GENETICS, GENOMICS AND BREEDING OF CONIFERS

Genetics, Genomics and Breeding of Crop Plants

Books in this Series:

Published or in Press:

- Jinguo Hu, Gerald Seiler & Chittaranjan Kole: *Sunflower*
- Kristin D. Bilyeu, Milind B. Ratnaparkhe & Chittaranjan Kole: *Soybean*
- Robert Henry & Chittaranjan Kole: *Sugarcane*
- Kevin Folta & Chittaranjan Kole: *Berries*
- Jan Sadowsky & Chittaranjan Kole: *Vegetable Brassicas*
- James M. Bradeen & Chittaranjan Kole: *Potato*
- C.P. Joshi, Stephen DiFazio & Chittaranjan Kole: *Poplar*
- Anne-Françoise Adam-Blondon, José M. Martínez-Zapater & Chittaranjan Kole: *Grapes*
- Christophe Plomion, Jean Bousquet & Chittaranjan Kole: *Conifers*
- Dave Edwards, Jacqueline Batley, Isobel Parkin & Chittaranjan Kole: *Oilseed Brassicas*
- Marcelino Pérez de la Vega, Ana María Torres, José Ignacio Cubero & Chittaranjan Kole: *Cool Season Grain Legumes*
- Yi-Hong Wang, Tusar Kanti Behera & Chittaranjan Kole: *Cucurbits*

GENETICS, GENOMICS AND BREEDING OF CONIFERS

Editors

Christophe Plomion
INRA
UMR BIOGECO
Cestas
France

Jean Bousquet
Centre d'étude de la forêt
Université Laval
Québec
Canada

Chittaranjan Kole
Department of Genetics and Biochemistry
Clemson University
Clemson, SC
USA

CRC Press
Taylor & Francis Group
an **informa** business
www.crcpress.com

6000 Broken Sound Parkway, NW
Suite 300, Boca Raton, FL 33487
270 Madison Avenue
New York, NY 10016
2 Park Square, Milton Park
Abingdon, Oxon OX14 4RN, UK

Science Publishers
Jersey, British Isles
Enfield, New Hampshire

Published by Science Publishers, an imprint of Edenbridge Ltd.

- St. Helier, Jersey, British Channel Islands
- P.O. Box 699, Enfield, NH 03748, USA

E-mail: info@scipub.net Website: www.scipub.net

Marketed and distributed by:

CRC Press
Taylor & Francis Group
an informa business
www.crcpress.com

6000 Broken Sound Parkway, NW
Suite 300, Boca Raton, FL 33487
270 Madison Avenue
New York, NY 10016
2 Park Square, Milton Park
Abingdon, Oxon OX14 4RN, UK

ISBN 978-1-57808-719-8

Cover photograph reproduced by kind courtesy of Didier Bert, Scientist Dendroecology, INRA Bordeaux, France

Library of Congress Cataloging-in-Publication Data

Genetics, genomics and breeding of conifers / editors, Christophe Plomion,
Jean Bousquet, Chittaranjan Kole. -- 1st ed.
p. cm. -- (Genetics, genomics and breeding of crop plants)
Includes bibliographical references and index.
ISBN 978-1-57808-719-8 (hardcover)
1. Conifers--Genetics. 2. Conifers--Genome mapping. 3. Conifers--Breeding. I. Plomion, Christophe. II. Bousquet, Jean, 1960- III.
Kole, Chittaranjan. IV. Series: Genetics, genomics and breeding of crop plants.
SD397.C7G45 2011
585'.2--dc22

2011006584

The views expressed in this book are those of the author(s) and the publisher does not assume responsibility for the authenticity of the findings/conclusions drawn by the author(s). Also no responsibility is assumed by the publishers for any damage to the property or persons as a result of operation or use of this publication and/or the information contained herein.

Preface to the Series

Genetics, genomics and breeding has emerged as three overlapping and complimentary disciplines for comprehensive and fine-scale analysis of plant genomes and their precise and rapid improvement. While genetics and plant breeding have contributed enormously towards several new concepts and strategies for elucidation of plant genes and genomes as well as development of a huge number of crop varieties with desirable traits, genomics has depicted the chemical nature of genes, gene products and genomes and also provided additional resources for crop improvement.

In today's world, teaching, research, funding, regulation and utilization of plant genetics, genomics and breeding essentially require thorough understanding of their components including classical, biochemical, cytological and molecular genetics; and traditional, molecular, transgenic and genomics-assisted breeding. There are several book volumes and reviews available that cover individually or in combination of a few of these components for the major plants or plant groups; and also on the concepts and strategies for these individual components with examples drawn mainly from the major plants. Therefore, we planned to fill an existing gap with individual book volumes dedicated to the leading crop and model plants with comprehensive deliberations on all the classical, advanced and modern concepts of depiction and improvement of genomes. The success stories and limitations in the different plant species, crop or model, must vary; however, we have tried to include a more or less general outline of the contents of the chapters of the volumes to maintain uniformity as far as possible.

Often genetics, genomics and plant breeding and particularly their complimentary and supplementary disciplines are studied and practiced by people who do not have, and reasonably so, the basic understanding of biology of the plants for which they are contributing. A general description of the plants and their botany would surely instill more interest among them on the plant species they are working for and therefore we presented lucid details on the economic and/or academic importance of the plant(s); historical information on geographical origin and distribution; botanical origin and evolution; available germplasms and gene pools, and genetic and cytogenetic stocks as genetic, genomic and breeding resources; and basic information on taxonomy, habit, habitat, morphology, karyotype, ploidy level and genome size, etc.

Classical genetics and traditional breeding have contributed enormously even by employing the phenotype-to-genotype approach. We included detailed descriptions on these classical efforts such as genetic mapping using morphological, cytological and isozyme markers; and achievements of conventional breeding for desirable and against undesirable traits. Employment of the in vitro culture techniques such as micro- and megaspore culture, and somatic mutation and hybridization, has also been enumerated. In addition, an assessment of the achievements and limitations of the basic genetics and conventional breeding efforts has been presented.

It is a hard truth that in many instances we depend too much on a few advanced technologies, we are trained in, for creating and using novel or alien genes but forget the infinite wealth of desirable genes in the indigenous cultivars and wild allied species besides the available germplasms in national and international institutes or centers. Exploring as broad as possible natural genetic diversity not only provides information on availability of target donor genes but also on genetically divergent genotypes, botanical varieties, subspecies, species and even genera to be used as potential parents in crosses to realize optimum genetic polymorphism required for mapping and breeding. Genetic divergence has been evaluated using the available tools at a particular point of time. We included discussions on phenotype-based strategies employing morphological markers, genotype-based strategies employing molecular markers; the statistical procedures utilized; their utilities for evaluation of genetic divergence among genotypes, local landraces, species and genera; and also on the effects of breeding pedigrees and geographical locations on the degree of genetic diversity.

Association mapping using molecular markers is a recent strategy to utilize the natural genetic variability to detect marker-trait association and to validate the genomic locations of genes, particularly those controlling the quantitative traits. Association mapping has been employed effectively in genetic studies in human and other animal models and those have inspired the plant scientists to take advantage of this tool. We included examples of its use and implication in some of the volumes that devote to the plants for which this technique has been successfully employed for assessment of the degree of linkage disequilibrium related to a particular gene or genome, and for germplasm enhancement.

Genetic linkage mapping using molecular markers have been discussed in many books, reviews and book series. However, in this series, genetic mapping has been discussed at length with more elaborations and examples on diverse markers including the anonymous type 2 markers such as RFLPs, RAPDs, AFLPs, etc. and the gene-specific type 1 markers such as EST-SSRs, SNPs, etc.; various mapping populations including F_2, backcross, recombinant inbred, doubled haploid, near-isogenic and pseudotestcross; computer software including MapMaker, JoinMap, etc. used; and different types of genetic maps including preliminary, high-resolution, high-density, saturated, reference, consensus and integrated developed so far.

Mapping of simply inherited traits and quantitative traits controlled by oligogenes and polygenes, respectively has been deliberated in the earlier literature crop-wise or crop group-wise. However, more detailed information on mapping or tagging oligogenes by linkage mapping or bulked segregant analysis, mapping polygenes by QTL analysis, and different computer software employed such as MapMaker, JoinMap, QTL Cartographer, Map Manager, etc. for these purposes have been discussed at more depth in the present volumes.

The strategies and achievements of marker-assisted or molecular breeding have been discussed in a few books and reviews earlier. However, those mostly deliberated on the general aspects with examples drawn mainly from major plants. In this series, we included comprehensive descriptions on the use of molecular markers for germplasm characterization, detection and maintenance of distinctiveness, uniformity and stability of genotypes, introgression and pyramiding of genes. We have also included elucidations on the strategies and achievements of transgenic breeding for developing genotypes particularly with resistance to herbicide, biotic and abiotic stresses; for biofuel production, biopharming, phytoremediation; and also for producing resources for functional genomics.

A number of desirable genes and QTLs have been cloned in plants since 1992 and 2000, respectively using different strategies, mainly positional cloning and transposon tagging. We included enumeration of these and other strategies for isolation of genes and QTLs, testing of their expression and their effective utilization in the relevant volumes.

Physical maps and integrated physical-genetic maps are now available in most of the leading crop and model plants owing mainly to the BAC, YAC, EST and cDNA libraries. Similar libraries and other required genomic resources have also been developed for the remaining crops. We have devoted a section on the library development and sequencing of these resources; detection, validation and utilization of gene-based molecular markers; and impact of new generation sequencing technologies on structural genomics.

As mentioned earlier, whole genome sequencing has been completed in one model plant (Arabidopsis) and seven economic plants (rice, poplar, peach, papaya, grapes, soybean and sorghum) and is progressing in an array of model and economic plants. Advent of massively parallel DNA sequencing using 454-pyrosequencing, Solexa Genome Analyzer, SOLiD system, Heliscope and SMRT have facilitated whole genome sequencing in many other plants more rapidly, cheaply and precisely. We have included extensive coverage on the level (national or international) of collaboration and the strategies and status of whole genome sequencing in plants for which sequencing efforts have been completed or are progressing currently. We have also included critical assessment of the impact of these genome initiatives in the respective volumes.

Comparative genome mapping based on molecular markers and map positions of genes and QTLs practiced during the last two decades of the last century provided answers to many basic questions related to evolution, origin and phylogenetic relationship of close plant taxa. Enrichment of genomic resources has reinforced the study of genome homology and synteny of genes among plants not only in the same family but also of taxonomically distant families. Comparative genomics is not only delivering answers to the questions of academic interest but also providing many candidate genes for plant genetic improvement.

The 'central dogma' enunciated in 1958 provided a simple picture of gene function—gene to mRNA to transcripts to proteins (enzymes) to metabolites. The enormous amount of information generated on characterization of transcripts, proteins and metabolites now have led to the emergence of individual disciplines including functional genomics, transcriptomics, proteomics and metabolomics. Although all of them ultimately strengthen the analysis and improvement of a genome, they deserve individual deliberations for each plant species. For example, microarrays, SAGE, MPSS for transcriptome analysis; and 2D gel electrophoresis, MALDI, NMR, MS for proteomics and metabolomics studies require elaboration. Besides transcriptome, proteome or metabolome QTL mapping and application of transcriptomics, proteomics and metabolomics in genomics-assisted breeding are frontier fields now. We included discussions on them in the relevant volumes.

The databases for storage, search and utilization on the genomes, genes, gene products and their sequences are growing enormously in each second and they require robust bioinformatics tools plant-wise and purpose-wise. We included a section on databases on the gene and genomes, gene expression, comparative genomes, molecular marker and genetic maps, protein and metabolomes, and their integration.

Notwithstanding the progress made so far, each crop or model plant species requires more pragmatic retrospect. For the model plants we need to answer how much they have been utilized to answer the basic questions of genetics and genomics as compared to other wild and domesticated species. For the economic plants we need to answer as to whether they have been genetically tailored perfectly for expanded geographical regions and current requirements for green fuel, plant-based bioproducts and for improvements of ecology and environment. These futuristic explanations have been addressed finally in the volumes.

We are aware of exclusions of some plants for which we have comprehensive compilations on genetics, genomics and breeding in hard copy or digital format and also some other plants which will have enough achievements to claim for individual book volume only in distant future. However, we feel satisfied that we could present comprehensive deliberations on genetics, genomics and breeding of 30 model and economic plants, and their groups in a few cases, in this series. I personally feel also happy that I could work with many internationally celebrated scientists who edited the book volumes on the leading plants and plant groups and included chapters authored by many scientists reputed globally for their contributions on the concerned plant or plant group.

We paid serious attention to reviewing, revising and updating of the manuscripts of all the chapters of this book series, but some technical and formatting mistakes will remain for sure. As the series editor, I take complete responsibility for all these mistakes and will look forward to the readers for corrections of these mistakes and also for their suggestions for further improvement of the volumes and the series so that future editions can serve better the purposes of the students, scientists, industries, and the society of this and future generations.

Science publishers, Inc. has been serving the requirements of science and society for a long time with publications of books devoted to advanced concepts, strategies, tools, methodologies and achievements of various science disciplines. Myself as the editor and also on behalf of the volume editors, chapter authors and the ultimate beneficiaries of the volumes take this opportunity to acknowledge the publisher for presenting these books that could be useful for teaching, research and extension of genetics, genomics and breeding.

Chittaranjan Kole

Preface to the Volume

Conifers are woody plants, the great majority being trees. They represent 650 species, some ranking as the largest, tallest, and longest living non-clonal terrestrial organisms on Earth. They are of immense ecological importance, dominating many terrestrial landscapes and representing the largest terrestrial carbon sink. They are evolutionary distinct from angiosperm trees on many accounts and with their extraordinary large genomes, they provide a different view of plant genome biology and evolution. They are also of great economic importance, as they are primarily used for timber and paper production worldwide. Domestication of some of these species was started about 60 years ago through traditional genetic improvement programs. It has resulted in advances in overall growth, wood quality, pest resistance and adaptation, but breeding still remains a slow process because of long generation intervals typical of most conifers and because most traits cannot be correctly evaluated at an early stage.

During the past 20 years, more and more sophisticated genomics tools have been developed to describe the extreme plasticity and variability of these species at different levels of integration (from genes up to phenotypes) and are now being integrated into breeding to accelerate the domestication process by a more precise exploitation of genetic diversity. Application of genomic-based science is also playing an important role in understanding the evolution, patterns of nucleotide variation and the molecular basis of quantitative traits and adaptation. Altogether, this new knowledge is also expected to help delineate more efficient gene conservation strategies.

This book will give the reader an in-depth review of the current state-of-the-art of genetic and genomic research conducted in conifers. Each chapter is the product of specialists in their field. Their goal was to report on the latest trends and findings and at the same time, promote awareness and make this knowledge accessible to the vast majority. Accordingly, the chapters are well documented and illustrated. Their contribution is greatly appreciated.

The book begins with an exhaustive description of the conifers in terms of classification, geographical distribution, life history and ecology, morphology and fossil history as well as phylogenetics (Chapter 1). It is followed by a chapter devoted to their economic importance and the development of conifer breeding programs worldwide, which lead to significant improvement of productivity and quality (Chapter 2). Chapter 3 deliberates on various classical and molecular cytogenetical tools useful to elucidate evolution, integrate physical and genetic maps, conserve species and assist in marker-based breeding. Chapter 4 describes the applications of neutral genetic markers from the perspectives of conservation genetics, phylogeography and gene flow studies. In Chapter 5, research efforts on linkage mapping, emerging gene maps as well as QTL detection and architecture are reviewed. An exhaustive review of investigations on candidate genes is provided in Chapter 6, from estimates of nucleotide diversity and recombination to new-generation selection signatures studies and the development of association mapping and outlier detection approaches. The ever-increasing applications of molecular markers into breeding from the management operations to selection strategies are considered in Chapter 7. Switching to more functional aspects, Chapters 8 and 9 review the current status of our understanding of transcriptome, proteome and metabolome modifications in responses to developmental changes and environmental constraints. The rapid advances in sequencing and cataloging the conifer gene space are also reported (Chapter 8). As a prerequisite for the sequencing of a conifer genome, insights into the characteristics of the large conifer genomes, especially with respect to the composition and evolution of transposable elements, are provided in Chapter 10. The book ends with refreshing views on the challenges faced by the conifer genomics community and how the pace of rapid advancement of the "omic" sciences might affect our understanding of conifer biology and the future use of conifer genetic resources (Chapter 11).

This book is a testimony to the substantial progress made in the field of conifer genetics and genomics and the definite value of conifers as a model system. Although the tools and concepts that are presented will continue to evolve rapidly, we hope this volume will provide a solid foundation for further development in conifer and more generally in forest tree genetics, genomics and breeding.

Christophe Plomion
Jean Bousquet
Chittaranjan Kole

Contents

List of Contributors

R. Alía
Department of Forest Ecology and Genetics, Center of Forest Research, INIA, 28040 Madrid, Spain.
Email: *alia@inia.es*

B. Andersson
Skogforsk (Sävar), Box 3, S-918 21 Sävar, Sweden.
Email: *bengt.andersson@skogforsk.se*

Francesca Bagnoli
Plant Protection Institute, CNR, Via Madonna del Piano 10, 50019 Sesto Fiorentino (FI), Italy.
Email: *bagnoli@ipp.cnr.it*

J.-C. Bastien
INRA—Centre de Recherche d'Orléans, 2163, Avenue de la Pomme de Pin, CS 400001 ARDON, F-45075 Orléans Cedex 2, France.
Email: *jean-charles.bastien@orleans.inra.fr*

J. Beaulieu
Natural Resources Canada, P.O. Box 10380, Stn. Sainte-Foy, Québec, QC G1V 4C7, Canada.
Email: *jeanbeau@nrcan-rncan.gc.ca*

Jean Bousquet
Centre d'étude de la forêt, Université Laval, Québec, Québec G1V 0A6, Canada.
Email: *jean.bousquet@sbf.ulaval.ca*

R.D. Burdon
Scion (NZ Forest Research Institute Ltd.), Private Bag 3020, Rotorua 3010, New Zealand.
Email: *rowland.burdon@scionresearch.com*

C. Burgarella
Department of Forest Ecology and Genetics, Center of Forest Research, INIA, 28040 Madrid, Spain.
Email: *concettaburgarella@hotmail.com*

Rebecca Dauwe
Department of Wood Science, Faculty of Forestry, 4030-2424 Main Mall, Vancouver, BC, V6T 1Z4, Canada.
Email: *rebecca.dauwe@u-picardie.fr*

Jeffrey F.D. Dean
Warnell School of Forestry and Natural Resources, University of Georgia, Athens, GA 30602, USA.
Email: *jeffdean@uga.edu*
Tel: +1-706-542-1710

Emanuele De Paoli
Istituto Agrario di San Michele all'Adige, Vie E. Mach 1, 38010 San Michele all'Adige, Italy.
Email: *emanuele.depaoli@iasma.it.*

S. Dillon
CSIRO Plant Industry, GPO Box 1600, Canberra, ACT 2601, Australia.
Email: *shannon.dillon@csiro.au*

W.S. Dvorak
North Carolina State University, Campus Box 8008, Raleigh, NC 27695-8008, USA.
Email: *w_dvorak@ncsu.edu*

A.J. Eckert
Section of Evolution and Ecology and Center for Population Biology, University of California at Davis, Davis, CA 95616, USA.
Email: *ajeckert@ucdavis.edu*

Bruno Fady
INRA, UR629, Ecologie des Forêts Méditerranéennes, Domaine Saint Paul, Site Agroparc, 84914 Avignon, France.
Email: *bruno.fady@avignon.inra.fr*

Silvia Fineschi
Plant Protection Institute, CNR, Via Madonna del Piano 10, 50019 Sesto Fiorentino (FI), Italy.
Email: *fineschi@ipp.cnr.it*

M.R. García-Gil
Umeå Plant Science Center, Swedish University of Agricultural Science, SE 901 83 Umeå, Sweden.
Email: *m.rosario.garcia@genfys.slu.se*

P.H. Garnier-Géré
INRA, UMR1202 Biodiversity Genes & Communities, 69 route d'Arcachon, 33612 Cestas Cedex, France.
Email: *pauline@pierroton.inra.fr*

David S. Gernandt
Departamento de Botánica, Instituto de Biología, Universidad Nacional Autónoma de México, A.P. 70-233, México Distrito Federal 04510, Mexico.
Email: *dgernandt@ibiologia.unam.mx*

S.C. González-Martínez
Department of Forest Ecology and Genetics, Center of Forest Research, INIA, 28040 Madrid, Spain.
Email: *santiago@inia.es*
Tel: +34 913471499

D. Grivet
Department of Forest Ecology and Genetics, Center of Forest Research, INIA, 28040 Madrid, Spain.
Email: *dgrivet@inia.es*

M. Heuertz
Department of Forest Ecology and Genetics, Center of Forest Research, INIA, 28040 Madrid, Spain.
and
Université Libre de Bruxelles, Faculté des Sciences, Behavioural and Evolutionary Ecology cp160/12, av. F.D. Roosevelt 50, 1050 Brussels, Belgium.
Email: *mheuertz@ulb.ac.be*

Nathalie Isabel
Natural Resources Canada, Canadian Forest Service, Laurentian Forestry Centre, 1055 du P.E.P.S., P.O. Box 10380, Stn Sainte-Foy, Québec, Québec G1V 4C7, Canada.
Email: *nisabel@cfl.forestry.ca*

Nurul Islam-Faridi
Forest Tree Molecular Cytogenetics, Southern Institute of Forest Genetics, US Forest Service.
and
Dept. of Ecosystem Science & Management, Texas A&M University, College Station, TX 77843, USA.
Email: *nfaridi@tamu.edu*

J.P. Jaramillo-Correa
Department of Forest Ecology and Genetics, Center of Forest Research, INIA, 28040 Madrid, Spain; and
Department of Evolutionary Ecology, Ecology Institute, Universidad Nacional Autónoma de México, Ciudad Universitaria, Tercer circuito Exterior, Apartado Postal 70-275, México, D.F.
Email: *jaramillo@miranda.ecologia.unam.mx*

J.N. King
British Columbia Forest Service, PO Box 9519 Stn Prov Govt, Victoria, B.C. V8W 9C2, Canada.
Email: *john.king@gov.bc.ca*

T. Kondo
Forest Tree Breeding Centre, 3809-1 Ishi, Juo, Hitachi, Ibaraki 319-1301, Japan.
Email: *kontei@affrc.go.jp*

J. Krakowski
British Columbia Ministry of Forests and Range, Box 335, Mesachie Lake, B.C. V0R2N0, Canada.
Email: *Jodie.Krakowski@gov.bc.ca*

Konstantin V. Krutovsky
Department of Ecosystem Science and Management, Texas A&M University, College Station, Texas 77843-2138, USA.
Email: *k-krutovsky@tamu.edu*

M. Lascoux
Program in Evolutionary Functional Genetics, Evolutionary Biology Centre, Uppsala University, 75326 Uppsala, Sweden.
Email: *Martin.Lascoux@ebc.uu.se*

S.J. Lee
Forest Research, Northern Research Station, Roslin, Midlothian, EH25 9SY, Scotland.
Email: *steve.lee@forestry.gsi.gov.uk*

Aaron Liston
Department of Botany and Plant Pathology, 2082 Cordley Hall, Oregon State University, Corvallis, Oregon 97331, USA.
Email: *listona@science.oregonstate.edu*

John J. Mackay
Center for Forest Research, Laval University, Québec City, Québec, Canada, G1V 0A6.
Email: *John.mackay@sbf.ulaval.ca*

Shawn D. Mansfield
Department of Wood Science, Faculty of Forestry, 4030-2424 Main Mall, Vancouver, BC, V6T 1Z4, Canada.
Email: *shawn.mansfield@ubc.ca*

S.E. McKeand
North Carolina State University, Campus Box 8002, Raleigh, NC 27695-8002, USA.
Email: *steve_mckeand@ncsu.edu*

Michele Morgante
Dipartimento di Scienze Agrarie ed Ambientali, Università di Udine, Via delle Scienze 208, 33100 Udine, Italy; and
Istituto di Genomica Applicata, Parco Scientifico e Tecnologico di Udine, Via Linussio 51, 33100 Udine, Italy.
Email: *michele.morgante@uniud.it*

T.J. Mullin
BioSylve Forest Science NZ Limited, 45 Krokoro Road, Lower Hutt 5012, New Zealand.
Email: *tim.mullin@biosylve.com*

D.B. Neale
Department of Plant Sciences, University of California at Davis, Davis, CA 95616, USA; and
Institute of Forest Genetics, Pacific Southwest Research Station, US Department of Agriculture Forest Service, Placerville, CA 95667, USA.
Email: *dbneale@ucdavis.edu*

C. Dana Nelson
USDA Forest Service, Southern Research Station, Southern Institute of Forest Genetics, 23332 Success Road, Saucier, MS 39574, USA.
Email: *dananelson@fs.fed.us*

Sylvie Oddou-Muratorio
INRA, UR629, Ecologie des Forêts Méditerranéennes, Domaine Saint Paul, Site Agroparc, 84914 Avignon, France.
Email: *sylvie.oddou@avignon.inra.fr*

L. Pâques
INRA—Centre de Recherche d'Orléans, 2163 Avenue de la Pomme de Pin, CS 400001 ARDON, F-45075 Orléans Cedex 2, France.
Email: *luc.paques@orleans.inra.fr*

Betty Pelgas
Natural Resources Canada, Canadian Forest Service, Laurentian Forestry Centre, 1055 du P.E.P.S., P.O. Box 10380, Stn Sainte-Foy, Québec, Québec G1V 4C7, Canada.
Email: *betty.pelgas@RNCan-NRCan.gc.ca*

Andrea Piotti
Department of Environmental Sciences, University of Parma, Viale Usberti 11/A, 43100 Parma, Italy.
Email: *andrea.piotti@nemo.unipr.it*

A. Raffin
INRA (Pierroton), 69 route d'Arcachon, 33612 CESTAS Cedex, France.
Email: *annie.raffin@pierroton.inra.fr*

Kermit Ritland
Department of Forest Sciences, University of British Columbia, Vancouver, British Columbia V6T 1Z4, Canada.
Email: *kermit.ritland@ubc.ca*

Andrew Robinson
Department of Wood Science, Faculty of Forestry, 4030-2424 Main Mall, Vancouver, BC, V6T 1Z4, Canada.
Email: *andrewrobinsonnz@gmail.com*

J.H. Russell
British Columbia Ministry of Forests and Range, Box 335, Mesachie Lake, B.C. V0R2N0, Canada.
Email: *john.russell@gov.bc.ca*

O. Savolainen
Department of Biology, University of Oulu, 90014 Oulu, Finland.
Email: *outi.savolainen@oulu.fi*

Federico Sebastiani
Plant Genetics Institute, CNR, Via Madonna del Piano 10, 50019 Sesto Fiorentino (FI), Italy.
Email: *federico.sebastiani@unifi.it*

T. Skrøppa
Norwegian Forest and Landscape Institute, Høgskoleveien 8, 1432 Ås, Norway.
Email: *tore.skroppa@skogoglandskap.no*

M. Stoehr
British Columbia Ministry of Forests, PO Box 9519, Stn Prov Govt, Victoria, B.C. V8W 9C2, Canada.
Email: *michael.stoehr@gov.bc.ca*

John V. Syring
Department of Biology, Linfield College, 900 SE Baker St., McMinnville, Oregon 97128, USA.
Email: *jsyring@linfield.edu*

Yoshihiko Tsumura
Forestry and Forest Products Research Institute, Tsukuba, Ibaraki 305-8687, Japan.
Email: *ytsumu@ffpri.affrc.go.jp*

Giovanni G. Vendramin
Plant Genetics Institute, CNR, Via Madonna del Piano 10, 50019 Sesto Fiorentino (FI), Italy.
Email: *giovanni.vendramin@igv.cnr.it*

Phillip L. Wilcox
Scion: New Zealand Forest Research Institute Ltd, Private Bag 3020, Rotorua 3046, New Zealand.
Email: *phillip.wilcox@scionresearch.com*

Ann Willyard
Biology Department, Hendrix College, 1600 Washington Ave, Conway, Arkansas 72032, USA.
Email: *willyard@hendrix.edu*

A. Yanchuk
British Columbia Ministry of Forests, PO Box 9519, Stn Prov Govt, Victoria, B.C. V8W 9C2, Canada.
Email: *alvin.yanchuk@gov.bc.ca*

Abbreviations

μ	Mutation rate
2-D	Two-dimensional
2-DE	Two-dimensional electrophoresis
ABC	Approximate Bayesian computation
AFA	Adaptive Force Acoustics
AFLP	Amplified fragment length polymorphism
AGP	Arabinogalactan protein
ANOVA	Analysis of variance
AT	Adenine-Thymine
ATRS	*Arabidopsis*-type telomere repeat sequence
BAC	Bacterial artificial chromosome
BHT	Butylated hydroxytoluene
BIC	Bayesian information criterion
BLUP	Best Linear Unbiased Prediction
bp	Base pairs
CAPS	Cleaved amplified polymorphic sequence
CCA	Canonical correlation analysis
CDA	Canonical discriminate analysis
cDNA	Complementary-DNA
CDS	Complete coding sequences
Ch	Chromosome
ChIP-Seq	Chromatin ImmunoPrecipitation coupled with next-generation sequencing
CID	Carbon isotope discrimination
cM	CentiMorgan
CMA	Chromamycin A_3
COLD-PCR	CO-amplification at lower denaturation temperature-PCR
COS	Conserved orthologous set
cpDNA	Chloroplast-DNA
cpSSR	Chloroplast-SSR
DAPI	4′, 6-Diamidino -2-phenylindole
DArT	Diversity array technology
DIGE	Differential in gel electrophoresis

DOE	Department of Energy (US)
DOP	Degenerate oligonucleotide primed
DUF	Domains of unknown function
EBV	Estimated breeding value
ECD	Electrochemical detector
eQTL	Expression-QTL
EST	Expressed sequence tag
ESTP	Expressed sequence tag polymorphism
ESU	Evolutionary significant units
FA	Factor analysis
FID	Flame ionization detector
FISH	Fluorescent/ce in situ hybridization
FL-cDNA	Full length-cDNA
Fst	Fixation index
GA	Gibberellin
GAB	Gene-assisted breeding
GAS	Gene-assisted selection
Gbp	Giga base pair
GBV	Genomic breeding value
GC	Gas chromatography
GC	Guanine-Cytosine
GC/MS	Gas chromatography-mass spectroscopy
GDP	Gross domestic product
GIS	Geographic information system
GLM	General linear model
GNP	Gross national product
GS	Genome Selection
GS	Genomic selection
Gst	Average amount of differentiation observed over multiple loci
G_{st}	Population differentiation statistic
GWS	Genome-wide selection
GWS	Genome-wide scan
HCA	Hierarchical cluster analysis
H_e	Expected heterozygosity statistic
HMM	Hidden Markov model
HMPR	Hypomethylated partial restriction
H_o	Observed heterozygosity statistic
HPLC	High pressure liquid chromatography
HSD	Honestly significant difference
HSP	heat-shock protein
HTS	High-throughput sequencing
IBF	Identity by function

ICAT	Isotope-coded affinity tags
IE	Isoelectric point
IEF	Isoelectric focusing
IHGSC	International Human Genome Sequencing Consortium
Indel	Insertion/deletion
INTA	Instituto Nacional de Tecnología Agropecuaria (Argentina)
IR	Inverted repeat
IS	Importance sampling
ISSR	Iinter-simple sequence repeat
iTRAQ	Isobaric tag for relative and absolute quantitation
ITS	Internal transcribed sequence
IUFRO	International Union of Forestry Research Organizations
JGI	Joint Genome Institute (US)
Kb	Kilobase(s)
Kbp	Kilo base pair
LC ESI-MS/MS	Liquid chromatography electrospray ionization tandem mass spectrometry
LC/ESI/MS	Liquid chromatography electrospray ionization mass spectrometry
LC/MS	Liquid chromatography-mass spectrometry
LD	Linkage disequilibrium
LDD	Long-distance dispersal
LG	Linkage group
LGM	Last glacial maximum
LSC	Large region of single copy genes
LTR	Long terminal repeats
MAB	Marker-assisted backcrossing
MAF	Minimum allele frequency
MALDI-TOF MS	Matrix assisted laser desorption/ionization time of flight mass spectrometry
MARG	Marker-assisted recovery of genotypes
MAS	Marker-aided/assisted selection
MAS	Magic angle spinning
Mbp	Mega base pair
MCMC	Markov Chain Monte Carlo
MDA	Multiple discriminate analysis
MFA	Microfibril angle
miRNA	Micro-RNA
MLM	Mixed linear model
MOE	Modulus of elasticity
MPB	Mountain pine beetle
MPK	Mitogen-activated protein kinase

Mr	Relative molecular weight
mRNA	Messenger-RNA
MS/MS	Mass spectrometry/mass spectrometry or Tandom mass spectrometry
MS^n	Mass spectrometry to the "n" th power
MSTFA	*N*-methyl-*N*-trimethylsilyltrifluoroacetamide
mtDNA	Mitochondrial-DNA
mtSSR	Mitochondrial-SSR
MudPIT	Multidimensional protein identification technology
MW	Molecular weight
MY	Million years
Mya	Million years ago
N	Census number
NCA	Nested clade analysis
NCBI	National Center for Biotechnology Information (US)
NCPA	Nested clade phylogeographic analysis
ncRNA	Non-coding RNA
nDNA	Nuclear-DNA
N_e	Effective population size
$N_e m_e$	Effective number of migrants per generation
NGS	Next-generation sequencing
NHGRI	National Human Genome Research Institute
NMR	Nuclear magnetic resonance spectroscopy
NRC	National Research Council (Canada)
nrITS	Nuclear ribosomal-ITS
N_s	Status number, a measure of effective population size
nSSR/nucSSR	Nuclear-SSR
ORF	Open reading frame
PAC	Product of approximating conditionals
PAGE	Polyacrilamide gel electrophoresis
PCA	Principal component analysis
PCR	Polymerase chain reaction
PCSR	Proximal CMA band-specific repeat
PET	paired-end tag
PGI	Plant Gene Indices
pI	Isoelectric point
PLSR	Partial least squares regression
PMF	Peptide mass fingerprinting
PUT	Putative transcripts
PVPP	Polyvinylpolypyrrolidone
QCI	Queen Charlotte Islands
QTL	Quantitative trait loci
QTN	Quantitative trait nucleotide

R&D	Research and development
RAPD	Random(ly) amplified polymorphic DNA
rDNA	Ribosomal-DNA
RFLP	Restriction fragment length polymorphism
RNA-Seq	Whole-transcriptome shotgun sequencing
RT-PCR	Reverse transcrtiptase-PCR
SAGE	Serial analysis of gene expression
SAMT	S-Adenosylmethionine transferase
SCAR	Sequence characterized amplified region
SD	Standard deviation
SDS	Sodium dodecyl sulfate
SE	Somatic embryogenesis
SFS	Site frequency spectrum
SGS	Spatial genetic structure
SILAC	Stable isotope labeling by amino acids in cell culture
siRNA	Small-interfering RNA
siRNA	Short interfering RNA
SNP	Single nucleotide polymorphism
SPF	Spruce-Pine-Fir lumber specification
SPME	Solid-phase microextraction
SSC	Small region of single copy genes
SSR	Simple sequence repeat
STS	Sequence tagged site
tasiRNA	Trans-acting siRNA
TBR	Tree bisection reconnection
TDT	Transmission disequilibrium test
TLP	TUBBY-like protein
TOF	Time-of-flight
U-HPLC	Ultra-HPLC
USD	US Dollar
USDA	United States Department of Agriculture
UTR	Untranscribed region
UV/Vis	Ultraviolet-visible spectrophotometry
VGN	Crucifer Genome Network
WGS	Whole-genome sequencing/shotgun

1

The Conifers (Pinophyta)

David S. Gernandt,[1,] Ann Willyard,[2] John V. Syring,[3] and Aaron Liston[4]*

ABSTRACT

Conifers (Pinophyta) are woody trees or shrubs with simple leaves, simple pollen cones, and compound or reduced ovulate cones. Despite their dominance in many terrestrial landscapes, the 670 species of extant conifers make up less than 0.3% of the species diversity of modern land plants. The fossil record of conifers, which extends to the Carboniferous, indicates that a much greater diversity is now extinct. Conifers occur on six of the seven continents and include both widely distributed, dominant species that form vast forests and narrow endemics. They rank as the largest, tallest, and longest living non-clonal terrestrial organisms on the Earth. *Pinus* is the largest extant genus with approximately 20 species distributed throughout the Northern Hemisphere. It is rivaled in diversity in the Southern Hemisphere and the tropics by *Podocarpus*, with approximately 105 species. Genetic diversity is often high in conifers, promoted by large population size, outcrossing reproductive systems, high mutation rates, and long distance dispersal of pollen and sometimes seeds. Estimates of ages and mutation rates in the group are expected to improve greatly as conceptual advances related to fossil interpretation converge with the enormous quantities of new sequence data being generated by genetic and phylogenetic studies of living species. Contrasting patterns of organellar and nuclear inheritance

[1]Departamento de Botánica, Instituto de Biología, Universidad Nacional Autónoma de México, A.P. 70-233, México Distrito Federal 04510, Mexico; e-mail: *dgernandt@ibiologia.unam.mx*
[2]Biology Department, Hendrix College, 1600 Washington Ave, Conway, Arkansas 72032, USA; e-mail: *willyard@hendrix.edu*
[3]Department of Biology, Linfield College, 900 SE Baker St., McMinnville, Oregon 97128, USA; e-mail: *jsyring@linfield.edu*
[4]Department of Botany and Plant Pathology, 2082 Cordley Hall, Oregon State University, Corvallis, Oregon 97331, USA; e-mail: *listona@science.oregonstate.edu*
*Correspondig author

make conifers an important system for studying pollen and seed flow, hybridization, lineage sorting, and gene coalescence.

Keywords: conifers; ecology; fossils; molecular clock; phylogeny; Pinophyta

1.1 Conifer Diversity

1.1.1 Classification and Phylogeny

1.1.1.1 Are Conifers Monophyletic?

Conifers are classified with seed plants, which include five living groups: conifers, cycads, *Ginkgo*, gnetophytes and angiosperms. The first four groups comprise the gymnosperms, which expose their ovules during pollination. There is wide acceptance for the rank of order Coniferales (also called Pinales), and conifers have often been recognized at the higher taxonomic ranks of class (Coniferae, Coniferopsida or Pinopsida) and division (Coniferophyta or Pinophyta). Living conifers are grouped in six families, 71 genera (Fig. 1-1), and ca. 670 species. In a recent global checklist (Farjon 2001), 69 genera and 630 species were recognized; we treat *Callitropsis* and *Xanthocyparis* as separate from *Cupressus* and recognize more species in Pinaceae and Cupressaceae.

Despite intensive study, the phylogenetic relationships among the major lineages of living and extinct seed plants remain ambiguous, with some DNA sequence analyses indicating that gnetophytes (*Ephedra*, *Gnetum*, and *Welwitschia*) are derived from conifers, rendering the conifers paraphyletic. Cladistic analyses of morphological characters (Crane 1985; Doyle and Donoghue 1986; Nixon et al. 1994) have recovered the gnetophytes and extinct gymnosperm groups like Bennettitales as more closely related to angiosperms than to other extant gymnosperms, thus supporting the Anthophyte hypothesis (Arber and Parkin 1907). Shared characters uniting these groups include "flower-like" reproductive structures, double fertilization (Friedman 1994), and the presence of vessels in their wood. In contrast, most molecular phylogenetic studies reject this hypothesis, placing gnetophytes either as sister to the conifers, the "gnetifer" hypothesis (Chaw et al. 1997), or sister to Pinaceae, within the conifers, the "gnepine" hypothesis (Bowe et al. 2000; Chaw et al. 2000). A close relationship between conifers and gnetophytes is supported by morphological characters such as simple leaves, compound ovulate cones, and wood anatomical characters that are also shared with *Ginkgo* such as tracheids with helical sculpturing intercalated with circular bordered pits, and the presence of a torus suspended by margo threads maintaining separation of the pits (Carlquist

1996). Nevertheless, results from molecular data have shown striking sensitivity to the choice of analytical method, characters, and taxonomic sampling (reviewed by Mathews 2009).

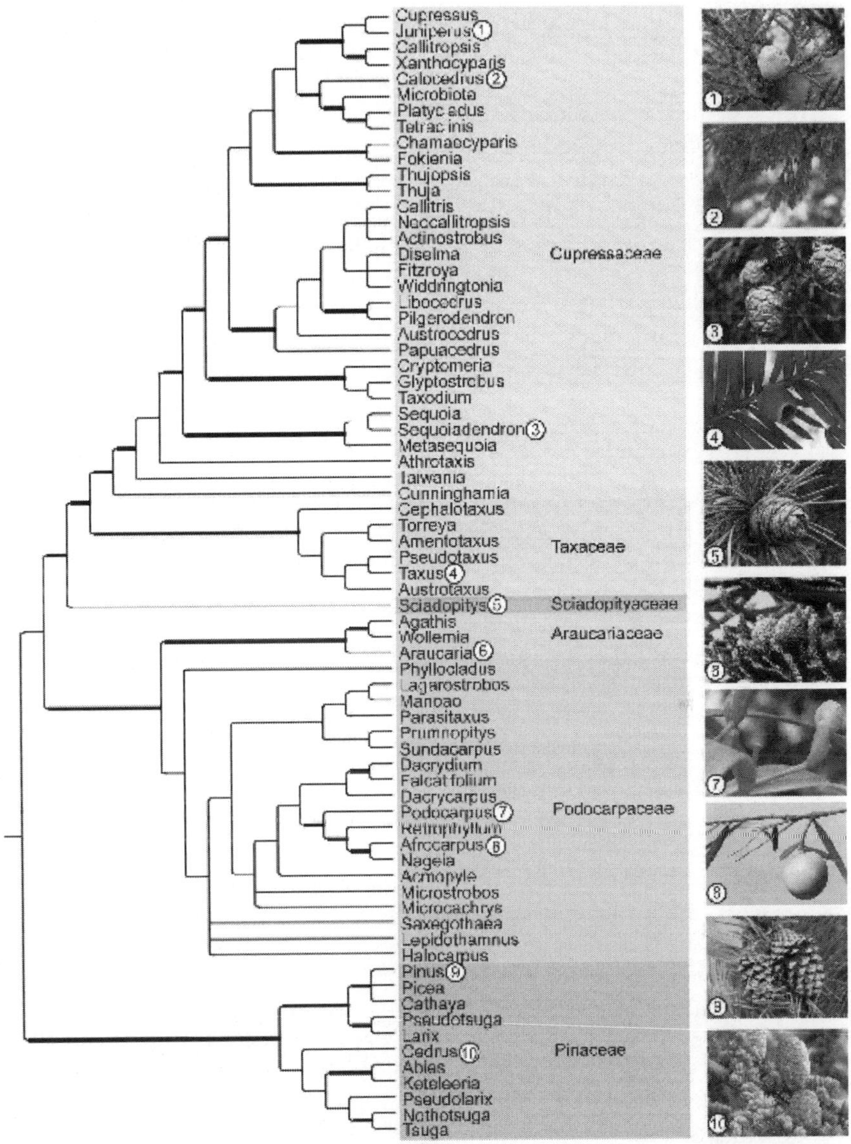

Figure 1-1 Conifer phylogenetic tree. A representation of our current understanding of intergeneric relationships.

Color image of this figure appears in the color plate section at the end of the book.

A recent phylogenetic analysis of 14 kbp of cpDNA for 38 taxa including 22 conifers representing all families failed to recover an association between Gnetales and conifers (Rai et al. 2008). This study included Sciadopityaceae and both subfamilies of Pinaceae, the most comprehensive taxonomic sampling of conifers to date. The chloroplast genome sequence of *Welwitschia* (McCoy et al. 2008) confirms that Gnetales possess the large inverted repeat that is present in most seed plants but lacking in conifers (Strauss et al. 1988; Wakasugi et al. 1994; Hirao et al. 2008). However, the *Pinus* "remnant" inverted repeat (495 bp including a duplicated *trn*I-CAU and partial *psb*A) could be derived from the inverted repeat of Gnetales (McCoy et al. 2008). The repeated *trn*I-CAU in *Cryptomeria* may be derived in a similar fashion (Hirao et al. 2008). These results are inconsistent with the hypothesis of a Gnetales–Pinaceae clade, but do not reject a sister group relationship between Gnetales and Pinophyta (the gnetifer hypothesis). For the purpose of this book, we exclude Gnetales from the conifers.

Another contentious issue in conifer classification has been the phylogenetic placement of the Taxaceae, most of whose members lack recognizable ovulate cones. Taxaceae has usually been considered a conifer family (Pilger 1926; Page 1990), but some botanists (Sahni 1920; Florin 1948) argued that it should be treated as a separate order, Taxales, principally because its ovules are borne terminally on lateral shoots rather than in cones. However, evidence from wood and leaf anatomy, embryological characters, and chloroplast, mitochondrial, and nuclear DNA sequences all unambiguously place Taxaceae within the conifers (Hart 1987; Chaw et al. 1993; Chaw et al. 2000; Quinn et al. 2002; Doyle 2006; Rai et al. 2008).

1.1.1.2 Relationships at the Level of Family and Genus

The six extant families of conifers are Pinaceae, Podocarpaceae, Araucariaceae, Sciadopityaceae, Taxaceae, and Cupressaceae (Table 1-1). Relationships among families and genera have become much clearer in recent years (Fig. 1-1). Molecular sequence data from the nuclear and chloroplast genomes have recovered Pinaceae as monophyletic and in a sister position to all other conifer families (Chaw et al. 1997; Stefanoviç et al. 1998; Quinn et al. 2002; Rai et al. 2008). The result is also supported by the loss of an intron in the mitochondrial *nad1* gene (Gugerli et al. 2001b) and in a morphological analysis of conifer genera (Hart 1987). Podocarpaceae and Araucariaceae have been recovered as sister groups consistently and with high branch support using nuclear and chloroplast data, but not with morphology (Hart 1987; Doyle 2006). Nuclear and chloroplast data strongly support a sister relationship between Taxaceae and Cupressaceae with Sciadopityaceae successively sister to them, which contrasts with previous morphological evidence uniting Cupressaceae and Sciadopityaceae (e.g., Hart 1987).

Table 1-1 Conifers of economic and/or ecological importance. This table is representative of conifer diversity, and not comprehensive.

Family	# Genera	Representative Genera or Subgenera	# Species	Representative Species	Native Range	Common Names	Notes	Genetic Resources
Pinaceae	11	*Abies*	50	*A. alba*	Europe	silver fir	forestry in Europe	nSSR
				A. balsamea	E Canada, NE USA	balsam fir	forestry in Canada, E USA	
		Cedrus	4	*C. atlantica*	NW Africa	Atlas cedar	forestry in NW Africa, horticulture	
		Larix	10	*L. decidua/L. sibirica/L. gmelinii*	N Eurasia	larch, Siberian larch, Dahurian larch	forestry in Europe & Russia	
				L. laricina	Canada, N USA	tamarack	forestry in Canada, E USA	
		Picea	34	*P. abies*	Europe	Norway spruce	forestry in Europe	nSSR, EST
				P. glauca	Canada, N USA	white spruce	forestry in Canada	EST
				P. mariana/P. rubens	E Canada, NE USA	black spruce, red spruce	forestry in Canada, E USA	
				P. sitchensis	W Canada, NW USA	Sitka spruce	forestry in Canada, W USA, NZ	EST
		Pinus subg. *Pinus*	ca. 80	*P. banksiana/P. contorta*	Canada, N USA	jack pine, lodgepole pine	forestry in Canada, Scandinavia	
				P. brutia/P. halepensis	Mediterranean	Brutia pine, Aleppo pine	forestry in Mediterranean, other arid zones	nSSR

Table 1-1 contd....

Table 1-1 contd....

Family	# Genera	Representative Genera or Subgenera	# Species	Representative Species	Native Range	Common Names	Notes	Genetic Resources
				P. caribaea	C America, Caribbean	Caribbean pine	subtropical forestry	nSSR
				P. densiflora	China, Korea, Japan	Japanese red pine	forestry in Japan, Korea	EST
				P. elliottii	SE USA	slash pine	forestry in SE USA, S Africa	nSSR
				P. kesiya	SE Asia		forestry in SE Asia	
				P. massoniana	China	Chinese red pine	forestry in SE Asia	
				P. merkusii	SE Asia	Sumatran pine	forestry in SE Asia	
				P. nigra/P. thunbergii	Eurasia, Japan	black pine, Japanese black pine	horticulture	
				P. pinaster	Mediterranean	maritime pine	forestry in Mediterranean, other arid zones	nSSR, EST
				P. pinea	Mediterranean	Italian stone pine	forestry in Mediterranean, other arid zones	EST
				P. oocarpa	Mexico		subtropical forestry	
				P. patula	Mexico	Mexican weeping pine	subtropical forestry	
				P. ponderosa	W Canada, W USA	ponderosa pine	forestry in W USA	

		P. radiata	California, USA	Monterey pine, radiata pine	forestry in Australia, NZ, Chile, S Africa	nSSR, EST
		P. resinosa	E Canada, NE USA	red pine	forestry in E USA	
		P. sylvestris	Europe	Scots pine	forestry in Europe, Russia	EST
		P. tabuliformis/P. yunnanensis	China, Korea	Chinese pine, Yunnan pine	forestry in China	
		P. taeda	SE USA	loblolly pine	forestry in SE USA, Australia	nSSR, EST
Pinus subg. *Strobus*	ca. 40	*P. armandii*	China, Japan	Armand pine	forestry in China	
		P. cembroides/P. edulis/P. monophylla	W USA, N Mexico	pinyon pine	local forestry, ecological genomics	
		P. albicaulis/P. lambertiana/P. monticola	W Canada, W USA	whitebark pine, sugar pine, western white pine	pathogen induced decline	
		P. cembra/P. sibirica/P. pumila	N Eurasia, Japan	stone pine	local forestry	
		P. krempfii	Vietnam	Krempf's pine	only pine with flat needles	
		P. longaeva	W USA	Great Basin bristlecone pine	oldest living tree	
Pseudotsuga	4	*P. menziesii*	Canada, W USA	Douglas-fir	forestry in Canada, USA, NZ	nSSR, EST
Tsuga	9	*T. canadensis*	E Canada, NE USA	Eastern hemlock	pest-induced dieback	

Table 1-1 contd....

Table 1-1 contd....

Family	# Genera	Representative Genera or Subgenera	# Species	Representative Species	Native Range	Common Names	Notes	Genetic Resources
				T. heterophylla	W Canada, W USA	Western hemlock (Canadian pine, Australia)	forestry in Canada	
Araucariaceae	3	*Agathis*	23	*A. australis*	New Zealand	kauri	historical forestry in NZ	
				A. damarra	SE Asia	East Indian kauri	forestry in SE Asia	
		Araucaria	19	*A. angustifolia*	Brazil, Paraguay, Argentina	Parana pine	forestry in Brazil	nSSR
				A. araucana	S Argentina, S Chile	monkey-puzzle tree	local forestry	nSSR
				A. bidwillii	NE Australia	bunya-bunya tree	forestry in Australia, horticulture	
				A. cunninghamii	NE Australia, New Guinea	hoop pine	local forestry	nSSR
				A. heterophylla	Norfolk Island	Norfolk Island pine	horticulture	
		Wollemia	1	*W. nobilis*	SE Australia	Wollemi pine	discovered in 1994, horticulture	
Podocarpaceae	19	*Afrocarpus*	6	*A. falcatus/A. gracilior*	E Africa, S Africa	yellow-wood	local forestry	

		Nageia	6	*N. nagi*	SE Asia	broad-leaved podocarpus	local forestry	
		Parasitaxus	1	*P. usta*	New Caledonia	corail	only parasitic gymnosperm	
		Podocarpus	105	*P. totara*	New Zealand	totara	forestry in NZ	
Sciadopityaceae	1	*Sciadopitys*	1	*S. verticillata*	Japan	koyamaki, umbrella pine	horticulture	
Taxaceae (includes Cephalotaxaceae)	6	*Taxus*	10	*T. baccata/T. cuspidata*	Europe, China, Korea, Japan	yew, Japanese yew	local forestry, horticulture, pharmaceuticals	EST
Cupressaceae	31	*Callitris*	16	*C. glaucophylla*	E Australia	white cypress pine	forestry in Australia	
		Callitropsis	18	*C. lusitanica*	Mexico, C America	Mexican cypress	forestry in C America, horticulture	
				C. macrocarpa	California, USA	Monterey cypress	horticulture	
				C. nootkatensis	W Canada, NW USA	Alaska yellow-cedar	forestry in Canada	
		Chamaecyparis	5	*C. obtusa*	S Japan, Taiwan	hinoki cypress	forestry in Japan, Taiwan	nSSR, EST
				C. lawsoniana	NW USA	Port Orford cedar	historical forestry, pathogen induced decline	
		Cryptomeria	1	*C. japonica*	China, Japan	sugi	forestry in Japan, China	nSSR, EST

Table 1-1 contd....

Table 1-1 contd....

Family	# Genera	Representative Genera or Subgenera	# Species	Representative Species	Native Range	Common Names	Notes	Genetic Resources
		Cunninghamia	1-2	*C. lanceolata*	China, Vietnam, Laos	China fir	forestry in China	EST
		Cupressus	12	*C. sempervirens*	Mediterranean	Italian cypress	horticulture	
		Fitzroya	1	*F. cupressoides*	S Argentina, S Chile	alerce	local forestry; tetraploid	
		Juniperus	67	*J. virginiana*	E Canada, E USA	Eastern redcedar	local forestry	
				J. communis	Circumboreal	common juniper	horticulture, flavoring (gin)	nSSR
		Metasequoia	1	*M. glyptostroboides*	China	dawn redwood	discovered in 1944, horticulture	
		Platycladus	1	*P. orientalis*	China, Korea, E Russia	Chinese arbor-vitae	forestry in China	
		Sequoia	1	*S. sempervirens*	California, USA	coast redwood	forestry in USA, tallest trees; hexaploid	
		Sequoiadendron	1	*S. gigantea*	California, USA	giant sequoia	largest trees, horticulture	
		Taiwania	1	*T. cryptomerioides*	China	Taiwania	local forestry	EST
		Taxodium	2	*T. distichum* / *T. mucronatum*	USA, Mexico, Guatemala	baldcypress	local forestry, horticulture	EST
		Thuja	5	*T. plicata*	W Canada, NW USA	western red cedar	forestry in Canada, USA	

Pinaceae is comprised of 10–11 genera (the separation of *Nothotsuga* from *Tsuga* is not universally accepted), and there is moderate support from all three genomes (Wang et al. 2000) and from morphology (Hart 1987; Gernandt et al. 2008) for dividing the family into two subfamilies, Pinoideae and Abietoideae. Relationships among the genera of subfamily Abietoideae (*Cedrus, Abies, Keteleeria, Tsuga, Nothotsuga,* and *Pseudolarix*) are not robust and the subfamily may actually be paraphyletic to Pinoideae (Wang et al. 2000; Gernandt et al. 2008).

Relationships among the approximately 19 genera of Podocarpaceae are less certain than in other families. Morphological, embryological, and molecular evidence indicate that *Podocarpus sensu lato* is notmonophyletic (Page 1989; Kelch 1997, 1998; Conran et al. 2000; Sinclair et al. 2002; Barker et al. 2004). *Podocarpus* is now restricted to ca. 105 species, with the designation of the genera *Afrocarpus, Dacrycarpus, Nageia, Parasitaxus, Prumnopitys, Retrophyllum,* and *Sundacarpus*. The genus *Phyllocladus* sometimes has been recognized as the separate family, Phyllocladaceae, based on the presence of cladodes and reduced, scale-shaped leaves. It also lacks an epimatium (a fleshy structure subtending the ovule probably homologous to the ovuliferous scale; Tomlinson and Takaso 2002). However, the epimatium is absent in other Podocarpaceae genera (e.g., *Microstrobus*) and the sister relationship between *Phyllocladus* and the rest of Podocarpaceae is not robust; recognition of the separate family Phyllocladaceae is thus unsupported.

Araucariaceae includes three extant genera. Early studies of phylogenetic relationships based on *rbcL* sequences recovered *Wollemia* as sister to *Agathis* and *Araucaria* (Setoguchi et al. 1998), but more recent studies with longer cpDNA data sets and more taxa have recovered *Wollemia* as sister to *Agathis* (Quinn et al. 2002; Knapp et al. 2007; Rai et al. 2008).

Cephalotaxus has often been separated from Taxaceae because its ovules are borne in recognizable cones. Molecular evidence has shown that it is the sister group to the other five genera of Taxaceae (Cheng et al. 2000; Quinn et al. 2002; Rai et al. 2008). Although recognition of Cephalotaxaceae would not render Taxaceae paraphyletic, its relatively modest genetic and morphological differentiation from Taxaceae are considered insufficient for recognition of a separate family. Its sister relationship with remaining members of the family is consistent with the hypothesis that the absence of compound ovulate cones in the other genera is due to a secondary loss. The remaining five genera of Taxaceae were divided into tribes Taxeae and Torreyeae, and this division is reflected in two clades inferred from DNA (Cheng et al. 2000) but not by morphology (Hart 1987).

Taxodiaceae (nine genera) was formerly recognized as separate from Cupressaceae, but the morphological differences are minor (alternate vs. opposite leaves in four ranks or whorled) and they possess similar cone morphology and karyotypes (Eckenwalder 1976). *Sciadopitys* was often classified in Taxodiaceae, despite its unique dimorphic shoots, presence of cladodes (photosynthetic branchlets) in place of leaves, and a chromosome number of $2n = 20$ (Farjon 2005). Molecular evidence has demonstrated that all Taxodiaceae genera except *Sciadopitys* are paraphyletic to Cupressaceae (Brunsfeld et al. 1994; Gadek et al. 2000; Kusumi et al. 2000; Quinn et al. 2002; Rai et al. 2008). *Sciadopitys* is now recognized as a monotypic family and the Cupressaceae has been expanded to include the other genera previously placed in Taxodiaceae. The 31 genera of Cupressaceae have been divided into seven subfamilies (Gadek et al. 2000).

The paraphyly of the genera formerly classified in Taxodiaceae with Cupressaceae clarifies the interpretation of ancestral states for this family. For example, *Cunninghamia*, the sister group to all other Cupressaceae, has three inverted ovules on each bract, while *Taiwania*, which is successively sister to the rest of Cupressaceae, has two ovules. This suggests that the proliferation of ovules on each bract-scale complex, the erect orientation of ovules in Cupressaceae and the reduced number of ovules per scale in some species of *Juniperus*, are more recently derived innovations (Farjon and Ortiz Garcia 2003).

1.1.2 Geographic Distribution

The natural range of conifers is from 55° south latitude on Tierra del Fuego in South America (*Pilgerodendron uviferum* Florin) (Veblen et al. 1995), to 75° north latitude deep within the Arctic Circle in Siberia (*Larix gmelinii* (Rupr.) Kuzen) (Farjon 2003). Many occur in extreme environments typified by high altitudes, high latitudes, and/or ecosystems with nutrient-poor soil (Stopes and Kershaw 1910; Richardson and Rundel 1998; Coomes et al. 2005). Between these extremes, ecological limitations on conifer distribution appear to be predominantly controlled by their ability to compete with angiosperms (Bond 1989; Coomes et al. 2005).

Although mostly absent from deserts, conifers are often found in environments with relatively high levels of evaporative stress, such as high light—low temperature (alpine tree line), high light—high temperature (semi-desert pinyon-juniper woodlands), and in temperate ecosystems with summer drought and winter rain where they compete well with deciduous angiosperms. The most extensive coniferous region in the world is the northern boreal forest, where *Picea, Abies, Pinus,* and *Larix* (Pinaceae) are dominant genera (Richardson and Rundel 1998). Conifer-dominated ecosystems are more frequent in the Northern Hemisphere, while in the Southern Hemisphere conifers are typically found either individually or as associates in mixed hardwood-conifer forests (Ogden and Stewart 1995). Geographic ranges of the species vary widely, from continent-wide (e.g., *Pinus sylvestris* Mill.) to narrow endemics only recently discovered (e.g., *Wollemia nobilis* W.G. Jones, K.D. Hill & J.M. Allen). Many species are rare and/or threatened with extinction (Farjon et al. 1999). In *Pinus*, geographic ranges have been shown to decrease with increasing proximity to the equator (Stevens and Enquist 1998), while species diversity increases dramatically along this same gradient (Farjon et al. 1993).

Overall, the Northern Hemisphere contains about 70% of total conifer diversity (Farjon 2001). Regions with high species diversity include California, Mexico, the Chinese provinces of Sichuan and Yunnan, and the eastern Himalayas, Japan, Taiwan, and New Caledonia (Farjon 2001). Pinaceae comprises 11 genera and 238 species distributed throughout Eurasia, North Africa, the Himalayas, and North and Central America. *Pinus*, with approximately 120 species, is the largest genus. The only member of Pinaceae that occurs naturally in the Southern Hemisphere is *Pinus merkusii* Jungh. & de Vriese, with a distribution that crosses the equator in Sumatra.

Podocarpaceae, with approximately 19 genera and 190 species, and Araucariaceae, with three genera and approximately 42 species, are distributed across the Southern Hemisphere and the tropics. Podocarpaceae occurs in Africa, South America, Australia, South and East Asia, Indonesia, and numerous other islands of the South Pacific. Other Podocarpaceae taxa occur north of the equator in East Africa, Japan, China, Central America, and Mexico. *Podocarpus*, with ca. 105 is the largest genus, and once better studied, may eventually be shown to be more diverse than *Pinus* (Farjon 2001, 2003). Araucariaceae occurs in South America, South and East Asia, Australia, and on islands throughout the South Pacific. The largest genera are *Agathis* (ca. 23 species) and *Araucaria* (19 species).

Sciadopitys verticillata Siebold & Zucc., the sole representative of Sciadopityaceae, is endemic to southern Japan. Taxaceae (6 genera and 24 species) occurs primarily in the Northern Hemisphere (North and Central America, Eurasia, and the Himalayas), but *Taxus sumatrana* Miquel de Laub. occurs south of the equator and the monotypic genus *Austrotaxus* is endemic to New Caledonia. *Taxus*, with 10 species, is the largest genus. Cupressaceae, with approximately 31 genera (approximately 18 monotypic) and 165 species (Little 2006), occurs on every continent except Antarctica. *Juniperus*, with ca. 67 species, is the largest genus.

1.1.3 Life History and Ecology

Most conifers are monopodal trees, and include the largest and longest living non-clonal organisms on Earth (Waring and Franklin 1979). Western North American ecosystems provide a striking array of the world's tallest and largest trees, including *Sequoia*, the tallest (maximum height 115 m), and *Sequoiadendron*, the most massive (>1,400 m^3). Other genera that attain extraordinary size in western North America include *Pseudotsuga, Picea, Abies, Pinus, Thuja*, and *Chaemaecyparis*. However, this habit is not geographically limited; *Agathis australis* Steud. and *Dacrycarpus dacrydioides* (A. Rich) de Laub. (New Zealand), *Fitzroya cupressoides* I.M. Johnst. (South America), *Cryptomeria japonica* D. Don (Japan) and *Taxodium mucronatum* Ten. (Mexico and Guatemala) are all remarkable. Other conifers are shrubs either throughout their range (e.g., *Microcachrys tetragona* Hook.f.) or at their altitudinal extremes where they may take on a Krummholz form (e.g., *Pinus albicaulis* Engelm., *Athrotaxis selaginoides* D. Don). *Parasitaxus usta* Vieill. de Laub. (Podocarpaceae) is the only parasitic conifer; it obtains carbon from the roots of *Falcatifolium taxoides* (Brongn. & Gris) de Laub. (Podocarpaceae) via a vesicular-arbuscular mycorrhizal association (Feild and Brodribb 2005).

Conifer forests achieve dominance in a variety of environments through a suite of structural characters (Waring and Franklin 1979). The leaves of most conifers are evergreen (retained for years, sometimes decades), and possess several modifications that reduce water loss while conducting photosynthesis under a wider range of conditions than most angiosperms. Conifer leaves are typically needle-like (Araucariaceae, Pinaceae, Podocarpaceae, Sciadopityaceae, and Taxaceae) or scale-like (Cupressaceae), conferring a high surface area to volume ratio and maximizing the diffusion of heat. The conical crowns, the separation between branch layers, the arrangement, density, and orientation of leaves on branches, the thickness of the cuticle covering the epidermis, and the distribution and degree to which the stomata are sunken in the epidermis are important in enhancing photosynthesis and limiting environmental stress (Smith and Brewer 1994). Roughly 20 species in five genera are deciduous (Pinaceae: *Larix, Pseudolarix*; Cupressaceae: *Glyptostrobus, Metasequoia*, and *Taxodium*).

Loehle (1988) estimated a typical life span of North American conifers at 400 years, while Enright and Ogden (1995) estimate 525 years for all Southern Hemisphere conifers. This is in stark contrast to the 250 years calculated for angiosperm trees (Loehle 1988). *Pinus longaeva* D.K. Bailey is the oldest recorded, non-clonal living organism in the world, with one living individual aged at ca. 4,700 years. A 9,550-year old *Picea abies* (L.) H. Karst has been recently reported from Sweden, but awaits publication in a peer-reviewed journal. There are a number of species with the potential to exceed 2,000 years (e.g., *Sequoiadendron giganteum* (Lindl.) J. Buchholz, *Lagarostrobos franklinii* (Hook.f.) Quinn, *Fitzroya cupressoides* I.M. Johnst.) (Lanner 2002).

Conifers have unisexual reproductive structures, with ovulate and pollen cones either on the same (monoecious) or different plants (dioecious). Other gymnosperm groups—Cycadales, Ginkgoales and Gnetales—are dioecious (only rarely monoecious), and dioecy occurs in genera of the families Araucariaceae, Podocarpaceae, Taxaceae, and Cupressaceae (Coulter and Chamberlain 1917; Sporne 1965). Only Pinaceae and Sciadopityaceae are exclusively monoecious. The reproductive cycle of most conifers is one to three years (Owens et al. 1998). The minimum age to first seed set is highly variable, but in natural populations of *Pinus*, ranges between ca. 10–25 years (Mirov 1967; Lanner 1998). However, the first seed crops and the seed from early producers are likely to be minimal in number with reduced viability (Lanner 1998). This makes it difficult to establish generation times, complicating calculations of per-generation mutation rates and effective population size.

While certainly less common than in angiosperms, conifers display a wide range of asexual forms of reproduction that allow them to maintain dominance at a site (Ogden and Stewart 1995). These include resprouting from basal lignotubers in *Sequoia sempervirens* Endl., from the root collar in *Pinus rigida* Mill., from epicormic buds on buried stems in *Actinostrobus acuminatus* Parl., and vegetative layering in *Picea* and *Phyllocladus aspleniifolius* (Labill.) Hook.f.

Outcrossing in conifers is promoted through dioecy, monoecy, and physical separation of the sexes on the plant. Self-fertilization is possible, but the effects of inbreeding depression are pronounced (Mirov 1967), leading to a reduction in seed set and growth (Keeley and Zedler 1998; Sorensen 2001). Nevertheless, it is possible that facultative selfing has proved beneficial by providing a means for conifers to disperse across the landscape, taking advantage of landscape disturbances and responding to changing climates. Prezygotic isolating mechanisms in conifers are limited (Williams et al. 2001), allowing for the potential of interspecific hybridization. However, in *Pinus*, the ability to hybridize is generally restricted to members of the same subsection, suggesting that barriers develop through time. Even amongst closely related species, some pairings never yield any progeny (Critchfield 1986). While studies documenting potential hybrid speciation exist (Ma et al. 2006), most interspecific hybridization is geographically restricted to regions of sympatry. Even so, introgression at the local level may prove important in the maintenance of intraspecific heterozygosity (Mirov 1967; Ledig 1998).

Seed dispersal most commonly occurs via wind, as in the dry, winged seeds of most Pinaceae, or a combination of birds and small mammals as in seeds surrounded by arils or epimatia (Taxaceae or Podocarpaeae), or the dry, wingless seeds of the "stone pines". In *Juniperus*, the unit of dispersal is the fleshy cone. Bird dispersal is more predominant in the

Southern Hemisphere due to the prevalence of the Podocarpaceae (Enright et al. 1995). Seed transport in excess of 22 km has been reported for bird dispersal in *Pinus* (Lanner 1998; Ledig 1998), while the range of pollen dispersal can be on the order of tens to hundreds of kilometers (Burczyk et al. 2004). Widespread distribution of the pollen acts mainly as a cohesive force reducing population differentiation, while occasional long-distance dispersal of the seed provides a means for species migration and population establishment (Ledig 1998). In at least some conifers, migration rates have been shown to be among the fastest of all tree species (Ledig 1998; Sannikov and Sannikova 2008).

Conifers are found in ecosystems that can exhibit tremendous biomass accumulation and some of the highest worldwide productivities (Franklin and Halpern 2000). Given their propensity to attain great heights, they are commonly canopy emergents. Conifers are generally early successional, light-demanding species unable to regenerate in mass under dense canopies. However, due to varying degrees of shade tolerance (Enright and Ogden 1995), some species occur in late successional forests where they are able to regenerate in the understory (e.g., members of Taxaceae, *Tsuga canadensis* Carriére, *Prumnopitys ferruginea* (D. Don) de Laub.) (Enright and Ogden 1995). Through periods of episodic recruitment following disturbance, coupled with their tremendous longevity, "relictual" conifer stands or individuals of early successional species can be found in mixed hardwood-conifer forests.

Disturbance is an integral component of succession for many conifers. Most shade-intolerant species have evolved strategies to take advantage of a variety of disturbance regimes (Agee 1998; Enright and Ogden 1995). Fire has probably been the most thoroughly studied disturbance (Veblen et al. 1995; Agee 1998) and has been an intensive selective force in the evolution of conifer life-history strategies. Fire strategies vary by species, and adaptations include cone serotiny and flammable foliage (e.g., *Pinus contorta* Dougl. ex Loudon), resprouting (e.g., *Widdringtonia cupressoides* Endl., *Sequoia sempervirens*, *Pinus rigida*), insulating bark (e.g., *Pinus ponderosa* Douglas ex P. Lawson & C. Lawson), and the seedling grass stage of several species of pines (e.g., *Pinus devoniana* Lindl., *P. palustris* Mill., *P. merkusii* Jungh. & de Vriese) (Keeley and Zedler 1998). Many conifers lacking these specific life history features are adapted to reinvade burned sites through the production of light, wind born seeds (Barnes 1991; Larson and Franklin 2005).

1.1.4 Cytology and Genetics

Conifer basic chromosome numbers vary from nine in Podocarpaceae to 22 in Pinaceae (*Pseudolarix*). The ancestral condition is likely to be 12 chromosomes (Flory 1936; Page 1990), however this has not been examined in a phylogenetic framework. Numbers can be conserved within genera, as in *Pinus* ($n = 12$), or they can be highly variable, as in *Dacrydium* or *Podocarpus* (Page 1990). Polyploidy has played a minor role in the evolution of conifers, the only naturally occurring cases are tetraploid *Fitzroya cupressoides* and hexaploid *Sequoia sempervirens* (Ahuja 2005).

Genetic diversity in conifers is generally high, promoted by large population sizes, long life spans, outcrossing reproductive systems, high mutation rates, and long distance dispersal of pollen, and sometimes seeds (Hamrick et al. 1992; Ledig 1998). Hamrick et al. (1992) estimated an average of 71.1% polymorphic loci and 16.9% expected heterozygosity across representative gymnosperms heavily favoring conifers. Ledig (1998) recognizes *Pinus* as one of the most variable of organisms with an average of 70.4% polymorphic loci and typical expected heterozygosity of 13 to 16%. Quiroga and Premoli (2007) reported 57.0% polymorphic loci and an expected heterozygosity of 14.8% in *Podocarpus parlatorei* Pilg., values within the reported range for other conifers. Some conifers do have low levels of genetic diversity. Most known examples are narrow endemics, including *Pinus torreyana* Carrière (Provan et al. 1999), *Picea chihuahuana* Martínez (Ledig et al. 1997), and *Picea omorika* (Pančić) Purk. (Ballian et al. 2006). In contrast, *Pinus resinosa* Aiton has low genetic diversity but a wide geographic distribution in eastern North America (Walter and Epperson 2005). Due to their outcrossing reproductive system, the ability of pollen to travel vast distances, and occasional long-distance seed dispersal, most species of conifers show little among population differentiation (Ledig 1998). Exceptions occur where drift is acting on small, fragmented populations (Ledig et al. 1997; Ge et al. 1998; Ballian et al. 2006).

As a result of their life history traits, conifers will generally have large effective population sizes (N_e), though variation by species is expected according to individual history (Syring et al. 2007a, b). Across *Pinus*, N_e estimates range from 1.7×10^4 in *P. flexilis* James to 1.2×10^5 in *P. lambertiana* Dougl. (Syring et al. 2007b). Values for three species of *Picea* are on the same order of magnitude as the higher *Pinus* estimates ($1.2–1.5 \times 10^5$) (Bouillé and Bousquet 2005). For comparison, reports from both inbreeding and outcrossing angiosperm species are typically less than 1.0×10^4 (Schoen and Brown 1991; Reusch et al. 2000). Large N_e promotes the retention of allelic diversity and has implications for phylogenetic analyses (see below). Because conifer species are less likely to form large, contiguous populations in the Southern Hemisphere (Enright 1995), it is tempting to assume that N_e will be larger for Northern Hemisphere species. However, geographic range is known to be a poor predictor of N_e (Syring et al. 2007b). Future estimates of N_e would prove informative.

1.2 Morphology and Fossil History

1.2.1 Morphology

Conifers are woody trees or shrubs with resin canals and single-veined simple leaves reduced to needles, scales, or blades. They have unisexual simple pollen cones and compound ovulate cones. Different interpretations have been proposed for the morphological and embryological characters that unite conifers (Hart 1987; Loconte and Stevenson 1990; Rothwell and Serbet 1994; Doyle 2006). Some characters are present in other plant groups, living and extinct, including uniseriate rays in the wood, also in *Ginkgo*, a torus in the tracheid pits, also in *Ginkgo*, *Gnetum*, and *Ephedra*, and simple needle-like leaves, also in *Ephedra*. The resin canals that are present in almost all conifers, variously distributed in leaves, shoots, roots, and/or seed coats, have been considered a synapomorphy (shared derived character); but the mucilage canals of *Gingko* are similar (Hart 1987). Other characters thought to be synapomorphies, such as the compound ovulate cone with inverted ovules, are lost, reduced, or otherwise modified in some genera (see below). Simple pollen cones with helically arranged scale-like microspophylls bearing free sporangia abaxially, are prevalent, but in Cupressaceae the microsporophylls are cyclical and borne terminally on lateral shoots (Stewart and Rothwell 1993). Characters such as five or fewer free nuclear divisions during embryogenesis and a stratified or tiered proembryo system are considered conifer synapomorphies (Hart 1987; Loconte and Stevenson 1990), but our knowledge of these characters is lacking in many living and fossil species.

Ovulate cones. The ovulate cone of conifers is generally interpreted as a compound inflorescence that includes a central axis that gives rise to fertile axillary shoots, often reduced to ovuliferous scales (Florin 1951). Cones of Cupressaceae, Sciadopityaceae, Araucariaceae, and Pinaceae have bract-scale complexes that are bilateral and dorsiventrally compressed. In Araucariaceae and some Cupressaceae, the bract-scale complex shows varying degrees of fusion, and in some taxa it is difficult to identify these structures (Tomlinson and Takaso 2002; Farjon and Ortiz Garcia 2003).

Independent, extreme reduction of the cone scale complex has taken place in Podocarpaceae and Taxaceae. The Podocarpaceae cone is composed of one or two ovules subtended by a scale that is often modified into an epimatium, which is in turn subtended by a bract (Tomlinson and Takaso 2002). Exceptionally, up to 15 ovule bearing complexes per cone can be present (*Prumnopitys*). In Taxaceae, the ovule is born terminally on its axis. In *Taxus*, the terminal ovule and aril are produced on a short secondary axis subtended by bracts, occasionally with indeterminate growth, while in *Torreya* they are produced on a primary axis (Tomlinson and Takaso

2002). In *Juniperus* (Cupressaceae), the cone scales are fleshy and fused into a bird dispersed "berry-like" structure. Multiple lineages of *Pinus* (Pinaceae) have cones with relatively few scales and enlarged, functionally wingless, bird-dispersed seeds. One of the two seeds per cone scale often aborts, presumably allowing for the more extensive growth of the surviving seed.

Wood. The exceptional size and height of many conifers with respect to other living organisms is due in part to the strength of their wood (secondary xylem), which is composed of thick walled vertical tracheids with bordered pits and lacks vessels. In addition to conducting water and nutrients, these cells provide much greater mechanical support than thin-walled parenchyma cells (Greguss 1955). The type of pitting, together with the arrangement of the horizontal rays, is diagnostic for conifer families. The horizontal rays have also undergone specialization, from homogeneous, thin walled parenchyma as seen in cycads, *Ginkgo*, and fossil conifer woods similar to modern Araucariaceae and Podocarpaceae, to heterogeneous, with variously pitted ray parenchyma and ray tracheids. Heterogeneous rays are found in two separate lineages, Cupressaceae (*Sequoia* and *Metasequoia*) and Pinaceae.

Pollen morphology and ovule orientation. The pollen grains of many conifers have air bladders, or sacci. The presence of air bladders facilitates pollen dispersal by wind, although their primary function is probably to orient pollen grains on pollen drops exuded on the micropyle of the ovulate cone, allowing germination towards the nucellar chamber (Doyle 1945; Tomlinson 1994). In many conifers, fertilization is facilitated by the absorption of the pollen drop, which draws the pollen inside the nucellus. Pollen drops appear to be functionally linked to ovule inversion and the presence of pollen sacs (Tomlinson and Takaso 2002). Families with ovules that are inverted during pollination (Pinaceae and Podocarpaceae) tend to have saccate pollen, and families with erect ovules during pollination (Araucariaceae, Sciadopityaceae, Taxaceae, and Cupressaceae) have nonsaccate pollen.

1.2.2 Fossil Record

Conifers have a rich fossil history, and evidently the living species represent only a fraction of past diversity (reviewed in Stockey 1982; Alvin 1988, Miller 1988; Rothwell and Scheckler 1988; Rothwell et al. 2005; Stockey et al. 2005). Gymnosperms were morphologically diverse during the Carboniferous (Pennsylvanian; ca. 300 Mya) and Permian (ca. 250 Mya). The sister group to conifers may be the Cordaitales, a diverse lineage of small woody shrubs or trees with large, helically arranged strap-shaped leaves and compound, monosporangiate ovulate and pollen cones known from Pennsylvanian to Permian compressions/impressions and permineralizations in Europe and North America (Grand'Eury 1877; Florin 1951; Crane 1985; Doyle and Donoghue 1986; Rothwell 1988; Rothwell et al. 2005; Hilton and Bateman 2006).

The earliest credible conifer dates from Middle Pennsylvanian (ca. 310 Mya) as fragments of shoots and leaves described from England as *Swillingtonia denticulata* Scott and Chaloner. Well preserved remains of Voltziales (the "walchian" conifers) appear by the Upper Pennsylvanian (ca. 300 Mya) and whole plant reconstructions based on attached fossils or on detached vegetative organs, pollen, and seed cones with matching cuticle variation have been proposed for taxa from North America, Eurasia, and South America (Rothwell et al. 2005). Representative Voltziales have been included in phylogenetic analyses linking them to modern conifers (Miller 1988; Rothwell and Serbet 1994; Hilton and Bateman 2006). Voltzialian conifers from the Permian and Triassic (ca. 300–200 Mya) have been described as "transition conifers" (Florin 1951) because they appear to show similarities to modern conifers. For example, *Pseudovoltzia* has been considered to be similar to *Cryptomeria* in possessing many ovules born on lobed scales (Miller 1982; Stockey et al. 2005).

Fossils that have been attributed to extant conifer families appear in the Triassic, including *Compsostrobus neotericus* Delevoryas and Hope (Pinaceae), *Rissikia media* (Tenison-Woods) Townrow (Podocarpaceae), *Stachyotaxus septentrionalis* Agardh (Palissyaceae, with similarities to Taxaceae), and *Parasciadopitys aequata* Yao, Taylor & Taylor (Cupressaceae or Sciadopityaceae). In the absence of whole plant reconstructions, these are best treated as members of the stem lineages that led to modern families, and thus they require caution when used for molecular clock calibrations (see below). Conifer diversity expanded during the Mesozoic (ca. 250–65 Mya), with fossils attributed to at least the stem lineages of all six extant families present by the Middle Jurassic (ca. 160 Mya) (Stewart and Rothwell 1993). Early fossils attributed to the crown of extant families are: *Araucarites phillipsii* Carruthers and *Brachyphyllum mamillare* Lindley & Hutton (Araucariaceae), *Elatides thomasii* Harris (Cupressaceae), *Pinus belgica* Alvin (Pinaceae), *Pseudoaraucaria heeri* Alvin (Pinaceae), *Podocarpus ryosekiensis* Kimura, Ohana & Mimota (Podocarpaceae), *Sciadopitys macrophylla* Manum (Sciadopityaceae), and *Paleotaxus redeviva* Florin (Taxaceae).

Among extant genera, only *Araucaria* is unequivocally represented in the Jurassic (ca. 200 Mya). *Pseudolarix* has been reported from the Upper Jurassic (ca. 145 Mya) (LePage and Basinger 1995) but these fossils lack anatomical confirmation. By the Lower Cretaceous (ca. 125 Mya), several extant genera of other families are known, including *Pinus* (Pinaceae), *Sciadopitys* (Sciadopityaceae), *Metasequoia* (Cupressaceae), and *Nageia* (Podocarpaceae).

1.2.3 Molecular Clock Calibration

Fixing the age of an evolutionary divergence in a molecular clock study can be used to estimate the divergence *time* of another node; and to estimate evolutionary *rates*. The assumptions required for a molecular clock introduce substantial limitations that should not be underestimated when interpreting results. The amount of divergence (shown as a branch length on a tree) encompasses two factors that are very difficult to tease apart: rate and time. For example, the same observed amount of relative divergence might have been created by a fast rate of evolution over a short time, a slow rate over a long time, or some combination of these factors. To convert a phylogenetic tree to a molecular clock, one or more of the nodes where lineages diverge must be fixed in time. The process of assigning a chronological date to a node is called calibrating, and the date is usually based on the fossil record. Unless the evolutionary rate has remained constant, the branch lengths must then be "fitted" so that all of the extant lineages that split from a common ancestor arrive at the tips of the tree at the same time—the present time if living organisms are sampled.

There are three important issues to consider when calibrating a molecular clock, and each adds to the cumulative level of uncertainty: accuracy of the age estimate for the fossil, methods to smooth uneven evolutionary rates among branches, and the association of a fossil to a particular node on the tree. Many fossil ages have been improved with more precise radiometric dating and by comparisons and refinements among worldwide strata (Ogg et al. 2008). Statistical models for rate smoothing are an active area of theoretical research (Sanderson 2002; Yang and Rannala 2006). Even if the uncertainty involved with fossil dating and rate smoothing are properly taken into account, placing the fossil at an inappropriate node can produce wildly inaccurate dates for other nodes (e.g., Grauer and Martin 2004; Magallon 2004). One of the biggest challenges is to determine if a fossil represents a species that would fit along the *stem* of a lineage or if it supports the most recent common ancestor of extant groups (called the *crown*). Calibrating with the oldest fossil is often described as a conservative choice, but placing the oldest known fossil at the crown node is not conservative and should only be done with justification. The abundance of conifer fossils with as-yet poorly understood relationships to living species can create extremely misleading molecular clocks (Willyard et al. 2007).

The neutral rate of evolution (often called μ) is an important parameter for many models of population and lineage dynamics. Estimation of μ requires some form of a molecular clock, whether explicit (i.e., by using a phylogeny) or implicit (i.e., using pairs of samples for which the phylogenetic relationship is assumed to be known). Regions of nucleotide sequences that are noncoding (e.g., intergenic spacers and introns) and third codon positions of protein coding regions can be used to estimate silent, (also called synonymous) relative substitution rates. With the addition of a fixed calibration point and a generation time, these relative rates can be converted to absolute rates. As with any use of a molecular clock, the assumption that the selected calibration age matches the assigned divergence has a major effect on the results. Dramatic lineage-specific (Gaut et al. 1996; Soltis et al. 2002) and locus-specific rate differences have been reported among plants (Senchina et al. 2003; Mower et al. 2007). Our knowledge of the true range of rate variation for conifers (and for all plants) is just beginning to unfold, but it seems prudent to examine the calibration assumptions if a rate estimation falls far outside of reported ranges.

As an example of a useful molecular clock, two alternative calibration points were examined in a combined analysis of cpDNA sequences and morphological characters in Pinaceae (Gernandt et al. 2008). *Pityostrobus bernissartensis* Alvin (Barremian/Aptian, ca. 123 Mya) was used to set a minimum age for the divergence of *Pinus* from the *Picea-Cathaya* clade. The oldest fossil record for an extant Pinaceae genus is *Pseudolarix erensis* Krassilov (LePage and Basinger 1995) (Mongolia, Upper Oxfordian, ca. 155 Mya), but this lacks the anatomical details required for confident determination. Regardless, using it to constrain the divergence of *Pseudolarix* from its sister group results in age estimates that are only slightly older than the *Pityostrobus* calibration.

The oldest representative of *Pinus*, *P. belgica* Alvin, has been placed at strikingly different nodes in published studies. This fossil has been associated with the divergence between *Pinus* and other modern Pinaceae genera (Wang et al. 2000; García-Gil et al. 2003; Willyard et al. 2007) or at the divergence of subgenera (Sokol and Williams 2005). Because Alvin (1960) further ascribed *P. belgica* to subsection *Pinus*, its age has also been applied to the divergence between representatives from the sections of subg. *Pinus* (Krupkin et al. 1996; Dvornyk et al. 2002; Brown et al. 2004; Eckert and Hall 2006). The association of the oldest *Pinus* fossil with a crown node has been shown to result in unreasonable age and rate estimates (see below) and a substantial distortion in biogeographical interpretations (Willyard et al. 2007). Some alternative calibrations that have been published for *Pinus* should also be approached with a generous dose of skepticism. For example, an origin for the genus *Pinus* of 195 Mya (Kutil and Williams 2001) was based on a presumed Jurassic origin (Miller 1977) that lacks explicit fossil evidence. This calibration scenario was bolstered with the assumption that both pine subgenera were already present by the Cretaceous. However, reassignment of a fossil from the Magothy Formation in Delaware from subgenus *Strobus* (Miller 1977) to subgenus *Pinus* (Miller and Malinky 1986) negates the support for a Jurassic origin for *Pinus*. Based on our current understanding, the appearance of subg. *Strobus* dates to the Upper

Cretaceous based on permineralized wood anatomy (Santonian, ca. 85 Mya, *Pinuxylon* sp.; (Meijer 2000), or to the mid Eocene based on either leaf anatomy (ca. 45 Mya, *Pinus similkameenensis*; (Miller 1973), or ovulate cones (ca. 43 Mya; Axelrod 1986). Because extant species of subgenus *Pinus* and subgenus *Strobus* are diagnosable by two vs. one fibrovascular bundle per leaf (Gernandt et al. 2005), foliage fossils may offer better support for *Pinus* than other fossil organs. The divergence of *Pinus* subgenera 47–48 Mya based on the presence of fossils representing both subgenera in the Eocene (Kossack and Kinlaw 1999) has been used to make reasonable estimates. A 45 Mya calibration also shows surprising agreement with a recent cpDNA-based estimate of seed plant divergences using entirely different calibration sources (Magallón and Sanderson 2005). When the divergence of Gnetophytes and Pinaceae is constrained with a fossil date of 216 Mya, estimated divergence dates for *Pinus* subgenera are ca. 50 Mya.

While reasonable estimates of mutation rate would benefit the conifer community, these rates are difficult to estimate, display notoriously wide variance even within genera (Willyard et al. 2007), and are lacking for many important taxa. The volatility of the estimates is mostly attributable to inconsistent calibration (above), complicated by the need to choose an arbitrary generation time in organisms with long and overlapping fertile lifespans. *Pinus* has received the most attention to date, and published synonymous mutation rate estimates for *Pinus* vary 50-fold. For the nuclear genome, the higher rates are angiosperm-like ($\mu = 2.8 \times 10^{-9}$ substitutions per site per year for *gypsy*-like retrotransposons (Kossack and Kinlaw 1999) and similar rates from antigenic distances (Prager et al. 1976). Other nuclear estimates are far slower than angiosperms (e.g., $\mu = 0.05 \times 10^{-9}$ for *Adh*; Dvornyk et al. 2002; also Brown et al. 2004; Sokol and Williams 2005; Ma et al. 2006; Pyhäjärvi et al. 2007; Gernandt et al. 2008). Bouillé and Bousquet (2005) also report nuclear genome values for three species of *Picea* ($\mu = 1.1$–1.7×10^{-8}) that are similar to the slower estimates for *Pinus*. Some absolute substitution rates calculated for the pine chloroplast (e.g., $\mu = 0.06 \times 10^{-9}$ in Krupkin et al. 1996) are among the slowest reported for any plant. These slow rates appear inconsistent with estimates of per-generation deleterious mutation rates, which are known to be at least 10-fold higher in pines than in self-compatible annual flowering plants (Karkkainen et al. 1996; Klekowski 1998). In contrast, nuclear genome estimates inferred from calibrating the split of *Pinus* subgenera at 45 Mya ($\mu = 1.3 \times 10^{-9}$) are of the same order of magnitude as those reported from seed plant phylogeny (Willyard et al. 2007 and references therein). When corrected for a mean generation time (e.g., 25 or 50 years), mutation rates in conifers may actually be 1.5–4.0 fold faster than annual angiosperms (Gaut et al. 1996; Koch et al. 2000; Clark et al. 2005).

1.3 Phylogenetics

1.3.1 Organellar Studies

Chloroplast DNA was the initial data source for comparative molecular genetic studies of plants, and it has had an enormous impact on the fields of phylogenetics and population genetics for almost 20 years. A pioneering study of chloroplast-based phylogeny was conducted in *Pinus* (Strauss and Doerksen 1990) and one of the first plant chloroplast genomes to be completely sequenced was *Pinus thunbergii* Parl. (Wakasugi et al. 1994). Another important early discovery was the absence of the ca. 25 Kbp inverted repeat in *Pinus* (Strauss et al. 1988) and other conifers (Raubeson and Jansen 1992), which typically have plastomes ca. 120,000–130,000 bp in length.

In contrast to most other seed plants, the chloroplast is predominantly paternally inherited in conifers (Mogensen 1996). However, small percentages (9 of 361 progeny from a controlled cross) of maternal chloroplast inheritance and heteroplasmy (3 of 80 open-pollinated seeds) have been documented in *Chamaecyparis obtusa* Siebold & Zucc. (Shiraishi et al. 2001). Early reports of limited maternal inheritance and heteroplasmy in *Pinus banksiana* Lamb. × *P. contorta* hybrids (Dong et al. 1992) and in *Pinus radiata* D. Don (Cato and Richardson 1996) were not as rigorously documented, and require confirmation. The prevalence of this phenomenon in other conifers remains to be evaluated.

The population dynamics of organellar DNA has important consequences. The effective population size (N_e) of organelles is one-half (in monoecious plants) to one-quarter (in dioecious plants) that of nuclear genes (Birky et al. 1983), resulting in more rapid coalescence than nuclear loci. Therefore, the retention of ancestral polymorphisms is predicted to be less common at organellar loci. In practice, this means that geographic partitioning of genetic diversity is often recovered, making plastid and mitochondrial DNA very popular in phylogeographic studies of conifers (reviewed in Petit et al. 2005). Note that most of these studies have only considered single species in isolation, and that broader taxonomic sampling may reveal alternative explanations for genetic differentiation. For example, Liston et al. (2007) found that very strong cpDNA differentiation between northern and southern populations of *Pinus lambertiana* was likely due to introgression from *Pinus albicaulis*. The haploid nature of organellar genomes makes them an effective tool for recognizing interspecific hybridization, including both recent (Senjo et al. 1999; Liston et al. 2007; Wachowiak and Prus-Glowacki 2008) and ancient (Wang and Szmidt 1994) events.

Reasons for the early popularity of chloroplast DNA include its haploid nature, high copy number per cell, stable structure, rarity of recombination, relatively small size, primer availability, and ease of amplification. A limitation for the utility of chloroplast DNA is its relatively slow substitution rate (Willyard et al. 2007). This was less of a problem when restriction fragment length polymorphism (RFLP) approaches were common in the 1980s and early 1990s, but as DNA sequencing became routine, the 0.5–3 Kb regions typically sequenced often have insufficient variation for comparisons at the intraspecific level and among closely related species. Another limitation that must be considered with uniparentally inherited chloroplast (and mitochondrial) markers is the potential for phylogenetic patterns to be obscured by natural interspecific hybridization and subsequent reticulate evolution (Little 2004; Liston et al. 2007).

The most variable part of the chloroplast genome is mononucleotide repeats, and these microsatellites have been characterized in *Pinus* and widely used in conifers (Petit et al. 2005). However, homology assessment is problematic with length variants, limiting phylogenetic utility of these data. Furthermore, analysis of 244 chloroplast microsatellites in 15 accessions of *Arabidopsis thaliana* (L.) Heynh. found a linear relationship between repeat length and mutation rate (Jakobsson et al. 2007), a correlation that is not accounted for in most estimates of population genetic parameters from microsatellites.

Recent advances in genomic analysis have the potential to revolutionize the phylogenetic and population genetic analyses of conifers (and other plants). Massively parallel sequencing technologies produce millions of base pairs in a single run, and thus are very well-suited for sequencing multiple chloroplast genomes in a single run (Cronn et al. 2008; Parks et al. 2009; Whittall et al. 2010). Obtaining dozens or even hundreds of nearly complete chloroplast genomes will permit definitive analyses of this organelle, unbiased by the short length of Sanger sequences or the unusual mutation patterns of microsatellites.

In contrast to chloroplasts, plant mitochondrial genomes are characterized by an extremely slow rate of sequence evolution (Wolfe et al. 1987; Mower et al. 2007) and structural complexity due to frequent intragenomic recombination and the presence of duplicated and/or rearranged subgenomic molecules. The mitochondrial genome of conifers has not been well-characterized. Beyond an estimate of 1,000 Kbp for the *Larix* mitochondrial genome (Kumar et al. 1995), little is known about its size and structure in conifers. Due to the low rate of nucleotide substitution, the most popular markers in conifer mtDNA studies are length variants (often minisatellites) in introns or other non-coding regions (Jaramillo-Correa et al. 2003).

Mitochondrial inheritance is generally considered to be strictly maternal in animals and most plants (Birky 2001), but see Ballard and Whitlock (2004) and McCauley and Olson (2008) for some exceptions. In contrast to this general pattern, there is ultrastructural evidence for paternal or biparental inheritance of mitochondria in all conifers except Pinaceae (Mogensen 1996; Wilson and Owens 2003). However, genetic confirmation of paternal inheritance has only been obtained for Cupressaceae (Neale et al. 1989, 1991; Kondo et al. 1998). Maternal mtDNA inheritance predominates in Pinaceae, but there is genetic evidence for a small amount (6 of 125 seedlings) of paternal mtDNA inheritance in controlled crosses between *Pinus banksiana* and *P. contorta* (Wagner et al. 1991). The discovery of recombinant mitotypes in a *Picea mariana* Britton, Sterns, & Poggenb. and *P. rubens* Sarg. hybrid zone is also indicative of biparental inheritance (Jaramillo-Correa and Bousquet 2005). However, these authors found no heteroplasmic individuals among the 834 trees analyzed, providing further evidence for the rarity of paternal mtDNA inheritance in Pinaceae.

Several studies have taken advantage of the contrasting maternal and paternal inheritance of cytoplasmic organelles in Pinaceae to compare historical patterns of gene flow via pollen and seed (Petit et al. 2005; see also Aizawa et al. 2007; Meng et al. 2007; Jaramillo-Correa et al. 2008). In addition, mtDNA has been used in studies of interspecific hybridization (Senjo et al. 1999; Ballard and Whitlock 2004; Jaramillo-Correa and Bousquet 2005) and phylogeny (Gugerli et al. 2001a; Semerikov and Lascoux 2003) in Pinaceae. Although studies within species and between hybridizing taxa have been very informative, the use of mitochondrial loci in plant phylogenetic studies requires caution. In addition to occasional biparental inheritance and structural complexity, interspecific horizontal gene transfer between mitochondrial genomes has been documented in angiosperms and the Gnetales (Richardson and Palmer 2007).

1.3.2 Nuclear Studies

1.3.2.1 Benefits of Low-copy Nuclear Markers

The use of low-copy nuclear loci, either in addition to or as an alternative to cpDNA, offer three primary benefits in phylogenetics. First, low-copy nuclear markers can sample a range of substitution rates and patterns (Yang 1998), and they can provide greater resolution than cpDNA and nuclear ribosomal DNA (nrDNA) for identifying temporally-compressed radiation events, relationships among closely-related species, and complex historical hybridization events (Harris and Disotell 1998; Springer et al. 2001; Cronn et al. 2002; Malcomer 2002; Cronn et al. 2003; Small et al. 2004; Syring et al. 2007b). Studies that include direct comparisons of the phylogenetic utility of low-copy nDNA to cpDNA in conifers confirm the differential in variability between these data sources. In *Pinus*, silent substitutions per site averaged 3.1-fold higher across eight nuclear loci (4,427 bp) compared with a sample of three cpDNA regions (3,318 bp) (Willyard et al. 2007). In a more taxon-rich sample across four loci, Syring et al. (2005) found divergence rates to be intermediate to the nuclear ribosomal internal transcribed spacers (nrITS) and cpDNA, with exons diverging 2.1-times faster than cpDNA, and introns diverging 1.3-times faster than nrITS. Both studies clearly illustrate the variability in the divergence rates among nDNA. Willyard et al. (2007) documented a ca. 3.3-fold difference (0.063 to 0.205 substitutions/sites) in silent substitution rates, while Syring et al. (2005) found an eight-fold difference between the slowest and fastest evolving regions. For gaining insight into the overall phylogenetic pattern in *Pinus*, the range in divergence amongst loci has proven beneficial in reconstructing a combination of both deep and shallow nodes (Syring et al. 2005, 2007b). The patterns of among-locus substitution rate variation reported in *Pinus* are typical of the broader Pinaceae (Wang et al. 2000; Gros-Louis et al. 2005) as well as the Cupressaceae (Kusumi et al. 2002).

Secondly, low-copy nDNA loci chosen from separate genetic linkage groups provide multiple independent markers for use in phylogenetics, in contrast to cpDNA markers which derive from a single linkage group. One of the benefits of this independence is that these loci can be used for resolving conflict between cpDNA and nrITS hypotheses. For example, conflicting topologies in *Pinus* using cpDNA (Wang et al. 1999; Gernandt et al. 2005) and nrDNA (Liston et al. 2003) have been partially resolved using multiple low-copy markers (Syring et al. 2005). Multiple independent markers have also revealed the complex phylogenetic history of *Pinus chiapensis* (Martínez) Andresen (Syring et al. 2007a).

Thirdley, the exponential increase of expressed sequence tags (ESTs) in GenBank (*http: //www.ncbi.nlm.nih.gov/dbEST/dbEST summary.html)* is beginning to provide a nearly limitless supply of low-copy nuclear loci for use in conifer phylogenetics. Nearly one million conifer EST markers were submitted to GenBank by the end of 2009. However, coverage across the conifers remains uneven (Table 1-1). Pinaceae has ca. 850,000, with ca. 470,000 markers from *Picea* (*Picea glauca* (Moench) Voss—ca. 300,000) and ca. 365,000 markers from *Pinus* (*Pinus taeda* L.—ca. 330,000). Cupressaceae and Taxaceae are poorly represented, and there are no EST entries for Araucariaceae, Podocarpaceae, or Sciadopityaceae.

1.3.2.2 Challenges and Pitfalls Working with Nuclear Loci

Despite the aforementioned benefits of low-copy nuclear markers, their use in conifer phylogenetics to date has been narrowly limited within the Pinaceae and Cupressaceae. While the tedious progress of nDNA marker development is partly responsible, biological factors, detailed below, present more fundamental challenges.

One of the more challenging aspects of working with low-copy nuclear genes is assessing orthology in the presence of high heterozygosity (Small et al. 2004; Syring et al. 2005). The haploid genome size (1C) of conifers ranges from 5.8–36.0 pg, which is roughly an order of magnitude higher than the size of most angiosperm genomes (Ohri and Khoshoo 1986; Davies et al. 1997; Grotkopp et al. 2004; Leitch et al. 2005). Large genome size acts as a hindrance to the development and application of low-copy nuclear markers for phylogenetic applications, as multigene families (Kinlaw and Neale 1997) and repetitive and retrotransposon DNA (Friesen et al. 2001) are abundant. Despite these obstacles, genomics and comparative mapping efforts (Brown et al. 2001; Devey et al. 2004; Krutovsky et al. 2004; Pelgas et al. 2006) reveal the scope of conserved synteny (linear gene order) among species, and aid in the detection of orthology. Comparative analysis of expressed sequence tag (EST) markers (Kirst et al. 2003; Ujino-Ihara et al. 2005) has revealed the conservation of genetic structure across highly divergent members of the seed plants, thereby providing aid in the choice of loci to address the level of interest.

Arguably the most daunting obstacle to working with nDNA in conifers is the slow coalescence (time to monophyly) of allele lineages within species, a property demonstrated to impact both repeated gene families (e.g., nrDNA, Gernandt et al. 2001; Campbell et al. 2005) and low copy loci (Bouillé and Bousquet 2005; Syring et al. 2007b). Bouillé and Bousquet (2005) demonstrated a striking case of non-coalescence across three low-copy nuclear genes in three species of *Picea*. Allelic coalescence among randomly selected alleles from these spruce species was estimated at 10–18 million years, values that overlapped with estimated divergence times (13–20 Mya) for the species studied. Explicit tests in *Pinus* subsects. *Strobus* and *Cembroides* demonstrated that allele lineages in 11 of 28 species tested failed to coalesce at one low-copy locus (Syring et al. 2007b). For these 11 species, coalescent expectations indicate that reciprocal monophyly will be more likely than paraphyly in 1.71 to 24.0 million years, and that complete genome-wide coalescence in these species may require up to 76.3 million years (Rosenberg 2003; Syring et al. 2007b).

The timing of speciation events and the historic effective population size are critical factors impacting the rate of allelic coalescence. Both factors interplay in determining whether species are genetically unique, and whether gene trees can accurately trace phylogenetic history. Because other conifer species share many life history traits with *Picea* and *Pinus* (e.g., highly outcrossing, long-lived trees with large effective population sizes), we should expect to encounter similar phylogenetic difficulties in related genera and families (Bouillé and Bousquet 2005; Syring et al. 2007b). Issues of non-coalescence highlight the importance of considering the magnitude of intraspecific diversity within the overall pattern of phylogenetic divergence (Fig. 1-2). While allelic non-coalescence may be highly problematic in the reconstruction of resolved phylogenies, it should be noted that this information is decidedly informative in elucidating the *process* of speciation (Syring et al. 2007b).

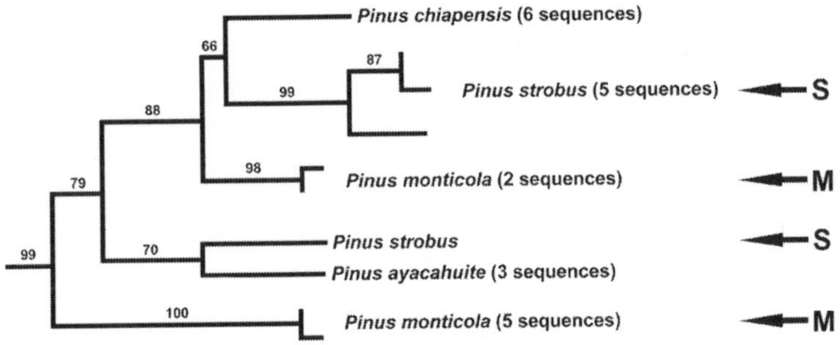

Figure 1-2 Example of allelic nonmonophyly. One of two most parsimonious trees derived from phylogenetic analysis of the cesA locus (see Syring et al. 2007b). Bootstrap values from 1,000 replicates are shown near nodes. Bold arrows indicate two cases where alleles have failed to coalesce, one in *Pinus strobus* (S) and the other in *P. monticola* (M). Note that one of the purposes of this particular study was to determine the sister species to *P. chiapensis*. If only a single sample had been sequenced per species then the choice of the individual could have dramatically affected the conclusions.

Low-copy nuclear loci (one to several copies) offer several distinct advantages over working with markers from large families, such as nrITS. Complications arising from the determination of orthology and allelic coalescence are exacerbated in large gene families. Florescence *in situ* hybridization has indicated the presence of 6–20 nrITS arrays across four genera of the Pinaceae (Liu et al. 2003), and ca. 4% of the *Picea* genome is apparently composed of nrITS copies (Bobola et al. 1992). The use of ITS in phylogenetic analyses requires the assumption that concerted evolution is fully homogenizing all of the variation present in the numerous gene copies. Multiple studies have indicated this is not the case in Pinaceae (Bobola et al. 1992; Gernandt and Liston 1999; Gernandt et al. 2001; Campbell et al. 2005). Differentiating between paralogy and incomplete lineage sorting then becomes impossible. Further, the rates of pseudogenization, recombination (particularly among ITS1 subrepeats), and PCR-mediated recombination can all be expected to be greater than in low-copy loci (Gernandt et al. 2001; Álvarez and Wendel 2003). Finally, despite greater variation in nrITS relative to cpDNA and to low-copy nDNA (Syring et al. 2005), the difficulty in assessing sequence homology severely hampers the construction of alignments. These considerations negate the usual benefits of high variability and ease of amplification in nrITS.

The preceding difficulties highlight some of the major challenges of incorporating nDNA into a phylogenetic analysis. A final difficulty is that the potential to amplify many independent loci raises the question of how these loci should be analyzed. To improve our phylogenetic resolution there is a temptation to use a total evidence approach or conditional combination approaches (Huelsenbeck et al. 1996; Sanderson and Shaffer 2002; Maddison and Knowles 2006). However, the retention of ancient allele lineages detected through population-level sampling reinforces the concept of locus independence and significantly reduces the options for traditional combination approaches. Coalescent-based approaches (Hudson 1990) may provide the greatest chance of resolving interlocus discrepancies driven by the process of incomplete lineage sorting, although effective multi-locus methods are still under development (Edwards et al. 2007; Wakeley 2009).

Acknowledgements

We thank Matt Parks and Rich Cronn for comments on the manuscript. Figure 1-1 was designed by Jesús Romero. Parts of the introduction, classification and phylogeny, and morphology sections were translated and modified from a chapter in "El árbol de la vida: sistemática y evolución de los seres vivos" coauthored by David Gernandt and Alejandra Vázquez Lobo and edited by Pablo Vargas Gómez and Rafael Zardoya for Editorial Reverté. This work was supported by National Science Foundation grants ATOL-0629508 to Sarah Mathews and DEB-0317103.

References

Agee JK (1998) Fire and pine ecosystems. In: DM Richardson (ed) Ecology and Biogeography of *Pinus*. Cambridge Univ Press, Cambridge, UK, pp 193–218.

Ahuja MR (2005) Polyploidy in gymnosperms: revisited. Silvae Genet 54: 59–69.

Aizawa M, Yoshimaru H, Saito H, Katsuki T, Kawahara T, Kitamura K, Shi F, Kaji M (2007) Phylogeography of a northeast Asian spruce, *Picea jezoensis*, inferred from genetic variation observed in organelle DNA markers. Mol Ecol 16: 3393–3405.

Álvarez I, Wendel JF (2003) Ribosomal ITS sequences and plant phylogenetic inference. Mol Phylogenet Evol 29: 417–434.

Alvin KL (1960) Further conifers of the Pinaceae from the Wealden Formation of Belgium. Instit Roy Sci Nat Belg Mem 146: 1–39.

Alvin KL (1988) On a new specimen of *Pseudoaraucaria major* Fliche (Pinaceae) from the Cretaceous of the Isle of Wight. Bot J Linn Soc 97: 159–170.

Arber E, Parkin J (1907) On the origins of angiosperms. Bot J Linn Soc 38: 29–80.

Axelrod DI (1986) Cenozoic history of some western American pines. Ann MO Bot Gard 73: 565.

Ballard JWO, Whitlock MC (2004) The incomplete natural history of mitochondria. Mol Ecol 13: 729–744.

Ballian D, Longauer R, Mikiæ T, Paule L, Kajba D, Gömöry D (2006) Genetic structure of a rare European conifer, Serbian spruce (*Picea omorika* (Panè.) Purk.). Plant Syst Evol 260: 53–63.

Barker NP, Muller EM, Mill RR (2004) A yellowwood by any other name: molecular systematics and the taxonomy of *Podocarpus* and the Podocarpaceae in southern Africa. S Afr J Sci 100: 629–632.

Barnes BV (1991) Deciduous forests of North America. In: E Röhrig, B Ulrich (eds) Ecosystems of the World, Temperate Deciduous Forests, vol. 7. Elsevier, New York, USA, pp 219–334.

Birky CW (2001) The inheritance of genes in mitochondria and chloroplasts: laws, mechanisms, and models. Annu Rev Genet 35: 125–148.

Birky CW, Maruyama T, Fuerst P (1983) An approach to population and evolutionary genetic theory for genes in mitochondria and chloroplasts, and some results. Genetics 103: 513–527.

Bobola MS, Smith DE, Klein AS (1992) Five major nuclear ribosomal repeats represent a large and variable fraction of the genomic DNA of *Picea rubens* and *P. mariana*. Mol Biol Evol 9: 125–137.

Bond WJ (1989) The tortoise and the hare: ecology of angiosperm dominance and gymnosperm persistence. Biol J Linn Soc 36: 227–249.

Bouillé M, Bousquet J (2005) Trans-species shared polymorphisms at orthologous nuclear gene loci among distant species in the conifer *Picea* (Pinaceae): implications for the long-term maintenance of genetic diversity in trees. Am J Bot 92: 63–73.

Bowe LM, Coat G, DePamphilis CW (2000) Phylogeny of seed plants based on all three genomic compartments: extant gymnosperms are monophyletic and Gnetales' closest relatives are conifers. Proc Natl Acad Sci USA 97: 4092–4097.

Brown GR, Kadel EE, Bassoni DL, Kiehne KL, Temesgen B, van Buijtenen JP, Sewell MM, Marshall KA, Neale DB (2001) Anchored reference loci in loblolly pine (*Pinus taeda* L.) for integrating pine genomics. Genetics 159: 799–809.

Brown GR, Gill GG, Kuntz RJ, Langley CH, Neale DB (2004) Nucleotide diversity and linkage disequilibrium in loblolly pine. Proc Natl Acad Sci USA 101: 15255–15260.

Brunsfeld SJ, Soltis PS, Soltis DE, Gadek PA, Quinn CJ, Streng DD, Ranker TA (1994) Phylogenetic relationships among the genera of Taxodiaceae and Cupressaceae: evidence from *rbcL* sequences. Syst Bot 19: 253–262.

Burczyk J, Lewandowski A, Chalupka W (2004) Local pollen dispersal and distant gene flow in Norway spruce (*Picea abies* [L.] Karst.). For Ecol Manag 197: 39–48.

Campbell CS, Wright WA, Cox M, Vining TF, Major CS, Arsenault MP (2005) Nuclear ribosomal DNA internal transcribed spacer 1 (ITS1) in *Picea* (Pinaceae): sequence divergence and structure. Mol Phylogenet Evol 35: 165–185.

Carlquist S (1996) Wood, bark, and stem anatomy of Gnetales: a summary. Int J Plant Sci 157: S58–S76.

Cato SA, Richardson TE (1996) Inter- and intraspecific polymorphism at chloroplast SSR loci and the inheritance of plastids in *Pinus radiata* D. Don. Theor Appl Genet 93: 587–592.

Chaw SM, Long H, Wang BS, Zharkikh A, Lie WH (1993) The phylogenetic position of Taxaceae based on 18S rRNA sequences. J Mol Evol 37: 624–630.

Chaw SM, Zharkikh A, Sung HM, Lau TC, Li WH (1997) Molecular phylogeny of extant gymnosperms and seed plant evolution: analysis of nuclear 18S rRNA sequences. Mol Biol Evol 14: 56–68.

Chaw SM, Parkinson CL, Cheng Y, Vincent TM, Palmer JD (2000) Seed plant phylogeny inferred from all three plant genomes: monophyly of extant gymnosperms and origin of Gnetales from conifers. Proc Natl Acad Sci USA 97: 4086–4091.

Cheng Y, Nicolson RG, Tripp K, Chaw S-M (2000) Phylogeny of Taxaceae and Cephalotaxaceae genera inferred from chloroplast *matK* gene and nuclear rDNA ITS region. Mol Phylogenet Evol 14: 353–365.

Clark RM, Tavaré S, Doebley J (2005) Estimating a nucleotide substitution rate for maize from polymorphism at a major domestication locus. Mol Biol Evol 22: 2304–2312.

Conran JG, Wood GM, Martin PG, Dowd JM, Quinn CJ, Gadek PA, Price RA (2000) Generic relationships within and between the gymnosperm families Podocarpaceae and Phyllocladaceae based on an analysis of the chloroplast gene *rbcL*. Aust J Bot 48: 715–724.

Coomes DA, Allen RB, Bentley WA, Burrows LE, Canham CD, Fagan L, Forsyth DM, Gaxiola-Alcantar A, Parfitt RL, Ruscoe WA, Wardle DA, Wilson DJ, Wright EF (2005) The hare,

the tortoise and the crocodile: the ecology of angiosperm dominance, conifer persistence and fern filtering. J Ecol 93: 918–935.
Coulter JM, Chamberlain CJ (1917) Morphology of Gymnosperms. Chicago Univ Press, Chicago, IL, USA.
Crane PR (1985) Phylogenetic analysis of seed plants and the origin of angiosperms. Ann MO Bot Gard 72: 716–793.
Critchfield WB (1986) Hybridization and classification of the white pines (*Pinus* section *Strobus*). Taxon 35: 647–656.
Cronn RC, Small RL, Haselkorn T, Wendel JF (2002) Rapid diversification of the cotton genus (*Gossypium*: Malvaceae) revealed by analysis of sixteen nuclear and chloroplast genes. Am J Bot 89: 707–725.
Cronn RC, Small RL, Haselkorn T, Wendel JF (2003) Cryptic repeated genomic recombination during speciation in *Gossypium gossypioides*. Evolution 57: 2475–89.
Cronn R, Liston A, Parks M, Gernandt DS, Shen R, Mockler T (2008) Multiplex sequencing of plant chloroplast genomes using Solexa sequencing-by-synthesis technology. Nucl Acids Res 36: e122.
Davies BJ, O'Brien IEW, Murray BG (1997) Karyotypes, chromosome bands and genome size variation in New Zealand endemic gymnosperms. Plant Syst Evol 208: 169–185.
Devey ME, Carson SD, Nolan MF, Matheson AC, Te Riini C, Hohepa J (2004) QTL associations for density and diameter in *Pinus radiata* and the potential for marker-aided selection. Theor Appl Genet 108: 516–24.
Dong J, Wagner DB, Yanchuk AD, Carlson MR, Magnussen S, Wang X-R, Szmidt AE (1992) Paternal chloroplast DNA inheritance in *Pinus contorta* and *Pinus banksiana*: independence of parental species or cross direction. J Hered 83: 419–422.
Doyle J (1945) Development lines in pollination mechanisms in the Coniferales. Proc Roy Dublin Soc 24: 43–62.
Doyle JA (2006) Seed ferns and the origin of angiosperms. J Torr Bot Soc 133: 169–209.
Doyle JA, Donoghue MJ (1986) Seed plant phylogeny and the origin of angiosperms: an experimental cladistic approach. Bot Rev 52: 321–431.
Dvornyk V, Sirviö A, Mikkonen M, Savolainen O (2002) Low nucleotide diversity at the *pal1* locus in the widely distributed *Pinus sylvestris*. Mol Biol Evol 19: 178–188.
Eckenwalder JE (1976) Re-evaluation of Cupressaceae and Taxodiaceae: a proposed merger. Madroño 23: 237–256.
Eckert AJ, Hall BD (2006) Phylogeny, historical biogeography, and patterns of diversification for *Pinus* (Pinaceae): phylogenetic tests of fossil-based hypotheses. Mol Phylogenet Evol 40: 166–182.
Edwards SV, Liu L, Pearl DK (2007) High-resolution species trees without concatenation. Proc Natl Acad Sci USA 104: 5936–5941.
Enright NJ (1995) Conifers of tropical Australasia. In*:* NJ Enright, RS Hill (eds) Ecology of the Southern Conifers. Smithsonian Institution Press, Washington DC, USA, pp 197–222.
Enright NJ, Ogden J (1995) The southern conifers—a synthesis. In*:* NJ Enright, RS Hill (eds) Ecology of the Southern Conifers. Smithsonian Institution Press, Washington DC, USA, pp 271–287.
Enright NJ, Hill RS, Veblen TT (1995) The southern conifers—an introduction. In*:* NJ Enright, RS Hill (eds) Ecology of the Southern Conifers. Smithsonian Institution Press, Washington DC, USA, pp 1–9.
Farjon A (2001) World Checklist and Bibliography of Conifers. Royal Botanic Gardens, Kew, UK.
Farjon A (2003) The remaining diversity of conifers. Acta Hort 615: 75–89.
Farjon A (2005) A Monograph of Cupressaceae and *Sciadopitys*. Royal Botanic Gardens, Kew, UK.
Farjon A, Page CN, Schellevis N (1993) A preliminary world list of threatened conifer taxa. Biodivers Conserv 2: 304–326.

Farjon A, Page CN, Brown MJ (1999) Conifers: Status Survey and Conservation Action Plan. Island Press, Washington DC, USA.
Farjon A, Ortiz García S (2003) Cone and ovule development in *Cunninghamia* and *Taiwania* (Cupressaceae sensu lato) and its significance for conifer evolution. Am J Bot 90: 8–16.
Feild TS, Brodribb TJ (2005) A unique mode of parasitism in the conifer coral tree *Parasitaxus ustus* (Podocarpaceae). Plant Cell Environ 28: 1316–1325.
Florin R (1948) On the morphology and relationships of the Taxaceae. Bot Gaz 110: 31–39.
Florin R (1951) Evolution in cordaites and conifers. Acta Hort Berg 15: 285–388.
Flory WS (1936) Chromosome numbers and phylogeny in the gymnosperms. J Arnold Arbor 17: 83–89.
Franklin JF, Halpern CB (2000) Pacific northwest forests. In*:* MG Barbour, WD Billings (eds) North American Terrestrial Vegetation. Cambridge Univ Press, New York, USA, pp 123–159.
Friedman W (1994) The evolution of embryogeny in seed plants and the developmental origin and early history of endosperm. Am J Bot 81: 1468–1486.
Friesen N, Brandes A, Heslop-Harrison JS (2001) Diversity, origin, and distribution of retrotransposons (*gypsy* and *copia*) in conifers. Mol Biol Evol 18: 1176–1188.
Gadek PA, Alpers DL, Heslewood MM, Quinn CJ (2000) Relationships within Cupressaceae sensu lato: a combined morphological and molecular approach. Am J Bot 87: 1044–1057.
García-Gil M, Mikkonen M, Savolainen O (2003) Nucleotide diversity at two phytochrome loci along a latitudinal cline in *Pinus sylvestris*. Mol Ecol 12: 1195–1206.
Gaut BS, Morton BR, McCaig BC, Clegg MT (1996) Substitution rate comparisons between grasses and palms: synonymous rate differences at the nuclear gene *Adh* parallel rate differences at the plastid gene *rbcL*. Proc Natl Acad Sci USA 93: 10274–10279.
Ge S, Hong D, Wang H, Liu Z, Zhang C (1998) Population genetic structure and conservation of an endangered conifer, *Cathaya argyrophylla* (Pinaceae). Int J Plant Sci 159: 351–357.
Gernandt DS, Liston A (1999) Internal transcribed spacer region evolution in *Larix* and *Pseudotsuga* (Pinaceae). Am J Bot 86: 711–723.
Gernandt DS, Liston A, Piñero D (2001) Variation in the nrDNA ITS of *Pinus* subsection *Cembroides*: implications for molecular systematic studies of pine species complexes. Mol Phylogenet Evol 21: 449–467.
Gernandt DS, Gaeda López G, Ortiz García S, Liston A (2005) Phylogeny and classification of *Pinus*. Taxon 54: 29–42.
Gernandt DS, Magallón S, Geada López G, Zerón Flores O, Willyard A, Liston A (2008) Use of simultaneous analyses to guide fossil-based calibrations of Pinaceae phylogeny. Int J Plant Sci 169: 1086–1099.
Grand'Eury FC (1877) Flore carbonifère du Département de la Loire et du centre de la France. Imprimerie Nationale, Paris, France.
Graur D, Martin W (2004) Reading the entrails of chickens: molecular timescales of evolution and the illusion of precision. Trends Genet 20: 80–86.
Greguss P (1955) Identification of Living Gymnosperms on the Basis of Xylotomy. Akadémiai Kiadó, Budapest, Hungary.
Gros-Louis M-C, Bousquet J, Pâques LE, Isabel N (2005) Species-diagnostic markers in *Larix* spp. based on RAPDs and nuclear, cpDNA, and mtDNA gene sequences, and their phylogenetic implications. Tree Genet Genomes 1: 50–63.
Grotkopp E, Rejmanek M, Sanderson MJ, Rost TL (2004) Evolution of genome size in pines (*Pinus*) and its life-history correlates: supertree analyses. Evolution 58: 1705–1729.
Gugerli F, Senn J, Anzidei M, Madaghiele A, Büchler U, Sperisen C, Vendramin GG (2001a) Chloroplast microsatellites and mitochondrial *nad1* intron 2 sequences indicate congruent phylogenetic relationships among Swiss stone pine (*Pinus cembra*), Siberian stone pine (*Pinus sibirica*), and Siberian dwarf pine (*Pinus pumila*). Mol Ecol 10: 1489–1497.

Gugerli F, Sperisen C, Büchler U, Brunner I, Brodbeck S, Palmer J, Qiu Y (2001b) The evolutionary split of Pinaceae from other conifers: evidence from an intron loss and a multigene phylogeny. Mol Phylogenet Evol 21: 167–175.

Hamrick JL, Godt MJW, Sherman-Broyles SL (1992) Factors influencing levels of genetic diversity in woody plant species. New For 6: 95–124.

Harris EE, Disotell TR (1998) Nuclear gene trees and the phylogenetic relationships of the mangabeys (Primates: Papionini). Mol Biol Evol 15: 892–900.

Hart JA (1987) A cladistic analysis of conifers: preliminary results. J Arnold Arbor 68: 269–307.

Hilton J, Bateman RM (2006) Pteridosperms are the backbone of seed-plant phylogeny. J Torr Bot Soc 133: 119–168.

Hirao T, Watanabe A, Kurita M, Kondo T, Takata K (2008) Complete nucleotide sequence of the *Cryptomeria japonica* D. Don. chloroplast genome and comparative chloroplast genomics: diversified genomic structure of coniferous species. BMC Plant Biol 8: 70.

Hudson RR (1990) Gene genealogies and the coalescent process. In: D Futuyma, J Antonovics (eds) Oxford Surveys in Evolutionary Biology. Oxford Univ Press, Oxford, UK, pp 1–44.

Huelsenbeck JP, Bull JJ, Cunningham CW (1996) Combining data in phylogenetic analysis. Trends Ecol Evol 11: 152–158.

Jakobsson M, Säll T, Lind-Halldén C, Halldén C (2007) Evolution of chloroplast mononucleotide microsatellites in *Arabidopsis thaliana*. Theor Appl Genet 114: 223–235.

Jaramillo-Correa JP, Bousquet J (2005) Mitochondrial genome recombination in the zone of contact between two hybridizing conifers. Genetics 171: 1951–1962.

Jaramillo-Correa JP, Bousquet J, Beaulieu J, Isabel N, Perron M, Bouillé M (2003) Cross-species amplification of mitochondrial DNA sequence-tagged-site markers in conifers: the nature of polymorphism and variation within and among species in *Picea*. Theor Appl Genet 106: 1353–1367.

Jaramillo-Correa JP, Aguirre-Planter E, Khasa DP, Eguiarte LE, Piñero D, Furnier GR, Bousquet J (2008) Ancestry and divergence of subtropical montane forest isolates: molecular biogeography of the genus *Abies* (Pinaceae) in southern Mexico and Guatemala. Mol Ecol 17: 2476–2490.

Kärkkäinen K, Koski V, Savolainen O, 1 (1996) Geographical variation in the inbreeding depression of Scots pine. Evolution 53: 111–119.

Keeley JE, Zedler PH (1998) Evolution of life histories in *Pinus*. In: DM Richardson (ed) Ecology and Biogeography of *Pinus*. Cambridge Univ Press, Cambridge, UK, pp 219–250.

Kelch DG (1997) The phylogeny of the Podocarpaceae based on morphological evidence. Syst Bot 22: 113–132.

Kelch DG (1998) Phylogeny of Podocarpaceae: comparison of evidence from morphology and 18S rDNA. Am J Bot 85: 986–986.

Kinlaw CS, Neale DB (1997) Complex gene families in pine genomes. Trends Plant Sci 2: 356–359.

Kirst M, Johnson AF, Baucom C, Ulrich E, Hubbard K, Staggs R, Paule C, Retzel E, Whetten R, Sederoff R (2003) Apparent homology of expressed genes from wood-forming tissues of loblolly pine (*Pinus taeda* L.) with *Arabidopsis thaliana*. Proc Natl Acad Sci USA 100: 7383–7388.

Klekowski E (1998) Mutation rates in mangroves and other plants. Genetica 102: 325–331.

Knapp M, Mudaliar R, Havell D, Wagstaff SJ, Lockhart PJ (2007) The drowning of New Zealand and the problem of *Agathis*. Syst Biol 56: 862–870.

Koch MA, Haubold B, Mitchell-Olds T (2000) Comparative evolutionary analysis of chalcone synthase and alcohol dehydrogenase loci in *Arabidopsis*, *Arabis*, and related genera (Brassicaceae). Mol Biol Evol 17: 1483–1498.

Kondo T, Tsumura Y, Kawahara T, Okamura M (1998) Paternal inheritance of chloroplast and mitochondrial DNA in interspecific hybrids of *Chamaecyparis* spp. Jap J Breed 48: 177–179.

Kossack DS, Kinlaw CS (1999) *IFG*, a gypsy-like retrotransposon in *Pinus* (Pinaceae) has an extensive history in pines. Plant Mol Biol 39: 417–426.

Krupkin AB, Liston A, Strauss SH (1996) Phylogenetic analysis of the hard pines (*Pinus* subgenus *Pinus*, Pinaceae) from chloroplast DNA restriction site analysis. Am J Bot 83: 489–498.

Krutovsky KV, Troggio M, Brown GR, Jermstad KD, Neale DB (2004) Comparative mapping in the Pinaceae. Genetics 168: 447–461.

Kumar R, Lelu MA, Small I (1995) Purification of mitochondria and mitochondrial nucleic acids from embryogenic suspension cultures of a gymnosperm, *Larix × leptoeuropaea*. Plant Cell Rep 14: 534–538.

Kusumi J, Tsumura Y, Yoshimaru H, Tachida H (2000) Phylogenetic relationships in Taxodiaceae and Cupressaceae sensu stricto based on *matK* gene, *chlL* gene, *trnL-trnF* IGS region, and *trnL* intron sequences. Am J Bot 87: 1480–1488.

Kusumi J, Tsumura Y, Yoshimaru H, Tachida H (2002) Molecular evolution of nuclear genes in Cupressaceae, a group of conifer trees. Mol Biol Evol 19: 736–747.

Kutil B, Williams C (2001) Triplet-repeat microsatellites shared among hard and soft pines. J Hered 92: 327–332.

Lanner RM (1998) Seed dispersal in *Pinus*. In: DM Richardson (ed) Ecology and Biogeography of *Pinus*. Cambridge Univ Press, Cambridge, UK, pp 281–295.

Lanner RM (2002) Why do trees live so long? Ageing Res Rev 1: 653–671.

Larson AJ, Franklin JF (2005) Patterns of conifer tree regeneration following an autumn wildfire event in the western Oregon Cascade Range, USA. For Ecol Manag 218: 25–36.

Ledig FT (1998) Genetic variation in *Pinus*. In: DM Richardson (ed) Ecology and Biogeography of *Pinus*. Cambridge Univ Press, Cambridge, UK, pp 251–280.

Ledig FT, Jacob-Cervantes V, Hodgskiss P, Eguiluz-Piedra T (1997) Recent evolution and divergence among populations of a rare Mexican endemic, Chihuahua spruce, following Holocene climatic warming. Evolution 51: 1815–1827.

Leitch IJ, Soltis DE, Soltis PS, Bennett MD (2005) Evolution of DNA amounts across land plants (Embryophyta). Ann Bot 95: 207–217.

LePage BA, Basinger JF (1995) Evolutionary history of the genus *Pseudolarix* Gordon (Pinaceae). Int J Plant Sci 156: 19–29.

Liston A, Gernandt DS, Vining TF, Campbell CS, Piñero D (2003) Molecular phylogeny of Pinaceae and *Pinus*. In: RR Mill (ed) Proc 4th Int Conifer Conf. Acta Hort, pp 107–114.

Liston A, Parker-Defeniks M, Syring JV, Willyard A, Cronn R (2007) Interspecific phylogenetic analysis enhances intraspecific phylogeographical inference: a case study in *Pinus lambertiana*. Mol Ecol 16: 3926–3937.

Little DP (2004) Documentation of hybridization between Californian cypresses: *Cupressus macnabiana × sargentii*. Syst Bot 29: 825–833.

Little DP (2006) Evolution and circumscription of the true cypresses (Cupressaceae: *Cupressus*). Syst Bot 31: 461–480.

Liu Z-L, Zhang D, Hong D-Y, Wang X-R (2003) Chromosomal localization of 5S and 18S–5.8S–25S ribosomal DNA sites in five Asian pines using fluorescence in situ hybridization. Theor Appl Genet 106: 198–204.

Loconte H, Stevenson DW (1990) Cladistics of the Spermatophyta. Brittonia 42: 197–211.

Loehle C (1988) Tree life history strategies: the role of defenses. Can J For Res/Rev Can Res For 18: 209–222.

Ma X-F, Szmidt AE, Wang X-R (2006) Genetic structure and evolutionary history of a diploid hybrid pine *Pinus densata* inferred from the nucleotide variation at seven gene loci. Mol Biol Evol 23: 807–816.

Maddison WP, Knowles LL (2006) Inferring phylogeny despite incomplete lineage sorting. Syst Biol 55: 21–30.

Magallon S (2004) Dating lineages: molecular and paleontological approaches to the temporal framework of clades. Int J Plant Sci 165: S7–S21.

Magallón SA, Sanderson M (2005) Angiosperm divergence times: the effect of genes, codon positions, and time constraints. Evolution 59: 1653–1670.
Malcomer ST (2002) Phylogeny of *Gaertnera* Lam. (Rubiaceae) based on multiple DNA markers: evidence of rapid radiation in a widespread, morphologically diverse genus. Evolution 56: 42–57.
Mathews S (2009) Phylogenetic relationships among seed plants: persistent questions and the limits of molecular data. Am J Bot 96: 228–236.
McCauley DE, Olson MS (2008) Do recent findings in plant mitochondrial molecular and population genetics have implications for the study of gynodioecy and cyto-nuclear conflict? Evolution 62: 1013–1025.
McCoy SR, Kuehl JV, Boore JL, Raubeson LA (2008) The complete plastid genome sequence of *Welwitschia mirabilis*: an unusually compact plastome with accelerated divergence rates. BMC Evol Biol 8: 130.
Meijer JJF (2000) Fossil woods from the Late Cretaceous Aachen Formation. Rev Palaeobot Palynol 112: 297–336.
Meng L, Yang RUI, Abbott RJ, Miehe G, Hu T, Liu J (2007) Mitochondrial and chloroplast phylogeography of *Picea crassifolia* Kom. (Pinaceae) in the Qinghai-Tibetan Plateau and adjacent highlands. Mol Ecol 16: 4128–4137.
Miller CN (1973) Silicified cones and vegetative remains of *Pinus* from the Eocene of British Columbia. Contrib Mus Paleont Univ Mich 24: 101–118.
Miller CN (1977) Mesozoic conifers. Bot Rev 43: 217–280.
Miller CN (1982) Current status of Paleozoic and Mesozoic conifers. Rev Palaeobot Palynol 37: 99–114.
Miller CN (1988) The origin of modern conifer families. In*:* CB Beck (ed) Origin and Evolution of Gymnosperms. Columbia Univ Press, New York, USA, pp 449–486.
Miller CN, Malinky J (1986) Seed cones of *Pinus* from the late Cretaceous of New Jersey, USA. Rev Palaeobot Palynol 46: 273–291.
Mirov NT (1967) The Genus *Pinus*. Ronald Press, New York, NY, USA.
Mogensen HL (1996) The hows and whys of cytoplasmic inheritance in seed plants. Am J Bot 83: 383–404.
Mower JP, Touzet P, Gummow JS, Delph LF, Palmer JD (2007) Extensive variation in synonymous substitution rates in mitochondrial genes of seed plants. BMC Evol Biol 7: 135.
Neale DB, Marshall KA, Sederoff RR (1989) Chloroplast and mitochondrial DNA are paternally inherited in *Sequoia sempervirens* D. Don Endl. Proc Natl Acad Sci USA 86: 9347–9349.
Neale DB, Marshall KA, Harry DE (1991) Inheritance of chloroplast and mitochondrial DNA in incense-cedar (*Calocedrus decurrens*). Can J For Res/Rev Can Rech For 21: 717–720.
Nixon KC, Crepet WL, Stevenson D, Friis EM (1994) A reevaluation of seed plant phylogeny. Ann MO Bot Gard 81: 484–533.
Ogden J, Stewart GH (1995) Community dynamics of the New Zealand conifers. In*:* NJ Enright, RS Hill (eds) Ecology of the Southern Conifers. Smithsonian Institution Press, Washington DC, USA, pp 81–119.
Ogg JG, Ogg G, Gradstein FM (2008) The Concise Geologic Time Scale. Cambridge Univ Press, Cambridge, UK.
Ohri D, Khoshoo TN (1986) Genome size in gymnosperms. Plant Syst Evol 153: 119–132.
Owens JN, Takaso T, Runions CJ (1998) Pollination in conifers. Trends Plant Sci 3: 479–485.
Page CN (1989) New and maintained genera in the conifer families Podocarpaceae and Pinaceae. Notes Roy Bot Gard Edinburgh 45: 377–395.
Page CN (1990) Coniferophytina (conifers and ginkgoids). In*:* KU Kramer, PS Green (eds) The Families and Genera of Vascular Plants, vol 1: Pteridophytes and Gymnosperms. Springer, Berlin, Germany, pp 278–361.
Parks M, Cronn R, Liston A (2009) Increasing phylogenetic resolution at low taxonomic levels using massively parallel sequencing of chloroplast genomes. BMC Biol 7: 84.

Pelgas B, Beauseigle S, Acheré V, Jeandroz S, Bousquet J, Isabel N (2006) Comparative genome mapping among *Picea glauca*, *P. mariana* × *P. rubens* and *P. abies*, and correspondence with other Pinaceae. Theor Appl Genet 113: 1371–1393.

Petit RJ, Duminil J, Fineschi S, Hampe A, Salvini D, Vendramin GG (2005) Comparative organization of chloroplast, mitochondrial and nuclear diversity in plant populations. Mol Ecol 14: 689–701.

Pilger R (1926) Coniferae. In: A Engler (ed) Die natürlichen Pflanzenfamilien nebst ihren Gattungen und wichtigeren Arten insbesondere der Nutzpflanzen. Engelmann, Leipzig, Germany, pp 121–149.

Prager EM, Fowler DP, Wilson AC (1976) Rates of evolution in conifers (Pinaceae). Evolution 30: 637–649.

Provan J, Soranzo N, Wilson N, Goldstein D, Powell W (1999) A low mutation rate for chloroplast microsatellites. Genetics 153: 943–947.

Pyhäjärvi T, García-Gil MR, Knürr T, Mikkonen M, Wachowiak W, Savolainen O (2007) Demographic history has influenced nucleotide diversity in European *Pinus sylvestris* populations. Genetics 177: 1713–1724.

Quinn CJ, Price RA, Gadek PA (2002) Familial concepts and relationships in the conifers based on *rbcL* and *matK* sequence comparisons. Kew Bull 57: 513–531.

Quiroga MP, Premoli AC (2007) Genetic patterns in *Podocarpus parlatorei* reveal the long-term persistence of cold-tolerant elements in the southern Yungas. J Biogeogr 34: 447–455.

Rai HS, Reeves PA, Peakal R, Olmstead RG, Graham SW (2008) Inference of higher-order conifer relationships from a multi-locus plastid data set. Can J Bot/Rev Can Bot 86: 658–669.

Raubeson LA, Jansen RK (1992) A rare chloroplast-DNA structural mutation is shared by all conifers. Biochem Syst Ecol 20: 17–24.

Reusch TBH, Stam WT, Olsen JL (2000) A microsatellite-based estimation of clonal diversity and population subdivision in *Zostera marina*, a marine flowering plant. Mol Ecol 9: 127–140.

Richardson AO, Palmer JD (2007) Horizontal gene transfer in plants. J Exp Bot 58: 1–9.

Richardson DM, Rundel PW (1998) Ecology and biogeography of *Pinus*: an introduction. In: DM Richardson (ed) Ecology and Biogeography of *Pinus*. Cambridge Univ Press, Cambridge, UK, pp 3–46.

Rosenberg NA (2003) The shapes of neutral gene genealogies in two species: probabilities of monophyly, paraphyly, and polyphyly in a coalescent model. Evolution 57: 1465–1477.

Rothwell GW (1988) Cordaitales. In: CB Beck (ed) Origin and Evolution of Gymnosperms. Columbia Univ Press, New York, USA, pp 273–297.

Rothwell GW, Scheckler SE (1988) Biology of ancestral gymnosperms. In: CB Beck (ed) Origin and Evolution of Gymnosperms. Columbia Univ Press, New York, USA, pp 85–134.

Rothwell GW, Serbet R (1994) Lignophyte phylogeny and the evolution of Spermatophytes: a numerical cladistic analysis. Syst Bot 19: 443–482.

Rothwell GW, Mapes G, Hernandez-Castillo GR (2005) *Hanskerpia* gen. nov. and phylogenetic relationships among the most ancient conifers (Voltziales). Taxon 54: 733–750.

Sahni B (1920) On certain archaic features in the seed of *Taxus baccata* with remarks on the antiquity of the Taxineae. Ann Bot 34: 117–133.

Sanderson MJ (2002) Estimating absolute rates of molecular evolution and divergence times: a penalized likelihood approach. Mol Biol Evol 19: 101–109.

Sanderson MJ, Shaffer HB (2002) Troubleshooting molecular phylogenetic analyses. Annu Rev Ecol Syst 33: 49–72.

Sannikov SN, Sannikova NS (2008) Outline of the hydrochory theory for some coniferous species. Dokl Bot Sci/Dokl Akad Nauk 418: 67–69.

Schoen DJ, Brown AH (1991) Intraspecific variation in population gene diversity and effective population size correlates with the mating system in plants. Proc Natl Acad Sci USA 88: 4494–4497.

Semerikov VL, Lascoux M (2003) Nuclear and cytoplasmic variation within and between Eurasian *Larix* (Pinaceae) species. Am J Bot 90: 1113–1123.

Senchina DS, Alvarez I, Cronn RC, Liu B, Rong J, Noyes RD, Paterson AH, Wing RA, Wilkins TA, Wendel JF, 4 (2003) Rate variation among nuclear genes and the age of polyploidy in *Gossypium*. Mol Biol Evol 20: 633–643.

Senjo M, Kimura K, Watano Y, Ueda K, Shimizu T (1999) Extensive mitochondrial introgression from *Pinus pumila* to *P. parviflora* var. *pentaphylla* (Pinaceae). J Plant Res 112: 97–105.

Setoguchi H, Asakawa Osawa T, Pintaud J-C, Jaffré T, Veillon J-M (1998) Phylogenetic relationships within Araucariaceae based on *rbcL* gene sequences. Am J Bot 85: 1507-1516.

Shiraishi S, Maeda H, Toda T, Seido K, Sasaki Y (2001) Incomplete paternal inheritance of chloroplast DNA recognized in *Chamaecyparis obtusa* using an intraspecific polymorphism of the *trnD-trnY* intergenic spacer region. Theor Appl Genet 102: 935–941.

Sinclair WT, Mill RR, Gardner MF, Woltz P, Jaffré T, Preston J, Hollingsworth ML, Ponge A, Möller M (2002) Evolutionary relationships of the New Caledonian heterotrophic conifer, *Parasitaxus usta* (Podocarpaceae), inferred from chloroplast *trnL-F* intron/spacer and nuclear rDNA ITS2 sequences. Plant Syst Evol 233: 79–104.

Small RL, Cronn RC, Wendel JF (2004) Use of nuclear genes for phylogeny reconstruction in plants. Aust Syst Bot 17: 145–170.

Smith WK, Brewer CA (1994) The adaptive importance of shoot and crown architecture in conifer trees. Am Nat 143: 165–169.

Sokol K, Williams C (2005) Evolution of a triplet repeat in a conifer. Genome 48: 417–426.

Soltis PS, Soltis DE, Savolainen V, Crane PR, Barraclough TG (2002) Rate heterogeneity among lineages of tracheophytes: integration of molecular and fossil data and evidence for molecular living fossils. Proc Natl Acad Sci USA 99: 4430–4435.

Sorensen FC (2001) Effect of population outcrossing rate on inbreeding depression in *Pinus contorta* var. *murrayana* seedlings. Scand J For Res 16: 391–403.

Sporne KR (1965) The Morphology of Gymnosperms: The Structure and Evolution of Primitive Seed-plants, 2nd edn. Hutchinson Univ, London, UK.

Springer MS, DeBry RW, Douady C, Amrine HM, Madsen O, de Jong WW, Stanhope MJ (2001) Mitochondrial versus nuclear gene sequences in deep-level mammalian phylogeny reconstruction. Mol Biol Evol 18: 132–43.

Stefanoviç S, Jager M, Deutsch J, Broutin J, Masselot M (1998) Phylogenetic relationships of conifers inferred from partial 28S rRNA gene sequences. Am J Bot 85: 688–688.

Stevens GC, Enquist BJ (1998) Macroecological limits to the abundance and distribution of *Pinus*. In: DM Richardson (ed) Ecology and Biogeography of *Pinus*. Cambridge Univ Press, Cambridge, UK, pp 183–190.

Stewart WN, Rothwell GW (1993) Paleobotany and the Evolution of Plants, 2nd ed. Cambridge Univ Press, Cambridge, UK.

Stockey RA (1982) The Araucariaceae: an evolutionary perspective. Rev Palaeobot Palynol 37: 133–154.

Stockey RA, Kvacek J, Hill RS, Rothwell GW, Kvacek K (2005) The fossil record of Cupressaceae s. lat. In: A Farjon (ed) A Monograph of Cupressaceae and *Sciadopitys*. Royal Botanic Gardens, Kew, Richmond, Surrey, UK, pp 64–68.

Stopes MC, Kershaw EM (1910) The anatomy of Cretaceous pine leaves. Ann Bot 24: 395–402.

Strauss SH, Doerksen AH (1990) Restriction fragment analysis of pine phylogeny. Evolution 44: 1081–1096.

Strauss SH, Palmer JD, Howe GT, Doerksen AH (1988) Chloroplast genomes of two conifers lack a large inverted repeat and are extensively rearranged. Proc Natl Acad Sci USA 85: 3898–3902.

Syring J, Willyard A, Cronn R, Liston A (2005) Evolutionary relationships among *Pinus* (Pinaceae) subsections inferred from multiple low-copy nuclear loci. Am J Bot 92: 2086–2100.

Syring J, del Castillo RF, Cronn R, Liston A (2007a) Multiple nuclear loci reveal the distinctiveness of the threatened, neotropical *Pinus chiapensis*. Syst Bot 32: 703–717.

Syring J, Farrell K, Businský R, Cronn R, Liston A (2007b) Widespread genealogical nonmonophyly in species of *Pinus* subgenus *Strobus*. Syst Biol 56: 163–181.
Tomlinson PB (1994) Functional morphology of saccate pollen in conifers with special reference to Podocarpaceae. Int J Plant Sci 155: 699–715.
Tomlinson PB, Takaso T (2002) Seed cone structure in conifers in relation to development and pollination: a biological approach. Can J Bot/Rev Can Bot 80: 1250–1273.
Ujino-Ihara T, Kanamori H, Yamane H, Taguchi Y, Namiki N, Mukai Y, Yoshimura K, Tsumura Y (2005) Comparative analysis of expressed sequence tags of conifers and angiosperms reveals sequences specifically conserved in conifers. Plant Mol Biol 59: 895–907.
Veblen TT, Burns BR, Kitzberger T, Lara A, Villalba R (1995) The ecology of the conifers of southern South America. In: NJ Enright, RS Hill (eds) Ecology of the Southern Conifers. Smithsonian Institution Press, Washington DC, USA, pp 120–155.
Wachowiak W, Prus-Glowacki W (2008) Hybridisation processes in sympatric populations of pines *Pinus sylvestris* L., *P. mugo* Turra and *P. uliginosa* Neumann. Plant Syst Evol 271: 29–40.
Wagner DB, Dong J, Carlson MR, Yanchuk AD (1991) Paternal leakage of mitochondrial DNA in *Pinus*. Theor Appl Genet 82: 510–514.
Wakasugi T, Tsudzuki J, Ito S, Nakashima K, Tsudzuki T, Sugiura M (1994) Loss of all *ndh* genes as determined by sequencing the entire chloroplast genome of the black pine *Pinus thunbergii*. Proc Natl Acad Sci USA 91: 9794–9798.
Wakeley J (2009) Coalescent Theory: An Introduction. Roberts & Company, Greenwood Village, CO, USA.
Walter R, Epperson BK (2005) Geographic pattern of genetic diversity in *Pinus resinosa*: contact zone between descendants of glacial refugia. Am J Bot 92: 92–100.
Wang X-R, Szmidt AE (1994) Hybridization and chloroplast DNA variation in a *Pinus* species complex from Asia. Evolution 48: 1020–1031.
Wang X-R, Tsumura Y, Yoshimaru H, Nagasaka K, Szmidt AE (1999) Phylogenetic relationships of Eurasian pines (*Pinus*, Pinaceae) based on chloroplast *rbcL, matK, rpl20, rps18* spacer, and *trnV* intron sequences. Am J Bot 86: 1742–1753.
Wang XQ, Tank DC, Sang T (2000) Phylogeny and divergence times in Pinaceae: evidence from three genomes. Mol Biol Evol 17: 773–781.
Waring RH, Franklin JF (1979) Evergreen coniferous forests of the Pacific Northwest. Science 204: 1380–1386.
Whittall JB, Syring J, Parks M, Buenrostro J, Dick C, Liston A, Cronn R (2010) Finding a (pine) needle in a haystack: chloroplast genome sequence divergence in rare and widespread pines. Mol Ecol 19 (Suppl 1): 100–114.
Williams CG, Zhou Y, Hall SE (2001) A chromosomal region promoting outcrossing in a conifer. Genetics 159: 1283–1289.
Willyard A, Syring J, Gernandt DS, Liston A, Cronn R (2007) Fossil calibration of molecular divergence infers a moderate mutation rate and recent radiations for *Pinus*. Mol Biol Evol 24: 90–101.
Wilson VR, Owens JN (2003) Cytoplasmic inheritance in *Podocarpus totara* (Podocarpaceae). Acta Hort 615: 171–172.
Wolfe KH, Li WH, Sharp PM (1987) Rates of nucleotide substitution vary greatly among plant mitochondrial, chloroplast, and nuclear DNAs. Proc Natl Acad Sci USA 84: 9054–9058.
Yang Z (1998) On the best evolutionary rate for phylogenetic analysis. Syst Biol 47: 125–133.
Yang Z, Rannala B (2006) Bayesian estimation of species divergence times under a molecular clock using multiple fossil calibrations with soft bounds. Mol Biol Evol 23: 212–226.

2

Economic Importance, Breeding Objectives and Achievements

T.J. Mullin,[1,*] B. Andersson,[2] J.-C. Bastien,[3,a] J. Beaulieu,[4] R.D. Burdon,[5] W.S. Dvorak,[6] J.N. King,[7] T. Kondo,[8] J. Krakowski,[9,c] S.J. Lee,[10] S.E. McKeand,[11] L. Pâques,[3,b] A. Raffin,[12] J.H. Russell,[9,d] T. Skrøppa,[13] M. Stoehr[14,e] and A. Yanchuk,[14,f]

ABSTRACT

This chapter reviews the historical context, economic importance, objectives and achievements to-date for many of the more important conifers undergoing domestication through genetic improvement programs around the world. These provide examples of the context in which genomic technologies will have an impact in forestry. Unlike many other crop plants and livestock animals, forest trees have only been exposed to a few cycles of breeding and selection, and most retain very large amounts of genetic variation in natural populations. These factors present both opportunities and hurdles in the effective application of genomic technologies to existing operational breeding programs.

Keywords: operational tree breeding, plantation programs, breeding, selection, genetic testing, genomic technologies, molecular markers, quatitative trait loci, *Pinus, Picea, Pseudotsuga, Larix, Cryptomeria, Chamaecyparis, Cupressus, Thuja, Cunninghamia, Sequoia*

2.1 Introduction

Although conifers are generally regarded as undomesticated trees, genetic improvement through breeding, selection and testing has had a significant impact on the productivity and quality of plantations established in a wide variety of species worldwide. Many conifers have been the target

---

For affiliations see at the end of this chapter on page 127.
*Corresponding author

of tree improvement efforts over the last 50 years, and many of these are now well into their second, third or even fourth cycle of breeding. In the context of these well-established programs, emerging genomic tools offer the greatest potential for immediate impact and deployment of benefits to production forests.

The purpose of this chapter is to describe the context in which genomics can have an impact on current breeding and reforestation of conifers. Descriptions are given for each species or group of species covering historical perspectives, economic importance, breeding objectives, and achievements to-date. In addition, some brief notes are given on the application of genomics technologies, particularly with respect to their current use, or lack thereof, in breeding and selection. While a wide range of species and programs are discussed (see Table 2-1), the list is not exhaustive, although we have attempted to capture some of the most important.

2.2 Pines—*Pinus* L.

The genus *Pinus* is the largest genus in the family Pinaceae and is widely distributed throughout the Northern Hemisphere, with as many as 100 recognized species (Richardson 1998). Many of these are of great economic importance for wood production and are the targets of intensive tree improvement programs, some of the more important of which are discussed here, organized as regional groups.

2.2.1 Northeastern North American Pines (Pinus strobus, P. resinosa, P. banksiana, and P. rigida)

2.2.1.1 Historical Perspective

Four pines grow in the northeast of North America, and all of them have played a major role in the development of this region. Eastern white pine (*Pinus strobus* L.) and red pine (*Pinus resinosa* Ait.) are both characteristic of the Great Lakes and St. Lawrence Forest Regions, where fire plays a role in the establishment of extensive stands (Whitney 1986). While jack pine (*Pinus banksiana* Lamb.) also occurs in the same region, it is primarily a boreal species that is also well adapted to forest fire. It bears serotinous cones, which allow the dispersion of quantities of viable seed following fire (Rudolph and Laidly 1990; Farrar 1995). Pitch pine (*Pinus rigida* Mill.) mainly grows in the Appalachians but in the northeast it can be seen on sandy soils of Pennsylvania, New Jersey, New York and Maine states. It can also be seen in isolated stands up to southern Quebec.

Eastern white pine and red pine were over harvested for many decades, owing to the huge size of the mature trees and their prized wood qualities.

Table 2-1 Species and programs discussed in this chapter.

Family/genus Species/group	Country programs discussed
Pines (*Pinus*)	
Northeastern North American pines (*Pinus strobus, P. resinosa, P. banksiana*, and *P. rigida*)	Canada, United States
Lodgepole Pine (*Pinus contorta*)	Canada, Sweden
Western White Pine (*Pinus monticola*)	Canada, Unites States
Southern Pines (*Pinus taeda, P. elliottii, P. palustris*, and *P. echinata*)	United States, Brazil, Argentina, China, South Africa, Swaziland, Zimbabwe, Australia
Maritime pine (*Pinus pinaster*)	France
Scots pine (*Pinus sylvestris*)	Sweden, Finland, Germany, France, Lithuania, Latvia, Poland, Spain
Radiata Pine (*Pinus radiata*)	New Zealand, Chile, Australia, Spain, South Africa
Spruces (*Picea*)	
Black spruce (*Picea mariana*)	Canada, United States
White spruce (*Picea glauca*)	Canada, United States
Red spruce (*Picea rubens*)	Canada
Sitka spruce (*Picea sitchensis*)	Canada, Great Britain
Norway spruce (*Picea abies*)	Norway, Sweden, Finland, Germany
Other Pinaceae	
Douglas-fir (*Pseudotsuga menziesii*)	Canada, United States, Germany, Belgium, Fance, Italy, Spain, United Kingdom, New Zealand, Argentina, Chile
Larches in Europe (*Larix* spp.)	France
Cypresses (Cupressaceae)	
Sugi (*Cryptomeria japonica*)	Japan
Other Cupressaceae including the whitecedars (*Chamaecyparis lawsoniana, C. nootkatensis, C. obtusa, C. pisifera*), the cypresses (*Cupressus lusitanica, C. macrocarpa, C, sempervirens*), the arborvitae (*Thuja plicata*), Chinese fir (*Cunninghamia lanceolata*), Coastal redwood (*Sequoia sempervirens*)	China, Japan, Korea, New Zealand, Greece, Italy, France, Canada, United States, Columbia, Mexico, El Salvador, Guatemala, Honduras, Kenya, Rwanda, Uganda, Tanzania, South Africa

They were especially suitable for ship masts. By the end of the 19th century, their extensive resources had been decimated, especially in eastern Canada (Daoust and Beaulieu 2004), where they were extensively used in shipbuilding for the British navy. The introduction of an exotic pathogen, the white pine blister rust (*Cronartium ribicola* J.C. Fisher) in the early 20th century decimated remaining eastern white pine stands and caused major losses to advance regeneration. As a result, there are today only scattered remnants of the magnificent natural stands that once covered eastern Canada. Reforestation efforts were engaged for many years to rebuild the pine reserves. Red pine has been one of the most extensively planted species in the northern United States and Canada for many decades. However, the virulence of pests such as blister rust, the white pine weevil (*Pissodes strobi* Peck) and scleroderris canker (*Gremmeniella abietina* Lagerberg Morelet) in large part explains the failures and the cutbacks in the reforestation programs of those two species as well as their reduced presence in the landscape.

The extensive commercial harvest of jack pine forests is more recent, due to poor access to remote stands. The smaller dimension of trees, relative to that of the eastern white and the red pines, made it less attractive to the first settlers. However, as settlement expanded in the 19th century, the need of lumber for house building increased and the utilization of jack pine consequently increased.

Pitch pine, which is a medium-sized tree, was also important during the days of wooden ships (Little and Garrett 1990) due to the large amount of resin its wood contains, which allows it to resist decay. However, it has not been as heavily harvested as the other pines.

2.2.1.2 Economic Importance

Eastern white pine wood is generally pale yellow/white and has medium strength so it can be easily machined. Much of the high grades are now reserved for lumber, while lower-grade material goes into pulp and paper. Its wood is excellent for doors, windows, panelling, moldings and cabinet work (Farrar 1995). Red pine is used primarily for the production of lumber, piling, poles, cabin logs, railway ties, posts, mine timbers, box boards, pulpwood, and fuel. While both species are still a significant resource for industry, their relative economic importance has decreased considerably over the years mainly due to the drastic reduction in supply and quality.

Eastern white pine was formerly one of the favored species for reforestation with annual production up to 40 million seedlings for fiber production and Christmas trees (Eckert and Kuser 1988). However, damage to plantations caused by white pine weevil and blister rust has considerably reduced the interest in this species by public and private landowners. Nevertheless, some reforestation continues with 5 million seedlings still planted annually in eastern Canada. In comparison, reduced demand from industry has caused annual shipments of red pine seedlings to fall to about 1 million, despite that damage from pests is not as serious as for eastern white pine.

Jack pine is now one of the most important commercial tree species in Canada and the Lake States. Its wood is moderately hard and heavy, and relative to other softwoods, of intermediate strength (Hosie, 1979). It is used in building construction as framing, sheathing, scaffolding and interior woodwork. Moreover, it has other commercial applications such as power poles, railroad ties, pilings, mine timbers (Cayford and McRae 1983), and boxes and crates. It is also a good source of wood chips for pulping and commercially important in the manufacture of newsprint in eastern Canada (Law and Valade 1994). Jack pine is extensively planted in Canada with annual shipments of about 80 million seedlings. The vast majority of these seedlings are planted in Ontario and Quebec in the boreal forest where most of the harvesting of jack pine forests occurs.

Pitch pine has a coarse-grained wood that is moderately strong. Southern sources have denser and higher strength and are extensively used in construction of factories, warehouses, bridges, docks, roof trusses, beams, posts, joists and piles. Other uses include interior finish, sheathing, subflooring, fencing, mine timbers, and railroad ties.

2.2.1.3 *Breeding Objectives*

Over the past 50 years, a number of private- and public-sector organizations have carried out research on the genetics of these pine species. This research demonstrated relatively little morphological variation for red pine in provenance trials (Fowler and Lester 1970). As the presence of moderate to large phenotypic variation is necessary to make good progress through breeding, no applied breeding program was initiated for red pine, such a program being not justifiable based on the extent of reforestation activity. For pitch pine, despite the presence of variation in phenotypic traits and its capacity to hybridize with shortleaf and loblolly pines, reforestation programs were not large enough to justify an investment in the development operational breeding activities. In contrast, research on the genetics of eastern white and jack pine showed extensive variation in various adaptive traits and reforestation programs were large enough to support applied breeding programs.

As for most of forest tree species, breeders' objectives for eastern white pine are to develop improved varieties that are adapted to ecological conditions where they are planted. This is made by delineating breeding zones and selecting and hybridizing superior genotypes for height growth, volume, stem straightness and crown shape for these zones. Moreover, due to its high susceptibility to white pine blister rust and white pine weevil, breeders aim to develop resistance to these pests in their varieties. As eastern white pine has not co-evolved with blister rust, gene variants conferring resistance to those harboring these variants are rare. This has prompted the development of hybrids with other white pine species to transfer resistance genes to the eastern white pine gene pool.

Jack pine breeders aim at developing, for each breeding, zone varieties that are improved for height growth and volume, cold hardiness and reduced branching. Jack pine is known to develop undesirable branch and form characteristics, especially on poor sites and when stand density is not sufficiently high. Variation in branch and form traits has been shown to be partially under genetic control and this can be exploited to improve the species. Jack pine is also sensitive to pests such as scleroderris canker and western gall rust (*Endocronartium harknessii* [J.P. Moore] Y. Hirat.) in the western part of its natural range. However, studies have reported resistance to various pests in the species (Yeatman and Teich 1969). More recently, tree breeders have also focussed on wood traits in order to maintain fiber attributes that give a competitive advantage to the industry using jack pine.

2.2.1.4 *Breeding Achievements*

2.2.1.4.1 Eastern White Pine

In eastern Canada, the first breeding program for eastern white pine was initiated in Ontario by Carl Heimburger in 1946 with the aim of developing blister rust resistant varieties. Plus-trees in natural stands, which were free of disease symptoms were propagated and used in the production of interspecific hybrids from rust-resistant species such as Himalayan white pine (Pinus wallichiana A.B. Jackson). The program was successful, resulting in the development of rust-resistant interspecific hybrids (Zsuffa 1981), although selected hybrids were not included in operational eastern white pine seed orchards in Ontario (Cherry et al. 2000).

In the late 1970s, an intensive plus-tree selection program in natural populations was launched by the Ontario Ministry of Natural Resources. By the late 1980s, the province had developed eight breeding populations and a network of 18 seed orchards covering over 130 ha. In the mid-1990s, new breeding activities were carried out with the initiation of a genecological study of eastern white pine seed sources from Ontario east of Lake Superior (Joyce et al. 2002).

In Quebec, the eastern white pine breeding program was initiated in the mid-1970s (Corriveau and Lamontagne 1977). From 1976 to 1986, about 150 plus-tree selections were made in natural stands and established by grafting to set up the first-generation breeding population. Production of full-sib families through controlled crosses was initiated in the early 1990s. Several experimental designs were established to evaluate the general- and specific-combining ability for a variety of traits. Six seed orchards were also established in the 1980s to produce seeds for the reforestation program.

Seeds were also collected in more than 100 natural stands in Quebec and others were obtained from collaborators of neighboring provinces and states to set up genecological tests including over 550 open-pollinated progenies from over 225 seed sources. Early growth and phenological trait assessments made it possible to study the genetic structure and patterns of genetic variation in these white pine populations (Beaulieu et al. 1996; Li et al. 1997a). Breeding values were also estimated for height, 12 years after plantation; 14% gain was expected through selection of the best 50 progenies (Daoust and Beaulieu 2004). A new breeding population was set up, grafting three plus-trees from each of the 50 selected families. Several series of new progeny tests were also established in the 1990s and 2000s including half- and full-sib families collected into the breeding orchards.

Recent efforts have focussed on the development of interspecific hybrids resistant to blister rust. The most promising material developed in Ontario for this purpose by Carl Heimburger and Louis Zsuffa was grafted and put in a breeding orchard in Quebec to facilitate controlled crosses. Since the early 2000s, over 130 control crosses have been made to create new hybrids. Seedlings were inoculated with blister rust, and after two inoculation phases, some of the hybrids appeared promising (G Daoust pers. comm. 2009). Somatic embryogenesis techniques are being used to propagate them (Klimaszewska et al. 2001). For the short-term, these hybrid somatic seedlings are being deployed to clonal trials, further testing resistance to blister rust.

There are also some seed orchard facilities established in the Atlantic region in Canada, by JD Irving Ltd in New Brunswick, the Department of Natural Resources in Nova Scotia, the Department of Environment, Energy and Forestry in Prince Edward Island. These seed orchards are now producing the genetically improved seed required for their reforestation programs.

Research on the genetics of eastern white pine in the eastern United States began in the early 1950s with interspecific hybridization experiments and early tests for resistance to the white pine weevil and the white pine blister rust (Kriebel 2004). From the 1950s to the 1980s, extensive cooperative tree improvement activities took place with range-wide and regional provenance trials providing information on geographical variation in adaptability and growth (Wright 1970; Kriebel 1982). Progeny testing was also carried out which allowed estimating inheritance of growth traits and the potential for genetic gain (Adams and Joly 1977; Kriebel 1978, 1983). Some of the progeny tests have been converted into seed orchards. Results of hybridization experiments demonstrated that the two most promising hybrids exhibiting desirable fiber attributes were *P. strobus* x *Pinus wallichiana* A.B. Jackson and *P. strobus* x *Pinus monticola* Douglas ex D. Don and their reciprocal crosses (Wright 1959; Kriebel 1972). Efforts to develop weevil and blister rust resistance in this species have not yet been successful but continue with application of genetic engineering technologies (Kriebel 2004). Ongoing genomics research at the CFS should lead to better understanding of the interaction between the host and disease, and to the development of efficient tools to select eastern white pine tolerant to blister in the future. Eastern white pine is also highly sensitive to sulfur dioxide and ozone, and genetic variation in tolerance to these air pollutants has been investigated (Karnosky and Houston 1979).

2.2.1.4.2 Jack Pine

Early studies of jack pine demonstrated considerable genetic variation in growth traits and insect and disease resistance (Jeffers and Nienstaedt 1972; Polk 1974; Canavera 1975; Yeatman 1975; Rudolph and Yeatman 1982), and breeding programs have since been established throughout much of its natural range in Canada and in the Lake States. Breeding programs were first based on selection of superior provenances, followed by selection of plus-trees within the best provenances to set up the first breeding and seed orchards.

In Atlantic Canada, the New Brunswick Tree Improvement Council initiated its jack pine breeding activities in the mid-1970s based on a seedling seed orchard strategy. About 850 plus-trees were selected in natural stands or provenance tests, and seed collected to establish open-pollinated family tests and seedling seed orchards (Simpson and Tosh 1997). Selected trees were superior for height, stem straightness and crown shape. About 43 ha of first-generation seed orchards were established from 1978 to 1986, with seed production starting in 1984. Prolific seed production allowed roguing of 50% of the families based on height and stem straightness while still meeting needs of the reforestation program. Selection of top-performing families and best phenotypes within these families for straightness and crown and branching traits was completed in 1997 in order to form the 400-parent second-generation breeding population (Tosh and McInnis 2000). Early results of realized gain tests established with seeds collected in the rogued first-generation seed orchards indicates that a 18 to 20% gain in volume (Weng et al. 2006) and a 25% gain in stem straightness could be achieved (Simpson and Tosh 1997).

In Quebec, the Ministry of Natural Resources and Wildlife has conducted a jack pine breeding program since the 1970s. Twelve seed orchards were established with plus-trees selected in both natural stands and provenance trials established in the 1950s, in collaboration with the Canadian Forest Service. Progeny tests were also associated with each of these seed orchards. In 2002, roguing of all first-generation seed orchards was completed based on family breeding values for height, stem straightness and tolerance to scleroderris canker. The gain in merchantable volume was estimated to 2.6 to 8.8 m^3/ha at 40 years (Beaudoin et al. 2004). Three hundred second-generation selections have since been made in the best families of the first-generation progeny tests across two breeding zones (M Desponts pers. comm.).

In Ontario, basic research on the genetics of jack pine was conducted primarily by the Petawawa National Forest Institute with collaborators between the 1950s and the 1970s (Yeatman 1974). The first breeding zones were delineated largely by administrative boundaries in the 1970s and 1980s and tree improvement activities were implemented at the regional level. More than 15 first-generation seed orchards were set up in the various regions, with about 50 family tests accompanying these seed orchards. Since then, all orchards have been rogued at least once. An advanced-generation breeding strategy was developed in 1994 and one of the jack pine programs was selected as a pilot for second-generation breeding. Since then, both an elite and an infusion population were assembled. Controlled crosses were carried out in the elite population and open-pollinated seedlots were collected from the trees making up the infusion population, with new progeny tests now in place to estimate breeding values of these second-generation selections (Ford et al. 2006).

Breeding activities conducted in Manitoba, Saskatchewan and Alberta have been on a smaller scale than those in the eastern Canadian provinces, although seed orchards provide most of the seed needed for the reforestation programs (Falk et al. 2004; Hansen et al. 2006). Some orchards have been rogued based on results of progeny tests. Selections for the establishment of second-generation seed orchards were made in the mid 2000s in Saskatchewan (Corriveau 2004), where the establishment of a new series of progeny trials is underway (Hansen et al. 2006).

In the United States, jack pine has been the subject of breeding programs in Minnesota, Michigan, Wisconsin and Maine. Seed orchards were established in each of those states and second-generation seed orchards in some cases (Stine et al. 1995).

Jack pine is an important commercial species, and there is no doubt that intensive breeding activities will be maintained for this species in the future. In Canada, the breeding plans of various agencies include wood properties among the selection criteria for advanced-generation breeding and consideration of marker-aided selection (MAS) for wood traits to help shorten breeding cycles. Accordingly, it is anticipated that genomic resources will be developed extensively for this species in the coming years.

2.2.2 Lodgepole Pine (*Pinus contorta*)

2.2.2.1 *Historical Perspective*

Lodgepole pine (*Pinus contorta* Dougl.) is one of the most important ecological and commercial hard pine species in western North America. With its large geographic range and economic importance, it has amassed a substantial utilization and management history as well as a very large body of research. The largest portion of its commercial range and management exists in the Canadian provinces of British Columbia and Alberta. It is also locally important in several of the northwest United States, but typically the volumes harvested are relatively small compared to British Columbia and Alberta.

In the 1970s and 80s, it became one of the most utilized species for reforestation in Sweden, and still is considered locally important in the northern latitudes of that country largely due to its superior growth rates over the native *Pinus sylvestris* L. (Elfving et al. 2001). It was also introduced to dozens of other countries, such as New Zealand, Argentina, and Great Britain, as its potential as a productive exotic conifer was explored around the world; however, the results have varied from important successes as an exotic to dramatic failures, e.g., becoming invasive (Ledgard 2001).

The first summary of the genetics of lodgepole pine was in the important "The Genetics Of" series, authored by William Critchfield (1980). Another large body of work on lodgepole pine was published by Koch (1987), which examined phenotypic variation for dozens of characteristics across the natural range of the species. Many other studies followed, ranging from mating systems in natural populations (e.g., Epperson and Allard 1984), biogeography of the species with molecular and quantitative studies (Wheeler and Guries 1982a, 1982b; Wheeler and Critchfield 1985; Yang et al. 1996; Godbout et al. 2008), leaf chemistry (von Rudloff and Lapp 1987), variation in quantitative traits of interest in genetic improvement programs (Xie et al. 2007), disease and pest resistance (Yanchuk et al. 1988; Wu et al. 2005), and more recently examination of the impacts of climate change on the potential adaptation and optimization of populations (Rehfeldt et al. 1999; Wang et al. 2006).

Although there are four subspecies recognized in the *Pinus contorta* complex, the largest and most important is var. *latifolia*, commonly referred to as "interior" pine. Shore pine, var. *contorta*, is the second largest component in terms of range, followed by var. *bolandaria* and the small outlier var. *murrayana* (tamarack pine), both of which are restricted to the southern part of the distribution in the United States. One unique biological characteristic of "*latifolia*" pine particularly is the serotinous "closed" seed cone, which is thought to have evolved as a regeneration strategy in response to fire. Lodgepole pine's basic adaptive strategy can be described as a genetic "specialist" (Rehfeldt 1988) as it typically has strong genetic clines, although these clines vary greatly across its natural range. All varieties exhibit special ecological distributions or "niches", which is not surprising considering the large range of lodgepole pine. Below, we focus on the most common of the three subspecies, var. *latifolia*.

2.2.2.2 *Economic Importance*

Harvest volumes for lodgepole pine in managed stands are in the order 350 m^3/ha, at a rotation age of 50–80 years. In the 1960s, lodgepole pine was not treated as a serious economic crop due to relatively small log diameters, but expansive natural monocultures (resulting mainly from regeneration after fire) and new processing technology in the '70s moved lodgepole to the forefront of economic importance in western Canada. Furthermore, rapid early growth and ease of establishment made lodgepole pine a favorite species for reforestation, with annual planting numbers in British Columbia in the order of 70–80 million trees.

The recent annual harvest of lodgepole pine in British Columbia (2006-2007) was over 35 million cubic meters, which represent approximately one-half of the annual allowable harvest in the province. These particularly high harvest levels have been due in part to increased salvage logging, as an attempt to obtain some remaining value of the millions of cubic meters pine being killed in an epidemic outbreak of *Dentroctonus ponderosae* (mountain pine beetle, MPB). It is expected that by the end of the MPB outbreak, approximately 80% of the mature lodgepole in British Columbia would have been killed. The devastating loss of a majority of the mature as well as young lodgepole stands from MPB attack represents a massive economic and social challenge to British Columbia, and is a sobering reminder of the drastic changes that climate, insect and diseases, combined with forest management practices can have on forests dominated by a single species. Spread of MPB into the neighboring province of Alberta may impact the species there as well.

In Sweden, reforestation with the introduced lodgepole pine peaked during the 1980s, with up to 40,000 ha planted annually (ca 20% of total reforestation). Planted area decreased during the early 1990s, levelling out at approximately 3,000 ha per year since then (Swedish Forest Agency, *http://www.svo.se*). In total, lodgepole pine now covers ca 600,000 ha (3% of commercial forest land). Harvest of lodgepole pine is still negligible, and mainly from thinnings, since planted stands have not yet reached rotation age for final cut.

2.2.2.3 Breeding Objectives and Achievements

In British Columbia, interest in breeding of lodgepole pine followed forest industry expansion into the interior, and provenance and progeny testing commenced in the late 1960s and early '70s. In 1969, one of the largest provenance tests for any conifer species in the world was established by Keith Illingworth, representing 153 seed sources planted across 60 sites in British Columbia and the Yukon (Ying et al. 1985). This network of trials has provided an enormously rich database for many questions related to breeding zone development, selection of superior provenances, and research on the effects of climate on adaptability and productivity. Data at age 32 years were the last that could be collected, as many of the test sites have been damaged extensively by MPB.

In the 1970s, significant breeding efforts were developed in Sweden, with the majority of the breeding population originating from the very high latitude natural populations in Canada (Ericsson 1994). Breeding commenced to develop 11 advanced-generation breeding groups that cover climatic differences in the country (Wilhelmsson and Andersson 1993), although the majority of investment is on nine of these.

Currently in British Columbia, eight breeding zones are recognized, and five of these now have second-generation tests in place, varying in age from 3–5 years. The initial population development for these breeding zones utilized provenance test data and incorporated local and superior non-local open-pollinated families, with 300–400 families per zone. Traits under selection have primarily been height at age 10, height growth with restrictions on wood density loss (due to small adverse genetic correlations) and disease traits such as western gall rust. Genetic gains in volume growth at rotation are currently predicted to be between 7 and 12% among the various breeding zones. Future breeding objectives may shift emphasis to a few other diseases and pests, and to attempt to address new concerns over climate change, adaptation and forest health. Genomic studies are underway to help elucidate gene expression in the mountain pine beetle system. The TRIA project (*www.thetriaproject.ca*) is hoping to utilize genomic tools to better understand the interactions between the genomes of bark beetles, fungal pathogens and host pine trees.

Lodgepole pine in British Columbia and Alberta will undoubtedly remain among the top two species in reforestation and forest management over the next rotation. Its ecological suitability and relatively fast growth rates, across many interior sites in its native range, will make it difficult for other species, native and non-native to substantially replace it on the landscape.

2.2.3 Western White Pine (*Pinus monticola*)

2.2.3.1 Historical Perspective

Western white pine (*P. monticola* Dougl. ex D. Don) is a member of the five-needle pines, which have long been an important part of the landscape of western North America, not just for their commercial and historic importance, but also for their aesthetic and ecological values. Other five-needle pines such as whitebark pine (*Pinus albicaulis* Engelm.) and limber pine (*Pinus flexilis* James) provide valuable tree cover for wildlife in exposed alpine country, food for birds and mammals and act as stabilizing elements for snow packs and soils in these steep and fragile environments. Sugar pine (*P. lambertiana* Dougl.) has also been a major commercial timber species.

The western five-needle pines have suffered from several serious problems, first with over-harvesting and then with issues arising from fire control removing the regeneration environment, browsing of young regeneration by ungulates, and mountain pine beetle (*Dendroctonus ponderosae* Hopkins). The most serious problem has been white pine blister rust caused by the exotic rust *Cronartium ribicola* J.C. Fischer, accidentally introduced to North America in the early part of the 1900s. All breeding efforts on North American white pines have targeted resistance to this pathogen. Arguably the largest effort has been made with western white pine, particularly in the "Inland Empire" program in Idaho.

2.2.3.2 Breeding Objectives

Richard Bingham initiated resistance breeding programs in Idaho as early as 1946 (Bingham 1983). McDonald et al. (2004) reviewed this and the other western regional programs (USDA Forest Service Regions 1, 5 and 6). Strongly influenced by Bingham's work in Region 1, the Region 6 program (Oregon and Washington) was started a decade later at the Dorena Tree Improvement Center near Cottage Grove, Oregon. Both Regions 1 and 6 started by rigorously selecting healthy survivors, then producing full-sib crosses, often in standing ortets (Bingham 1983). A Phase II program in both regions had less exacting candidate tree selection but followed up with inoculation and screening of open-pollinated progeny. Region 1 introduced this Phase II program in 1965 and has screened over 3,000 candidate trees (McDonald et al. 2004). This open-pollinated testing phase started in Region 6 in 1971 and Dorena has since screened over 4,900 western white pine and 4,500 sugar pine candidate parent trees (R Sniezko pers. comm.).

An early program in British Columbia screened ramets (grafted cuttings) from canker-free field selections following the protocols developed for eastern white pine in Wisconsin and Minnesota (King and Hunt 2004).

This early effort was abandoned, but in 1983 a program based on the USDA western regional Phase II programs of inoculation was initiated. Open-pollinated seedlots were screened from 300 widely distributed candidate parent trees from the coast and 300 from the interior regions of British Columbia (Hunt 2004).

All three of these western white pine programs were influenced by that initiated by Bingham, but were regionally adjusted. For example, in British Columbia, where the rust severity is generally less than in the United States, canker-free parents with intact lower branches were selected as candidate trees. In the Inland Empire, stand infections could average more than 150 cankers per tree so while most selected candidates had fewer than three cankers, canker-free trees were so rare that disease-free status could not used as a criterion (McDonald et al. 2004).

2.2.3.3 *Breeding Achievements*

Although white pine blister rust is an exotic pathosystem in North America, two important inheritable forms of resistance have been noted: major gene (R-gene) and multigenic "partial resistance" (Kinloch 2003). Although these may not always be distinguishable in the observed phenotypic distribution of resistance, progress has still been made by selecting the phenotype based on early field survival and slow canker growth. More information on the underlying genetic mechanisms will ultimately have implications for the effectiveness, practicality and durability of resistance. Breeding program activities are shifting from open-pollinated screening to controlled (full-sib and backcross) breeding to gain a more thorough understanding of what controls the phenotypic expression and durability of resistance.

As for eastern white pine, genomics resources and tools are expected to be developed in order to select western white pines tolerant to blister rust. A suite of candidate genes is already available in white pine to test for associations with "partial resistance", and a further association study, utilizing a diallel population composed of selected Oregon and British Columbia selections, is expected to be initiated later in 2010.

2.2.4 *Southern Pines (Pinus taeda, P. elliottii, P. palustris, and P. echinata)*

2.2.4.1 *Historical Perspective and Economic Importance*

In the southern United States, the "South", 10 species of southern yellow pines (*Pinus* sp.) are common across many forest ecosystems. In the late 1800s and early 1900s when commercial forestry started, longleaf pine (*P. palustris* Mill.) and shortleaf pine (*P. echinata* Mill.) were the most important commercial species in the South. When plantation forestry developed in the middle of the 20th century, loblolly pine (*P. taeda* L.) and slash pine (*P. elliottii* Engelm.) were the species of choice for planting and continue to be the most commercially important timber species today both in the United States (Wear and Greis 2002; McKeand et al. 2003; Sampson 2004) and in other countries (Zobel et al. 1987; Bridgwater et al. 1997). Both loblolly and slash pines have also been used extensively as exotic species in plantation forestry programs in Australia, China, southern Africa, and southern South America (Zobel et al. 1987; Bridgwater et al. 1997).

The silvical characteristics of loblolly and slash pine have some important distinctions. Loblolly pine is broadly adapted to a wide range of sites and is limited primarily by winter cold and drought. When the best genetic material is planted and given the necessary resources to grow, mean annual increments for loblolly pine of 20 m^3 ha^{-1} yr^{-1} are readily achieved (Allen et al. 2005). Slash pine typically does best on wet, poorly drained soils in the lower coastal plains of the Southeast (Baker and Langdon 1990; Lohrey and Kossuth 1990).

In 2007, 1.1 billion trees were planted in the South, with loblolly pine (840 million) accounting for 77.4%, and slash pine (126 million) 11.6% of the planting (McNabb 2007). On average, for each of the past five years, approximately 500,000 ha of loblolly pine and 80,000 ha of slash pine were planted in the region, all with genetically improved seedlings (McKeand et al. 2003).

There are more than 5 million ha of loblolly and slash pines planted outside of the United States. The majority of these are found in Brazil and China, with lesser amounts in Argentina, Australia, South Africa (and surrounding countries), and Uruguay. The first plantings of the southern pines in China were in the 1920s (Bridgwater et al. 1997). Through the early 1990s, slash pine dominated, because of its ability to survive on poor sites. Since the 1990s, large plantation areas of loblolly and slash pines have been established in southern and eastern China, with commercial quantities of seeds coming from seed stands and orchards in the United States and Zimbabwe.

The first introductions and commercial plantings of the southern pines occurred in northern Argentina, southern Brazil in the 1940s and '50s (Bridgwater et al. 1997; C Peirano pers. comm.). Brazil has the largest plantation area of southern pines in Latin America with 1.6 million ha established, 1.2 million ha of loblolly and 360,000 ha of slash pine (D Chaves pers. comm.). The current planting rate is approximately 30,000 ha per year. Argentina has approximately 400,000 and 270,000 ha of loblolly and slash pine established, respectively, primarily in the subtropical northern provinces of Missiones and Corrientes. Slash pine and the hybrid, *P. elliottii x P. caribaea* var. *hondurensis* is especially preferred in Corrientes where low-lying areas have a tendency to be wet during the year. Improved silviculture practices, like bedding, might make *P. taeda* more attractive as a species on these sites in the future. The forestry plantation programs in Uruguay are relatively new. Currently, 144,500 ha of loblolly pine have been established with the prospect of much greater expansion (J Posse pers. comm. 2008).

The southern pines have been grown in South Africa since the 1890s with expansion into other countries in southern and eastern Africa since the late 1920s (Poynton 1977). In general, loblolly pine has always been a secondary species to *P. patula* in most highland areas of the region. Its main disadvantages are its relatively poor stem form (Poynton 1977), its propensity to produce reaction wood at the base in some environments (van der Sijde et al. 1985), and its adaptability to a limited number of sites in the southern African environment. Recently, however, it has gained more attention as an alternate to *P. patula* on some sites because of its resistance to Pitch canker (*Fusarium circinatum*) and as sawmillers learn how to identify and handle the reaction-wood problem.

Slash pine remains the most widely planted of the southern pines in southern Africa, especially in South Africa, Swaziland and Zimbabwe, on the drier and/or colder sites, for both pulpwood and sawtimber (Bridgwater et al. 1997). In South Africa and Swaziland, 192,000 and 10,000 ha of slash pine and 27,500 and 2,000 ha of loblolly pine have been established, respectively (DWAF 2006). Lesser amounts occur in southern and eastern Africa as far north as the equator (Bridgwater et al. 1997; Poynton 1977).

Loblolly pine and slash pine were introduced into Queensland, Australia in 1917 and 1925, respectively (Bridgwater et al. 1997). Loblolly pine was eventually found to be poorly adapted to the region but slash pine did well on the excessively wet sites. On sites that are better drained, the *P. elliottii* x *P. caribaea* var. *hondurensis* hybrid and pure *Pinus caribaea* are now preferred with 62,120 and 50,345 ha established, respectively (I Last pers. comm. 2008). The seeds for the pure species and hybrids all come from advanced generation breeding orchards.

2.2.4.2 *Breeding Objectives*

In the southern United States, deployment practices such as planting only the best open-pollinated families to the best sites are resulting in dramatic increases in productivity. Increased resistance to fusiform rust disease, caused by the fungus *Cronartium quercuum* (Berk) Miyabe ex Shirai f. sp. *fusiforme*, especially in slash pine, has also had major impacts on plantation yields (Vergara et al. 2004). In the early 2000s, 59% of the loblolly and 43% of the slash were annually deployed as open-pollinated families by companies and small landowners (McKeand et al. 2003). In the last 10 years, seed orchard managers have had great success in developing methods to mass produce full-sib families for operational planting. The gains from improved quality and yield are very impressive when both the female and male parents are selected (e.g., Bramlett 1997; Bridgwater et al. 1998; Jansson and Li 2004; McKeand et al. 2008). From 2000 to 2007, over 94 million full-sib family seedlings were planted in the South (McKeand et al. 2008), and annual deployment of mass control-pollinated seedlings has grown to 35–40 million (4–5% of the total planting). Propagation of selected clones has also become a reality via somatic embryogenesis (e.g., Pait 2005), with over 10 million somatic seedlings planted to-date, and the numbers increasing annually.

From the beginning of tree improvement programs in the region (see Schmidtling et al. 2004 for a summary of tree breeding in the southern United States), the focus has been on selecting, breeding, testing, and planting trees that provide landowners with the greatest return on their investments (e.g., Zobel 2005). Historically, the greatest emphasis for both loblolly pine and slash pine was on volume production; more wood production for both pulp and solid wood products. For slash pine, a critical trait for volume production is resistance to fusiform rust disease. Slash pine is extremely susceptible to rust, and gains in rust resistance have been a major success (Vergara et al. 2004). Stem form traits (straightness, forking, and small, flat-angle branches) were also important criteria. In fact, the most dramatic improvement made in loblolly pine was the improvement in straightness.

Outside of the United States, tree breeding programs for the southern pines have been in existence in some countries since the late 1950s and early '60s (Poynton 1977; Mullin et al. 1978). The breeding objectives in these exotic environments are similar to those in the United States. Breeding for volume has been important in most countries, but stem straightness and branch characteristics have received very high priority in tree improvement programs especially in southern and eastern Africa to maximize recovery rates at local sawmills. The gains from selection for form traits are apparent when African bred loblolly pine is grown in compartments adjacent to genetic material from other countries. In the last 10 years especially, breeding programs have also concentrated on improving wood quality traits.

Most breeding programs for exotic southern pine plantations are one to two cycles behind the most advanced programs in the United States. There is still general usage of open-pollinated seeds from clonal orchards for operational planting, but this is gradually giving way to more use of seeds from control crosses with seedlings multiplied by vegetative propagation. Cutting programs of several million seedlings per year are common in industrial nurseries at some localities. The more advanced programs in southern Latin America are also actively involved in testing genetic material produced from somatic embryogenesis.

2.2.4.3 Breeding Achievements

For loblolly pine originating from the largest improvement program, second-generation seed orchards currently produce 77% of the seed while third-cycle orchards produce 12% (NCSUCTIP 2008). Estimated gains in volume production for open-pollinated families at rotation age from the second-generation improvement program pine vary from 13 to 21% over non-improved checks depending on the region in the southern United States. From rogued second-generation orchards, gain estimates vary from 26 to 35% (Li et al. 1999). Third-cycle orchards are expected to produce volume gains of 30 to 40% over non-improved.

If only the best family is planted in a region, the gains could be as high as 50–60%. With mass controlled pollination (Bramlett 1997), gains in stem form and sawtimber potential can be as much as 100% over the non-improved check which is twice as great as that from open-pollinated families (McKeand et al. 2008). Resistance to fusiform rust resistance has also been greatly improved in loblolly pine. There are individual families that have less than 10% infection when non-improved checks have 50% infection (Isik et al. 2008).

In the southern United States, breeding programs have put much more emphasis on improving traits that are important to solid wood products. Volume production is still the most important trait, but selection against stem defects such as excessive sweep, forking, ramicorn branching, and large, steep branches has become more prevalent. With the development of rapid screening techniques for wood quality traits such as bending strength (e.g., Jones et al. 2005; Roth et al. 2007) and wood density (Isik and Li 2003), breeders are now incorporating these traits into selection indices to improve value.

For slash pine, realized gain in stand yield for first-generation averaged about 10%, or an extra inside-bark volume of 25 m^3/ha at age 25 years (Vergara et al. 2004). For fusiform rust resistance, which is critical for productivity in slash pine stands, the 25% realized gain for rust resistant material compared to rust susceptible material obtained at age 16 was conservatively extrapolated to a 25-year-old rotation-age gain of 51.4 m^3/ha in inside-bark volume (Vergara et al. 2007).

Gain estimates from programs outside the United States are not readily available. The average productivity of first-generation loblolly pine in southern Latin America is approximately 28–30 m^3/ha/year. Estimates for growth of third-generation material on the best locations (deep fertile soils) in southern Latin America are 55 m^3/ha/year (over bark). Growth rates of this magnitude are already being seen in some operational plantings of second-generation material established on good sites in Santa Catarina, Brazil and measured at 16 years of age. Serious diseases have not yet affected the southern pines in Latin America or in southern and eastern Africa, but insect attacks are becoming more problematic in Southeast Asia (H Wang pers. comm. 2008).

The southern pines, and in particular loblolly pine, have long been the subject of marker studies, particularly for wood quality traits and disease resistance. Most recently, the breeding programs for loblolly and slash pine have entered a collaboration with the Conifer Translational Genomics Network (CTGN), in the hopes of making genomics-assisted selection an operational reality (*http://www.pinegenome.org/ctgn/*). Led by UC Davis, with additional funding from the USDA and the US Forest Service, the CTGN will genotype up 7,500 trees and analyze genetic variation at about 7,000 loci previously identified as single nucleotide polymorphisms (SNPs). Phenotypic information will be associated with SNP variation, focussing on stem volume, fusiform rust resistance, wood quality, and stem form, with the goal to develop selection tools.

2.2.5 *Maritime Pine (Pinus pinaster)*

A dozen pine species are native to Europe, and several of them are cultivated to more or less a large extent, such as European black pine (*Pinus nigra* Arn.) and Scots pine (*P. sylvestris* L.) as well as the Mediterranean pines of southern Europe: Turkish pine (*P. brutia* Ten.), Canary Island pine (*P. canariensis* C. Sim.), Aleppo pine (*P. halepensis* Mill.), Maritime pine (*P. pinaster* Ait.), and stone pine (*P. pinea* L.). These species have long been used for wood production and in some cases for resin production (Maritime pine), or even edible seed production (stone pine). Some are the subject of breeding and plantation in different countries, usually where they are also present in natural stands: for example European black pine and sub-species in Spain, France, Greece, and Turkey, Scots pine in many countries corresponding to its natural area (the largest of the genus *Pinus*), Turkish pine in Turkey, Aleppo pine in Greece and Israel, stone pine in Spain, and Maritime pine in France, Portugal and Spain. Of these Mediterranean pines, Maritime pine has been the most extensively planted and has also been introduced as an exotic outside Europe, in areas such as southwestern Australia. Breeding of Maritime pine in southwestern France started in the 1960s, after several species and provenances trials had shown that the local "pin des Landes" was the best adapted and the fastest growing tree in the Aquitaine soil and climatic conditions.

2.2.5.1 *Historical Perspective*

Between the end of the 18th century and today, the Maritime pine stands in southwestern France (Aquitaine region) expanded from a natural forest of

250,000 ha, located along rivers and the Atlantic coast, to a cultivated area of one million ha. Such a progression was the result of the determination of the local land owners and public authorities to stabilize coastal dunes, drain 700,000 ha of marshes, and plant a new forest. Once the forest was settled, the challenges of nature and of a changing economic background had to be addressed repeatedly (Riou-Nivert 2002). Between 1939 and 1950, fire destroyed 400,000 ha. In the 1950s, the resin market collapsed, due to international competition and the emergence of oil by-products. Production objectives were reoriented towards timber, supported by the progress in silviculture and the breeding of improved varieties, marketed in the early 1980s. During the winter of 1985, an intense cold wave in southwestern France destroyed 30,000 ha of Maritime pine plantations from Spanish and Portugese provenances, which had been established during the major reforestation effort following the fires of the 1940s. The genetic origin of seed source stands is now systematically verified by a terpene test, and seed harvest from non-local stands is forbidden. A hurricane in December 1999 felled more than 100,000 ha in Aquitaine: 28 million cubic meters were levelled (*http://agreste.agriculture.gouv.fr/*). Once more, the Maritime pine forest resource had to be reconstituted. Reforested areas increased, reaching 23,000 ha/year, while 100% of plantations have been established with seedlings from second-generation seed orchards (GPMF 2002).

2.2.5.2 *Economical Importance*

Maritime pine is by far the most planted tree species in France where it represents 10% of the forest area and 24% of wood harvest (French Ministry of Agriculture, *http://agreste.agriculture.gouv.fr/*). Average productivity is about 10 m^3/ha/y, but can reach 20–25 m^3/ha/y on the best sites. The rotation age is typically 45 years and is decreasing with improved varieties. Today, some 8.5 million m^3 are harvested annually, most of which is processed locally, 60% as saw-timber, and 40% as industrial round wood. Forest management and primary wood processing represents 40,000 jobs in Aquitaine and are an essential economic activity in this region, bringing a turnover that is greater than that of Bordeaux wines.

2.2.5.3 *Breeding Objectives and Strategies*

2.2.5.3.1 Provenance Choice and Plus-tree Selection

The breeding program started in the 1960s, when early provenance trials had already shown the superiority of the local Landes provenance for growth and cold resistance (Illy 1966). Aquitaine is the most northern region of the species' natural distribution, which is otherwise localized on the

Atlantic coast of Spain and Portugal and around the Mediterranean basin (Spain, southeastern France, Italy, Tunisia, Algeria, and Morocco). Cold resistance was identified as an important issue, especially when lowest night temperatures in Aquitaine can reach –10°C or –15°C every few winters (-20°C in February 1985). The local provenances were thus chosen to build up a breeding population, despite their form defects: trunk flexuosity and poor branching. A total of 380 plus-trees were selected phenotypically on the coastal sand dunes of Aquitaine, based on height and diameter, and visual scoring of stem form. This first phenotypic selection proved to be efficient for improving growth and stem straightness, as shown by a progeny test comparing plus-trees progeny with their non-selected neighbor-tree progenies on two locations after 10 years old (Danjon 1995). In addition, genetic variation among provenances and performance of crosses between provenances were explored (Harfouche and Kremer 2000; Harfouche et al. 2000). Among all tested combinations, Landes x Corsica families proved to be the best material for growth and form in Aquitaine conditions. A few hundred clones from the Corsica provenance were selected in provenance trials located in Aquitaine, based on growth, stem straightness, branch quality, *pyralis* resistance (*Dioryctria sylvestrella*), and cold resistance. The objective of this second population is to produce improved Landes x Corsica varieties for better stem straightness and branch quality.

2.2.5.3.2 Breeding Strategy and Selection Criteria

Development of the breeding program followed a classical recurrent-selection approach, with a main population composed of the Landes plus-trees. In the first two cycles of selection, factorial mating designs with four to six crosses per parent, or hierarchical (nested) mating designs with two crosses per parent were used to produce the next generation. Forward selection was based on an individual-family index, with total height, diameter at breast height and stem deviation from vertical at 10 years as selection criteria (Baradat and Pastuszka 1992; Durel 1992). For these three traits, narrow-sense heritability in the base population was moderate to low (0.19, 0.14 and 0.16, respectively) (Bouffier et al. 2008b), and an adverse genetic correlation exists between growth and stem straightness (-0.2 between diameter and straightness). Following this strategy, the main population has cycled through three generations, with more than 4,500 individuals selected, and 5,000 families tested over 500 ha of trials (GPMF 2002).

The changes in genetic variance accompanying selection and breeding has been studied in this material, using an individual genetic model to estimate heritability and additive coefficient of variation over three generations for traits under selection (Bouffier et al. 2008b). For growth traits, the results showed a clear decrease (30 to 40%) in genetic variation from the Aquitaine natural resource to the selected plus-tree founder breeding population, and stabilization of variance from founder breeding population to the next-generation breeding population. The pattern was different for stem straightness, and difficult to interpret due to different measurement methods over time. It was concluded that the recurrent selection strategy based on one main population could sustain several generations of breeding and selection, considering the level of additive coefficients of variation for the selected traits and a status number of the breeding population (population effective size, Lindgren et al. 1996) close to 100.

For the next generation of the breeding population, the focus is on a reduction of population census size and better management of pedigrees, to optimize selection efficiency while producing regularly renewed varieties with increasing genetic gains. Eight unrelated sublines were assembled within the breeding population based on pedigrees and breeding values, allowing the deployment of unrelated selections to clonal seed orchards. Status number is used as an indicator of genetic diversity. Double-pair mating designs are used to produce material for progeny tests and the base of the next generation, while polycross testing is performed for parental ranking. Trials are replicated on several contrasting sites, usually with single-tree plots and a large number of replications per site.

Including new selection criteria is also a focus: studies on pests and diseases resistance (Jactel et al. 1996; Kleinhentz et al. 1998; Burban et al. 1999; Lung-Escarmant and Guyon 2004), wood quality (Pot et al. 2002; Bouffier et al. 2008c, 2009), and drought tolerance (Dubos et al. 2003; Dubos and Plomion 2003; Nguyen-Queyrens and Boucher-Lannat 2003; Eveno et al. 2008) are ongoing. Some new criteria have already been included in the selection process: rust resistance (*Melampsora pinitorca*) is tested on future seed orchard parents through a cut-shoot assessment (Desprez-Loustau 1990), wood density is evaluated at the family level in progeny trials with an IML-Resi tool (Bouffier et al. 2008a), and branch quality is scored visually in progeny tests (GPMF 2002).

Breeding forest tree varieties in the context of changing climate is another challenge. Models predicting the evolution of climate in southwestern France during the next decades show an elevation of air temperature and a seasonal shift in precipitation distribution from spring and summer to winter, which will likely result in decreased forest productivity (Loustau et al. 2005). Although these are hypotheses, interest to improve drought resistance has been increasing. Current varieties of Maritime pine in Aquitaine were selected, tested and used in one breeding zone. Selection is aimed at producing multipurpose varieties adapted to the different soil types of Aquitaine, including dry, semi-humid and humid podzol soils. In the near future, seed orchards could be rogued to favor clones that are better adapted to the drier sites as and when their progeny tests are assessed in a changing climate. As for future varieties, different strategies are considered: locating progeny trials in more southern and drier sites as an anticipation of future climate, infusing new diversity into the breeding population either by selecting better adapted trees in the local provenance, using the national network for Maritime pine natural genetic resources conservation, either by selecting adapted inter-provenance combinations (Landes x Portugal and Landes x Morocco progenies are already being tested), as well as introducing new selection criteria for drought resistance, e.g., water-use efficiency (Brendel et al. 2002), resistance to cavitation (Lopez et al. 2005), and molecular markers for these traits.

2.2.5.4 *Breeding Achievements*

Three generations of seed orchards have been produced. For economic and technical reasons, the deployment strategy for Maritime pine varieties in Aquitaine is based mainly on open-pollinated orchards.

The first-generation orchards were seedling seed orchards based on a very large number of full-sib families, corresponding to the progeny tests of plus trees, and were rogued after genetic assessment. These orchards demonstrated genetic gains of 10–15 % in volume and stem straightness at about age 15 years (GPMF 2002).

Second-generation orchards were characterized by a reduced genetic base and greater genetic gain, compared to those in the first generation. They were based on a few tens (usually around 30) of backward-selected clones, either as classical grafted clonal orchard or as a randomized plantation of polycross families obtained by controlled pollination between selected clones (Baradat et al. 1992). The polycross family seed orchards were planted over 180 ha and are open-pollinated. The expected genetic gain was estimated from progeny trials at age 13 years to be 30% for both criteria over unimproved material. Since the hurricane of 1999, when annual reforested areas of Maritime pine in Aquitaine increased from 15,000 to 23,000 ha, 70% have originated from second-generation orchards.

Third-generation seed orchards have been established over 180 ha, and should enter in production by 2010–2015. They are either clonal or polycross family seed orchards. This third generation also includes a Landes x Corsica variety, to be produced by controlled pollination.

In the future, seed orchards will have to be renewed more rapidly, to better respond to likely climate change and developments in the marketplace. Recent adaptations in the Maritime pine breeding program, such as the optimization of population management through sublining and of selection efficiency with BLUP evaluations, are expected to be augmented with marker-assisted selection for complex traits such as wood quality and drought resistance.

2.2.6 Scots Pine (*Pinus sylvestris*)

2.2.6.1 Historical Perspective

Scots pine (*Pinus sylvestris* L.) has a wide natural geographic distribution, the most extensive in the genus *Pinus* and in the family Pinaceae (e.g., Boratyński 1991). The species ranges from Scotland and Spain in the west to the far east of Russia (Siberia), and from Spanish Sierra Nevada mountains in south to the northernmost part of Scandinavia. It occurs on a variety of soils in very diverse climates, in pure as well as in mixed forest stands. In northern Europe and Asia, Scots pine is a dominant species of the boreal forest (Willis et al. 1998). It has also been introduced to North America as an exotic species, initially both for ornamentals, Christmas trees, and timber production, but now grown primarily for Christmas trees.

Scots pine is the most intensively studied tree species from the standpoint of provenance variation. Provenance studies had already started by the end of the 19th century (reviewed by Langlet 1971). In 1907, an international provenance study with pine from different climatic regions was established by IUFRO members, and this was followed by several others (e.g., Giertych 1991). The aim of early provenance research was to reveal the possible use of seed from different origins with respect to germination, survival, and growth. Langlet (1936) undertook an extensive study of physiological variation in Scots pine from 582 localities in Sweden in the 1930s. He demonstrated a genetically controlled clinal variation in physiological traits related to cold hardiness. Eiche (1966) established a large provenance series in the early 1950s, from which much valuable information has been extracted, elucidating genetic parameters for many traits. Eiche (1966) demonstrated hereditary adaptation of provenances and the possibility to improve survival in plantations suffering from cold damage by transferring provenances from north to south. This pioneering work has been followed by numerous population genetic studies in the same or new field experiments in Sweden (e.g., Remröd 1976; Eriksson et al. 1980; Persson 1994) and in other countries (e.g., Giertych and Mátyás 1991).

2.2.6.2 Economic Importance

Scots pine is one of the most commercially important Eurasian forest trees and widely used in plantation programs in temperate zones (Volosyanchuk 2002). It has major economic significance throughout its natural range (Mikola 1991), both for high-quality sawn products and for pulp and paper, with Russia, Finland, and Sweden comprising the largest areas for timber production. In addition, Scots pine has also been widely planted for timber production beyond its natural distribution in western Europe, Eurasia, and North America, and to a small extent even in Mexico and New Zealand (Boratyński 1991).

In Sweden, commercial Scots pine forests occupy 12 million ha of productive forest land (ca 50%), with a total stocking of about 1,100 million m^3, an annual cut of 30 million m^3, and annual planting of 120 million seedlings (ca 32% of total seedling production). According to the Swedish Forest Agency (*http://www.svo.se*), the value of forest product exports in 2007 totalled 127,000 million SEK (ca US$18 billion), or 11 % of total exports and 4 % of GNP. More than 100,000 people are employed in the forestry sector (2.2 % of all workers). Based on its share of total harvest volume, Scots pine contributes roughly 30–40% to these figures.

In Finland, there are 13.6 million ha of Scots pine dominated forest, representing 65% of the forest area and 50% of the standing volume (FFRI 2008). About 55,000 ha annually are artificially regenerated with pine, where direct seeding is used on more than half of the area (requiring 20-times as much seed as planting).

According to Russian Federal Agency of Forestry (A. Fedorkov, pers. comm.), Scots pine in Russia covers 117 million ha (42 million in Europe and 75 million in Asia). It is the second most dominant species with a standing inventory of 15,000 million m^3, or 20% of total standing volume.

In other European countries, areas dominated with Scots pine are considerably smaller, but still contribute significantly to total production and are considered economically important.

2.2.6.3 Breeding Objectives

Much information on genetic parameters for many traits is available from a large number of investigations carried out over many years (e.g., reviewed by Giertych and Mátyás 1991; Eriksson 2008). Significant genetic variation and heritability has been shown for both growth traits (e.g., Haapanen 2001), stem quality and wood properties (e.g., Ståhl and Ericson 1991; Persson et. al 1995) and adaptive traits (e.g., Persson and Andersson 2003), demonstrating good potential for improvement through breeding. Hannrup (1999) grouped Scots pine traits, where phenology traits generally show high values for both additive genetic variation and heritability, growth traits show large genetic variation but low heritability, and morphological traits such as wood density and tracheid length show little variation but high heritability. Further, genetic correlations between height at different sites within climatic regions are usually high (Haapanen 1996; Zhelev et al. 2003; Persson et al. 2006), indicating limited genotype-environment interaction.

Although Scots pine is considered a cold-hardy species, withstanding short vegetative growth periods and very low winter temperatures, regeneration at high latitudes and altitudes is at risk of mortality due to cold damage. Because of its large natural range, Scots pine is host to many different pests (Stephan 1991). Genetic variation in resistance to fungi (e.g., *Melampsora pinitorqua, Cronartium ssp., Phacidium infestans, Gremmeniella abietina*) has been shown (e.g., Quencez and Bastien 2001), while similar results are lacking for insect resistance.

Overall breeding objectives for Scots pine in Sweden are to improve value production, while maintaining sufficient genetic diversity and preparedness for climatic change, through a multiple-population breeding strategy (Danell 1993; Wilhelmsson and Andersson 1993). Target traits are grouped in selection traits for improved (i) adaptation/survival, (ii) yield, and (iii) stem and wood quality. Selection indices based on genetic variation, correlation between assessed traits and goal-traits, and economic weights are used to identify predictors of highest economic yield in different geographic areas (populations). The sustainability of such programs over 10 generations, with spruce as the model species, has been validated in simulation studies by Rosvall et al. (1998).

Finland has a breeding program of similar size and structure as that in Sweden (Haapanen and Mikola 2008), with the main objectives to improve growth and branching (branch size and angle). Parallel objectives can be found in other programs, for example: in Russia, where yield and stem quality are targeted (A Fedorkov pers. comm.); height, survival and stem shape in the Czech Republic (O Ivanek pers. comm.); growth traits, stem straightness, and branch quality in Latvia (Jansons et al. 2008); height growth, stem quality, and frost resistance in Turkey (Bilir and Ulusan 2008).

Historically, there has been an extensive Scots pine breeding program for the northeastern part of Germany, focussed on preservation of genetic variability, adaptation to site conditions and climate changes, resistance to biotic and abiotic hazards, yield and quality, and transfer of valuable genetic material into practice (Kohlstock and Schneck 1992). The program included progeny-tested orchards as well as plans for cross breeding (two-clone orchards) to utilize specific combining ability. However, interest in tested regeneration material of Scots pine has declined, and the future of Scots pine breeding in Germany is uncertain (V Schneck pers. comm.).

In some other countries, breeding objectives reflect concern about the impacts of biotic stress. Although productivity and stem quality are important, particular emphasis is put on resistance to pathogens in France (C Bastien pers. comm.), and tree health in Lithuania (D Danusevicius pers. comm.), while genetic conservation and genetic variability are stressed in Spain (R Alia pers. comm.) and in Poland (J Kowalczyk pers. comm.).

2.2.6.4 Breeding Achievements

Generally for Scots pine, genetic gain from breeding programs are realized through crops from clonal or seedling seed orchards. Deployment of Scots pine to plantations is entirely from seed, as vegetative propagation is currently difficult and costly on a large scale. A worldwide review in 1991 showed that there were ca 10,000 ha of Scots pine orchards established (Mátyás 1991), indicating high expectations of improved regeneration stock through breeding. The former Soviet Union contributed half of the total orchard area, with Finland, Sweden, China, and Poland as other major contributors.

The level of genetic gain obtained depends on both testing accuracy and selection intensity, both of which vary among countries and programs. In addition, estimates of genetic gains are usually available only from a restricted number of trials, which usually introduces an upward bias due to unaccounted G x E variance, etc. This makes it difficult to generalize on breeding achievements; however, some rather comprehensive results and estimates of breeding accomplishments are available.

In Sweden, the first round of improved regeneration stock from phenotypically selected trees in clonal orchards started to accumulate in the 1980s. Gains, as predicted from large series of progeny trials with unselected control lots as comparison, showed superiority in height (9.2%), breast height diameter (5.4%), and volume (18.9%) at age 27 (Andersson et al. 2007). Calculations based on growth and site-index functions indicated that the height superiority corresponds to a 10% difference in volume production at full rotation (80–100 years). There were minor changes in survival (–1.4%), ramicorn frequency (–1.0%), and stem break frequency (1.3%). Jansson (2007) found 11.7% superiority in volume per ha at age 30 for progenies from phenotypically selected trees in south Sweden, estimated from five trials with block-plots. Based on genetic parameters from numerous Scots pine trials and realized selection intensities, Rosvall et al. (2002) estimated genetic gains for the third round (1.5 generation) of Scots pine orchards currently under establishment in Sweden. Figures varied between 23 to 27 % predicted gain at rotation-age volume production per ha, and included both initial phenotypic selection gains and gains from selection of genetically tested material. In addition to gain in growth, a gain of 5–13% in survival was estimated for those orchards intended for climatically harsh sites. Since selection indices include also pest resistance, stem and branching characters, etc., improvements are also expected here, but no precise estimates are available.

In Finland, Haapanen (M Happanen pers. comm.) reports 15–20% predicted gain in stem volume at age 12–20 years for bulked open-pollinated first-generation orchard seed lots, in comparison with local wild seed lots. In addition, improved branch quality (smaller branch diameter) was observed. Establishment of tested 1.5-generation orchards started in 1997 with expected gains of 25–30 % in early stem volume.

The genetic quality (breeding value) for height and stem form of seed orchards in Britain have been predicted (Lee 1999). Existing, first-generation seed orchards with phenotypically selected trees, were 8–12% superior in height and 0–3% superior in stem form compared to unimproved seed from registered stands, at age 10 years. New orchards with top-performing progeny-tested clones are predicted to give genetic gains of 14–20% in height and 5–19% in stem form, depending on how the traits are weighted.

In Latvia, Scots pine breeding is at the beginning of the second cycle. Orchard seed (almost 100%) from both the first-generation and 1.5-generation orchards is used in operational forestry. Genetic gains from open-pollinated 1.5-generation mother trees are predicted to be 10–14% in height and diameter at age 21–36 years (Jansons et al. 2008).

In summary, many programs still utilize improved stock from first-generation orchards with phenotypically selected trees. The superiority of this stock is around 10% in early height (20% in early volume), which corresponds roughly to 10% in full-rotation volume (Andersson et al. 2007). In addition, some improvement in stem and branch quality is achieved. These gains should be rather accurate for first-generation untested orchards over various countries and programs, as the plus-tree selection was carried out in similar ways (Pihelgas 1991). Depending on pollen contamination rates, predictions should be somewhat reduced to give realized gains.

Orchards with tested clones (1.5-generation) are also coming into production. Depending on the size of the breeding program, and thereby the selection intensity, and the weighting of traits, the superiority generally varies between 15–30% for early height or full-rotation volume, although realized gains would be somewhat reduced by outside-orchard pollen contamination. Adaptive traits (survival), stem and branch quality, and resistance to fungus were also targeted, are expected to yield additional gains.

Marker-aided or genomic selection approaches are not used in operational Scots pine breeding. However, microsattelites are used for paternity identification in applied research projects, e.g., investigations on mating patterns and contamination rates in seed orchards (Torimaru et al. 2010). Although numerous research projects on MAS using simple sequence repeats (SSRs) and SNPs are in progress, large-scale genome-wide association or evaluation using dense SNP maps are considered to have the best potential for assisting Scots pine breeding in the future.

2.2.7 *Radiata Pine* (Pinus radiata)

Within its native California, radiata pine (*Pinus radiata* D. Don) is a comparatively obscure species, prized much more for its amenity value and producing Christmas trees than as a timber species. Elsewhere, it has become an extremely important commercial species (Scott 1960; Burdon 2002). Plantations occupy over 4 million ha, roughly 500 times the present natural distribution of the species. Its very rapid growth, ease of collecting and storing seed, easy handling in the nursery, amenability to transplanting, modest edaphic requirements (typical of true pines), and the versatility of its wood, make it the utility softwood of choice almost wherever it will grow satisfactorily. Climatic conditions that exceed its tolerances include severe winter cold, heavy snowfalls, damp heat, and severe drought especially combined with high temperatures, such that a mild oceanic climate suits it best. Also, it demands higher soil fertility than many pines. The site tolerances reflect a natural habitat that is a highly localized variant of a Mediterranean climate, with summer sea fogs caused by a cold ocean current. The limitations mean that the successful plantings are very predominantly within the Southern Hemisphere, New Zealand, Chile and Australia being the largest growers, and Spain being the only Northern Hemisphere country with major plantings.

2.2.7.1 *Historical Perspective*

While known to the Spanish who colonized Mexico and California, radiata pine only became known to mainstream European plant collectors and botanists with the collection of herbarium specimens and seed by David Douglas in 1833, although the species was named from herbarium specimens collected separately by John Coulter. This was during a fashion for collecting and exchanging the many newly discovered conifer species from North America and northern India. Material from the Douglas collection was distributed in Britain and later to British colonies (Shepherd 1990). Early introductions were on a specimen-tree scale.

In New Zealand, the first confirmed introduction was in 1859, via Australia. Its good growth, over the length of the country, soon led to fresh seed importations, and larger-scale plantings. The country evidently became self-sufficient for seed by the early 1880s. Later on, New Zealand became a major seed exporter, notably to South Africa and parts of Australia. Only after 1921 did New Zealand make a massive commitment to the species for timber, to make good depletion of the native timbers. This led to a planting boom during 1925–1935. After World War II major processing industries were established with pulp and paper mills as well as sawmills. From around 1960, the plantations were seen as the base for major exports, leading to a second planting boom beginning in the 1960s and peaking during the 1970s and '80s.

The first introduction to Australia was slightly earlier than in New Zealand. Local shortages of softwoods led to establishment of plantations, in addition to plantings for shelter and amenity. Commercial plantings began in South Australia in 1885; Victoria and New South Wales followed suit in the 1920s and 1930s, with major expansion after World War II. While there are pulp mills and reconstituted wood plants, Australia continues to place a major emphasis on producing light structural timber from the species. Radiata pine, despite some early problems with micronutrient deficiencies, has become by far the pre-eminent softwood plantation species in southeastern Australia, and with significant plantings in western Australia.

The first introduction to Chile was much later, in 1887, near Concepción. It soon became popular in that locality, but it was only towards 1940 that it was used for large-scale afforestation. Rapid expansion of processing plants occurred in the 1960s and 1970s, with considerable emphasis on pulping. Major planting began again in the 1970s, with strong financial encouragement from the government.

Introduction to Spain reportedly occurred in the 1860s. Plantings are close to the north coast, predominantly near the western end of the Pyrenees, in the Basque Autonomous Region. The plantings are almost all dispersed among large numbers of very small owners.

In South Africa, seed importations began earlier than in New Zealand and Australia, but the species is confined to the climate of a coastal strip in western Cape Province, where it is favored for light structural timber. The species has also been tried in many other countries, with many failures. Initial success was often published enthusiastically, unlike the subsequent failures. Even so, there are various countries where some plantings have performed acceptably, although good statistics are elusive.

2.2.7.2 Economic Importance

The area over which the species is managed is not huge compared with some other species. Few forest tree species, however, are grown more intensively as plantation crops. Worldwide, it is thought there are over 4.1 million ha planted to radiata pine, with around 1.6 million and 1.4 million ha in New Zealand and Chile, respectively (MAF 2007; INFOR 2007). A total of 730,000 ha are established in Australia (ABARE 2007) and close to 300,000 ha in Spain (DGB 2005; cited by Crecente-Campo 2009).

The versatility of the radiata pine wood rates highly, even among pine timbers, despite shortcomings of its corewood. It can be used for solid-wood products, and both chemical and mechanical pulps. The solid-wood products cover both light structural and appearance-grade lumber, and the latter can be quite high-value. While at present its long-fiber kraft pulp rates as a commodity, it now tends to be a by-product rather than a primary one. The mechanical pulps are now valued for magazine papers.

In New Zealand, radiata pine has since around 1925 been the mainstay for replacing dwindling supplies of native timber species that are not readily domesticated. It has since become the basis of major export industries, with a total annual roundwood harvest of nearly 20 million m^3, making the country the 12th largest producer of coniferous roundwood and the third largest exporter of coniferous logs. The contribution of the forestry sector, including derived industries, which is around 95% based on radiata pine, is estimated at around 3.5% of the GDP (over US$2.5 billion), and 10% of the country's total export receipts, based on 7% of the land area (MAF 2007). This has been despite a depressed state of the sector, due to a strong New Zealand dollar, an historical focus of the corporates on producing commodity products and on log exports, and some correction of over-harvesting, factors that obscure the species' full contribution to wealth. In addition to producing wood, *P. radiata* makes important contributions to soil conservation and provision of shelter, the shelter plantings containing an additional timber resource.

In Chile, radiata pine is also the mainstay of a large forestry sector, albeit less pre-eminent than in New Zealand. Annual roundwood harvest is some 25 million m^3 (INFOR 2007), the country being the ninth largest producer of coniferous roundwood. Contributions to GDP of primary production from the species are estimated at around US$ 2.3 billion, ca 2% GDP (G Ortiz pers. comm.). The species also greatly dominates forestry exports (ca. US$3.9 billion) (loc cit). The contribution has been helped by very active government encouragement to reduce the economy's extreme exposure to the world market for copper. Many of the plantings have also rehabilitated severely degraded land.

In Australia, radiata pine is also less pre-eminent in the forestry sector than in New Zealand, being 73% of total softwood plantation area (Wu et al. 2007), and it is oriented very much towards local markets. The species probably contributes around 11 million m^3 to the annual roundwood harvest. In the solid-wood area, conifers comprise around three-quarters of the sawtimber production (ABARE 2007), and are widely favored for light construction, for which locally grown radiata pine is generally well suited.

2.2.7.3 Breeding Objectives

Given the general attractiveness of the species for domestication, and strong indications of great genetic variation, it was a logical subject for breeding (Burdon 2004). Research on its genetic variation began in Australia in the late 1930s. Active breeding work, however, began there only in the 1950s, and the effort was long fragmented among the states. In New Zealand, an intensive breeding program was mounted in the early 1950s, from the Forest Service's Forest Research Institute in a large, centralized operation. South Africa began breeding work at the same time, but with the comparative small area that suits the species, the breeding program has remained relatively minor. In Chile, an abortive start was made on breeding in the early 1970s, but was started afresh, adapting the United States Industry/University Cooperative model, in 1976. In the Basque Autonomous Region of Spain a breeding program began around 1990.

Among breeding objectives the basic, common features have been general health and vigor. Beyond these, breeding objectives have varied among the programs. This largely reflects the phenotypic plasticity of the species, whereby environmental effects mean that different traits are prime candidates for genetic improvement on different sites.

In New Zealand, early breeding largely addressed improving tree form on the fertile pumice-land sites that carried a major portion of the plantation estate. Apart from dominant crown status and general health, trees were selected very intensively for stem straightness, and light, wide-angled branching. The specification for branching generally led to choice of trees with a "multimodal" or "short-internode" branching habit. Favorable genetic correlations between this habit, growth rate, and general tree form, led in 1968 to explicit choice of a short-internode "ideotype", by way of indirect selection to help improve growth and form and to control branch size. In this context, pruning butt logs was done to produce high-quality appearance-grade timber. Selection of an alternative, "long-internode" ideotype was pursued in a side program (Shelbourne et al. 1986), for assuring clear-cuttings of timber without having to prune, albeit at a cost of potential genetic gain in growth and form. Early market acceptance, however, was almost nil. An offshoot of the main, short-internode breeding program was selecting for resistance to *Dothistroma pini*, which is now become an almost universal selection trait, along with resistance to needle cast associated with *Cyclaneusma minus*. A portfolio of different breeds, representing different breeding goals (Jayawickrama and Carson 2000), has become a distinctive feature of New Zealand's radiata pine breeding program. This largely finesses the question of assigning explicit economic weights to different traits, and reflects the combination of the species' environmental plasticity, the diversity of end products in New Zealand creating a complex production system, and the associated difficulty of assigning economic weights. Shorter rotations and more aggressive thinning regimes have exposed shortcomings in wood quality, leading to a recent focus on genetic improvement of stiffness and stability in service.

In Australia and South Africa, where tree form was often better because of lower soil fertility, selection tended to focus less on tree form and more on vigor. In recent years, the selection in Australia has shifted more to wood density and stiffness, reflecting the importance of structural timber combined with how the widespread use of fertilizer tended to compromise timber stiffness. The emphasis on structural timber means a simpler production system than in New Zealand, which has encouraged efforts to derive explicit economic weights for breeding-goal traits (e.g., Ivković et al. 2006a, b). However, complex patterns of pronounced genotype-site interaction in Australia (e.g., Wu and Matheson 2005) pose a continuing challenge in breeding for good local adaptation.

Chile followed New Zealand in pursuing the short-internode ideotype, despite good prospects of growing satisfactory trees of a long-internode ideotype (Burdon 1978). As with other main growers of the species, there is now an increasing focus on genetic improvement of wood properties. The structure of the breeding program, being originally company-based, produced a built-in regionalization.

2.2.7.4 *Breeding Achievements*

Breeding achievements have depended not only on efficient selection and testing for appropriate breeding objectives, but also on early and efficient delivery of genetic gain in planting stock. Open-pollinated clonal seed orchards were initially used for delivering genetic gain. In New Zealand, the first orchard planting was in 1958 and the first orchard seed was produced in 1968. Despite a difficult learning experience, the country was self-sufficient for seed-orchard seed by 1986. In Australia, there were mixed fortunes with early seed orchards, and major losses of orchards from fire in 1983. Chile benefited from the New Zealand experience on how to site and manage the orchards, most of the orchards coming into full production faster than their early New Zealand counterparts.

From around 1980, delivery systems for genetic gain have been changing. Use of controlled pollination, to capture more genetic gain, has become possible through vegetative multiplication of top-ranked controlled crosses, or through large-scale controlled pollination in orchards that produce seed close to the ground. More recently, clonal forestry, i.e., mass-propagation of well-characterized clones, has been commercialized, to capture non-additive in addition to the additive gene effects and achieve greater crop uniformity, although technical challenges remain.

Despite good combinations of variability and heritability, the important task of demonstrating and quantifying genetic gain accurately is not straightforward. Available growth data come mainly from quite young trees, whose performance needs to be projected into harvest-age production and quality and value of the logs and their end-products. However, special genetic-gain trials, and growth modelling (e.g., Carson et al. 1999), have allowed projection of stem volume gains at different crop ages. For tree-form traits, available data often involve subjective scores, which pose their own problems of quantification.

Notwithstanding these difficulties, major genetic gains have been achieved (Wu et al. 2007; Burdon et al. 2008; F Drppelmann pers. comm.), even in the first generation of breeding. In New Zealand, considerable gains have been achieved in growth rate and, on many sites, massive improvements in tree form. This has allowed large reductions in initial stocking and tending costs, and will mean better wood recovery through reduced logging waste. However, the major genetic shift towards a short-internode habit brings an increased dependence on pruning to improve clearwood yield.

In keeping with the intensive breeding programs, radiata pine has in recent years become the subject of considerable genomic research (see references in Plomion et al. 2007; Wilcox et al. 2007; Burdon and Wilcox 2011), although many of the findings remain unpublished. The species has been involved in several comparative genomics studies (op cit), which have all indicated close synteny and colinearity among various pines. Searches for quantitative trait loci (QTL) (e.g., Carson et al. 1997; Devey et al. 2004; Cato et al. 2006) have suggested a general paucity of large-effect QTL. This, combined with generally minimal population-wide linkage disequilibrium (e.g., Kumar et al. 2003), has led to a shift in emphasis towards association genetics for pursuing the option of genome-based selection. Also, at least one genomic study (Kuang et al. 1998) has endorsed the hypothesis that very imperfect and variable effective self-fertility is due to genetic load in the form of deleterious recessive genes.

2.3 Spruces (*Picea* A. Dietr.)

The genus *Picea* is a member of the family Pinaceae, with about 40 species distributed throughout the cooler parts of the Northern Hemisphere. Within their natural ranges, some of these are extremely important economically and have been the focus of reforestation and breeding efforts. A few have also become important far outside their natural range.

2.3.1 Black Spruce (*Picea mariana*)

2.3.1.1 *Historical Perspective*

Black spruce (*Picea mariana* [Mill.] B.S.P.) is one of the most widely distributed and hardiest of boreal forest conifers in North America (Farrar 1995). It ranges from northern Massachusetts to northern Labrador on the Atlantic coast, and west across Canada to the west coast of Alaska (Viereck and Johnston 1990). It is also one of the most planted tree species in Canada. It harbors large amounts of genetic variation in quantitative traits, which is an indication of the adaptive capacity of its populations (Khalil 1984). Various patterns of clinal variation have been reported for germination rate, survival rate, phenology, juvenile growth and hardiness (e.g., Dietrichson 1969; Morgenstern 1969; Corriveau 1981; Park and Fowler 1988; Beaulieu et al. 1989a; Morgenstern and Mullin 1990; Parker et al. 1994; Beaulieu et al. 2004).

Natural introgressive hybridization between black spruce and red spruce has been documented in the sympatric zone that mainly takes place in southern Quebec, New Brunswick, Nova Scotia and New England (e.g., Perron and Bousquet 1997). As both are close species, they are known to introgress, but introgressed populations are generally found on disturbed sites (Morgenstern 1996).

Due to this phenomenon, more attention must be paid to seed source transfer in order to make sure that they are well adapted to the environmental conditions of the recipient site.

Knowledge of patterns of genetic variation as well as of the strength of genetic control on characters is of fundamental importance if a breeding program and reforestation efforts are to succeed. While selection and breeding were begun in some Canadian provinces in the 1960s, large-scale tree improvement programs have been initiated in most of the provinces since the mid-1970s (Park et al. 1993).

2.3.1.2 *Economic Importance*

Black spruce is a medium-size tree that can reach on poor-drained sites average heights of about 20 m and diameters of 30 cm whereas on well-drained sites, it can reach up to 30 m high and 60 cm in diameter (Farrar 1995). It is one of the most important species in Canada and northern United States for manufacturing high quality pulp and paper and solid wood products, including framing material, millwork, crating and piano sounding boards (Alden 1997). Historically, it has also provided specialized products such as healing salves from spruce gum, beverages, aromatics and binding material (Viereck and Johnston 1990).

2.3.1.3 Breeding Objectives

As for many other commercial species, selection is primarily based upon stem growth, wood quality traits and tolerance to biotic and abiotic adverse factors. Contrary to other spruces and pines, the stem is generally straight and the crown form is fairly uniform in black spruce. Consequently, less emphasis was put on selection criteria for these traits.

2.3.1.4 Breeding Achievements

While black spruce programs have been established in most jurisdictions where the species grows naturally, their importance and progress varies among them. New Brunswick, for instance, is now in a position to initiate a third breeding cycle. Tree improvement activities are carried out by the New Brunswick Tree Improvement Council, formed by the New Brunswick Department of Natural Resources, the Canadian Forest Service and six large industrial companies (Tosh and Fullarton 2006). Since the first breeding activities in the mid-1970s, two generations of seed orchards were set up by the Council. The first-generation was established as seedling seed orchards between 1980 and 1987, with open-pollinated seed collected from plus-trees selected in natural stands. Progeny tests accompanied the seed orchards and data collected in these tests allowed improvement of the orchards by roguing. A second-generation series was established between 1989 and 1997 by grafting selections into orchards. Since then, all polycross and controlled crosses needed to evaluate the general combining and specific combining abilities of the selected trees were made and all the tests are now in place (Tosh and Fullarton 2006) and third-generation selection in the older full-sib tests will begin in the near future. A portion of the annual reforestation stock requirement is now produced using somatic embryogenesis techniques from elite crosses.

Nova Scotia, Prince Edward Island and Newfoundland and Labrador have also established first-generation seed orchards that now supply the current seed demand for reforestation. A second-generation program is currently conducted by the industrial partners of the Nova Scotia Tree Improvement Working Group (Frame and Steeves 2006). First-generation seed orchards in this province as well as in Prince Edward Island have been rogued to increase expected gain (MacKinnon et al. 1997).

In Quebec, breeding activities also began in the 1970s. Five breeding zones were delineated using 16-year data collected on a range-wide provenance trial replicated on four sites (Beaulieu et al. 1989a). First-generation breeding populations were assembled by selecting superior phenotypes in the best performing provenances. Controlled crosses were carried out and full-sib families were vegetatively multiplied to establish progeny tests. In the 1980s, a network of 24 seedling seed orchards and 42 open-pollinated progeny tests were established using plus-tree selections made in natural stands. Roguing of first-generation seed orchards is now complete, as well as the establishment of the five second-generation clonal seed orchards with elite trees selected from progeny tests accompanying first-generation seed orchards and those established with full-sib families from controlled crosses in the first-generation breeding population. About 3 million rooted cuttings are also produced annually for the reforestation program using seeds form tested full-sib families. Breeding activities for the development of the third-generation program are now underway (Beaudoin et al. 2004).

In Ontario, first-generation breeding zones were created in the early 1970s, and as for most provinces, largely delineated by administrative boundaries (Ford et al. 2006). First-generation seed orchards were set up and they have now been rogued. The installation of genecological trials in the 1990s allowed including information on variation in adaptive traits in the delineation of biologically-sound second-generation breeding zones. The northeastern region has been selected to develop a pilot second-generation breeding program. Selection of superior genotypes was done in first-generation open-pollinated progeny tests and controlled crosses were made. Second-generation progeny tests were set up in accordance with a nucleus breeding system, with the breeding population substructured into elite and infusion populations (Cherry and Joyce 1998).

In Manitoba, breeding zones have been delineated and breeding work is achieved in a collaborative mode by Manitoba Conservation and three forest companies. First-generation seed orchards and open-pollinated progeny tests were established in the 1990s and the early 2000s, and some of the seed orchards have been rogued (Falk et al. 2006). Alberta is also breeding black spruce and has first-generation seed orchards in place.

In the United States, first-generation seedling seed orchards have also been set up in Maine and Vermont (Carter and Simpson 1985; Carter et al. 1988) to produce the genetically improved seed for their reforestation programs.

The development of genomic resources for black spruce is a focus of research at the CFS and Genome Canada's "Arborea" project, with the aim to develop MAS for adaptive traits such as growth and phenology as well as for wood quality traits in the context of shorter rotations. The objective is to more rapidly develop varieties that can better sustain climate-change conditions with faster generation turnover, and to optimize the forest products value chain.

2.3.2 White Spruce (*Picea glauca*)

2.3.2.1 Historical Perspective

White spruce (*Picea glauca* (Moench) Voss.) has a transcontinental range, from Newfoundland and Labrador west across Canada along the northern tree limit to Hudson Bay, the Northwest Territories, Yukon and Alaska, and is adapted to a wide range of soil and climatic conditions (Nienstaedt and Zasada 1990). It is a medium-sized tree that can reach up to 25 m in height and 60 cm in diameter (Farrar 1995). Research on the genetic variation of white spruce began in the 1930s in both Canada and the United States where it has demonstrated great potential for genetic improvement (Niensteadt and Teich 1971). Results from provenance trials have clearly indicated the superiority of white spruce populations originating from the Lower Ottawa Valley and adjacent areas in most of the regions where they were tested. This has had a great influence on the composition of the breeding populations. In some regions of British Columbia, white spruce grows with Sitka spruce (*Picea sitchensis* [Bong.] Carr.). It also grows with Engelmann spruce (*Picea engelmannii* Parry ex Engel.) in the same province, as well as in Alberta, the Northwest Territories and Yukon, and natural hybrids occur (Farrar 1995). White and Engelmann spruce have been shown to be the extreme forms of a clinal pattern of variation associated with altitude (Roche 1969) and hybrids are known as the "interior spruce complex". Hence, interior and white spruces from eastern regions of Canada have been considered to be sufficiently different to warrant separate breeding programs.

2.3.2.2 Economic Importance

White spruce is one of the most important commercial species in the boreal forest of North America. It is used extensively for manufacturing of pulpwood and solid-wood products. It is used for framing material, general millwork, boxes and crates and piano sounding boards (Farrar 1995; Alden 1997). Historically, its wood was used for fuel, its bark to cover summer dwellings, its branches for bedding and its resin for medicinal purposes by aboriginal people (Nienstaedt and Zasada 1990). Due to its high survival rate, capacity to adapt to various ecological conditions and fast growth rate, it is one of the most planted species in Canada, especially in British Columbia, Alberta and Quebec.

2.3.2.3 Breeding Objectives

In white spruce, selection has focussed primarily on increasing economic value by improving stem growth and straightness, volume, as well as crown form and branch size, while maintaining a broad genetic base for adaptability and pest resistance, especially to the white pine weevil (*Pissodes strobi* [Peck]) in western Canada. More recently, emphasis has been put on wood physical properties.

2.3.2.4 Breeding Achievements

White spruce breeding programs started in most jurisdictions in the 1970s and are now well established. Breeding strategies vary from region to region, but generally use progeny testing and recurrent selection combined with clonal seed orchards to produce seed for reforestation programs.

In New Brunswick and Nova Scotia, two distinctly different programs have been implemented for the first generation: (1) seedling seed orchards to capture genetic variation at the provenance level, and (2) clonal seed orchards to capture within provenance variation (Fowler 1986). The former were established between 1978 and 1982 on 8.6 ha using open-pollinated seed from the Lower Ottawa Valley to develop improved varieties adapted to the Maritimes (Carter and Simpson 1985). For the clonal seed orchards, plus-trees were selected in natural stands in each province. They were established between 1985 and 1987 and covered 9 ha. Polycrosses and pair mating were used to generate the seed for progeny tests to estimate both the general and specific combining abilities of selected parent trees, as well as selection plantations in which candidates were selected for the second-generation. Roguing of first-generation seed orchards has been completed and forward selection to establish second-generation clonal seed orchards is underway, with 4.1 ha established to-date (Tosh et al. 2009).

In Newfoundland and Labrador, clonal seed orchards were established in the early 1990s. Since then, polycrosses have been performed and selection plantations established in the early 2000s (English and Linehan 2000).

White spruce breeding activities in Quebec began in the early 1970s to support a major reforestation program. Between 1972 and 1990, over 360 million seedlings were planted (Beaulieu 1994). In the mid-1990s, about 70 million seedlings were planted annually on both private and public lands, but this has now declined to about 25 million. As in the Maritimes, breeding populations in Quebec were developed using various sources of superior material. First, analysis of data collected in genecological tests established in the late 1970s and early '80s allowed delineation of two large breeding zones based on patterns of genetic variation observed and existing ecological classification (Li et al. 1993; Beaulieu 1996; Li et al. 1997b). Provenance trials set up in the 1950s and '60s provided a first pool of tested material for the selection of superior genotypes to build the first-generation breeding populations and seed orchards. The volume production of the best provenances at age 25 years was 20–50% better than local seed sources. About 100 plus-trees were selected from these provenances and grafted in the early 1980s, together with additional selections made in the genecological tests. Polycross and pair matings were made and progeny tests established in the mid 1990s. Controlled crosses were completed and a new series of progeny tests was established in the early 2000s. Second-generation orchards have already begun to provide seed for reforestation programs, with some material vegetatively bulked up as rooted cuttings.

In Manitoba, three breeding zones were delineated and open-pollinated family tests as well as first-generation clonal seed orchards were established (Falk et al. 2004), some of which have been rogued (Falk et al. 2006). In Alberta, first-generation clonal seed orchards were established between 1982 and 1989 for each of three breeding zones, with accompanying open-pollinated progeny trials. These orchards have been rogued several times since their inception (Hansen et al. 2009).

In British Columbia, the interior spruce breeding program was structured in two phases. The first began in the mid-1960s and addressed the needs of three regions. The second began in the mid-1970s and focussed on other regions where interior spruce was important. First-generation seed orchards established at this time have now been rogued and provide much of the planting stock. Controlled crosses have also been made and full-sib second-generation tests are in place. Selection of superior genotypes has begun and the establishment of a breeding orchard has been initiated (Carlson et al. 2009).

In Maine, private forest companies have established small first-generation clonal seed orchards in the 1970s whereas 10 ha of clonal and seedling seed orchards were set up in the 1960s in the State of New York (Carter and Simpson 1985). Breeding programs have also been undertaken by the USDA Forest Service in Minnesota, Wisconsin and Michigan in the 1970s (Nienstaedt and Teich 1971).

Major research efforts have been underway in Canada through several projects in recent years, to develop genomic resources and implement molecular breeding for adaptive and wood quality traits in both eastern and western white spruces. Various schemes are being deployed to identify informative gene SNPs including QTL/gene co-localization studies, association genetic approaches and genomic selection. Association studies are also underway to evaluate the possibility of using candidate genes for early selection in the spruce terminal weevil resistance programs (also involving Sitka spruce) in the British Columbia Forest Service breeding programs.

2.3.3 Red Spruce (Picea rubens)

2.3.3.1 Historical perspective

Red spruce (*Picea rubens* Sarg.) is a common spruce in the Maritime provinces of Canada and southward into the Appalachian Mountains of the United States (Farrar 1995). It is also present in Quebec and Ontario but only in the southern regions. While it is an important forest species, it is not widely planted. As introgressed hybrids contribute substantially to observed variation in phenotypic traits and adaptation (Morgenstern et al. 1981), seed-source transfer has been carefully monitored. Research on the genetics of red spruce began in the early 1950s with the involvement of the Canadian Forest Service and the collaboration of the eastern Canadian provinces and the federal and state research organizations of the United States (Holst 1955). The aim was to identify superior populations that could be used directly in reforestation programs.

2.3.3.2 Economic Importance

When utilized for structural products, red spruce is not distinguished from other spruces and is processed in a group called SPF (Spruce, Pine and Fir). Its wood physical properties are in the range of those of white and black spruces (Jessome 1977) and its wood is mainly used for lumber, flakeboard, plywood, and pulpwood. Other marginal uses are for poles piling, boatbuilding and cooperage stocks, as well as sounding boards for a variety of musical instruments (Blum 1990).

2.3.3.3 Breeding Objectives

Selection in red spruce, has focussed on improving stem growth and straightness, volume as well as crown form and branch size while maintaining a broad genetic base for adaptability and pest resistance, especially to the spruce budworm (*Choristoneura fumiferana* [Clem.]) and the yellow-headed spruce sawfly (*Pikonema alaskensis* [Roh.]).

2.3.3.4 Breeding Achievements

Nova Scotia initiated its red spruce breeding program in 1976 and the selection of plus-trees was completed in 1985 (Fowler 1986). Clonal seed orchards were established to provide the genetically improved stock for the reforestation program. In Quebec, first-generation clonal seed orchards were also established in the 1980s using plus-trees selected in natural stands as well as in range-wide provenance trials established in the late 1950s by the Canadian Forest Service (Morgenstern et al. 1981). Reforestation of red spruce in Quebec has not been extensive, due to low productivity when planted on open sites and its susceptibility to winter drying and frost damage (Morgenstern et al. 1981; Beaulieu et al. 1989b), so advanced-generation breeding had not been continued. On the other hand, New Brunswick responded to an increased interest in planting red spruce by reviving its program in 2004. Second-generation red spruce clonal seed orchards had been established in 1999, and by 2007 they occupied 3.6 ha (Tosh and Fullarton 2009).

As red spruce is highly susceptible to winter dessication and given the relative paucity of breeding resources for this species, red spruce breeders in the future may resort to genomic approaches to identify adaptive polymorphisms and select trees that are more tolerant to winter drying and frost damage. The current development of gene catalogs and SNP directories for white spruce and black spruce should help accelerate the application of MAS in red spruce.

2.3.4 *Sitka Spruce (Picea sitchensis)*

2.3.4.1 *Historical Perspective*

In its native range, Sitka spruce (*Picea sitchensis* (Bong.) Carr.) occupies a narrow strip on the north Pacific coast of North America, extending for 2,900 km from 61°N latitude in south-central Alaska to 39°N in northern California. Throughout this tremendous north-south range, Sitka spruce is a coastal species, occupying islands of the Alexander Archipelago in Alaska and the Queen Charlotte Islands (QCI) in British Columbia, and, with the exception of river valleys, rarely reaching more than a few kilometres from the coast along a narrow strip on the mainland (Harris 1990).

While its natural range is not extensive and the species' economic importance ranks far below that of other western conifers, Sitka spruce is a keystone species in some of the most productive ecosystems of North America, particularly in the QCI (Peterson et al. 1997). Nevertheless, in the Pacific Northwest United States, British Columbia and Alaska, Sitka spruce is not a preferred species for reforestation and in fact is often considered unacceptable. This is because it is attacked by the white pine or terminal weevil (*Pissodes strobi* Peck), which repeatedly kills the emerging leader of young plantation trees. The weevil is a native insect that occurs across Canada and the northern United States. Sitka spruce is particularly susceptible to this pest; damage is so severe that young plantation trees often become stunted and bushy as terminal leaders are repeatedly killed and young trees fail to achieve apical dominance. This has reduced planting to offshore islands such as the QCI and Alaska.

Outside its natural range, Sitka spruce has played an important role in plantation forestry, particularly in northern Europe (Hermann 1987). In Great Britain, Sitka spruce is the most widely planted conifer, accounting for nearly 700,000 ha of forest or 30% of the total forest estate (Forestry Commission 2003). The species is well suited to areas of high rainfall and lower quality agricultural soils that predominate in the north and west of Britain. It is planted from Cornwall in the southwest England (latitude 51° N), through Wales and northwestern England, across northeastern England and southern Scotland and up into the Scottish Highlands (latitude 58°N).

Although the species was originally described by Archibald Menzies in 1792, it was not introduced into Britain until 1831 by David Douglas. By the time the British Forestry Commission (the State Forestry Service) was formed in 1919, experience from sample trees planted in arboreta and on large estates had established the species as fast-growing, hardy in exposed conditions and capable of growing on site types which at the time were mainly planted with Norway spruce (*Picea abies* [L.] Karst.). The superior growth of Sitka ultimately led to an increase in its popularity through the 1930s and beyond as the forest estate expanded under the then-government policy of afforestation.

2.3.4.2 Economic Importance

Wood from Sitka spruce offers unique qualities for manufacture of the highest quality sounding boards for many musical instruments, and its outstanding strength-to-weight ratio made it strategically important during both World Wars for construction of aircrafts (Brazier 1987). Although a relatively minor species in its native range, Sitka spruce is now hugely important to British forestry and wood utilization industries. The main objective of growing Sitka is to generate construction-grade timber that will displace material imported mainly from Scandinavia and the Baltic states, although smaller material also feeds the pulp and particle board industries which have become well established in Britain.

Annual growth in Britain averages 12 to 26 m^3/ha/yr, translating to rotation lengths of 50 years down to 35 years, depending on the site. Around 32 million plants are sold annually within Britain to plant over 12,000 ha, predominately for restock harvested forest land. Sitka spruce is also a primary plantation species in Brittany (France) and Ireland, where productivity of stands is similar to or greater than that in Britain (Vaudelet 1982; Serrière-Chadoeuf 1986; Guyon 1995; Thompson et al. 2005).

2.3.4.3 Breeding Objectives

Within its native range, breeding has focussed on developing robust resistance to the white pine weevil. This program is based on investigations of the extent and nature of genetic resistance to the pest, with the goal of restoring the Sitka spruce component of the regenerated coastal forests. The mechanisms for resistance are very likely complex, with the density of sclereid cells and resin canals thought to be important. In some genotypes, a strong resistance factor, almost a "total resistance", was also observed, but its mechanism is unknown (Alfaro et al. 2002). The evidence is that this resistance is stable, viable over a wide area and appears durable.

In Britain, the main objective is to increase the end-of-rotation value to the construction grade industry, relative to that achieved using unimproved seed imported from the Pacific Northwest. Trees are selected which combine good growth rate, with improved stem straightness and branching qualities, and better wood stiffness. Wood stiffness is a complex trait involving wood density, microfibril angle and other internal characteristics such as proportion of compression wood. Under current practice, only wood density is screened as a surrogate for wood stiffness.

2.3.4.4 Breeding Achievements

Breeding efforts in British Columbia have focussed on quantifying resistance to weevil, based on statistically testable data (King et al. 2004), and development of methodology for rapid screening (five years) using artificial infestations (Alfaro et al. 2008). Many populations, families and individuals have now been screened to ascertain which have resistance that is durable and useable in the breeding program (King and Alfaro 2009). The best individuals and families have been established into seed orchards which are now producing seeds with a high degree of resistance (King and Alfaro 2004). New guidelines for the deployment of resistant Sitka spruce have been proposed, which include recommending Sitka spruce as not only an acceptable but even the preferred species for many coastal sites (Heppner and Turner 2006).

The breeding of Sitka spruce in Britain has followed the classical breeding theory: selection of the best origin, selection of plus trees from stands in forests, followed by testing of selected plus trees through comparative half-sib progeny tests, subsequent measurement of trials and then re-selection of a breeding and production populations based on multi-trait index selection. Samuel et al. (2007) summarized the processes involved in identifying provenances best suited for planting in Britain. The general conclusion was that material from around the QCI (54° N) was most suitable for the bulk of Britain although in the milder areas of southwest England and Wales, Washington sources (48° N) or even Oregon material (45° N) were well adapted.

Plus-tree selection in Britain commenced during the early 1960s (Fletcher and Faulkner 1972) and progressed through into the early 1980s. Over 1,800 candidate trees of predominately QCI origin were selected. Progeny tests were established with open-pollinated seeds, with each candidate evaluated in replicated trials established on an average of three sites, and compared against standard controls of unimproved QCI and Washington origin (Lee 2001). The trials were measured regularly for height and later stem diameter, stem straightness and wood density. The best 340 plus tress were identified, based on a multi-trait index combining 15-year stem diameter, straightness and wood density, and these used as first-generation breeding parents. In the second generation, the program expects to stratify the breeding population into six sublines of equal mean genetic value, and to apply positive assortative mating within sublines (Lee 2001).

Improved planting stock has been available from the Sitka spruce breeding program since the early 1990s. Improved stock can be derived either from seedlings raised from seed collected in progeny-tested clonal seed orchards, or as rooted cuttings derived from stock plants originating from controlled pollinations. The controlled pollination of selected seed parents uses a polymix of 20 or so unrelated pollens, again from selected trees. Predictions of genetic gain have been impressive, up to around 20% for both stem diameter and stem straightness with minimal loss in wood density. More recently, these half-sibling family mixtures used in the production of stock plants and ultimately rooted cuttings, have given way to full-sibling families (Lee 2006). Sawmill studies involving trees from some of the earlier half-sib progeny tests have suggested end-of-rotation gains for volume of around 25% relative to unimproved QCI material (Lee and Matthews 2004), and an increase of high-end value sawlogs of up to 130% (Mochan et al. 2008). Improved material is in high-demand and this is now entirely satisfied from home-produced improved sources.

Despite having contrasting breeding objectives, groups in Canada and Britain are collaborating to develop geneomic resources and identify markers associated with a range of economically important traits, including disease and insect resistance and wood density. In particular, the British effort has invested in MAS, with an initial objective to identify a suite of DNA-based markers, which could be used in the laboratory as surrogates for direct field selection. Three large clonal trials were planted in 2004 on climatically contrasting sites across Britain. Each trial contains the same material; 1,500 clones from each of three full-sib families, along with the usual QCI control (used also in the Canadian program). It is hoped that the tests will enable the identification of QTLs contributing to wood density, stem and branch quality (Lee et al. 2007).

Research also continues to develop somatic embryogenesis and cryopreservation of Sitka spruce. If successful, this will prove instrumental in harnessing the material identified in the British MAS program for quick deployment to the field (Lee et al. 2004), and for confirmation and delivery of stable weevil resistance (El-Kassaby et al. 2001).

2.3.5 *Norway Spruce* (Picea abies)

2.3.5.1 *Historical Perspective*

Norway spruce (*Picea abies* [L.] Karst.) is one of the most abundant and economically most important forest tree species in Europe. Its natural geographic range covers 31 degrees of latitude from the Balkan Peninsula to its northernmost extension near Khatanga River, Siberia. Longitudinal range is from the French Alps to the Sea of Okhotsk in eastern Siberia. The vertical distribution is from sea level to altitudes above 2,300 m in the Italian Alps. Its natural range in Europe is to a large extent in the boreal and in the mountainous region of the temperate zone. The species is, however, widespread outside this range, particularly in western and central Europe. This is due to the fact that the proportion of Norway spruce has been substantially increased in Europe by reforestation and afforestation in order to establish forests for timber production. This process started in particular at the beginning of the 19th century when many forests in Europe had been affected by forest devastation due to overexploitation and soil degradation. The species can easily be established artificially outside its natural range, in particular in the rather oceanic climate in western Europe that seems to provide a physiological optimum for Norway spruce. It has been regenerated artificially in areas naturally occupied by European beech, oak and other broadleaved tree species. To some extent, Norway spruce has also been planted in North America, especially in eastern Canada. Due to the wide distribution of Norway spruce and considerable differentiation in provenances, it is not possible to define very distinct site requirements for the species.

The first provenance trials with Norway spruce were established in the late 1800s in Austria and were followed by several series of national and international experiments (König 2005). The most important of these are the two IUFRO series of 1938 and 1964, which together comprise more than 1,100 provenances and were planted at more than 40 locations. The field experiments have revealed certain genetic-geographic variation patterns with regard to growth and have clearly demonstrated that the local provenance as a rule is not the best (König 2005). A considerable increase in growth rate can therefore be obtained by judicious transfer of provenances. In the case of extreme environmental conditions, like in northern Scandinavia and at high altitudes in the Alps, large losses can result as a consequence of inappropriate provenance transfers.

2.3.5.2 *Economic Importance*

The most extensive coverage of Norway spruce is found in Sweden and Austria, where the species covers more than 25% of the total land area and more than 40% of the forest area (Spiecker 2000). A large coverage of Norway spruce, with 15–25% of the total land area and more than 25% of the forest land, can also be found in Finland, Norway, Czech Republic, and Slovakia. In Switzerland and Germany, the species covers 10–15% of the total land and more than 30% of the forest land. All these countries are in the natural range of the species, but with plantations also outside the areas where it occurs naturally. This is also the case in the western part of Europe; in Belgium, the Netherlands, Denmark, Great Britain, Ireland and most parts of France.

The highest volume production of Norway spruce is found in pure plantations and often outside its natural range (Schmidt-Vogt 1977; von Teuffel et al. 2004). On average, the annual increment of Norway spruce in Europe during the last 20-year period has been about 7.3 m^3/ha (von Teuffel et al. 2004), but growth rates are much higher in several countries where Norway spruce is planted as an exotic. Norway spruce accounts for 40% of the total increment in Nordic forests, making it a very important commercial tree species in this region. There has been a considerable increase in the growth rate of Norway spruce in Europe during the last 40–50 years, which could be due to several factors such as changes in land use, forest management, natural disturbances, climate changes and nitrogen deposition (Spiecker 2000). However, in recent decades, some problems have been exposed due to its susceptibility to air pollution, wind, snow, ice and storms, and also to certain fungi and weevils. The use of maladapted provenances has resulted in damage and reduced yield in plantations. These negative factors have made Norway spruce less popular in reforestation, in particular outside its natural range.

Norway spruce produces large volumes per unit area of straight timber that is suitable for structural applications, panelling and furniture. Its relatively fine branching and long, lean and straight fibers makes it particularly attractive as raw material for the pulp and paper industry. It is therefore a widely used and valuable tree species for the forest industry in Europe.

2.3.5.3 Breeding and Breeding Objectives

The genetic variability in Norway spruce has been studied in provenance and progeny trials, often planted at several sites, and by genetic markers such as isozymes and DNA markers. The most pronounced patterns of variation demonstrated in provenance trials relate to the populations' responses to climatic conditions. In northern Europe, these patterns of variability often relate to latitude and longitude, and to the degree of continentality, and will sometimes vary clinally. They are expressed as variation in budflush and duration of the annual growth period in spring, and the corresponding cessation of growth and development of frost-hardiness in autumn. These annual growth patterns have implications for frost-hardiness, growth potential and wood-quality traits, and are important for proper choice of reforestation materials. At the same time, there is large variability for the same traits within natural populations. In central Europe, the regional variation patterns are less clear, owing to a long history of planting and provenance transfers.

Breeding of Norway spruce was initiated in several European countries in the late 1940s (Danell 1991; Mikola 1993). The work typically started with the selection of phenotypically superior "plus trees" in natural stands (Skrøppa 1982; Gabrilavicius and Pliura 1993; Mikola 1993). Mature trees were selected that had superior height and diameter growth and stem and branch quality, compared to neighboring trees in the stand. These were established by grafting onto rootstocks in clonal archives and seed orchards. Each grafted seed orchard was composed of a rather large number of selected clones (50–500), with the intention of seed production for one geographic region. The seed orchards generally start to flower 10–15 years after grafting, although the periodicity and amount of flowering are very much dependent on climatic conditions at the orchard site. To promote flowering, orchards have often been located on warmer sites, relative to those from where the parents originated and where the orchard seed is intended for use.

It was soon realized that the selection of plus trees in natural Norway spruce stands is not an efficient method to identify superior genotypes. It is necessary to test the genetic value of each parent, based on an evaluation of their offspring. In the Nordic countries, this is done in progeny tests planted at several sites where assessments are made of survival, height and diameter growth and quality traits. The progeny tests are sometimes supplemented with tests where seedlings are grown under controlled conditions in growth chambers and measurements made of physiological traits. On the basis of several traits, a subset of the original parents is selected for further breeding. Seeds for operational planting can be collected selectively in the orchard, the orchard thinned or a new orchard established with the best progeny-tested parents.

In other countries, breeding programs were based on materials selected from populations with high adaptive potential exhibited in comparative provenance trials. The best individuals from families of the best provenances were selected to produce seeds in orchards, or to create a breeding population through controlled crosses. Some of these programs also targeted mass production of rooted cuttings of tested clones (Birot 1982; van de Sype and Roman-Amat 1989; Kleinschmit 1993). Of major concern in the breeding strategies have been the breeding objectives; the sizes of breeding and production populations required to maintain genetic diversity; test design and efficiency; and identification of suitable regions where the orchard seed should be recommended for use.

The principal breeding objectives in most programs are to improve the value of production in future spruce stands and to mitigate risk under variable environmental conditions. The selection criteria needed to achieve these goals will vary among different breeding populations, based on the varying regional conditions. Under the severe conditions in the northern boreal forest, adaptation to the climatic conditions is crucial. Frost hardiness in artificial freezing tests, the timing of flushing in spring and survival, vitality and lack of injuries in field tests are therefore important target traits. Spring frost events may also occur at more southern latitudes, and selection for late bud flushing may also be important here. Selection for yield is mostly based on height or diameter growth. Some programs aim to keep stem and wood quality at the present level, while others also want to select for improvement of quality traits. Another important target for breeding has been resistance to root rot (*Heterobasidion annosum*), but research efforts have not yet succeeded in developing reliable techniques for selection of resistant materials. In the last decade, adaptation to changing climate conditions has been an increasing concern. In Sweden, this objective has been addressed by establishing a system of multiple breeding populations, which are bred for adaptation to different combinations photoperiod and temperature conditions, including combinations that lie outside of what is normal under the present climatic conditions (Andersson 2002).

2.3.5.4 Breeding Achievements

The regeneration of Norway spruce forests is based both on natural regeneration and planting, with the former often preferred where it is feasible. While seed orchards are common in many countries, the bulk of Norway spruce seeds are still collected in natural or planted stands. Each seed lot should be identified by the geographic origin of the stand, and in several countries it is required that the seed stand should be selected for superior performance. The relative amounts of seeds from forest stands and from genetically improved seed harvested in seed orchards vary considerably among countries and regions within countries. In the Nordic region, there has been a considerable increase in the use of seed orchard seed during the last five-year period. In Norway, 77% of the 300 kg Norway spruce seeds sold in 2007 in the southeastern region originated from seed orchards. The nearly 12 tons of Norway spruce seeds that were produced in Swedish seed orchards in 2006 will be produce 1.2 billion plants, sufficient to regenerate 450,000 ha (Almqvist et al. 2008).

Genetic tests have shown that the productivity of Norway spruce stands established with seedlings originating from untested first-generation seed orchards is about 10% higher than those from unselected material of the same provenance (Andersson 2002). The difference in quality is less, but even here there has been some improvement. Genetic thinning of these orchards could increase the gain further. In Sweden, a second round of seed orchards was established using a mix of untested and tested parent trees. The gain in volume production from these orchards is estimated to be in the range of 12–25% (Rosvall 2001). In a third round of seed orchards, based on a new generation of tested parents from the breeding populations, a gain of some 35% is anticipated (Rosvall 2001).

A comparison of production and economics of Norway spruce stands in southern Sweden established with genetically improved and unimproved seedlings showed that the increased gain in volume production resulted from earlier thinnings and shorter rotation age (Rosvall et al. 2004). A 68% increase in the present value of improved planting stock could be expected, based on the realistic assumption of a 22% increase in volume growth and a 10-year reduction in rotation age.

The use of clonal forestry based on rooted cuttings was popular in the 1970s in Germany, Denmark and Sweden, but now occurs only on a small scale. The same is true for bulk propagation of rooted cuttings from selected full-sib families. Clonal forestry based on somatic embryogenesis has potential to become a valuable tool for intensive wood production, and methods for somatic embryogenesis in Norway spruce are now to a point where operational testing and deployment programs can be launched (Devillard and Högberg 2004).

Marker-aided or genomic selection has not yet been applied to breeding of Norway spruce. A list of "recommended" nuclear microsatellites has been established for the species, and research is underway using SNPs to identify candidate genes for the terminal bud set (M Lascoux pers. comm. 2009). Meanwhile, a project in Sweden is sequencing the Norway spruce genome; its results will facilitate the development of genetic markers and dissection of complex traits, and likely lead to applications in breeding (PK Ingvarsson pers. comm. 2009).

2.4 Other Important Pinaceae

2.4.1 Douglas-fir (*Pseudotsuga menziesii*)

Douglas-fir (*Pseudotsuga menziesii* [Mirb.] Franco) is an important timber species in western North America, where it is valued for its high wood and timber quality, fast growth and broad resistance to diseases and insects. Along the Pacific coast, the coastal variety *P. menziesii* var. *menziesii* extends in a continuous fashion from latitude 37° to 53°, while the interior variety *P. menziesii* var. *glauca* (Beissn.) Franco ranges from the 19° (from the mountains of central Mexico along the Rocky Mountains) to latitude 55° (Hermann and Lavender 1990). In the southern part of its interior range, Douglas-fir distribution is non-continuous. In the western United States, Douglas-fir grows on roughly 17 million ha (Smith et al. 2001), while in Canada it grows on 4.5 million ha (Hermann and Lavender 1999). It can grow from sea level up to 3,000 m on the slopes of the Rockies (Howe et al. 2006). Due to its desirable economic characteristics and its wide ecological niche, this highly adaptable and plastic species has been introduced to Europe and several countries of the southern hemisphere (New Zealand, Argentina and Chile), where it is a major commercial conifer species.

2.4.1.1 *Historical Perspective*

Douglas-fir has been a major species in North America since the mid-Pleistocene, establishing itself as a keystone species over large parts of its range (Lipow et al. 2003). During the last 200 years through to the early 1900s, forests in the Pacific Northwest of the United States and in British Columbia were mainly clearcut followed by slash burning and natural regeneration. This mode of reforestation favored the establishment of Douglas-fir on the dry soils of clearcuts, often replacing other species such as western hemlock (*Tsuga heterophylla* [Raf.] Sarg.) and western redcedar. In the 1950s, planting following clear cutting and slash burning reduced rotation ages and became the reforestation method of choice (Curtis et al. 2004). Only small relics of old-growth Douglas-fir are still present throughout its natural range.

Douglas-fir was introduced in Europe in 1827 by the Scottish botanist David Douglas. Initially planted as an ornamental, it was utilized for forest plantations by the end of the 19th century. Its position, compared to other forest species, remained modest until the middle of the 20th century when it became a major reforestation species in western Europe, mainly with the support of post-war national or regional forest grants. Today's plantations in western Europe exceed 700,000 ha, representing the largest area of Douglas-fir outside its natural range.

Douglas-fir was first introduced to New Zealand in 1859 (Miller and Knowles 1994). It has been used in plantations since the early 1900s and is economically the second-most planted species. The use of Douglas-fir in New Zealand initially declined in the mid 1960s after the fungal disease Swiss needle cast (*Phaeocrytopus gaumanii*) became established in the North Island. Interest and enthusiasm for Douglas-fir is now keenest in the South Island where growing conditions are more favorable, and wherever Swiss needle blight has not had a significant impact on stand health and productivity. In the South Island, there are large areas where Douglas-fir has distinct advantages as the primary commercial species due to its good growth and tolerance of winter climatic extremes.

Unknown Douglas-fir provenances were introduced to Argentina in the early 20th century on Victoria Island in Nahuel Huapi Lake in northwest Patagonia. Growth pattern studies associate this land race with Californian origins (Rehfeldt and Gallo 2001). However, only in 1940 was the first plantation made by the State Forests Institute, and it was not until the 1970s that the provincial government began programs with the objective to identify appropriate areas for intensive forest plantations (Buamscha 2002). The current area of Douglas-fir in Argentina ranges from Neuquen province (latitude 40° 15′ S) to Chubut province (latitude 43° 13′S).

The introduction of Douglas-fir to Chile was similar to that in Argentina. While first introduced early in the 20th century, it is only in 1940 that the first plantations were made, principally in the Cautín province (IX Region). Site conditions in the south of Chile, from Cautín (latitude 38° S) to Llanquihue (latitude 41° S) are favorable for the growing of Douglas-fir (Siebert et al. 2003).

2.4.1.2 *Economic Importance*

Douglas-fir is one of the most valuable and productive timber species. In the western United States alone, over 27.8 million m^3 of Douglas-fir lumber was produced in 2002 (Howe et al. 2006) while in British Columbia in 2006/07 Douglas-fir contributed roughly 12% of the provincial allowable annual cut (10.2 million m^3 of a total of 83.6 million m^3). In addition, in 1999, Douglas-fir accounted for one third of all log exports in the United States (Howard 2001) and 60% of all log exports in British Columbia in 2003/04.

France and Germany represent more than half and nearly one fourth of the European Douglas-fir area (400,000 ha and 160,000 ha, respectively). Other European countries where Douglas-fir is important are: United Kingdom (50,000 ha), Spain (35,000 ha), Belgium (20,000 ha) and Italy (12,000 ha). As a result of its plasticity and its high volume yield, Douglas-fir tends presently to replace Norway spruce in middle-elevation regions.

In New Zealand, Douglas-fir is planted on more than 113,000 ha (MAF undated) and is suited primarily as structural and framing timber because of its good stiffness and stability characteristics. Its timber falls within the density range of radiata pine, but has longer fibers and greater strength. A major advantage over radiata pine is that wood density and stiffness does not decrease seriously near the pith, so that framing timber can be sawn from much smaller logs including thinnings.

Currently, Douglas-fir occupies over 8,000 ha in the Argentinean Patagonian region (Jovanovski et al. 2005) and 15,000 ha in Chile (Siebert et al. 2003). This could be considered insignificant compared to the potential that Douglas-fir has in the region, and a shortage of wood supplies is foreseen for the future years. The price of Douglas-fir wood is similar and competes with that of native species such as southern beech (*Nothofagus* spp. Bl.) and Chilean cedar (*Austrocedrus chilensis* [D. Don] Pichi-Serm. & Bizzarri).

2.4.1.3 Breeding Objectives

Within its native range in North America, the main goal of the tree improvement programs is improvement of stem volume in genotypes adapted to their target environment (Howe et al. 2006). In the United States, small breeding zones were identified initially based on the seedling studies of Campbell (1979), who found that adaptive traits were finely shaped by local conditions. This led to the delineation of small breeding zones, 60,000 ha in size and elevational ranges up to 300 m (Silen and Wheat 1979). However, results from provenance studies established by Ching (1965) showed that local populations were rarely the best performers and provenance-site interactions were not important at age 25 years (White and Ching 1985). This led to the formation of much larger breeding zones (Stonecypher at al. 1996).

In British Columbia, early Douglas-fir improvement started with intraspecific (racial) crosses of parents growing in drastically different environments (Orr-Ewing 1966). A contrasting approach was also explored via the recurrent selfing of genotypes to the S_3 generation (Orr-Ewing 1976). However, practical tree improvement started with the diallel program of Chris Heaman, using six-parent partial diallels in 8 series planted on 11 sites per series (Yeh and Heaman 1987).

Early provenance evaluations in Germany indicated strong differentiation of populations in Douglas-fir provenances. To obtain more complete information about Douglas-fir variability, adaptability and physiology, IUFRO started in 1967 a systematic and representative collection of 182 indigenous provenances, covering the whole natural range. These were distributed to 59 institutions in 36 countries. This provenance collection, planted over more than 100 sites, has been the base of a vast number of biosystematics studies and provided several European institutes with genetic resources to start or diversify their breeding activity (Kleinschmit and Bastien 1992).

In 1985, six European countries (Belgium, France, Germany, Italy, Spain, and United Kingdom) agreed to collect a base population from a broad genetic base of superior provenances in previous IUFRO tests to provide accurate genetic parameter estimates for further breeding. This base population, made of 1,000 open-pollinated progenies harvested at low elevation from United States Pacific Northwest has been evaluated in field tests, covering 270 ha in western Europe and straddling over 10 degrees of latitude. Selection criteria were: adaptedness, expressed as survival, bud flush and bud set (frost damage avoidance), stem quality, volume growth, and wood quality (despite adverse genetic correlations with growth) (Rozenberg et al. 2001).

In New Zealand, Douglas-fir improvement started with provenance trials of large numbers of provenances in 1957 and 1959 from the United States Pacific Northwest and northern California (Shelbourne et al. 2007). Before the provenance trial results were known, a breeding population was established based on plus-tree selections from 35–50 year-old stands, probably from Washington provenances planted during the Depression in Kaingaroa Forest in the Central North Island. Parents were selected and grafted, and open-pollinated progeny tests established with little delay in the early 1970s. However, early test results from the 1957 and 1959 tests at age 13 years, showed superior growth of Californian and southern Oregon provenances, causing the breeding program to stall for the following 14 years (Shelbourne et al. 2007).

In 1988, in the wake of high log prices, industry interest revived and a new breeding program was started in New Zealand with 186 selections (superline) composed largely of better coastal fogbelt provenances in the 1959 provenance trials and material of Fort Bragg origin. Plus-trees were grafted in an archive and it was planned to progeny test these clones by polycross and use pair crossing for forward selection. This strategy failed to deliver sufficient seed or crosses, and has recently been revised to rely on an open-pollinated testing strategy in the clonal archive for generation turnover and breeding value estimation. It is intended that relatedness among selections will be assessed by DNA pedigree analysis (Shelbourne et al. 2007).

In Argentina, the growth potential of the land race is high. The principal objectives of the breeding program initialized in 1998 by INTA Bariloche were therefore: (1) to increase growth and improve form by selections from the land race; (2) to supply improved seed from seed production areas; (3) to broaden the genetic base from fast-growing Washington and

Oregon populations; (4) to assess genetic diversity of the land race; and (5) to maintain adaptability

In Chile, the breeding program has objectives similar to those in Argentina, nevertheless the propagation procedures are much more developed; for example, rooting cutting propagation, management of donor plants (hedges), and evaluation of flowering induction techniques.

2.4.1.4 Breeding Achievements

In coastal British Columbia, forward selections from the diallel program were grouped into sublines consisting of 10 to 15 parents in a total of 32 sublines. Each parent is progeny tested using a standard polymix and, at the same time, four to six full-sib families with a common parent are tested in 5 x 5 family blocks on two sites for the purpose of forward selection. This complementary testing is to be carried out in four series. Forward selections from the first series have been grafted for third-generation orchard establishment (Stoehr et al. 2008). The primary selection trait was height growth, while a secondary trait was wood density. For interior Douglas-fir, control crossing for second-generation testing is underway. Rotation-age volume gains in selections from first-generation open-pollinated tests were above 25%.

In the United States Pacific Northwest, realized genetic gains of elite crosses (between selected first-generation parents) in realized gain trials were close to the predicted values based on progeny tests, i.e., 6% for height, 8% for diameter and 28% for tree volume (St. Clair et al. 2004). Crossing and testing for second-generation orchards are underway (Howe et al. 2006).

Until recently, Douglas-fir plantations in Europe were established primarily with seeds collected from North America. As a consequence of the IUFRO provenance experiments, the European Community sponsored four missions to North America to check the status of the original IUFRO seed-collection stands. In order to preserve valuable Douglas-fir genetic resources in Europe, more than 1,000 ha of *ex situ* conservation plantations have been established in France, Germany and Belgium. The outstanding performance of Douglas-fir has justified the establishment of 34 seed orchards (163 ha) in seven countries of the European Union (26 (109 ha) in the "Qualified" category and 8 (54 ha) in the "Tested" category). The largest orchard plantings are France and Germany, with 8 orchards (98 ha) and 9 orchards (35 ha), respectively.

In New Zealand, selections from the two unrelated "superlines" will be grafted into open-pollinated orchards. Recently the economic importance of timber stiffness has been recognized, with stiffness as well as yield and log quality established as objective traits. Wood density or stiffness were not criteria in the selection of parents in the "superlines", and to remedy this a number of selections are now being undertaken for wood stiffness, diameter and stem straightness in both second-generation land-race stands of Fort Bragg, Californian origin and in existing progeny trials (H Dungey pers. comm.). Seedlots from the Fort Bragg land race have proven to be top performers for volume growth in the 1959- and 1996-planted trials.

In South America, first-generation testing and breeding are underway and early results have led to the establishment of seed orchards and seed production areas in Argentina (Gallo et al. 2005).

Detailed marker association studies have been limited to cloned (rooted cuttings) seedlings from a single full-sib family to identify QTLs for several adaptive traits, such as spring bud flush and spring and fall frost hardiness (Jermstad et al. 2001a, b). For spring flushing, there was congruence in QTL presence and linkage group location from year-to-year, but not between test sites, suggesting that different suites of genes are governing growth initiation in different environments. Significant QTLs were also found for spring and fall cold-hardiness, but their locations revealed that different genes are responsible for the two cold-hardiness traits. In a follow-up study, significant QTL x treatment interactions have been detected in the same genetic background (Jermstad et al. 2003), indicating that QTLs as tools for selection is still in a developmental state in Douglas-fir. In a more recent study, again in the same family, Wheeler et al. (2005) showed that, with a larger sample size, several QTLs for adaptive traits can be classified as candidate genes.

2.4.2 Larches in Europe (*Larix*)

The genus *Larix* Mill. is composed of 10 or so species distributed across the Northern Hemisphere, three in North America, six in Asia and one in Europe, with additional subspecies and natural hybrids often recognized (Schmidt 1995). Important breeding programs are well established for various species, including western larch (*Larix occidentalis* Nutt.) in British Columbia and the United States Pacific Northwest (Jaquish et al. 1995) and for tamarack (*Larix laricina* [Du Roi] K.Koch) in Quebec and the Canadian Maritime Provinces (Fowler et al. 1995). In Europe, the focus of genetic improvement is on European larch (*Larix decidua* Mill.), the exotic Japanese larch (*Larix kaempferi* [Lamb.] Carr.), and their hybrid, Dunkheld larch (*Larix × eurolepis* Henry, syn. *Larix marschlinsii* Coaz.); these European efforts are discussed in greater detail in the following sections.

2.4.2.1 *Historical Perspective*

Natural European larch forests are limited to mountainous areas, including the Alps, the Sudetan Mountains, the Tatras and hills in central Poland. Some relic populations also exist in the Romanian Carpathian Mountains. The native range of European larch is highly discontinuous and of small size but due to human pressure release on farmland, it has naturally extended upwards and downwards the mountains as in the French Alps since nearly a century. Yet, most of larch planting was outside its native range in northern and western Europe such as in Belgium, Denmark, France, Ireland, Scotland, Germany, Poland, etc.

Due to its fast growth, excellent stem form and durable wood, larch has long attracted attention from European foresters who attempted to move European larch from its native mountain range towards northern-Europe lowlands. These introductions, mostly from alpine seed sources, proved unsuccessful; after one or two decades of satisfactory growth, many plantations suffered dieback caused by larch canker (*Lachnellulla willkommii* Hartig).

Interest then shifted to an exotic species, Japanese larch, but either it proved poorly adapted to more continental sites because of its sensitivity to summer drought or it exhibited poor stem form (crookedness), due probably to a more extended photoperiod. Its popularity was thus short-lived.

Research on geographic variation and the general expansion of conifer tree improvement programs in the 1950s heightened awareness of the importance of genetic variation and the use of well-adapted seed sources. Breeders also resumed work on the hybrid larch, which had been discovered much earlier in 1900 occurring as open-pollinated offspring of Japanese larch growing near European larch on an estate in Dunkeld, Scotland (Larsen 1937).

2.4.2.2 Economic Importance

European larch forests now cover over 1 million ha in Europe with more than half of them established outside the species' native range. As such, while it plays a major economic role at regional levels such as in the Alps, larch appears as a minor species among other European conifers. Other exotic species like Douglas-fir and Sitka spruce have been adopted rapidly, while expansion of larch plantations has been relatively modest. A major reason for this has been a frequent lack of seed, due to highly irregular seed crops.

Planting of larch may yet expand in the future, as it processes exceptional soil and climate adaptability (wind resistance), juvenile growth (probably the fastest-growing conifer in Europe), and desirable wood properties (among the best coniferous wood in terms of not only physical and mechanical properties, but also appearance and natural durability). It can be used either in pure or mixed stands (e.g., with some broadleaf species), in afforestation/reforestation, but also in agro-forestry. Light tolerant, it is appreciated as a nurse species for shade-tolerant species, and being a fast grower, it allows an early economic return before slower-growing species in mixture (mostly broadleaves) come into commercial production. The most valuable use of its wood is as lumber for indoor use (flooring, wall panelling, carpentry), but also for traditional outdoor uses like in carpentry (bridges, towers), wall panelling, roof tiling, etc.

2.4.2.3 Breeding Objectives

Reforestation in the native range of European larch is often based on natural regeneration. Artificial regeneration by planting is also used but, for conservation purposes, only local reproductive material is planted, obtained mostly from selected seed stands and more rarely from seed orchards. In these areas, use of exotic larches is often prohibited. Population conservation is a priority with large protected areas being delimited for *in situ* conservation. Most of these forests established in (high) mountains and steep slopes play a protection role before timber production.

Intensively managed commercial stands are established outside the native range of European larch, in much more favorable environmental conditions (lower elevations, milder climate). Larch forests are in these areas established by planting, with clear-cutting at rotation age; natural regeneration is rarely practiced, and exotics are welcome. Wood production is the main target but management approaches vary; those in France and United Kingdom usually favoring fast growth, short rotations (40–50 years), limited environmental risks and fast economical returns, while elsewhere in central Europe, management is over longer rotations (90–120 years) seeking larger volumes and higher-value wood.

Larch breeding programs expanded rapidly across Europe in the 1940s, and have typically targeted reforestation in these more productive lowland sites. Most of the effort is on pure European larch and its hybrid with Japanese larch; breeding of pure Japanese larch is not pursued. Others like Eurasian larches, such as Siberian larch (*Larix russica* [Endl.] Sab. ex Trautv., syn. *Larix sibirica* Ledeb.), and the North American tamarack are sometimes established in Scandinavian countries (Martinsson and Lesinski 2007), but are not the objects of breeding work.

For both European and hybrid larches, breeding objectives are similar and include growth, stem form (crookedness), branching and resistance to larch canker. Late frost damage is rarely a concern and, even if phenology is currently assessed, it is seldom used as a selection criterion. In more advanced programs, some wood properties like wood density and modulus of elasticity (MOE; an indicator of wood strength) are included. Research is also ongoing for traits such as heartwood formation and quality, and drought tolerance.

As breeding zones are not (yet clearly) defined at national levels for larch, breeding is usually for whole countries, except in the native range and some areas unsuited for the species. Stable varieties across these large areas are required and genotype-environment interaction is used as a selection criterion.

For European larch, short-term, low-input breeding is used with the final aim to release first-generation seed orchard varieties. For hybrid larch as well, with very few exceptions, the strategy is generally restricted to first-generation hybrids, identifying outstanding varieties combining favorable parental traits. The French program for hybrid larch is an exception, where for the last 20 years breeding has been strongly linked to research on sources and prediction of interspecific heterosis, as well as conditions required to benefit from F_2-hybrids.

While abundant flowering and seed crops of European larch are irregular, production of improved varieties from seed orchards works rather well, especially on continental sites. In contrast, production of first-generation hybrids, either by sexual or asexual means, has remained problematic, which seriously impedes rapid expansion of hybrid larch plantations. Over the last two decades, research work has focussed on the improvement of propagation systems.

2.4.2.4 *Breeding Achievements*

2.4.2.4.1 European Larch

The first significant step towards larch breeding was achieved when results from international IUFRO provenances trials of European larch became available about 30 years ago (Schober 1985). Clearly, populations from Central Europe (*sudetica* and *polonica*) performed well all over Europe and were the least sensitive to canker, while some populations from the Alps were characterized by a better stem form, but having lower vigor and a high sensitivity to canker. Planting programs responded to these results, favoring selected seed stands of Sudetan and central Poland.

Most of European larch breeding populations were established with Sudetan and central Poland selections, with the addition to some landrace origin parents in some programs, and first-generation clonal seed orchards established with these materials. Commercial crops are now available from most and the evaluation of their genetic value is currently in progress, with improvement by roguing (1.5-generation orchards) planned. Recent results from genetic diversity studies and connected progeny trials focussed on best stands of Sudetan/central Poland origins will be exploited to establish second-generation orchards, emphasizing stem straightness of progeny-tested clones.

2.4.2.4.2 Interspecific Hybrids

The interspecific hybridization between European larch and Japanese larch is intended to combine favorable traits of both species: juvenile growth and larch canker resistance of Japanese larch, with stem straightness and fine branching of European larch. Thousands of hybrid combinations have been created in Europe by control crossing or by open-pollination in seed orchards. Overall, hybrids have shown superiority over their pure parental species for phenology, growth, stem form, and branching (Pâques 1989, 2002a); for example, gains in total height compared to the parental controls was from –5 to +140%.

While it might be that hybrids benefit from the canker resistance of Japanese larch, many interspecific hybrids fail to combine expected parental properties in a favorable way. For several traits, it has been shown that levels of heterosis can vary widely and can be either positive or negative (Pâques 2002a). The simple and supposedly low-cost strategy has clearly shown its limits and has finally proven to be costly due to the low rate of successful combinations. The work of C.S. Larsen, who conducted an active larch hybridization program in Denmark in the 1940s, had already shown the importance of the choice of parents and of their improvement prior to interspecific recombination. Many successful hybrid varieties still used in Europe today rely on this early work. A systematic approach has been used in France over the last 20 years to better understand the genetics of heterosis and the role of parental species, and to developing predictors of heterotic combinations (Pâques 2002b). Improvement of parental breeding populations integrates this knowledge. In parallel, composite breeding is explored as an alternative strategy: second-generation hybrids have been created to study levels of heterosis and the impacts of inbreeding depression (Pâques 2007).

Application of MAS has been proposed by Arcade et al. (2002) who found several significant QTLs for ring wood-density traits (effects ranging from 3.4–6.2% of the phenotypic variance). Generally, given its short rotation and early expression of important traits, the motivation to apply MAS as an early selection tool for larches is perhaps less than with other longer-rotation-species. Markers have, however, found a place in assessment of genetic distance in relation to heterosis and of allelic dosage in hybrids.

For more than 60 years, parents from the best hybrid combinations have been established in interspecific hybrid seed orchards with various layout designs (alternating rows of species clones, tree by tree, etc.), number of clones (bi-clonal orchards to multi-clonal orchards in excess of 200 clones), and clone origins. Unfortunately, commercial crops from these open-pollinated seed orchards have remained well below expected yield and quality. In addition to low seed set, the proportion of hybrids in seed lots is highly variable from orchard to orchard and from year to year (< 20% to > 60%) as first revealed by isozyme markers or more recently by cytoplasmic DNA markers (Acheré et al. 2004). Poor climatic conditions during pollination (frost damage, snow, alternating warm and cold days, etc.) and differences in parental phenology are the main causes of these failures.

2.4.2.4.3 Mass Propagation

Several approaches to overcome poor seed production in hybrid orchards have been tested, including the improvement of generative reproduction conditions, as well as the development of better vegetative propagation techniques. Among the various methods developed for generative reproduction, supplemental mass pollination by electrostatic dusting of female clones (kept in a separate orchard) is the most promising (higher seed set with up to 95% hybrid fertilization) and has been put into practice in France (Philippe et al. 2006). While clonal propagation by cuttings was extensively tested with disappointing results due to rapid ageing of donor plants, "bulk" propagation by cuttings of young seedlings from selected families looks more promising (Verger and Pâques 1993; Le Pichon et al. 2001) and is being implemented on a pilot scale. Recent and significant progress in somatic embryogenesis and cryo-conservation of larch (Lelu-Walter and Pâques 2009) offer new opportunities, in combination with cutting propagation and deployment, as maintenance of juvenility of donor materials is possible.

2.5 Cypresses (Cupressaceae)

Cypresses, the Cupressaceae including the former Taxodiaceae and Cunninghamiaceae families (Gadek et al. 2000), comprise a diverse group of species with a worldwide distribution. The family has species on every continent except Antarctica, and occurs across a wide range of climatic and edaphic environments. Many of the up to 30 genera are monotypic and a significant portion of the 140 or so species have localized, relict distributions.

The Cupressaceae is the most important family in horticulture, with thousands of varieties in existence. Many of these are also used for forestry. Of these, *Cryptomeria japonica* D. Don. (sugi, or Japanese cedar) is by far the most commonly planted and has the longest history of genetic improvement. We discuss this species separately, followed with a more general discussion of the other important members of the Cupressaceae.

2.5.1 Sugi (*Cryptomeria japonica*)

2.5.1.1 Historical Perspective

During a very long history of cultivation, many varieties of sugi have been developed. Miyajima (1983) classified these into two types; those cultivars that have been improved artificially, and those representing unimproved geographic races. The first cultivars were selected in the 16th century, and forest plantations first established in the early 18th century. At this time, the main cultivars were selected by foresters on Kyushu Island. Many cultivars have been developed subsequently, and most have been maintained vegetatively by cuttings. Because these cultivars have been cultivated for a long period of time, various characteristics such as growth performance, wood quality, rooting ability and flowering are well understood.

2.5.1.2 Economic Importance

Sugi is one of the most important timber species in Japan, favored for its straight bole and rapid growth. It has been planted over 4.53 million ha, and comprises 45% of the artificial forest in Japan. Total log production in Japan in 2007 was 29 million m^3, and almost half of this was sugi. Approximately 80% of houses built using the post-and-beam construction method use a pre-cut system. Therefore, there is an increasing need for high quality products, with good performance in terms of dryness, dimensional stability, and strength. For these reasons, the market share of kiln-dried lumber is increasing. Sugi wood is also durable and easily worked, and is typically used for buildings, bridges, ships, and furniture. A recent topic of wood industry is the development of new laminated wood products, using *Pseudotsuga menziesii* for outer-layers and sugi as inner-layers.

2.5.1.3 Breeding Objectives

Initially, the main breeding objective was to improve growth. Later, other breeding objectives, such as resistance to the sugi bark borer, *Semanotus japonicus*, and resistance to snow damage were included. Although sugi grows well, the wood has lower strength than that of imported timbers, and the high moisture content of heartwood prevents efficient kiln-drying. More recent breeding objectives have included improving wood strength and lowering the moisture content of heartwood.

It is said that 16% of Japanese people suffer from allergies due to sugi pollen, and addressing this problem has become an objective for breeders. Two strategies have been proposed to ameliorate this problem; one is to select varieties with lower pollen production, and the other is to select for low-allergenic pollen. The two major allergy proteins in the pollen of sugi have been documented; Cry j 1 and Cry j 2. Efficiency of CO_2 fixation is a further breeding objective to address issues associated with global warming.

2.5.1.4 Breeding Achievements

Systematic breeding of sugi began in the late 1950s. The Forestry Agency of the Japanese Ministry of Agriculture and Forestry has established a network of tree breeding stations throughout the country, so that all climatic conditions are represented. Over 3,600 plus trees have been selected from four breeding regions, excluding the Hokkaido Breeding Region, which is a cool-temperate area. As mentioned above, the main breeding objective initially was to improve growth, and volume production gains of 15% over local varieties were achieved. Based on progeny trials, 50 clones or families showing superior growth were selected, and a further 25 clones or families selected for bole straightness. The second-generation population was established using controlled crosses among these superior trees.

The breeding program also addressed the problem of the sugi bark borer, whose larvae feed on bark and xylem. An inoculation test was established, and 61 resistant clones have been identified and released for deployment. Another problem is that trees can become crooked in regions with heavy snowfall, due to the pressure exerted by snow load. Eight clones and 19 families that grow straight in these regions have been developed.

Research on wood quality using 563 sugi selected clones showed that the coefficient of variation was greater than 30% in heartwood moisture content, and 17.5% in MOE (Hirakawa et al. 2003). Furthermore, broad-sense heritability of MOE and heartwood moisture content was high, 0.597 to 0.857, and 0.53 to 0.57, respectively (Fujisawa et al. 1992, 1995). These results indicate that further improvement of these characters can be achieved by breeding. Fujisawa (1998) discussed quality management of fast-growing material, and recommended clonal forestry to attain high wood quality and to decrease variation of wood quality.

Kuramoto et al. (2000) analyzed QTLs associated with wood strength using a linkage map in the F_1 progeny of two cutting cultivars of sugi. Effective QTLs were associated with MOE and wood density. Several QTLs for MOE were detected in the linkage maps of parent cultivars. Because these QTLs explained approximately 45% of the total phenotypic variances in one parent cultivar, they were deemed appropriate for use in breeding programs.

Evaluation of wood quality is time-consuming and labor-intensive, so simple testing methods for standing trees are required to execute large-scale selection. There was a high correlation between stress-wave propagation velocity in the longitudinal direction and the MOE (Ikeda et al. 2000). Kamaguchi et al. (2000) proposed a non-destructive measurement to estimate heartwood moisture content, where vibration of the tree trunk is measured after lateral impact. Using these two simple methods to evaluate MOE and the heartwood moisture content, forward selection of second-generation candidates has been implemented.

As there is a large variation in production of male strobili, 131 plus trees bearing fewer male strobili were selected as "low pollen", and new seed orchards established with this material. Male sterility is an equivocal answer to the pollen allergy problem. Taira et al. (1993) first reported male-sterile trees, and found that this characteristic is controlled by a single recessive gene (Taira et al. 1999). To date, approximately 20 male-sterile trees have been identified. Genetic modification has been proposed to introduce this character, as it is difficult to introduce it into a population using traditional breeding methods. Goto et al. (1999) found that the major allergenic protein, Cry j 1, varied markedly among trees, and the DNA sequences of the gene encoding the protein have been reported (Griffith et al. 1993). This gene was located on a linkage map (Goto et al. 2003) and some Cry j 1 isoforms with different binding properties to monoclonal antibodies were found (Goto et al. 2004).

Volume, wood density, and carbon content of the wood have been evaluated as components of CO_2 fixation. Using such components high CO_2 fixation variety will be developed.

2.5.2 *Other Cupressaceae*

2.5.2.1 Historical Perspective

In addition to sugi, several other species in the Cupressaceae are commercially important with associated breeding programs, including some of the highest value timber species. These include: the whitecedars, such as Port Orford-cedar (*Chamaecyparis lawsoniana* [A.Murr.] Parl.) from the Pacific Northwest of the United States, yellow-cedar (*Chamaecyparis nootkatensis* (D. Don) Farjon & D.K. Harder[1]) from western North America, and Hinoki (*Chamaecyparis obtusa* [Sieb. & Zucc.] Endl.) from southern Japan and Taiwan; the cypresses such as Mexican cypress (*Cupressus lusitanica* Mill.) from Mexico, Monterey cypress (*Cupressus macrocarpa* Hartweg ex Gord.) from western California, and Italian cypress (*Cupressus sempervirens* L.)

[1]taxonomic authorities plan to resolve nomenclature for this species in 2011; current synonyms include *Callitropsis* and *Xanthocyparis* (Little et al. 2004).

from the Mediterranean region; the arborvitae such as western redcedar (*Thuja plicata* Donn ex D. Don.) from western North America; and Chinese fir (*Cunninghamia lanceolata* (Lambert) Hooker) from China and Vietnam. Many Cupressaceae species have had historical ties to indigenous people, both for spiritual value and traditional uses. Western redcedar, known as the "Tree of Life", has had a rich history with aboriginal cultures because of its multitude of traditional practical and aesthetic values.

2.5.2.2 Economic Importance

Many Cupressaceae species are highly prized for their aromatic and durable heartwood, as well as dimensional stability. Logs are typically higher in demand and price than most other commercial conifers. For example, western redcedar in British Columbia accounted for approximately 30% of the volume harvested on the coast and economic value to the provincial government (the primary owner of forest lands) was over 50% greater than coastal Douglas-fir (US$164/m^3 vs. $99/m^3) for 2006 and 2007. The high demand for logs drives enhanced reforestation programs for many Cupressaceae species.

Chinese fir is an important timber species in China with over 400,000 ha of plantations established annually (Minghe and Ritchie 1999a). The highly durable wood is used in construction, bridge and ship building, and coffin making. In Japan, hinoki and sugi together comprise approximately 70% of forest plantations. Hinoki wood is lemon-scented and rot-resistant, and is uniquely used in palace and temple construction. Coastal redwood (*Sequoia sempervirens* [D. Don] Endl.) is a high-value conifer species endemic to California, with approximately 0.5 million ha of commercial, second-growth forests (Olson et al. 1990). The heartwood is highly valued for its beauty, light weight, and resistance to decay.

Both Mexican and Monterey cypresses have been widely domesticated away from their native ranges in Central and North America, mostly in warm temperate and subtropical regions including New Zealand, southern Europe, and South America. Mexican cypress is a fast-growing, drought-tolerant tree that is used for saw logs, pulp, wind breaks, and as an ornamental. Following the introduction and spread of exotic canker diseases, Mexican cypress has become the most widely planted member of the Cupressaceae for tree improvement, supplanting the preferred Monterey cypress, whose logs are used for boats and furniture.

Approximately 12 million western redcedar and yellow-cedar trees are planted annually in British Columbia. The heartwood of western redcedar is very resistant to decay and has high dimensional stability. The wood is used for outdoor construction, including posts, decking, shingles, and siding. Yellow-cedar is used in finish carpentry, such as exterior siding, shingles, decking, exposed beams, glue-laminated beams, panelling, cabinetry and boat building. Both species are prized by artisans for carving.

2.5.2.3 *Breeding Objectives*

Despite the high value and use, only a handful of economically important Cupressaceae species have been the focus of tree improvement activities, although many species have been studied and evaluated for natural levels of population variability for a range of traits.

2.5.2.3.1 Asia-Pacific Region

The oldest tree improvement program in the world is most likely that of Chinese fir in China. Clonal forestry has been practiced for over 800 years (Minghe and Ritchie 1999a, b). More recently, recurrent selection programs have been developed in a number of provinces using both open-pollinated and full-sib crosses, together with wind-pollinated seed orchards (Zhuowen 2003). Selection has focussed on growth and wood density. Breeding programs are well-established for sugi (see previous section above for details) and Hinoki, the two most important conifer species for plantations in Japan. Tree breeding efforts for Hinoki in the first generation have utilized open-pollinated families from over 1,000 seed parents, and selections have been made for growth, bole straightness and heartwood color. Second-generation matings are integrating the different trait selections through a factorial design (T Kondo pers. comm.). Sawara cypress (*Chamaecyparis pisifera* [Sieb. & Zucc.] Endl.) is also the subject of improvement activities and plantation forestry, primarily in temperate montane areas. In Korea, Hinoki has been used in breeding programs since 1965, with selections for growth, form, and insect and disease resistance. The first orchards began producing seed in the 1980s, with advanced-generation orchards being developed (Kang 2007). The introduced species Mexican cypress and Monterey cypresss both have breeding programs in New Zealand using open-pollinated families for both first- and second-generation progeny trials, along with clonal trials in the second generation. The major target traits have been vigor and stem straightness, and, in the case of Monterey cypress, resistance to stem canker disease (*Seiridium* spp.).

2.5.2.3.2 Europe

Italian cypress is the Cupressaceae species with the most intensive tree improvement program in Europe. It has both cultural and economic importance. Since about 1927, the fungal pathogen *Seiridium cardinale*, indigenous to California, has spread rapidly throughout the global range of cypress and related species, including Europe, Asia, Africa, and Australasia, causing widespread and increasing mortality. The main impact of *Seiridium* fungi is death from stem canker, also termed cypress blight. Breeding programs have thus targeted selection and breeding for disease resistance, using clonal and sexual recombination to increase gains. To date, research programs in Greece, Italy, and France have conducted the largest body of work in breeding and improvement for canker resistance (Santini et al. 1997; Papageorgiou et al. 2005; P Raddi pers. comm.).

2.5.2.3.3 North America

Breeding of western redcedar in British Columbia has focussed on growth, heartwood durability, and mammalian damage resistance. Selections from first-generation polycrossed trials are currently being bred for advanced-generation testing. Minimal inbreeding depression and ease of vegetative propagation have facilitated selfing and cloning as tools for testing and deployment of populations. Short generation intervals have allowed for relatively quick advancement. The breeding program for yellow-cedar, also in British Columbia, has tested clones from partial diallels in the first generation. Forward selections based on vigor and stem form are being bred for second-generation testing. Port Orford-cedar, indigenous to California and Oregon, has a breeding program to develop resistance to the introduced pathogen *Phytophthora lateralis*. Both putative dominant single (major) gene and quantitative resistance mechanisms have been identified (Sniezko et al. 2004). Since this species has short generations and is amenable to vegetative and sexual reproduction, improvement has been rapid, yielding several hundred resistant first-generation F_1 selections available for deployment.

2.5.2.3.4 Central and South America

Mexican cypress, indigenous to Mexico, El Salvador, Guatemala, and Honduras, has been introduced throughout Central and South America for timber production (Cornelius et al. 1996), where it has naturalized in some areas. Privately established tree improvement trials began selections in Colombia in the 1970s (Ladrach 1983), with breeding programs starting in 1977. The major traits of interest are volume growth, disease and insect resistance, stem form, crown form, and wood quality. Most progeny trials and deployed seedlings are open-pollinated, but some controlled crossing has been successful.

2.5.2.3.5 Africa

Introduced Cupressaceae, particularly Mexican cypress and Monterey cypress, have been grown in plantation forests of tropical and subtropical countries for decades. Monterey cypress was a preferred species in Kenya for its greater yields, but due to its susceptibility to *Seiridium* canker disease, has widely been replaced by Mexican cypress (Roux et al. 2005). Stem taper, stem form, wood grain angle, stem branches, and susceptibility to key diseases were considered in the selection of trees for the Kenyan tree improvement program. The introduced cypress aphid (*Cinara cupressi*), has been spreading throughout the region since 1986 severely damaging stands, and is now the subject of selection and breeding for resistance, along with cypress canker (Ciesla 1991; Mugasha et al. 1997). Beginning in the late 1960s under individual corporate and government programs, South Africa has reaped considerable economic benefits from a comprehensive tree improvement program, including Mexican cypress. The predominance of the private sector in forest management and research in tree improvement, however, has limited the availability of information on these programs (Denison 2001).

2.5.2.4 Breeding achievements

2.5.2.4.1 Asia-Pacific Region

Improved clones of Chinese fir have been deployed for close to 800 years for reforestation. Prior to the 1950s, reforestation by either stump cuttings or rooted cuttings accounted for 80% of the plantings in Chinese fir (Minghe and Ritchie 1999a). An increased emphasis on seedling-based forestry resulted in only 65 million rooted cuttings planted in 1991, accounting for 5.4% of the total annual planting stock. First- and second-generation orchards have been established with this species as well, and in 1984, 30% of the planting demand was meet through orchard seed (Jusheng 1985), with volume gains from the second-generation orchard predicted to be 40% over unimproved seedlots (Zhuowen 2003). The majority of seed used for reforestation of Hinoki comes from over 334 ha of first-generation open-pollinated orchards (McKeand and Kurinobu 1998). Operational open-pollinated orchards of Hinoki are established on a regional basis, with some in advanced generations. In addition to orchard seed, clones are also available for reforestation throughout the breeding regions of Japan, although supply is limited (McKeand and Kurinobu 1998).

2.5.2.4.2 Europe

In Greece, dozens of canker-resistant clones of *Cupressus lusitanica* are available for deployment, but base levels of resistance in testing populations remain below 5% (Santini et al. 1997).

2.5.2.4.3 North America

All of the approximately 8 million western redcedar plants required annually in the primary breeding zone for this species in British Columbia are from improved first-generation orchard seed with expected volume gains at rotation of 7–10%. Selections from the relatively young breeding program is taking advantage of high breeding values to create operational full-sib family seedlots with 15–20% volume gain when available. Currently, yellow-cedar clonal planting stock is delivering up to 20% volume gain at rotation. For Port Orford-cedar, gains in *Phytophtora* resistance range from double to over 6 times that of wild populations from the same breeding zone (8–29% natural resistance vs. 27–63% selected resistance (Elliott 2006). All three of the above species have short generations, and are amenable to vegetative propagation and sexual reproduction including selfing, resulting in rapid improvement.

2.5.2.4.4 Central and South America

The Colombian program in the first generation of selection yielded early gains (age 3, relative to a rotation of 16 years) of 13% in height and 50% in volume, but effectively no difference in stem or crown form (Ladrach 1983).

2.5.2.4.5 Africa

South Africa, Kenya, Rwanda, Uganda, Tanzania, and other countries have established a network of Mexican cypress plantations for wood production from selected material that have been assessed for variability in growth and yield parameters. Most are from open-pollinated selections but seed production areas are now widely used to produce seed that can be transferred across cooperating countries on suitable sites, based on provenance studies. In Kenya, the Kenya Tree Seed Centre is the central repository and distribution center for forest seed and clone banks, and also manages a network of seed orchards. Since the early 1960s, plus-trees, provenance, and progeny testing have resulted in advanced-generation gains of approximately 30% for Mexican cypress (Bernard 2001) over unimproved yields.

2.5.2.4.6 Summary

The combination of substantial additive variation for economic traits, ease of grafting and cloning, precocious reproduction, and wide range of ecological adaptations make the Cupressaceae an ideal taxonomic group that has demonstrated many successes through tree improvement. Gains can be substantial when selecting for one or several traits, with limited or no trade-offs between growth and disease or insect resistance. Although the wood is generally soft, rapid fiber production supports a diverse range of forest products. The horticultural sector has long sought value in the Cupressaceae, and the emerging non-timber forest products sector is increasing its utilization for distilled oils, phytochemicals, bark, chips, and green foliage. To date, prospects for marker-aided selection are limited, given the lack of correlations identified between traits of interest and molecular markers or QTLs for this taxon; however, short generations and high gains from breeding programs indicate that phenotypic selection for quantitative traits, supported by genetic and biochemical data is a viable system for efficient improvement.

2.6 Concluding Remarks

Conifers are the target of major tree breeding efforts worldwide. While much progress has been made through conventional approaches to breeding, tree breeders face enormous challenges with long generation turnover times, costly field testing, and relatively undomesticated genetic resources. In this chapter, we have attempted to describe the varied circumstances and state of the art for breeding of many of the more important conifers.

Advances in molecular technologies could have an enormous impact on the rate of progress and achievements made by tree breeding programs. To succeed, new technologies must be carefully integrated into the context of existing programs so that they respond to opportunities and build on gains already realized. Markers are already playing an important role in understanding the patterns of variation and genetic basis for some traits, as well as assisting in the positive identification of individuals and their pedigree. However, with a few notable exceptions, markers for individual large-effect QTLs for most economic traits have not been discovered, which dampens somewhat the prospects for marker-assisted evaluation. This is consistent with experience in marker-aided breeding with livestock animals, where effort is now focussing on genome-wide scans and approaches to genomic selection (e.g., Meuwissen et al. 2001). Learning from the animal-breeding experience suggests that forward-looking tree breeding programs will be archiving pedigreed DNA samples with associated phenotypic records, in anticipation of the availablity of affordable chips that will permit scanning of very dense SNP maps in important conifer genomes.

Acknowledgments

The authors are greatful to their colleagues who shared information freely and in particular to Heidi Dungey, Richard Sniezko, James Turner, and Don Zobel for their contributions.

References

ABARE (Australian Bureau of Agricultural and Resource Economics) (2007) Australian forest and wood products statistics, September and December quarters 2007, Canberra, Australia.

Acheré V, Faivre-Rampant P, Pâques LE, Prat D (2004) Chloroplast and mitochondrial molecular tests identify European x Japanese larch hybrids. Theor Appl Genet 108: 1643–1649.

Adams WT, Joly RJ (1977) Analysis of genetic variation for height growth and survival in open-pollinated progenies of eastern white pine. In: F Chech, D Schmitt (eds) Proc 25th Northeast Forest Tree Improvement Conf, Orono, ME, USA, pp 117–131.

Alden, HA (1997) Softwoods of North America. USDA For Serv For Prod Lab Gen Tech Rep FPL-GTR-102.

Alfaro RI, Borden JH, King JN, Tomlin ES, McIntosh RL, Bohlmann J (2002) Mechanisms of resistance in conifers against shoot infesting insects: the case of the white pine weevil *Pissodes strobi* (Peck) (Coleoptera: Curculionidae). In: MR Wagner, KM Clancy, F Lieutier, TD Paine (eds) Mechanisms and Deployment of Resistance in Trees to Insects. Klubbert Publ, The Netherlands, pp 101–126.

Alfaro RI, King JN, Brown RG, Buddingh SM (2008) Screening of Sitka spruce genotypes for resistance to the white pine weevil using artificial infestations. For Ecol Manag 255: 1749–1758.

Allen HL, Fox TR, Campbell RG (2005) What is ahead for intensive pine plantation silviculture in the South? South J Appl For 29: 62–69.

Almqvist C, Simonsen R, Wennström U, Rosenberg O (2008) Norway spruce seed orchards—forestry's golden nugget. Skogforsk, Resultat Nr 3–2008.

Andersson B (2002) Forest tree breeding in Sweden. Nordiske GENressurser 2002 Nordic Council of Ministers, Copenhagen, Denmark, pp 36–38.

Andersson B, Elfving B, Persson T, Ericsson T, Kroon J (2007) Characteristics and development of improved *Pinus sylvestris* in northern Sweden. Can J For Res 37: 84–92.

Arcade A, Faivre-Rampant P, Pâques LE, Prat D (2002) Localisation of genomic regions controlling microdensitometric parameters of wood characteristics in hybrid larches. Ann For Sci 59: 607–615.

Baker JB, Langdon, OG (1990) Loblolly pine. In: RM Burns, BH, Hokala (eds) Silvics of North America, vol. 1: Conifers. Agriculture Handbook 654. USDA Forest Service, Washington, DC, USA, pp 497–512.

Baradat P, Pastuszka P (1992) Le pin maritime. In: A Gallais, H Bannerot (eds) Amélioration Végétale des Espèces Cultivées, INRA Éditions, INRA, France, pp 695–709.

Baradat P, Durel C-E, Pastuszka P (1992) The polycross seed orchard: an original concept. In: Proc IUFRO–AFOCEL Symp on Mass Production Technology for Genetically Improved Fast Growing Forest Tree Species, vol II. Éditions AFOCEL, Bordeaux, France, pp 53–62.

Beaudoin R, Desponts M, Mottet M-J, Périnet P, Perron M, Rainville A (2004) Tree improvement progress by the Direction de la recherche forestière. In: JD Simpson (ed) Proc 29th

Canadian Tree Improvement Association, part 1, Kelowna, BC, Canada, July 26–29, pp 38–42.

Beaulieu J (1994) L'amélioration génétique et le reboisement. In: AL D'Aoust, R Doucet (eds) Compte rendu du Colloque no 112 de l'ACFAS. La régénaration de la zone de la forêt mixte. Ressour. nat. Canada, Serv can for, Ressour nat Québec, Dir Rech For Montréal, Quebec, 19 May pp 107–133.

Beaulieu J (1996) Breeding program and strategy for white spruce in Quebec. Can For Serv Inf Rep LAU-X-117E.

Beaulieu J, Corriveau A, Daoust G (1989a) Phenotypic stability and delineation of black spruce breeding zones in Quebec. Can For Serv Inf Rep LAU-X-85E.

Beaulieu J, Corriveau A, Daoust G (1989b) Productivité et stabilité phénotypique de l'épinette rouge au Québec. For Chron 65: 42–48.

Beaulieu J, Plourde A, Daoust G, Lamontagne L (1996) Genetic variation in juvenile growth of *Pinus strobus* in replicated Quebec provenance-progeny tests. For Genet 3: 103–112.

Beaulieu J, Perron M, Bousquet J (2004) Multivariate patterns of adaptive genetic variation and seed source transfer in *Picea mariana*. Can J For Res 34: 531–545.

Bernard KN (2001) State of forest genetic resources in Kenya. Report prepared for the sub-regional workshop FAO/IPGRI/ICRAF on the conservation, management, sustainable utilization and enhancement of forest genetic resources in Sahelian and North-Sudanian Africa. Ouagadougou, Burkina Faso, 22–24 September 1998. Co-published by Food and Agriculture Organization of United Nation (FAO), Sub-Saharan Africa Forest Genetic Resources Programme of the International Plant Genetic Resources Institute (IPGRI/SAFORGEN), Danida Forest Seed Centre (DFSC) and International Centre for Research in Agroforestry (ICRAF). Working paper FGR/18E.

Bilir N, Ulusan D (2008) Seed orchards and seed collection stands of Scots pine in Turkey. In: D Lindgren (ed), Seed Orchards, Proc from a Conf at Umeå, 26–28 Sept 2007, Sweden, pp 25–36.

Bingham RT (1983) Blister rust resistant western white pine for the Inland Empire: the story of the first 25 years of the research and development program. USDA Forest Service, Gen Tech Rep INT-146, Ogden, UT, USA.

Birot Y (1982) Breeding strategies with Norway spruce in France with particular reference to the criteria and methods of selection. In: Proc IUFRO Joint Meeting of Working Parties on Genetics about Breeding Strategies including Multiclonal Varieties, Sensenstein, Germany, pp 10–21.

Blum BM (1990) Red spruce—*Picea rubens* Sarg. In: RM Burns, BH Honkala (eds) Silvics of North America, vol 1, Conifers. Agriculture Handbook 654. USDA Forest Service, Washington DC, USA, pp 250–259.

Boratyńsky A (1991) Range of natural distribution. In: M Giertych, C Mátyás (eds) Genetics of Scots Pine. Elsevier, Amsterdam, Netherlands, pp 19–30.

Bouffier L, Charlot C, Raffin A, Rozenberg P, Kremer A (2008a) Can wood density be efficiently selected at early stage in maritime pine (*Pinus pinaster* Ait.)? Ann For Sci 65: 106–114.

Bouffier L, Raffin A, Kremer A (2008b) Evolution of genetic variability for selected traits in breeding populations of maritime pine. Heredity 101: 156–165.

Bouffier L, Rozenberg P, Raffin A, Kremer A (2008c) Wood density variability in successive breeding populations of maritime pine. Can J For Res 38: 2148–2158.

Bouffier L, Raffin A, Rozenberg P, Meredieu C, Kremer A (2009) What are the consequences of growth selection on wood density in the French maritime pine breeding programme? Tree Genet Genomes 5: 11–25.

Bramlett DL (1997) Genetic gain from mass controlled pollination and topworking. J For 95(3): 15–19.

Brazier JD (1987) Man's use of Sitka spruce. In: DM Henderson, R Faulkner (eds) Proc Roy Soc Edinburgh Sect B (Biol Sci) 93 (Sitka Spruce, Proc Symp, Royal Botanic Garden, Edinburgh, 3–6 Oct 1986): 213–221.

Brendel O, Pot D, Plomion C, Rozenberg P, Guehl J-M (2002) Genetic parameters and QTL analysis of $\delta^{13}C$ and ring width in maritime pine. Plant Cell Environ 25: 945–953.

Bridgwater FE, Barnes RD, White T (1997) Loblolly and slash pines as exotics. In: T White, D Huber, G Powell (eds) Proc 24th South. Forest Tree Improvement Conf., Gainesville, FL, USA, pp 18–32.

Bridgwater FE, Bramlett DL, Byram TD, Lowe WJ (1998). Controlled mass pollination in loblolly pine to increase genetic gains. For Chron 74(2): 185–189.

Buamscha MG (2002) Nursery practices with exotic conifers in Patagonia, Argentina, and some reason to afforest the region with these species. In: RK Dumroese, LE, Riley, TD Landis (eds) National Proc: Forest and Conservation Associations-1999, 2000, and 2001. Proc RMRS-P-24. USDA Forest Service, Rocky Mountain Research Station, Ogden, UT, USA, pp 169–171.

Burban C, Petit RJ, Carcreff E, Jactel H (1999) Rangewide variation of the maritime pine bast scale *Matsucoccus feytaudi* Duc. (Homoptera: Matsucoccidae) in relation to the genetic structure of its host. Mol Ecol 8: 1593–1602.

Burdon R (1978) Mejoramiento Genetico [Forestal en Chile] (Genetic Improvement of Forest Trees in Chile). Project Working Document, FAO Project CHI/7.

Burdon RD (2002) *Pinus radiata* D. Don. In: CAB (compil) Pines of Silvicultural Importance. CABI Publ, Wallingford, UK, pp 359–279.

Burdon RD (2004) Genetics of *Pinus radiata*. In: J Burley, J Evans, J Youngquist (eds) Encyclopaedia of Forest Sciences. Elsevier Academic Press, Oxford, UK; San Diego, CA, USA, pp 1507–1516.

Burdon RD, Wilcox PL (2011) Integrating molecular markers in breeding. In: C Plomion, J Bousquet, C Kole (eds) Genetics, Genomics and Breeding of Conifers. Science Publ Enfield, NH, USA (in press).

Burdon RD, Carson MJ, Shelbourne CJA (2008) Achievements in forest tree improvement in Australia and New Zealand 10: *Pinus radiata* in New Zealand. Aust For 71(4): 263–280.

Campbell RK (1979) Genecology of Douglas-fir in a watershed in the Oregon Cascades. Ecology 60: 1036–1050.

Canavera DS (1975) Variation among the offspring of selected Lower Michigan jack pines. Silvae Genet 24: 12–15.

Carlson M, L'Hirondelle S, Jaquish B, King JN, O'Neill G, Russell J, Stoehr M, Ukrainetz N, Xie C-Y, Yanchuk A (2009) British Columbia Ministry of Forests and Range, Forest Genetics Research and Tree Breeding Program. In: JD Simpson (ed) Proc 31st Meeting Canadian Tree Improvement Association, Part 1, Quebec City, Quebec, Canada, 25–29 Aug 2008, pp 86–93.

Carson SD, Corbett GE, Lee JR, Wilcox PL, Richardson TE (1997). Marker-wood density association identified in a selectively genotyped population of *Pinus radiata*. In: RD Burdon, JM Moore (eds) IUFRO '97 Genetics of Radiata Pine, Proc NZ FRI-IUFRO Conf, 1–4 Dec and Workshop 5 Dec, Rotorua, New Zealand. FRI Bulletin No 203, p 345.

Carson SD, García OP, Hayes JD (1999 Realised gain and prediction of yield with genetically improved *Pinus radiata* in New Zealand. For Sci 45: 186–200.

Carter KK, Simpson JD (1985) Status and outlook for tree improvement programs in the northeast. North J Appl For 2: 127–131.

Carter KK, DeHayes DH, Demeritt ME Jr, Eckert RT, Garrett PW, Gerhold HD, Kuser JE, Steiner KC (1988) Tree Improvement in the Northeast: interim summary and recommendations for selected species. Univ Maine, Agri Exp Stn, Tech Bull 131, Orono, ME, USA.

Cato SA, Pot D, Kumar S, Douglas J, Gardner RC, Wilcox PL (2006) Balancing selection in a dehydrin gene associated with increased wood density and decreased radial growth in *Pinus radiata* (Abstract). In: Proc Plant Anim Genome XIV Conf, San Diego, CA, USA.

Cayford JH, McRae DJ (1983) The ecological role of fire in jack pine forests. In: RW Wein, DA MacLean (eds) The Role of Fire in Northern Circumpolar Ecosystems. John Wiley, Chichester, UK, pp 183–199.

Cherry ML, Joyce DG (1998) Implementation of a second-generation breeding program for black spruce in Ontario. Abstract 269. In: Frontiers of Forest Biology, Proc Joint Meeting of the NA For Biol Workshop/Western For Genet Assoc, Victoria, BC, Canada.
Cherry M, Lu P, Sinclair B (2000) Ontario Forest Research Institute. In: JD Simpson (ed) Proc 27th Can Tree Improve Assoc, part 1, 15–17 Aug, Sault Ste Marie, ON, Canada, pp 81–82.
Ching KK (1965) Early growth of Douglas-fir in a reciprocal planting. For Res Lab Res Paper 3, Oregon State Univ, Corvallis, OR, USA.
Ciesla WM (1991) Cypress aphid: a new threat to Africa's forests. Unasylva 167(42/4).
Cornelius J, Apedile L, Mesén F (1996) Provenance and family variation in height and diameter growth of *Cupressus lusitanica* Mill. at 28 months in Costa Rica. Silvae Genet 45: 82–85.
Corriveau AG (1981) Variabilité spatiale et temporelle de la croissance juvenile des provenances d'épinette noire au Québec. In: DFW Pollard, DG Edwards, CW Yeatman (eds) Proc 18th Canadian Tree Improvement Association Meeting, Duncan, BC, 17–20 Aug , Can For Serv, Petawawa Nat For Inst, Chalk River, ON, Canada, pp 181–187.
Corriveau AG, Lamontagne Y (1977) L'amélioration génétique du pin blanc au Québec. Fisheries and Environment Canada, Inf Rep LAU-X-31.
Corriveau L (2004) Weyerhaeuser—Tree improvement in Saskatchewan. In: JD Simpson (ed) Proc 29th Meeting Canadian Tree Improvement Association, part 1, 26–29 July, Kelowna, BC, Canada, pp 73–74.
Crecente-Campo F, Marshall P, Rodríguez-Soalleiro R (2009) Modeling non-catastrophic individual-tree mortality for *Pinus radiata* plantations in northwestern Spain. For Ecol Manag 257: 1452–1550.
Critchfield WB (1980) The Genetics of Lodgepole Pine. USDA For Serv Research Paper W-37
Curtis RO, Marshall DD, DeBell DS (2004) Silvicultural options for young-growth Douglas-fir forests: the Capitol Forest Study—establishment and first results. USDA For Ser Gen Tech Rep PNW-GTR-598.
Danell Ö (1991) Survey of past, current and future Swedish forest tree breeding. Silva Fenn 25: 241–247.
Danell Ö (1993) Breeding programs in Sweden. 1 General approach. In: SJ Lee (ed) Progeny Testing and Breeding Strategies, Proc Nordic Group of Tree Breeding, Oct, Edinburgh, Forestry Commission, UK, pp 128(i)–128(v).
Danjon F (1995) Observed selection effects on height growth, diameter and stem form in maritime pine. Silvae Genet 44: 10–19.
Daoust G, Beaulieu J (2004) Genetics, breeding, improvement and conservation of *Pinus strobus* in Canada. In: RA Sniezko, S Samman, SE Schlarbaum, HB Kriebel (eds) Breeding and Genetic Resources of Five-needle Pines: Growth, Adaptability and Pest Resistance, 23–27. July 2001, Merford, OR, USA. IUFRO Working Party 2.02.15. Proc RMRS-P-32, USDA Forest Service, Rocky Mountain Research Station, Fort Collins, CO, USA, pp 3–11.
Denison NP (2001) Tree improvement: what has South Africa achieved? S Afr For J 190: 1–2.
Desprez-Loustau M-L (1990) A cut-shoot bioassay for assessment of *Pinus pinaster* susceptibility to *Melampsora pinitorqua*. Eur J For Pathol 20: 386–391.
Devey ME, Groom KA, Nolan MF, Bell JC, Dudinski MF, Old KM, Matheson AC, Moran GF (2004) Detection and verification of quantitative trait loci for resistance to *Dothistroma* needle blight in *Pinus radiata*. Theor Appl Genet 108: 1056–1063.
Devillard C, Högberg A (2004) Somatic embryogenesis—tomorrow's spruce plant production for intensive forestry. Skogforsk, Resultat Nr 7-2004.
DGB (Dirección General para la Biodiversidad) (2005) Annuario de Estadistica Forestal 2005. Ministerio de Media Ambiente, Madrid, Spain.
Dietrichson J (1969) Genetic variation of cold damage, gowth rhythm and height growth in 4-year-old black spruce (*Picea mariana* [Mill.] B.S.P.). Medd Nor. Skogforsoksves 97: 112–128.
Dubos C, Plomion C (2003) Identification of water-deficit responsive genes in maritime pine (*Pinus pinaster* Ait.) roots. Plant Mol Biol 51: 249–262.

Dubos C, Le Provost G, Pot D, Salin F, Lalane C, Madur D, Frigerio J-M, Plomion C (2003) Identification and characterization of water-deficit responsive genes in hydroponically grown maritime pine (*Pinus pinaster* Ait.) seedlings. Tree Physiol 23: 169–179.

Durel CE (1992) Gains génétiques attendus après sélection sur index en seconde génération d'amélioration du pin maritime. Rev For Fr 44 (4): 341–355.

DWAF (South African Department of Water Affairs and Forestry) (2006) Report on the Commercial Timber Resources and Primary Roundwood processing in South Africa 2005/2006. Department of Water Affairs and Forestry, Johannesburg, SA.

Eckert RT, Kuser JE (1988) Eastern white pine (*Pinus strobus* L.). In: Tree Improvement in the Northeast: Interim Summary and Recommendations for Selected Species. Tech Bull No 131, Univ of Maine, Maine Agri Exp Sta, Orono, ME, USA, pp 31–34.

Eiche V (1966) Cold damage and plant mortality in experimental provenance plantations with Scots pine in northern Sweden. Stud For Suec 36.

Elfving B, Ericsson T, Rosvall, O (2001) The introduction of lodepole pine for wood production in Sweden—a review. For Ecol Manag 141: 15–29.

El-Kassaby YA, King JN, Ying CC, Yanchuk AD, Alfaro RI, Leal I (2001) Somatic embryogenesis as a delivery system for specialty products with reference to resistant Sitka spruce. In: Proc 26th Biennial South For Tree Improve Conf, Univ of Georgia, Athens, GA, USA, pp 154–168.

Elliott L (2006) Availability of Resistant Port-orford-cedar Seed. USDA Forest Service, Umpqua National Forest, Cottage Grove, OR, USA.

English B, Linehan B (2000) The status of tree improvement in Newfoundland and Labrador. In: JD Simpson (ed) Proc 27th Canadian Tree Improvement Association, Part 1, 15–17 Aug 2000, Sault Ste Marie, Ontario, Canada, pp 18–19.

Epperson BK, Allard RW (1984) Allozyme analysis of the mating system in lodgepole pine populations. J Hered 75: 212–214.

Ericsson T (1994) Lodgepole pine (*Pinus contorta* var. *latifolia*) breeding in Sweden—results and prospects based on early evaluations. Ph.D. thesis, Swedish Univ of Agri Sci, Dept of Forest Genetics and Plant Physiology, Umeå, Sweden.

Eriksson G, Andersson S, Eiche V, Ifver J, Persson A (1980) Severity index, transfer effects on survival and volume production of *Pinus sylvestris* in Northern Sweden. Stud For Suec 156.

Eveno E, Collada C, Guevara MA, Léger V, Soto A, Diaz L, Léger P, Gonzalez-Martinez SC, Cervera MT, Plomion C, Garnier-Gére PH (2008) Contrasting patterns of selection at *Pinus pinaster* Ait. drought stress candidate genes as revealed by genetic differentiation analyses. Mol Biol Evol 25: 417–437.

Falk G, McGimpsey G, Broughton K (2004) Manitoba's tree improvement program. In: JD Simpson (ed) Proc 29th Canadian Tree Improvement Association, part 1, 26–29 July 2004, Kelowna, BC, Canada, pp 70–72.

Falk G, McGimpsey G, Broughton K (2006). Manitoba's tree improvement program. In: JD Simpson (ed) Proc 30th Canadian Tree Improvement Association, part 1, 24–27 July, Charlottetown, PEI, Canada, pp 58–60.

Farrar JL (1995) Trees in Canada. Fitzhenry and Whiteside—Canadian Forest Service, Markham, Ontario, Canada.

FFRI (Finnish Forest Research Institute) (2008) Finnish Statistical Yearbook of Forestry 2008. Official Statistics of Finland. Finnish Forest Research Institute.

Fletcher AM, Faulkner R (1972) A Plan for the Improvement of Sitka Spruce by Selection and Breeding. Forestry Commission Research and Development Paper 85, Forestry Commission, Edinburgh, UK.

Ford R, Atack C, Laine M, Gordon S (2006) Northeast Seed Management Association (NeSMA). In: JD Simpson (ed) Proc 30th Meeting Canadian Tree Improvement Association, part 1, 24–27 July, Charlottetown, PEI, Canada, pp 52–53.

Fowler DP (1986) Strategies for the Genetic Improvement of Important Tree Species in the Maritimes. Can For Serv, Inf Rep M-X-156.

Fowler DP, Lester DT (1970) The Genetics of Red Pine. USDA Forest Service, Research Paper WO-8.

Fowler DP, Park YS, Loo-Dinkins J (1995). *Larix laricina*—silvics and genetics. In: WC Schmidt, KJ McDonald (eds) Ecology and Management of *Larix* Forests: A Look Ahead. Procs of an Int Symp 5–9 Oct 1992, Whitefish, Montana. USDA Forest Service Gen Tech Rep INT-GTR-319, Intermountain Research Station, Ogden, UT, USA, pp 54–57.

Frame H, Steeves D (2006) Cooperative tree breeding in Nova Scotia. In: JD Simpson (ed) Proc 30th Canadian Tree Improvement Association, part 1, 24–17 July, Charlottetown, PEI, Canada, p 13.

Fujisawa Y (1998) Forest tree breeding systems for a forestry based on the concept of quality management considering the production of high level raw materials. Bull For Tree Breed Center 15: 31–107. [In Japanese with English summary]

Fujisawa Y, Ohta S, Nishimura K, Tajima M (1992) Wood characteristics and genetic variations in Sugi (*Cryptomeria japonica*). Clonal differences and correlations between locations of dynamic moduli of elasticity and diameter growths in plus-tree clones. Mokuzai Gakkaishi 38: 638–644. [In Japanese with English summary]

Fujisawa Y, Ohta S, Nishimura K, Toda T, Tajima M (1995) Variation in moisture contents of heartwood among clones and test stands in sugi (*Cryptomeria japonica*). Mokuzai Gakkaishi 41: 249–255. [In Japanese with English summary]

Gabrilavicius R, Pliura A (1993) Breeding of Norway spruce in Lithuania. In: V Rhone (ed) Norway Spruce Provenances and Breeding. Proc IUFRO S2.2-11 Symp, Latvia, 1993, pp 193–199.

Gadek PA, Alpers DL, Heslewood MM, Quinn CJ (2000) Relationships within Cupressaceae *sensu lato*: a combined morphological and molecular approach. Am J Bot 87: 1044–1057.

Gallo L, Martinez-Meier A, Azpilicueta MM, Marchelli P, Mondino V (2005) Subprograma coníferas y otras especies en la región Patagónica. In: M Marcó et al. (eds) Mejores árboles para más forestadores: el programa de producción de material de propagación mejorado y el mejoramiento genético en el Proyecto Forestal de Desarrollo. Secretaría de Agricultura, Ganadería, Pesca y Alimentos, Buenos Aires, Argentina, pp 95–116.

Giertych M (1991) Provenance variation in growth and phenology. In: M Giertych, C Mátyás (eds) Genetics of Scots Pine. Elsevier, Amsterdam, The Netherlands, pp 87–101.

Giertych M, Mátyás C (eds) (1991) Genetics of Scots Pine. Elsevier, Amsterdam, The Netherlands.

Godbout J, Fazekas A, Newton CH, Yeh FC, Bousquet J (2008) Glacial vicariance in the Pacific Northwest: Evidence from a lodgepole pine mitochondrial DNA minisatellite for multiple genetically distinct and widely separated refugia. Mol Ecol 17: 2463–2475.

Goto Y, Kondo T, Yasueda H (1999) The variation of Cry j 1 in pollen among Japanese cedar plus trees selected in Kanto Breeding Region. Jap J Palynol 45: 149–152. [In Japanese with English summary]

Goto Y, Kondo T, Kuramoto N, Ide T, Yamamoto K, Inaoka K, Yasueda H (2003) Mapping the gene encoding Cry j 1: a major *Cryptomeria japonica* pollen allergen. Silvae Genet 52: 97–99.

Goto Y, Kondo T, Ide T, Yasueda H, Kuramoto N, Yamamoto K (2004) Cry j 1 isoforms derived from *Cryptomeria japonica* trees have different binding properties to monoclonal antibodies. Clin Exp Allergy 34: 1754–1761.

Griffith I J, Lussier BS, Garman R, Koury R, Yeung H, Pollock J (1993) cDNA cloning of Cry *j* I, the major allergen of *Cryptomeria japonica* (Japanese cedar). J Allergy Clin Immunol 91: 339.

GPMF (Groupe Pin Maritime du Futur) (2002) Le progrès génétique en forêt. INRA, UE Hermitage, F33610 Cestas, France.

Guyon A (1995) Peuplements forestiers littoraux de Bretagne: description et gestion. Rev For Fr 47: 255–262.

Haapanen M (1996) Impact of family-by-trial interaction on the utility of progeny testing methods for Scots pine. Silvae Genet 45: 130–135.

Haapanen M (2001) Time trends in genetic parameter estimates and selection efficiency for Scots pine in relation to field testing method. For Genet 8: 129–144.

Haapanen M, Mikola J (2008) Metsänjalostus 2050—pitkän aikavälin metsänjalostusohjelma. Metlan työraportteja/Working Papers of the Finnish Forest Research Institute 71.

Hannrup B (1999) Genetic parameters of wood properties in *Pinus sylvestris* (L.). Ph.D. Thesis, Swedish Univ Agri Sci, Umeå, Sweden, Silvestria 94.

Hansen CR, Dhir NK, Barnhardt L, Quinn J, Rweyongeza D, Palamarek D, Andriuk C, Antoniuk N, DeCosta T, Mochulski MA (2006) Genetics and Tree Improvement Program 2004–2006 Alberta Sustainable Resource Development. In: JD Simpson (ed) Proc 30th Canadian Tree Improvement Association, part 1, 24–26 July , Charlottetown, PEI, Canada, pp 69–72.

Hansen CR, Dhir NK, Barnhardt L, Rweyongeza D, Quinn J, Palamarek D, DeCosta T, Andriuk C, Antoniuk N, Mochulski MA (2009) Genetics and tree improvement program, 2006–2008. Alberta Sustainable Resource Development. In: JD Simpson (ed) Proc 31st Canadian Tree Improvement Association, part 1, 25–29 Aug 2008, Quebec City, Quebec, Canada, pp 69–74.

Harfouche A, Kremer A (2000) Provenance hybridization in a diallel mating scheme of maritime pine (*Pinus pinaster* Ait.). I. Means and variance components. Can J For Res 30: 1–9.

Harfouche A, Bahrman N, Baradat P, Guyon JP, Petit RJ, Kremer A (2000) Provenance hybridization in a diallel mating scheme of maritime pine (*Pinus pinaster* Ait.). II Heterosis. Can J For Res 30: 10–16.

Harris AS (1990) *Picea sitchensis* (Bong.) Carr. Sitka spruce. In: RM Burns, BH Honkala (eds) Silvics of North America, vol 1, Conifers. Agriculture Handbook 654, USDA Forest Service, Washington DC, USA, pp 260–267.

Heppner D, Turner, J (2006) British Columbia's coastal forests: Spruce weevil and western spruce budworm forest health Stand Establishment Decision Aids. BC J Ecosyst Manag 7(3): 45–49.

Hermann RK (1987) North American tree species in Europe: transplanted species offer good growth potential on suitable sites. J For 85(12): 27–32.

Hermann RK, Lavender, DP (1990) *Pseudotsuga menziesii* (Mirb.) Franco. Douglas-fir. In: RM Burns, BH Honkala (eds) Silvics of North America, vol 1, Conifers. Agriculture Handbook 654, USDA Forest Service, Washington DC, USA, pp 527–540.

Hermann RK, Lavender DP (1999) Douglas-fir planted forests. New Forest 17: 53–70.

Hirakawa Y, Fujjisawa Y, Nakada R, Yamashita K (2003) Wood properties of sugi clones selected from plus trees in Kanto Breeding Region. Bull of FFPRI 2: 31–41. [In Japanese with English summary]

Holst M (1955) Forest tree breeding and genetics at Petawawa Forest Experiment Station 1954. Proc 3rd Commercial Forest Tree Breeding in Canada. Appendices J-1 to J-12.

Hosie RC (1979) Native Trees of Canada. 8th edn. Fitzhenry & Whiteside, Don Mills, ON, Canada

Howard JL (2001) US Timber Production, Trade Consumption and Price Statistics 1965 to 1999. USDA Forest Service, Res Pap FPL-RP-595, Forest Products Lab, Madison, WI, USA.

Howe GT, Jayawickrama K, Cherry M, Johnson GR, Wheeler NC (2006) Breeding Douglas-fir. Plant Breed Rev 27: 245–353.

Hunt RS (2004) Blister-rust-resistant Western White Pine for British Columbia. Natural Resources Canada, Can For Serv Info Rep BC-X-397, Victoria, BC, Canada.

Ikeda K, Oomori S, Arima T (2000) Quality evaluation of standing trees by stress-wave propagation method and its application III. Application to sugi (*Cryptomeria japonica*) standing plus trees. Mokuzai Gakkaishi 46: 558–565. [In Japanese with English summary].

Illy G (1966) Recherches sur l'amélioration génétique du pin maritime. Ann Sci For 23: 757–948.

INFOR (Instituto Forestal) (2007) Estadisticas Forestales Chilenas 2006. INFOR Boletin Estadistico 117, Santiago, Chile (*fide* German Ortiz).

Isik F, Li B (2003) Rapid assessment of wood density of live trees using the Resistograph for selection in tree improvement programs. Can J For Res 33: 2426–2435.
Isik F, Amerson HV, Whetten RW, Garcia S, Li B, McKeand SE (2008) Resistance of *Pinus taeda* families under artificial inoculations with diverse fusiform rust pathogen populations and comparison with field trials. Can J For Res 38: 2687–2696.
Ivković M, Wu HX, McRae TA, Powell MB (2006a) Developing breeding objectives for radiata pine structural wood production. I Bioeconomic model and economic weights. Can J For Res 36: 2920–2931.
Ivković M, Wu HX, McRae TA, Matheson AC (2006b) Developing breeding objectives for radiata pine structural wood production. II Sensitivity analyses. Can J For Res 36: 2932–2942.
Jactel H, Kleinhentz M, Marpeau-Bezard A, Marion-Poll F, Menassieu P, Burban C (1996) Terpene variations in Maritime pine constitutive oleoresin related to host tree selection by *Dioryctria sylvestrella* Ratz (Lepidoptera: Pyralidae). J Chem Ecol 22 (5): 1037–1050.
Jansons Ā, Baumanis I, Haapanen M (2008) Parastās priedes (*Pinus sylvestris* L.) klonu atlase Kurzemes zonas 2. kārtas sēklu plantācijas izveidei un sagaidāmais ģenētiskais ieguvums (Selection of clones and genetic gain for second-stage seed orchards of Scots pine (*Pinus sylvestris* L.) in western part of Latvia). Mežzinātne 17: 88–116. [in Latvian with English summary]
Jansson G (2007) Gains from selecting *Pinus sylvestris* in southern Sweden for volume per hectare. Scand J For Res 22: 185–192.
Jansson G, Li B (2004) Genetic gains of full-sib families from disconnected diallels in loblolly pine. Silvae Genet 53: 60–64.
Jaquish B, Howe G, Fins L, Rust M (1995) Western larch tree improvement programs in the Inland Empire and British Columbia. In: WC Schmidt, KJ McDonald (eds) Ecology and Management of *Larix* Forests: A Look Ahead. Proc of an Int Symp, 5–9 Oct 1992, Whitefish, Montana. USDA Forest Service, Gen Tech Rep INT-GTR-319, Intermountain Research Station, Ogden, UT, USA, pp 452–460.
Jayawickrama KJS, Carson MJ (2000) A breeding strategy for the New Zealand Radiata Pine Breeding Cooperative. Silvae Genet 49: 82–90.
Jeffers RM, Nienstaedt H (1972) Precocious flowering and height growth of jack pine full-sib families. In: Proceedings Meeting Working Party on Progeny Testing (IUFRO) Macon, GA, USA, pp 19–33.
Jermstad KD, Bassoni DL, Jech KS, Wheeler NC, Neale DB (2001a) Mapping of quantitative trait loci controlling adaptive traits in coastal Douglas-fir. I Timing of vegetative bud flush. Theor Appl Genet 102: 1142–1151.
Jermstad KD, Bassoni DL, Wheeler NC, Anekonda TS, Aitken SN, Adams WT, Neale DB (2001b) Mapping of quantitative trait loci controlling adaptive traits in coastal Douglas-fir. II Spring and fall cold-hardiness. Theor Appl Genet 102: 1151–1158.
Jermstad KD, Bassoni DL, Jech KS, Ritchie GA, Wheeler NC, Neale DB (2003) Mapping of quantitative trait loci controlling adaptive traits in coastal Douglas fir. III Quantitative trail loci-by-environment interactions. Genetics 165: 1489–1506.
Jessome AP (1977) Strength and related properties of woods grown in Canada. For Tech Rep 21, Can For Serv, Eastern For Prod Lab, Ottawa, ON, Canada.
Jones PD, Schimleck LR, Peter GF, Daniels RF, Clark A III (2005) Nondestructive estimation of *Pinus taeda* L. wood properties for samples from a wide range of sites in Georgia. Can J For Res 35: 85–92.
Jovanovski A, Davel M, Mohr-Bell D (2005) Densidad basica de la madera de *Pseudotsuga menziesii* (Mirb.) Franco en la Patagonia. Invest Agrar: Sist Recur For 14(2): 153–160.
Joyce DG, Lu P, Sinclair RW (2002) Genetic variation in height growth of eastern white pine (*Pinus strobus*) in Ontario. Silvae Genet 51: 136–142.
Jusheng H (1985) A brief account of forest tree improvement in China. Forest Genetic Resources Information (FAO) 14: 2–6.
Kamaguchi A, Nakao T, Kodama Y (2000) Non-destructive measurement of heartwood moisture content in sugi (*Cryptomeria japonica* D. Don) standing tree by lateral impact vibration method. Mokuzai Gakkaishi 46: 13–19. [In Japanese with English summary]

Kang K-S (2007) Conservation and utilization of genetic materials through tree breeding and seed orchard management in Korea. In: Proceedings of a Symposium: A New Era for the Conservation of Forest Genetic Resources, 13 June, Forest Seed Research Center, Suanbo, Korea, pp 38–53.

Karnosky DF, Houston DB (1979) Genetics of air pollution tolerance of trees in the northeastern United States. In: Proc 26th Northeastern Tree Improvement Conf, 25–26 July 1978, University Park, PA, USA, pp 161–178.

Khalil MAK (1984) All-range black spruce provenance study in Newfoundland: performance and genotypic stability of provenances. Silvae Genet 33: 63–71.

King JN, Alfaro RI (2004) Breeding for resistance to a shoot weevil of Sitka spruce in British Columbia, Canada. In: C Walter, M Carson (eds) Plantation Forest Biotechnology for the 21st Century. Research Signpost Publ, Kerala, India, pp 119–128.

King JN, Alfaro RI, Cartwright C (2004) Genetic resistance of Sitka spruce (*Picea sitchensis*) populations to the white pine weevil (*Pissodes strobi*): distribution of resistance. Forestry 77: 269–278.

King JN, Hunt R (2004) Five needle pines in British Columbia, Canada: Past present and future. In: RA Sniezko, S Samman, SE Schlarbaum, HB Kriebel (eds) Breeding and Genetic Resources of Five-needle Pines: Growth, Adaptability, and Pest Resistance, 23–27 July 2001, Medford, OR, USA. IUFRO Working Party 2.02.15. USDA Forest Service, RMRS-P-32, Rocky Mountain Research Station, Fort Collins, CO, USA, pp 12–19.

King JN, Alfaro RI (2009) Developing Sitka spruce populations for resistance to the white pine weevil: summary of the research and breeding program. Tech Rep 50, Ministry of Forests and Range, Forest Science Program, British Columbia.

Kinloch BB Jr (2003) White pine blister rust in North America: past and prognosis. Phytopathology 93: 1044–1047.

Kleinhentz M, Raffin A, Jactel H (1998) Genetic parameters and gain expected from direct selection for resistance to *Dioryctria sylvestrella* Ratz. (Lepidoptera: Pyralidae) in *Pinus pinaster* Ait., using a full diallel mating design. For Genet 5: 147–154.

Kleinschmit J (1993) 25 years Norway spruce breeding in lower Saxony, Germany. In: V Rhone (ed) Norway Spruce Provenances and Breeding. Proc IUFRO S2.2–11 Symp, Latvia, pp 213–218.

Kleinschmit J, Bastien J-C (1992) IUFRO's role in Douglas-fir (*Pseudotsuga menziesii* (Mirb.) Franco) tree improvement. Silvae Genet 41: 161–173.

Klimaszewska K, Park YS, Overton C, MacEacheron I, Bonga JM (2001) Optimized somatic embryogenesis in *Pinus strobus* L. *In Vitro* Cell Dev Biol Plant 37: 392–399.

Koch P (1987) Gross characteristics of lodgepole pine trees in North America. USDA Forest Service, Inter Mtn Res Stn, Gen Tech Rep INT-227, Ogden, UT, USA.

Kohlstock N, Schneck H (1992) Scots pine breeding (*Pinus sylvestris* L.) at Waldsieversdorf and its impact on pine management in northeastern German lowland. Silvae Genet 41: 174–180.

König A (2005) Provenance research: evaluating the spatial pattern of genetic variation. In: TH Geburek, J Turok (eds) Conservation and Management of Forest Genetic Resources in Europe. Arbora Publ, Zvolen, pp 275–333.

Kriebel HB (1972) Embryo development and hybridity barriers in the white pines (Section Strobus). Silvae Genet 21: 39–44.

Kriebel HB (1978) Genetic selection for growth rate improvement in *Pinus strobus*. Genetika (Belgrade, Yugoslavia) 10(3): 269–276.

Kriebel HB (1982) Recommended Genetic Selections of some Forest Trees for Ohio. Ohio State Univ Res Bul 1148, Ohio Agri Res and Dev Center, OH, USA.

Kriebel HB (1983) Breeding eastern white pine: a world-wide perspective. For Ecol Manag 6: 263–279.

Kriebel HB (2004) Genetics and breeding of five-needle pines in the eastern United States. In: RA Sniezko, S Samman, SE Schlarbaum, HB Kriebel (eds) Breeding and Genetic Resources of Five-needle Pines: Growth, Adpatability and Pest Resistance, 23–27 July 2001, Merford,

OR, USA. IUFRO Working Party 2.02.15. USDA Forest Service, Rocky Mountain Research Station, Proc RMRS-P-32, Fort Collins, CO, USA, pp 20–27.

Kuang H, Richardson TE, Carson SD, Bongarten BC (1998) An allele responsible for seedling death in *Pinus radiata* D. Don. Theor Appl Genet 96: 640–644.

Kumar S, Echt CS, Wilcox PL, Richardson TE (2003) Testing for linkage disequilibrium in the New Zealand radiata pine breeding population. Theor Appl Genet 108: 292–298.

Kuramoto N, Kondo T, Fujisawa Y, Nakata R, Hayashi E, Goto Y (2000) Detection of quantitative trait loci for wood strength in *Cryptomeria japonica*. Can J For Res 30: 1525–1533.

Ladrach WE (1983) Ten years of industrial tree improvement in Colombia. In: Proc 17th Southern Forest Tree Improvement Conf, Atlanta, GA, USA, pp 8–22.

Langlet O (1936) Studier over tallens fysiologiska variabilitet och dess samband med klimatet. Meddelanden från statens skogsförsöksanstalt 29(4): 219–406.

Langlet O (1971) Two hundred years of gene ecology. Taxon 20: 653–722.

Larsen CS (1937) The employment of species, types and individuals in forestry. Roy Vet Agri Coll Yearbook (Copenhagen) 1937: 74–154.

Law KN, Valade JL (1994) Status of utilization of jack pine (*Pinus banksiana*) in the pulp and paper industry. Can J For Res 14: 2078–2084.

Le Pichon C, Verger M, Brando J, Le Bouler H (2001) Itinéraires techniques pour la multiplication végétative en vrac du mélèze hybride (*Larix x eurolepis* Henry). Rev For Fr LIII: 111–124.

Ledgard N (2001) The spread of lodgepole pine (*Pinus contorta*, Dougl.) in New Zealand. For Ecol Manag 141: 43–57.

Lee SJ (1999) Genetic Gain from Scots Pine: Potential for New Commercial Seed Orchards. Information Note, Forestry Commission, Edinburgh, UK.

Lee SJ (2001) Selection of parents for the Sitka spruce breeding population in Britain and the strategy for the next breeding cycle. Forestry 74: 129–143.

Lee SJ (2006) It's a family affair. Forestry and British Timber, Dec 2006: 14–16.

Lee SJ, Matthews R (2004) An Indication of the Likely Volume Gains for Improved Sitka Spruce Planting Stock. Forestry Commission Information Note 55, Forestry Commission, Edinburgh, UK.

Lee SJ, Cottrell J, John A (2004) Advances in Biotechnology: Powerful Tools for Tree Breeding and Genetic Conservation. Forestry Commission Information Note 50, Forestry Commission, Edinburgh, UK.

Lee SJ, A'Hara S, Cottrell J (2007) The use of DNA technology to advance the Sitka spruce breeding programme. In: Forest Research Annual Report and Accounts 2005–6, Forestry Commission, Edinburgh, UK, pp 30–36.

Lelu-Walter MA, Pâques LE (2009) Simplified and improved somatic embryogenesis of hybrid larches (*Larix* x *eurolepis* and *Larix* × *marschlinsii*). Perspectives for breeding. Ann For Sci 66: 104. DOI: 10.1051/forest/2008079.

Li B, McKeand SE, Weir RJ (1999) Tree improvement and sustainable forestry—impact of two cycles of loblolly pine breeding in the USA. For Genet 6: 229–234.

Li P, Beaulieu J, Corriveau A, Bousquet J (1993) Genetic variation in juvenile growth and phenology in a white spruce provenance-progeny test. Silvae Genet 42: 52–60.

Li P, Beaulieu J, Daoust G, Plourde A (1997a) Patterns of adaptive genetic variation in eastern white pine (*Pinus strobus*) from Quebec. Can J For Res 27: 199–206.

Li P, Beaulieu J, Bousquet J (1997b) Genetic structure and patterns of genetic variation among populations in eastern white spruce (*Picea glauca*). Can J For Res 27: 189–198.

Lindgren D, Gea LD, Jefferson PA (1996) Loss of genetic diversity monitored by status number. Silvae Genet 45: 52–59.

Lipow SR, Johnson GR, St. Clair JB, Jayawickrama KJ (2003) The role of tree improvement programs for *ex situ* gene conservation of coastal Douglas-fir in the Pacific Northwest. For Genet 10: 111–120.

Little DP, Schwarzbach AE, Adams RP, Hsieh C-F (2004) The circumscription and phylogenetic relationships of *Callitropsis* and the newly described genus *Xanthocyparis* (Cupressaceae). Am J Bot 91: 1872–1881.

Little S, Garrett PW. 1990. *Pinus rigida* Mill.—Pitch pine. In: RM Burns, BH Honkala (eds) Silvics of North America, vol 1: Conifers. Agriculture Handbook No. 654, USDA Forest Service, Washington, DC, USA, pp 456–462.

Lohrey RE, Kossuth SV (1990) Slash pine. In: RM Burns, BH Honkala (eds) Silvics of North America, vol 1: Conifers. Agriculture Handbook No. 654, USDA Forest Service, Washington DC, USA, pp 338–347.

Lopez OR, Kursar TA, Cochard H, Tyree MT (2005) Interspecific variation in xylem vulnerability to cavitation among tropical tree and shrub species. Tree Physiol 25: 1553–1562.

Loustau D, Bosc A, Colin A, Davi H, François C, Dufrêne E, Déqué M, Cloppet E, Arrouays D, Le Bas C, Saby N, Pignard G, Hamza N, Granier A, Bréda N, Ciais P, Viovy P, Ogée J, Delage F (2005) Modeling climate change effects on the potential production of French plains forests at the sub regional level. Tree Physiol 25: 813–823.

Lung-Escarmant B, Guyon D (2004) Temporal and spatial dynamics of primary and secondary infection by *Armillaria ostoyae* in a *Pinus pinaster* plantation. Phytopathology 94: 125–131.

MAF (Ministry of Agriculture and Forestry, New Zealand) (2007) A National Exotic Forest Description as on April 2007. Wellington, NZ.

MAF (Ministry of Agriculture and Forestry, New Zealand) (undated) New Zealand Forestry Industry Facts and Figures 2007/08. Wellington, NZ.

MacKinnon WJ, Glen WM, Myers MN (1997) Tree Improvement on Prince Edward Island. In: JD Simpson (ed) Proc 26th Meeting Canadian Tree Improvement Association, part 1, 18–21 Aug , Sainte-Foy, Quebec, Canada, pp 19–20.

Martinsson O, Lesinski J (2007) Siberian larch. Forestry and Timber in a Scandinavian Perspective. JiLU Jämtlands County Council Institute of Rural Development.

Mátyás C (1991) Seed orchards. In: M Giertych, C Mátyás (eds) Genetics of Scots Pine. Elsevier, Amsterdam, The Netherlands, pp 125–145.

McDonald G, Zambino P, Sniezko RA (2004) Breeding rust-resistant five-needle pines in the western United States: lessons from the past and a look to the future. In: RA Sniezko, S Samman, SE Schlarbaum, HB Kriebel (eds) Breeding and Genetic Resources of Five-needle Pines: Growth, Adaptability, and Pest Resistance, 23–27 July 2001, Medford, OR, USA. IUFRO Working Party 2.02.15. USDA Forest Service, RMRS-P-32 Rocky Mountain Research Station, Fort Collins, CO, USA, pp 28–50.

McKeand S, Kurinobu S (1998) Japanese tree improvement and forest genetics. J For 96(4): 12–17.

McKeand S, Mullin T, Byram T, White T (2003) Deployment of genetically improved loblolly and slash pine in the South. J For 101(3): 32–37.

McKeand SE, Gerwig DM, Cumbie WP, Jett JB (2008) Seed orchard management strategies for deployment of intensively selected loblolly pine families in the southern US. In: D Lindgren (ed) Seed Orchards, Proc from a Conf, 26–28 Sept 2007, Umeå, Sweden, pp 177–182.

McNabb K (2007) Forest Tree Seedling Production in the South for the 2006–2007 Planting Season. Technical Note 07–02, Southern Forest Nursery Management Cooperative, Auburn Univ, AL, USA.

Meuwissen THE, Hayes BJ, Goddard ME (2001) Prediction of total genetic value using genome-wide dense marker maps. Genetics 157: 1819–1829.

Mikola J (1991) Utilization of improved material: a survey. In: M Giertych, C Mátyás (eds) Genetics of Scots Pine. Elsevier, Amsterdam, The Netherlands, pp 265–275.

Mikola J (1993) Breeding of Norway spruce in Finland: problems and remedies. In: V Rhone (ed) Norway Spruce Provenances and Breeding. Proc IUFRO S2.2-11 Symp, Latvia, pp 231–239.

Miller JT, Knowles FB (1994) Introduced Forest Trees in New Zealand: Recognition, Role and Seed Source. 14 Douglas-fir. FRI Bull No 124 Part 14, Rotorua, NZ.

Minghe L, Ritchie GA (1999a) Eight hundred years of clonal forestry in China: I. traditional afforestation with Chinese fir (*Cunninghamia lanceolata* (Lamb.) Hook.). New For 18: 131–142.

Minghe L, Ritchie GA (1999b) Eight hundred years of clonal forestry in China: II. Techniques for mass production of rooted cuttings of Chinese fir (*Cunninhamia lanceolata* (Lamb.) Hook.). New For 18: 143–159.

Miyajima H (1983) Variety. In: K Sakaguchi (ed) Suginosubete, Zenkoku-rinngyou-kairyou-fukyu-kyoukai, Tokyo, Japan, pp 126–140 (in Japanese).

Mochan S, Lee S, Gardiner B (2008) Benefits of Improved Sitka Spruce: Volume and Quality of Timber. Forestry Commission Research Note 3, Forestry Commission, Edinburgh, UK.

Morgenstern EK (1969) Genetic variation in seedlings of *Picea mariana* (Mill.) B.S.P. II. Variation patterns. Silvae Genet 18: 161–167.

Morgenstern EK (1996) Geographic Variation in Forest Trees. Genetic Basis and Applications of Knowledge in Silviculture. UBC Press, Vancouver, BC, Canada.

Morgenstern EK, Mullin TJ (1990) Growth and survival of black spruce in the range-wide provenance study. Can J For Res 20: 130–143.

Morgenstern EK, Corriveau AG, Fowler DP (1981) A provenance test of red spruce in nine environments in eastern Canada. Can J For Res 11: 124–131.

Mugasha AG, Chamshama SAO, Nshubemuki L, Iddi S, Kindo AI (1997) Performance of thirty two families of *Cupressus lusitanica* at Hambalawei, Lushoto, Tanzania. Silvae Genet 46: 185–192.

Mullin LJ, Barnes RD, Prevôst MJ (1978) A Review of the Southern Pines in Rhodesia. Research Bulletin No 7, Rhodesia Forestry Commission, Harare, Zimbabwe.

NCSUCTIP (North Carolina State University Cooperative Tree Improvement Program) (2008) 52nd Annual Report, North Carolina State University Cooperative Tree Improvement Program. Department of Forestry and Environmental Resources, Raleigh, NC, USA.

Nguyen-Queyrens A, Bouchet-Lannat F (2003) Osmotic adjustment in three-year-old seedlings of five provenances of maritime pine (*Pinus pinaster*) in response to drought. Tree Physiol 23: 397–404.

Nienstaedt H, Teich A (1971) The Genetics of White Spruce. USDA Forest Service, Res Pap WO-15, Washington DC, USA.

Nienstaedt H, Zasada JC (1990) *Picea glauca* (Moench) Voss. White spruce. In: RM Burns, BH Honkala (eds) Silvics of North America, vol 1: Conifers. Agriculture Handbook 654. USDA Forest Service, Washington DC, USA, pp 204–226.

Olson DF Jr, Roy DF, Walters GA (1990). *Sequoia sempervirens* (D. Don) Endl. Redwood. In: RM Burns, BH Honkala (eds) Silvics of North America, vol 1: Conifers. Agriculture Handbook 654. USDA Forest Service, Washington DC, USA, pp 541–551.

Orr-Ewing AL (1966) Inter- and intraspecific crossing in Douglas-fir (*Pseudotsuga menziesii* (Mirb.) Franco). Silvae Genet 15: 121–126.

Orr-Ewing AL (1976) Inbreeding Douglas fir to the S_3 generation. Silvae Genet 25: 179–183.

Pait J (2005) Clonal forestry: out of the lab, finally. In: SE McKeand, B Li (eds) Proc 28th Southern Forest Tree Improvement Conf, 21–23 June 2005, North Carolina State Univ, Raleigh, NC, Publ 50, Southern Forest Tree Improvement Committee, USA, p 16.

Papageorgiou AC, Finkeldey R, Hattemer HH, Xenopoulos S (2005) Genetic differences between autochthonous and breeding populations of common cypress (*Cupressus sempervirens* L.) in Greece. Eur J For Res 124: 119–124.

Pâques LE (1989) A critical review of larch hybridization and its incidence on breeding strategies. Ann Sci For 46: 141–153.

Pâques LE (2002a) Heterosis in interspecific hybrids between European and Japanese larch. In: LE Pâques (ed) Improvement of Larch (*Larix* sp.) for Better Growth, Stem Form and Wood Quality, Proc of an Int Symp, 16–21 Sept, Gap (Hautes-Alpes)—Auvergne & Limousin, France, pp 155–161.

Pâques LE (2002b) Larch tree improvement programme in France. In: LE Pâques (ed) Improvement of Larch (*Larix* sp.) for Better Growth, Stem Form and Wood Quality,

Proc of an Int Symp, 16–21 Sept, Gap (Hautes-Alpes)—Auvergne & Limousin, France, pp 104–118.

Pâques LE (2007) Can F2-hybrids be a reasonable alternative to F1-hybrids for plantation of hybrid larch (*Larix* x *eurolepis*)? Influence of consanguinity levels on performance. In: M Perron (ed) Integrated Research Activities for Supply of Improved Larch to Tree Planting: Tree Improvement, Floral Biology and Nursery Production. LARIX 2007, Int Symp IUFRO Working Group S2.02.07 (Larch Breeding and Genetic Resources) Proc/Actes, 16–21 Sept, Saint-Michel-des-Saints and Québec City, Canada, p 25.

Park YS, Fowler DP (1988) Geographic variation of black spruce tested in the Maritimes. Can J For Res 18: 106–114.

Park, YS, Simpson JD, Adams GW, Morgenstern EK, Mullin TJ (1993) An updated breeding strategy for black spruce (*Picea mariana*) (Mill.) B.S.P. in New Brunswick. In: YS Park, GW Adams (eds) Workshop on Breeding Strategies of Important Tree Species in Canada, Aug 18, Fredericton, NB, Canada, Nat Resour Can Inf Rep M-X-186E, pp 41–54.

Parker WH, van Niejenhuis A, Charette P (1994) Adaptive variation in *Picea mariana* from northwestern Ontario determined by short-term common environment tests. Can J For Res 24: 1653–1661.

Perron M, Bousquet J (1997) Natural hybridization between *Picea mariana* and *P. rubens*. Mol Ecol 6: 725–734.

Persson B (1994) Effects of provenance transfer on survival in nine experimental series with *Pinus sylvestris* (L.) in northern Sweden. Scand J For Res 9: 275–287.

Persson B, Persson A, Ståhl EG, Karlmats U (1995) Wood quality of *Pinus sylvestris* progenies at various spacings. For Ecol Manag 76: 127–138.

Persson T, Andersson B (2003) Genetic variance and covariance patterns of growth and survival in northern *Pinus sylvestris*. Scand J For Res 18: 332–343.

Persson T, Andersson B, Ericsson T (2006) Contrasting covariance patterns between growth and survival in northern *Pinus sylvestris*. Paper III. In: T Persson (ed) Genetic Expression of Scots Pine Growth and Survival in Varying Environments. Doctoral Thesis, Dept For Gen Plant Phys, Swedish Univ Agricultural Sciences, Umeå, Sweden.

Peterson EB, Peterson NM, Weetman GF, Martin PJ (1997) Ecology and Management of Sitka Spruce, Emphasizing its Natural Range in British Columbia. UBC Press, Vancouver, BC, Canada.

Philippe G, Baldet B, Heois B, Ginisty C (2006) Reproduction Sexuée des Conifères et Production de Semences en Vergers à Graines. Centre National du Machinisme Agricole, du Génie Rural, des Eaux et des Forêts (CEMAGREF), Antony, France.

Pihelgas E (1991) Seed stands and plus trees. In: M Giertych, C Mátyás (eds) Genetics of Scots Pine. Elsevier, Amsterdam, The Netherlands, pp 117–123.

Plomion C, Chagné D, Pot D, Kumar S, Wilcox PL, Burdon RD, Prat D, Peterson DG, Paiva J, Chaumeil P, Vendramin GG, Sebastiani F, Nelson CD, Echt CS, Savolainen O, Kubisiak TL, Cervera MT, de María N, Islam-Faridi MN (2007) Pines. In: C Kole (ed) Genome Mapping and Molecular Breeding in Plants, vol 7: Forest Trees, Springer, Berlin, Heidelberg, Germany; New York, USA, pp 29–92.

Polk RB (1974) Heritabilities of some first-order branching traits in *Pinus banksiana* Lamb. In: Proc 8th Central States Tree Improvement Conf, pp 33–39.

Pot D, Chantre G, Rozenberg P, Rodrigues JC, Jones GL, Pereira H, Hannrup B, Cahalan C, Plomion C (2002) Genetic control of pulp and timber properties in maritime pine (*Pinus pinaster* Ait.). Ann For Sci 59: 563–575.

Poynton RJ (1977) Tree Planting in Southern Africa, vol 1: The Pines. Department of Forestry, Republic of South Africa, Pretoria, South Africa.

Quencez C, Bastien C (2001) Genetic variation within and between populations of *Pinus sylvestris* L. (Scots pine) for susceptibility to *Melampsora pinitorqua* Rostr (pine twist rust). Heredity 86: 36–44.

Rehfeldt GE (1988) Ecological genetics of *Pinus contorta*; a synthesis. Silvae Genet 37: 131–135.

Rehfeldt GE, Gallo L (2001) Introduction of ponderosa pine and Douglas-fir to Argentina using quantitative traits for retrospective identification and prospective selection of provenances. New For 21: 35–44.

Rehfeldt GE, Ying CC, Spittlehouse DL, Hamilton DA Jr (1999) Genetic responses to climate in *Pinus contorta*: niche breadth, climate change, and reforestation. Ecol Monogr 69: 375–407.

Remröd J (1976) Choosing Scots Pine (*Pinus sylvestris* L.) Provenances in Northern Sweden—Analyses of Survival, Growth, and Quality in Provenance Experiments Planted 1951. Roy Coll For, Dept For Genet, Res Notes 19. [In Swedish with English summary]

Richardson DM (ed) (1998) Ecology and Biogeography of *Pinus*. Cambridge Univ Press, Cambridge, UK.

Riou-Nivert P (2002) Le pin maritime, seigneur d'Aquitaine. Forêt Entr 148: 47–51.

Roche L (1969) A genecological study of the genus *Picea* in British Columbia. New Phytol 68: 505–554.

Rosvall O (2001) New seed orchards give high genetic gain. Skogforsk, Resultat Nr 2-2001.

Rosvall O, Lindgren D, Mullin TJ (1998) Sustainability robustness and efficiency of a multigeneration breeding strategy based on within-family clonal selection. Silvae Genet 47: 307–321.

Rosvall O, Jansson G, Andersson B, Ericsson T, Karlsson B, Sonesson J, Stener L-G (2002) Predicted genetic gain from existing and future seed orchards and clone mixes in Sweden. In: M Haapanen, J Mikola (eds) Integrating tree breeding and forestry. Proc Nordic Group for Management of Genetic Resources of Trees Meeting, Mekrijärvi, Finland, 23–27 March 2001. Research Paper 842, Finnish Forest Research Institute, Vantaa, Finland, pp 71–85.

Rosvall O, Jaconson S, Karlsson B, Lundström A (2004) Ökad Produktion—Trots Ökan Naturvård. Skogforsk, Redogörelse Nr 1-2004.

Roth BE, Li X, Huber DA, Peter GF (2007) Effects of management intensity, genetics and planting density on wood stiffness in a plantation of juvenile loblolly pine in the southeastern USA. For Ecol Manag 246: 155–162.

Roux J, Mekeb G, Kanyic B, Mwangid L, Mbagae A, Huntera GC, Nakabongea G, Heatha RN, Wingfield MJ (2005) Diseases of plantation forestry trees in eastern and southern Africa. S Afr J Sci 101: 409–413.

Rozenberg P, Franc A, Bastien C (2001) Improving models of wood density by including genetic effects: a case study in Douglas-fir. Ann For Sci 58(4): 385–394.

Rudolph TD, Yeatman CW (1982) Genetics of Jack Pine. USDA Forest Service, Research Paper WO-38, Washington DC, USA.

Rudolph TD, Laidly PR (1990) *Pinus banksiana* Lamb.—jack pine. In: RM Burns, BH Honkala (eds) Silvics of North America, vol 1: Conifers. Agriculture Handbook 654. USDA Forest Service, Washington DC, USA, pp 280–293.

Sampson RN (2004) Southern forests: yesterday, today, and tomorrow. In: HM Rauscher, K Johnsen (eds) Southern Forest Science: Past, Present, Future. Gen Tech Rep SRS-75, USDA Forest Service Southern Research Station, Asheville, NC, USA, pp 5–14.

Samuel CJA, Fletcher AM, Lines R (2007) Choice of Sitka Spruce Seed Origins for Use in British Forests. Forestry Commission Bulletin 127, Forestry Commission, Edinburgh, UK.

Santini A, Panconesi A, Di Lonardo V, Raddi P (1997) 20 years of research on genetic improvement of cypress for resistance to bark canker: problems and results. In: Proc 10th Congr Mediterranean Phytopathological Union, 1–5 June, Montpellier, France, pp 603–607.

Schmidt WC (1995) Around the world with *Larix*: An introduction. In: WC Schmidt, KJ McDonald (eds) Ecology and Management of *Larix* Forests: A Look Ahead. Proc of an Int Symp, 5–9 Oct 1992, Whitefish, Montana. Gen Tech Rep INT-GTR-319, USDA Forest Service, Intermountain Research Station, Ogden, UT, USA, pp 6–18.

Schmidtling RC, Robison TL, McKeand SE, Rousseau RJ, Allen HL, Goldfarb B (2004) The role of genetics and tree improvement in southern forest productivity. In: HM Rauscher, K

Johnsen (eds) Southern Forest Science: Past, Present, Future. Gen Tech Rep SRS-75. USDA Forest Service, Southern Research Station, Asheville, NC, USA, pp 97–108.
Schmidt-Vogt H (1977) Die Fichte. Band I: Taxonomie, Verbreitung, Morphologie, Ökologie, Waldgesellschaften. Paul Parey, Hamburg, Berlin, Germany.
Schober R (1985) Neue Ergebnisse des II. Internationalen Lärchenprovenienzversuches von 1958/59 nach Aufnhmen von Teilversuchen in 11 europäischen Ländern und den USA. Schriften aus der Forstlichen Fakultät der Universität Göttingen, Germany.
Scott CW (1960) *Pinus radiata*. FAO Forestry and Forest Products Studies No 14. FAO, Rome, Italy.
Serrière-Chadoeuf I (1986) Production et sylviculture de l'épicéa de Sitka en France. Rev For Fr 38 (Amélioration génétique des arbres forestiers): 140–148.
Shelbourne CJA, Burdon RD, Carson SD, Firth A, Vincent TG (1986) Development Plan for Radiata Pine Breeding. New Zealand Forest Service, Forest Research Institute Spl Publ
Shelbourne CJA, Low CB, Gea LD, Knowles RL (2007) Achievements in forest tree genetic improvement in Australia and New Zealand. 5: Genetic improvement of Douglas-fir in New Zealand. Aust For 70(1): 28–32.
Shepherd RW (1990) Early importations of *Pinus radiata* to New Zealand and distribution in Canterbury to 1885: Implications for genetic makeup of *Pinus radiata* stocks, part I, Hort in NZ 1(1): 33–38, part II, Ibid 1(2): 28–35.
Siebert H, von Einsiedel S, Freiin Truchsess A (2003) Mejoramiento de la calidad fustal en plantaciones de *Pseudotsuga menziesii* al crecer en asociación con *Acacia melanoxylon*. BOSQUE 24(3): 75–83.
Silen RR, Wheat JG (1979) Progressive tree improvement program in coastal Douglas-fir. J For 77(2): 78–83.
Simpson JD, Tosh K (1997) The New Brunswick Tree Improvement Council is 20 years old. For Chron 73(5): 572–577.
Skrøppa T (1982) Breeding strategies with Norway spruce in south-eastern Norway. In: Proc IUFRO Joint Meeting of Working Parties on Genetics about Breeding Strategies including Multiclonal Varieties, Sensenstein, Germany, pp 1–9.
Smith WB, Vissage JS, Darr DR, Sheffield RM (2001) Forest resources of the United States, 1997. Gen Tech Rep NC-219, USDA Forest Service, North Central Research Sation, St. Paul, MN, USA.
Sniezko RA, Hansen EM, Kolpak SE (2004) Simply inherited resistance to *Phytophthora lateralis* in Port-orford-cedar: greenhouse testing. In: BW Geils (ed) Proc 51st Western International Forest Disease Work Conf, 18–22 Aug 2003. Grants Pass, OR, USDA Forest Service, Rocky Mountain Research Station, Flagstaff, AZ, USA, p 87 (poster).
Spieker H (2000) Growth of Norway spruce (*Picea abies* (L.) Karst.) under changing environmental conditions in Europe. In: E Klimo, H Hager, J Kulhavý (eds) Spruce Monocultures in Central Europe—Problems and Prospects. EFI Proc No 33, pp 1–26.
St. Clair JB, Mandel NL, Jayawickrama KJS (2004) Early realized genetic gains for coastal Douglas-fir in the northern Oregon Cascades. West J Appl For 19(3): 195–201.
Ståhl EG, Ericson B (1991) Inheritance of wood properties. In: M Giertych, C Mátyás (eds) Genetics of Scots Pine. Elsevier, Amsterdam, The Netherlands, pp 231–241.
Stephan BR (1991) Inheritance of resistance to biotic factors. In: M Giertych, C Mátyás (eds) Genetics of Scots Pine. Elsevier, Amsterdam, The Netherlands, pp 205–218.
Stine RA, Miller LK, Wyckoff G, Li B (1995) A Comprehensive Tree Improvement Plan for Minnesota. Staff Paper Series No 102, Univ of Minnesota, Dep of Forest Resources, St. Paul, MN, USA.
Stoehr M, Yanchuk A, Xie C-Y, Sanchez L (2008) Gain and diversity in advanced generation coastal Douglas-fir selections for seed production populations. Tree Genet Genomes 4: 193–200.
Stonecypher RW, Piesch RF, Helland GG, Chapman JG, Reno HJ (1996) Results from genetic tests of selected parents of Douglas-fir (*Pseudotsuga menziesii* (Mirb.) Franco) in an applied tree improvement program. For Sci Monogr 32.

Taira H, Teranishi H, Kenda Y (1993) A case study of male sterility in sugi (*Cryptomeria japonica*). J Jap For Soc 75: 377–379. [In Japanese with English summary]

Taira H, Saito M, Furuta Y (1999) Inheritance of the trait of male sterility in *Cryptomeria japonica*. J For Res 4: 271–273. [In Japanese with English summary]

Thompson AD, Lally M, Pfeifer A (2005) Washington or Queen Charlotte Islands? Which is the best provenance of Sitka spruce (*Picea sitchensis*) for Ireland? Irish For 1: 19–33.

Torimaru T, Wang, X-R, Fries A, Andersson B, Lindgren D (2010) Evaluation of pollen contamination in an advanced Scots pine seed orchard. Silvae Genet (in press).

Tosh K, McInnis B (2000) New Brunswick Tree Improvement Council update. In: JD Simpson (ed) Proc 27th Canadian Tree Improvement Association, part 1, 15–17 Aug, Sault Ste. Marie, ON, Canada, pp 28–30.

Tosh K, Fullarton MS (2006) Tree improvement progress by the New Brunswick Department of Natural Resources. In: JD Simpson (ed) Proc 30th Canadian Tree Improvement Association, Part 1, 24–27 July, Charlottetown, PEI, Canada, pp 17–18.

Tosh K, Fullarton MS (2009) Tree improvement progress by the New Brunswick Department of Natural Resources. In: JD Simpson (ed) Proc 31st Canadian Tree Improvement Association, part 1, 25 -29 Aug 2008, Quebec City, Quebec, Canada, pp 18–19. DOI: 10.1051/forest/2008079.

Tosh K, Fullarton MS, Weng Y (2009) New Brunswick Tree Improvement Council update In: JD Simpson (ed) Proc 31st Canadian Tree Improvement Association, part 1, 25–29 Aug 2008, Quebec City, Quebec, Canada, pp 20–21. DOI: 10.1051/forest/2008079.

Van de Sype H, Roman-Amat B (1989) Analyse d'un test multiclonal d'épicéa commune (*Picea abies* (L.) Karst.). Variabilité génétique. Ann Sci For 46: 15–29.

Van der Sijde HA, Shaw MJP, van Wyk G (1985) Reaction wood in *Pinus taeda*—a preliminary report. S Afr For J 133: 27–32.

Vaudelet JC (1982) Conseils aux reboiseurs de la Bretagne septentrionale. Inform For 1: 33–52.

Vergara R, White TL, Huber DA, Shiver BD, Rockwood DL (2004) Estimated realized gains for first-generation slash pine (*Pinus elliottii* var. *elliottii*) tree improvement in the southeastern United States. Can J For Res 34: 2587–2600.

Vergara R, White TL, Huber DA, Schmidt RA (2007) Realized genetic gains of rust resistant selections of slash pine (*Pinus elliottii* var. *elliottii*) planted in high rust hazard sites. Silvae Genet 56: 231–242.

Verger M, Pâques LE (1993) Multiplication végétative du mélèze hybride (*Larix x eurolepis* Henry) par bouturage en vrac. Ann For Sci 50: 205–215.

Viereck LA, Johnston WF (1990) *Picea mariana* (Mill.) B.S.P.—black spruce. In: RM Burns, BH Honkala (eds) Silvics of North America, vol 1: Conifers. Agriculture Handbook 654. USDA Forest Service, Washington DC, USA, pp 227–237.

Volosyanchuk RT (2002) *Pinus sylvestris* L. In: CABI International (comp) Pines of Silvicultural Importance. CABI Publ, Wallingford, Oxon, UK, pp 449–466.

von Rudloff E, Lapp MS (1987) Chemosystematic studies in the genus Pinus. VI General survey of the leaf oil terpene composition of lodgepole pine. Can J For Res 17: 1013–1025.

von Teuffel K, Heinrich B, Baumgarten M (2004) Present distribution of secondary Norway spruce in Europe. In: H Spieker, J Hansen,E Klimo, JP Skovsgaard, H Sterba, K von Teuffel (eds) Norway Spruce Conversion—Options and Consequences. EFI Research Report 18, pp 11–34.

Wang T, Hamann A, Yanchuk A, O'Neill G, Aitken SN (2006) Use of response functions in selecting lodgepole pine populations for future climates. Global Change Biol 12: 2404–2416.

Wear DN, Greis JG (2002) Southern Forest Resource Assessment—Summary Report. Gen Tech Rep SRS-54, Southern Research Station, USDA Forest Service, Asheville, NC, USA.

Wear DN, Carter DR, Prestemon J 2007. The U.S. South's Timber Sector in 2005: A Prospective Analysis of Recent Change. Gen Tech Rep SRS-99, Southern Research Station, USDA Forest Service, Asheville, NC, USA.

Weng YH, Tosh K, Fullarton MS (2006) Realized genetic gains obtained in first generation programs for jack pine in New Brunswick, Canada. In: JA Loo, JD Simpson (eds) Proc 30th Canadian Tree Improvement Association, part 2. Symp Canada's Forests–Enhancing Productivity, Protection & Conservation, 24–27 July 2006, Charlottetown, PEI, Canada, p 41.

Wheeler NC, Guries RP (1982a) Population structure, genic diversity, and morphological variation in *Pinus contorta* Dougl. Can J For Res 12: 595–606.

Wheeler NC, Guries RP (1982b) Biogeography of lodgepole pine. Can J Bot 60: 1805–1814.

Wheeler NC, Critchfield WB (1985) The distribution and botanical characteristics of lodgepole pine: biogeographical and management implications. In: DM Baumgartner (ed) Lodgepole Pine: The Species and its Management. Washington State Univ, Pullman, WA, USA, pp 1–13.

Wheeler NC, Jermstad KD, Krutovsky K, Aitken SN, Howe GT, Krakowski J, Neale DB (2005) Mapping of quantitative trait loci controlling adaptive traits in coastal Douglas-fir. IV Cold-hardiness QTL verification and candidate gene mapping. Mol Breed 15: 145–156.

White TL, Ching KK (1985) Provenance study of Douglas-fir in the Pacific Northwest region. IV. Field performance at age 25 years. Silvae Genet 34: 84–90.

Whitney GG (1986) Relation of Michigan's presettlement pine forests to substrate and disturbance history. Ecology 67: 1548–1559.

Wilcox PL, Echt CE, Burdon RD (2007) Gene-assisted selection: applications of association genetics for forest tree breeding. In: NC Oraguzie, EHA Rikkerink, SE Gardiner, HN De Silva (eds) Association Mapping in Plants, Springer, New York, USA, pp 211–247.

Wilhelmsson L, Andersson B (1993) Breeding programmes in Sweden: 2. Breeding of Scots pine (*Pinus sylvestris*) and lodgepole pine (*Pinus contorta* ssp. *latifolia*). In: SJ Lee (ed) Progeny Testing and Breeding Strategies, Proc Nordic Group of Tree Breeding, Oct, Forestry Commission, Edinburgh, UK, pp 135–145.

Willis KJ, Bennet KD, Birks HJB (1998) The late quarternary dynamics of pines in Europe. In: DM Richardson (ed) Ecology and Biogeography of *Pinus*, Cambridge Univ Press, Cambridge, UK, pp 107–121.

Wright JW (1959) Species hybridization in the white pines. For Sci 5: 210–222.

Wright JW (1970) Genetics of eastern white pine. Res Pap WO-9, USDA Forest Service, Washington DC, USA.

Wu HX, Matheson AC (2005) Genotype by environment interaction in an Australia-wide diallel mating experiment: implications for regionalised breeding. For Sci 51: 1–11.

Wu HX, Ying CC, Ju H-B (2005) Predicting site productivity and pest hazard in lodgepole pine using biogeoclimate system and geographic varianble in British Columbia. Ann For Sci 62: 31–42.

Wu HX, Eldridge KG, Matheson AC, Powell MP, McRae TA (2007) Achievements in forest tree improvement in Australia and New Zealand 8: Successful introduction and breeding of radiata pine in Australia. Aust For 70: 215–225.

Xie C-Y, Carlson M, Murphy J (2007) Predicting individual breeding values and making forward selections from open-pollinated progeny test trials for seed orchard establishment of interior lodgepole pine (*Pinus contorta* ssp. *latifolia*) in British Columbia. New For 33: 125–138.

Yanchuk AD, Yeh FC, Dancik BP (1988) Variation of stem rust resistance in a lodgepole pine provenance-family plantation. For Sci 34: 1067–1075.

Yang RC, Yeh FC, Yanchuk AD (1996) A comparison of isozyme and quantitative genetic variation in *Pinus contorta* spp. *latifolia* by Fst. Genetics 142: 1045–1052.

Yeatman CW (1974) The Jack Pine Genetics Program at Petawawa Forest Experiment Station 1950–1970. Can For Serv Publ No 1331, Ottawa, ON, Canada.

Yeatman CW (1975) A progeny test of Ottawa Valley jack pine—6 year results. In: Proc 9th Central States Forest Tree Improvement Conf, pp 71–84.

Yeatman CW, Teich AH (1969) Genetics and breeding of jack and logepole pines in Canada. For Chron 45: 428–433.

Yeh FC, Heaman JC (1987) Estimating genetic parameters of height growth in seven-year-old Douglas-fir from disconnected diallels. For Sci 33: 946–957.

Ying CC, Illingworth K, Carlson M (1985) Geographic variation in lodgepole pine and its implication for tree improvement in British Columbia. In: DM Baumgartner (ed) Lodgepole Pine: The Species and its Management, Coop Ext Serv, Wash State Univ, Pullman, WA, USA, pp 45–53.

Zhelev P, Ekberg I, Eriksson G, Norell L (2003) Genotype environment interactions in four full-sib progeny trials of *Pinus sylvestris* (L.) with varying site indices. For Genet 10: 93–102.

Zhuowen Z (2003) Studies on the pollination characteristics and pollination level of Chinese fir seed orchard. Silvae Genet 53: 7–11.

Zobel BJ (2005) Our roots: the start of tree improvement in the South. In: SE McKeand, B Li (eds) Proc 28th Southern Forest Tree Improvement Conf, 21–23 June, North Carolina State Univy, Raleigh, NC, Publ 50, Southern Forest Tree Improvement Committee, USA, pp 1–5.

Zobel BJ, van Wyk G, Ståhl P (1987) Growing Exotic Forests. John Wiley, NY, USA.

Zsuffa L (1981) Experiences in breeding *Pinus strobus* L. for resistance to white pine blister rust. In: Proc XVII IUFRO World Congr, Division 2, Ibaraki, Japan, pp 181–183.

---

[1]BioSylve Forest Science NZ Limited, 45 Korokoro Road, Lower Hutt 5012, NEW ZEALAND;
e-mail: *tim.mullin@biosylve.com*

[2]Skogforsk (Sävar), Box 3, S-918 21 Sävar, Sweden; e-mail: *bengt.andersson@skogforsk.se*

[3]INRA-Centre de Recherche d'Orléans, 2163, Avenue de la Pomme de Pin, CS 400001 ARDON, F-45075 Orléans Cedex 2, FRANCE;
[a]e-mail: *jean-charles.bastien@orleans.inra.fr*
[b]e-mail: *luc.paques@orleans.inra.fr*

[4]Natural Resources Canada, P.O. Box 10380, Stn. Sainte-Foy, Québec, QC G1V 4C7, Canada;
e-mail: *jeanbeau@nrcan-rncan.gc.ca*

[5]Scion (NZ Forest Research Institute Ltd.), Private Bag 3020, Rotorua 3010, New Zealand;
e-mail: *rowland.burdon@scionresearch.com*

[6]North Carolina State University, Campus Box 8008, Raleigh, NC 27695-8008, USA;
e-mail: *w_dvorak@ncsu.edu*

[7]British Columbia Forest Service, PO Box 9519 Stn Prov Govt, Victoria, B.C. V8W 9C2, Canada;
e-mail: *john.king@gov.bc.ca*

[8]Forest Tree Breeding Centre, 3809-1 Ishi, Juo, Hitachi, Ibaraki 319-1301, Japan;
e-mail: *kontei@affrc.go.jp*

[9]British Columbia Ministry of Forests, and Range, Box 335, Mesachie Lake, B.C. V0R2N0, Canada;
[c]e-mail: *Jodie.Krakowski@gov.bc.ca*
[d]e-mail: *john.russell@gov.bc.ca*

[10]Forest Research, Northern Research Station, Roslin, Midlothian, EH25 9SY, Scotland; e-mail: *steve.lee@forestry.gsi.gov.uk*

[11]North Carolina State University, Campus Box 8002, Raleigh, NC 27695-8002, USA;
e-mail: *steve_mckeand@ncsu.edu*

[12]INRA (Pierroton), 69 route d'Arcachon, 33612 CESTAS Cedex, France;
e-mail: *annie.raffin@pierroton.inra.fr*

[13]Norwegian Forest and Landscape Institute, Høgskoleveien 8, 1432 Ås, Norway;
e-mail: *tore.skroppa@skogoglandskap.no*

[14]British Columbia Ministry of Forests, PO Box 9519 Stn Prov Govt, Victoria, B.C. V8W 9C2, Canada;
[e]e-mail: *michael.stoehr@gov.bc.ca*
[f]e-mail: *alvin.yanchuk@gov.bc.ca*

3

Cytogenetics

M. Nurul Islam-Faridi[1,*] and *C. Dana Nelson*[2]

ABSTRACT

Fluorescent *in situ* hybridization (FISH) has become the most important tool in molecular cytogenetics for positioning and characterizing DNA sequences on chromosomes. Although numerous genetic linkage and quantitative trait loci maps have been reported in conifer species, little progress has been made on developing standard karyotypes capable of identifying individual chromosomes. Standard karyotypes based on a reference set of cytological landmarks will greatly facilitate the integration of genetic and physical maps and their comparisons between species. Such information has the potential to significantly boost our knowledge of genome evolution and to guide marker-based tree breeding and species conservation. To date 25 species of *Pinus* (the most widely studied genus in the conifers), four of *Picea*, six of *Larix* and one each of *Abies* and *Pseudotsuga* have been karyotyped using FISH. A total of 19 loci (10 major and 9 intermediate to minor) of 18S rDNA have been reported in *Pinus taeda*, and this is the highest number of 18S rDNA loci observed in any plant species. Different patterns of 5S and 18S rDNA sites have also been reported among the Pinaceae genera, with *Pinus* showing the most variation.

Keywords: Pinales, Pinaceae, *Pinus*, Fluorescent *in situ* hybridization, Ribosomal DNA, Plant telomere repeat, Karyotype

3.1 Introduction

The conifers (Order Pinales = Coniferales, Division Pinophyta) consist of at least five families and approximately 600 species (Whetton 2005;

[1]US Forest Service, Southern Research Station, Southern Institute of Forest Genetics, Forest Tree Molecular Cytogenetics Laboratory, Texas A & M University, College Station, Texas 77843, USA; e-mail: *nfaridi@tamu.edu*

[2]US Forest Service, Southern Research Station, Southern Institute of Forest Genetics, Harrison Experimental Forest, Saucier, Mississippi 39574, USA.

*Corresponding author

Gymnosperm Database 2010) within more than 60 genera. Of these families, Pinaceae Lindl. is the most widespread ecologically and the most important economically. In addition, most of the genetics, genomics and cytogenetics research have been carried out in this family. In view of this, this chapter will concentrate on the cytogenetics of the Pinaceae conifers.

The Pinaceae consists of about 230 species within 11 genera growing mostly in the temperate to boreal zones of the Northern Hemisphere (Kozubov and Muratova 1986; Farjon 1990). The genera with the largest numbers of species include: *Pinus* (pines, ~110 species), *Abies* (true firs, ~50), *Picea* (spruces, ~30), *Larix* (larches, ~10). Among these genera, *Pinus* has been the most studied cytologically followed by *Picea*, *Larix* and *Abies*. All the genera of the family Pinaceae are diploid with $n = x = 12$ except *Pseudotsuga menziesii* and *Pseudolarix amabilis*. *Pseudotsuga menziesii* is $n = x = 13$, with an uncertain origin for the extra chromosome; although, Silen (1978) hypothesized that one of the progenitor species' ($2n = 2x = 24$) chromosomes broke into two pieces eventually forming two chromosomes. The other exception, *Pseudolarix amabilis*, is the only known polyploid species in the Pinaceae with $2n = 4x = 44$ (Sax and Sax 1933; Stiff 1952).

3.2 Traditional Cytogenetics

Pinaceae genomes are quite large compared to other plant species with 1C DNA contents ranging from 9.5 to 36 pg (e.g., see *http://data.kew.org/cvalues/*). Being diploid with $n = x = 12$, the Pinaceae chromosomes are very large and mostly indistinguishable based on morphometric characteristics such as chromosome length and centromeric index. Numerous attempts have been made to develop karyotypes of species within each of the major genera of Pinaceae since Sax and Sax (1933) first reported chromosome numbers and basic morphology. Little progress has been made using traditional cytological tools, including Giemsa C-banding, chromamycin A_3-banding, and chromosome length measurements, because insufficient variation in chromosome lengths and centromeric indices and inconsistency in banding patterns have made it difficult to unambiguously identify individual chromosomes (Saylor 1961, 1983; Borzan and Papes 1978; MacPherson and Filion 1981). In particular the number and locations of secondary constrictions and C-banding patterns were not consistent enough to develop robust karyotypes (Morgenstern 1962; Mergen and Burley 1964). Focusing on primary constriction locations, Burley (1965) observed inconsistencies in chromosome arm lengths and concluded that accurate analysis of karyotypes in the Pinaceae would be difficult.

3.3 Molecular Cytogenetics

Fluorescent *in situ* hybridization (FISH) has proven to be a powerful technique for localizing major DNA sequence-based features on plant chromosomes (Fig. 3-1). FISH analysis in addition to traditional cytogenetic methods is providing important details of the structural organization of plant genomes (Heslop-Harrison 1991; Leitch and Heslop-Harrison 1992; Leitch et al. 1992), bringing high-quality karyotypes into view. Various cytological markers are necessary to differentiate individual chromosomes in Pinaceae since most of the chromosomes are metacentric or nearly metacentric (e.g., 11 for *Pinus*, 10 or 11 for *Picea*) and similar in length. These markers when well placed on specific chromosomes provide the basis for the development of cytomolecular maps (Heslop-Harrison 1991), where connections can be made between gene positions on genetic (linkage groups) and physical maps and their corresponding positions within karyotypes (chromosomes). In several instances the cytomolecular maps do not correspond to their respective genetic maps due to heterogeneity of recombination (Tanksley et al. 1995; Islam-Faridi et al. 2002; Kim et al. 2005). Recombination can be suppressed in chromosomal regions carrying heterochromatic or repetitive DNA making the genetic map unresolved in these areas.

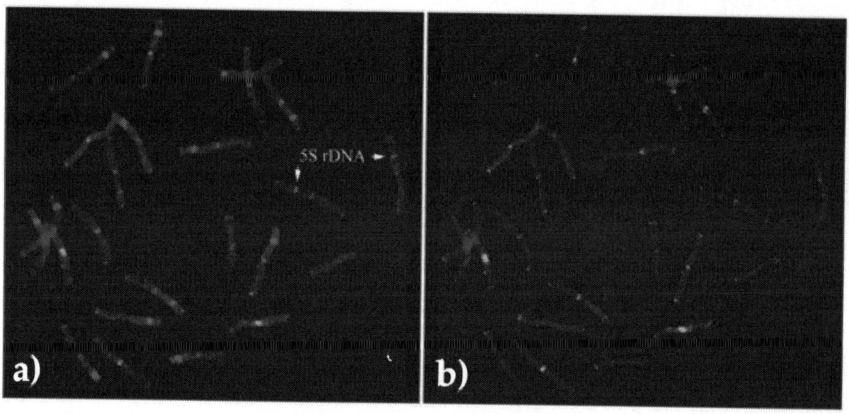

Figure 3-1 Fluorescent *in situ* hybridization images of ribosomal DNAs (18S-28S rDNA, 5S rDNA) and telomere (ATRS) probes on *Pinus echinata* somatic metaphase chromosomes: a) superimposed images of DAPI, Cy3 (red signals, 18S and 5S rDNA sites) and FITC (green signals, ATRS sites) filters; b) super imposed images of DAPI and FITC filters.

Color image of this figure appears in the color plate section at the end of the book.

3.3.1 Ribosomal DNA Gene Families

Ribosomal gene families (18S rDNA and 5S rDNA, with many variations in terminology for the 18S family including 18S-28S, 18S-25S, 18S-5.8S-26S and 45S) provide valuable cytological landmarks for karyotyping and studying the relationships between species and genera. The locations of specific ribosomal gene families in different species can be syntenic, to each other or be located on different chromosomes or on opposite arms of the same chromosomes, as has been found in the Magnoliophyta (i.e., angiosperms) species (Maluszynska and Heslop-Harrison 1993; Hanson et al. 1996; Cerbah et al. 1998; Zoldoš et al. 1999; Taketa et al. 2001; Kulak et al. 2002). To date 25 species of *Pinus*, four of *Picea*, six of *Larix*, and one each of *Abies* and *Pseudotsuga* have been karyotyped using FISH with 18S and 5S rDNA probes, as well as telomere DNA repeat sequence and unique repetitive DNA sequence probes and DNA binding fluorochrome banding patterns (discussed below). Different patterns of rDNA chromosomal distributions have been observed in these species (Table 3-1). From four to 10 interstitial 18S rDNA sites and from one to four 5S rDNA sites have been observed in *Pinus* spp. (Doudrick et al. 1995; Hizume et al. 2002a; Cai et al. 2006; Islam-Faridi et al. 2007). Physical locations of 18S rDNA alone can be used to identify some individual chromosomes. For example in *Pinus taeda*, although similar in length, Chromosome 5 (Ch5) can easily be distinguished from Ch6, since Ch5 harbors an intermediate intensity 18S rDNA site in the centromeric region. A major intensity 18S rDNA site located much further away from the centromere than any of the other 18S sites in the karyotype identifies Ch10. Further, Ch3 and Ch7 can be identified from similar sized Ch2 and Ch4 and Ch8 and Ch9, respectively, based on the differences in 18S rDNA sites near the centromere along with other cytological markers (see Table 4 in Islam-Faridi et al. 2007).

Structural rearrangements in genomes reflect the divergence of species, but the morphometric-determined karyotypes of Pinaceae are very similar making it difficult to differentiate the species, even among less related groups such as genera or even subspecies. In contrast, FISH with rDNA probes shows enough variation among species within sections of *Pinus* and even within subsections to differentiate their karyotypes (Hizume et al. 1992, 2002a; Lubaretz et al. 1996; Liu et al. 2003; Cai et al. 2006). Various distributional patterns of the 18S and 5S rRNA gene families in Pinaceae genomes clearly show differences among genera and as well between more closely related groups within genera (Fig. 3-2). The genus *Pinus* shows more variation in 18S and 5S rDNA distribution than the other Pinaceae genera, supporting conclusions based on the fossil record (Wang et al. 2000) that *Pinus* is most likely the oldest genus in the Pinales, and fitting well with recently proposed molecular dating of the time of divergence of Pinaceae genera (Wang et al. 2000).

Table 3-1 Number and distribution of 18S and 5S rDNA loci in *Pinaceae* genera.

Genera	18S rDNA loci		5S rDNA loci			References
	Interstitial	Centromeric	Interstitial	Syntenic with18S	Opposite arm of 18S	
Pinus						
Subgenus: *Pinus*	5–10	0–3	1–2	0–1P	0–1	Doudrick et al. 1995 Hizume et al. 2002a Islam-Faridi et al. 2007 Bogunic et al. 2010 Liu et al. 2003
Subgenus: *Stobus*	4–10		2–4	1–3D		Cai et al. 2006
Picea	5–8		1	1P		Lubaretz et al. 1996 Brown and Carlson 1997
Larix	2–3		1			Siljak-Yakolev et al. 2002 Lubaretz et al. 1996 Liu et al. 2006 Zhang et al. 2010
Abies*	5		2	1D		Puizina et al. 2008
Pseudotsuga*	2		1 1		1	Hizume et al. 1996 Amarasinghe and Carlson 1998

P = 5S rDNA is proximal to 18S.
D = 5S rDNA is distal to 18S.
*one species only.

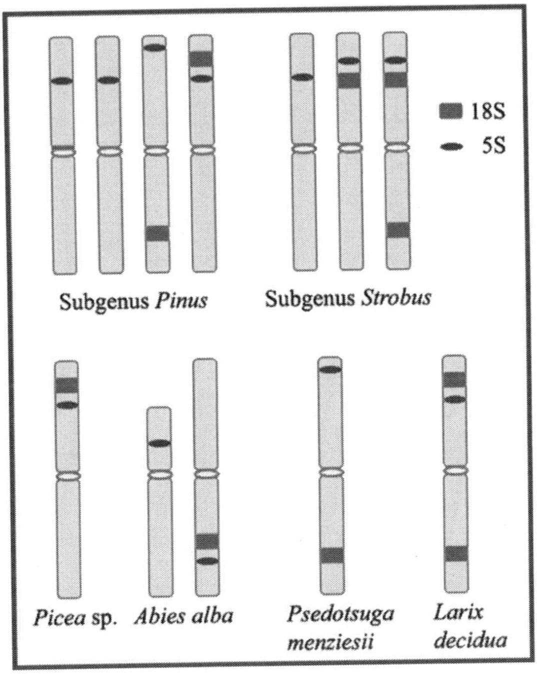

Figure 3-2 Diagrammatic representation of 18S and 5S rDNA loci in different Pinaceae genera. All 18S and 5S rDNA patterns reported in the less extensively studied genera (i.e., *Picea, Abies, Pseudotsuga* and *Larix*) are present in the more extensively studied genus *Pinus* (subgenera, *Pinus* and *Strobus*).

Color image of this figure appears in the color plate section at the end of the book.

3.3.2 Telomere Repeat Sequences

The plant telomere DNA repeat sequence ($(TTTAGGG)_n$) was first reported in *Arabidopsis thaliana* (Richards and Ausubel 1988), and later found in most plant species including green algae and bryophytes (Ganal et al. 1991; Fuchs et al. 1995). Fuchs et al. (1995) first reported an abundance of *Arabidopsis*-type telomere repeat sequence (ATRS) in the interstitial and centromeric regions of *Pinus* chromosomes. Later Hizume et al. (2002a) and Islam-Faridi et al. (2007) used ATRS along with 18S and 5S rDNA and a unique repetitive sequence clone PCSR (proximal CMA band-specific repeat) to characterize and differentiate individual chromosomes of four Eurasian *Pinus* (*P. sylvestris, P. densiflora, P. thunbergii* and *P. nigra*) and one North American (*P. taeda*) species, respectively. Each of these data sets allowed the construction of graphical ideograms providing starting points for cytomolecular maps. Three of the *P. taeda* chromosomes (Ch6, Ch9 and Ch11) had high-copy number (i.e., strong FISH signals) centromeric ATRS sites, while three (Ch4, Ch7 and Ch10) had low-copy centromeric ATRS sites, except that one Ch4 homolog had no ATRS site. The other six chromosomes had intermediate ATRS signals in their centromeric regions suggesting intermediate copy numbers (Islam-Faridi et al. 2007). When carefully characterized (i.e., positioned and quantified), ATRS has been found to be an excellent cytomolecular marker. Furthermore, when ATRS is used in combination with other markers including the rDNA gene families and DNA specific fluorochrome banding (discussed below), informative karyotypes can be developed where individual chromosomes are differentiated from each other and subdivided for more detailed analysis (Fig. 3-3, for details see Islam-Faridi et al. 2007, also see *http://www.srs.fs.usda.gov/sifg/sifg/pinustaeda.html*). In addition it has been suggested that telomere-like DNA sequences located at interstitial and centromeric regions are likely results of chromosomal rearrangements due to inversions, chromosome translocations and fusions (Meyne et al. 1990; Biessmann and Mason 1994; Fuchs et al. 1995; Schubert et al. 1995). Understanding these ATRS patterns among and within genera will provide further information about the divergence of the Pinaceae as well as about the remarkable conservation of their basic karyotypes.

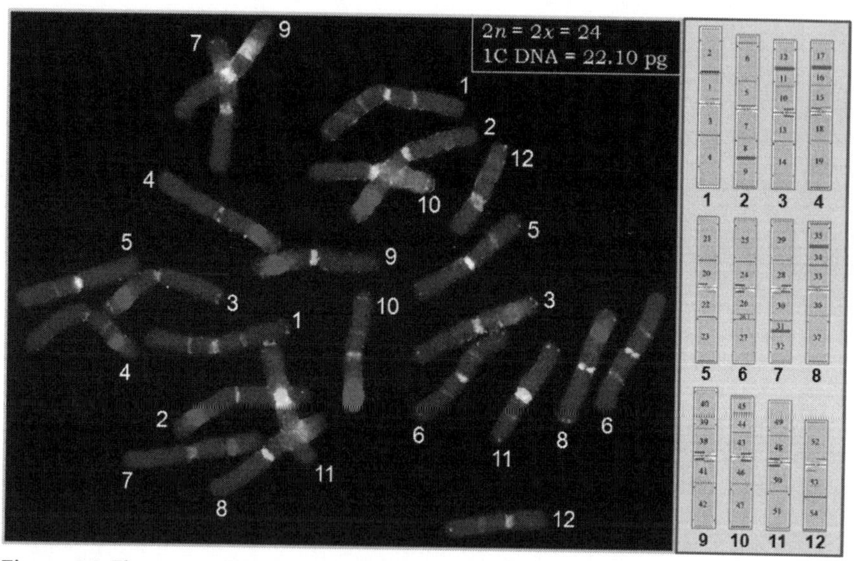

Figure 3-3 Fluorescent *in situ* hybridization image of *Pinus taeda* somatic metaphase chromosomes probed with 18S-28S rDNA (red signals) and *Arabidopsis*-type telomere repeat sequence (green signals). Numbers from 1 to 12 enumerate homologous chromosome pairs. The ideogram of *Pinus taeda* in the right hand side box is based on 108 readings of each measurement (see Islam-Faridi et al. 2007 for details).

Color image of this figure appears in the color plate section at the end of the book.

3.3.3 CMA and DAPI Fluorochrome Banding

Variations observed in fluorescent chromomycin A_3 (CMA) and 4′, 6-diamidino-2-phenylindole (DAPI) banding have facilitated further differentiation of individual chromosomes between and within Pinaceae genera. Heterochromatin regions with high GC-rich or AT-rich DNA sequences bind strongly with CMA or DAPI fluorochromes, respectively (Schweizer 1976), making them useful cytomolecular markers. CMA and DAPI banding were first adapted for use in the Pinaceae by Hizume's Lab (Hizume et al. 1983, 1989). Interestingly, CMA bands always coincide with 18S rDNA sites, but not the other way round. For example, the numbers of CMA bands reported in *Pinus* and *Picea* species are higher than the numbers of 18S rDNA sites, while the numbers and CMA bands and 18S rDNA sites are the same in *Abies alba*, *Larix decidua* and *Pseudotsuga menziesii* (Hizume et al. 1996; Lubaretz et al. 1996; Puizina et al. 2008). Various degrees of DAPI band intensities (minor/weak to major/strong fluorescent signals) are consistently observed primarily at the centromeric regions of *Pinus* species (Fig. 3-4). Both CMA and DAPI banding patterns have been used to identify homologous chromosomes pairs in Pinaceae and these data have facilitated karyotype development and comparison (Hizume et al. 1989, 1990, 2002b; Doudrick et al. 1995; Lubaretz et al. 1996; Siljak-Yakovlev et al. 2002; Islam-Faridi et al. 2007).

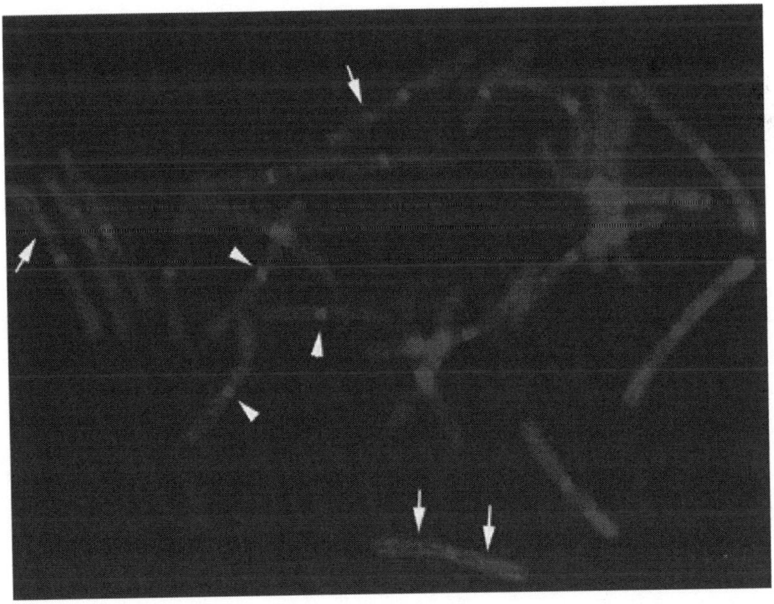

Figure 3-4 An inverted image of DAPI stained *Pinus taeda* chromosomes showing proximal (i.e., centromeric) (triangles) and interstitial (arrows) DAPI (AT-rich) bands.

3.3.4 Unique Repetitive Sequences

Various unique repetitive DNA sequences have been used in identifying individual chromosomes in Pinaceae species. Centromeric-associated (i.e., proximal) DNA sequences were identified by dissecting out centromeric and near-centromeric regions of *P. densiflora*, amplifying (using degenerate oligonucleotide primed (DOP)-PCR) the isolated DNA and cloning the amplified products (Hizume et al. 2001). Of the 31 clones obtained, six contained highly repetitive DNA sequences and showed localized FISH signals on *Pinus* chromosomes. Clone PDCD501 (later named PCSR) when used in FISH, was found to be localized to the proximal CMA bands of 10 of the 12 chromosomes of *Pinus densiflora* (Hizume et al. 2001). Clone PDCD159, on the other hand, hybridized at the proximal DAPI band of the remaining two chromosomes of *Pinus densiflora*. Four other clones strongly hybridized to the secondary constrictions (i.e., major 18S rDNA sites) and produced weak signals at the centromeric regions, and one of the four clones showed homology to the 26S rRNA gene. Clearly these repetitive clones are useful as FISH probes providing data to study the evolution of the genus *Pinus* (Hizume et al. 2001) as well as to enhance the existing karyotypes by improving the resolution between chromosomes. Demonstrating these attributes, the PCSR clone along with rDNA and ATRS markers were used for comparative karyotypic analyses of four *Pinus* species (Hizume et al. 2002a).

Three unique repetitive sequence clones (PATR140, PAF1 and 1PABCD6) were identified in *Picea abies* (Vischi et al. 2003) using a randomly sheared genomic library. Using these clones as FISH probes, Vischi et al. (2003) unambiguously identified all 12 chromosomes of *P. abies*. Clone PAF1 was found to be associated with the 18S rDNA gene. From the same library, Sarri et al. (2008) identified three clones (PAF1, PAG004P22F and PAG004E03C) containing satellite DNA sequences. These three clones were also used as FISH probes to assess their physical distribution across the chromosomal complement with their patterns being such that each chromosome could be uniquely identified. Sarri et al. (2008) suggest that the structural organization of centromeres is not simple and it varies within species. However, significant discrepancies were noticed when comparing this *Picea abies* karyotype with the one developed by Vischi et al. (2003). For example Ch10 and Ch6 of the earlier karyotype (Vischi et al. 2003) corresponds with Ch3 and Ch7of the later one (Sarri et al. 2008), respectively, and the chromosome arm identities of the secondary constrictions are also reversed.

3.4 Conclusions

Well developed karyotypes that uniquely distinguish each chromosome and chromosome arm and their graphical representation as ideograms are critical tools in moving towards cytomolecular mapping for genome-based applications. We and others have noted frequent instances where the ideograms within genera or even within species differ for no identified or apparent reason (Hizume et al. 2002a; Liu et al. 2003; Islam-Faridi et al. 2007), except where the differences appear to be due to variable chromosome preparations and under sampling in data collection. To overcome this limitation, it will be important for the conifer and especially Pinaceae community to develop standardized, reference karyotypes for each major taxa. Within *Pinus* the appropriate level of taxonomic discrimination at present would seem to be at the subsection level. Standardization should include high-quality chromosome preparations with minimal chromosome distortion and absence of cell wall debris along with multiple cells measured from multiple genotypes per taxa. Reference markers should include the major and minor 18S and 5S rDNA sites, major and intermediate ATRS bands (both centromeric and interstitial), and chromosome specific probes as they are identified. The later class of markers is now more readily within grasp as large-scale genome projects are progressing in *Pinus* and *Picea* including the development and analysis of bacterial artificial chromosome (BAC) libraries.

FISH has revolutionized plant cytogenetics and continues to yield new insights into the genomes of Pinaceae genera and species. Implementing FISH at a large-enough scale to cover the many Pinaceae taxa and to be confident of probe locations and intensities is still a challenge. Continued improvements in chromosome preparation techniques will aid in developing the large-scale capabilities needed to meet the challenge along with continued enhancements in microscopy technology and computerized digital analyses. Implementing standardization methods across laboratories and developing a standardized set of reference markers will facilitate the development of well-referenced ideograms and cytomolecular maps. These maps will be invaluable for whole genome mapping and sequencing. Recent advances in flow sorting (e.g., Li et al. 2004; Šafář et al. 2004) and laser capture microdissection (e.g., Zhou and Hu 2007) offer great promise in providing Pinaceae cytogeneticists new tools for isolating and analyzing individual chromosomes, chromosome arms or pieces of arms.

Acknowledgement

We thank colleagues Kostya Krutovsky, Tom Byram and George Hodnett for valuable comments and suggestions regarding an earlier draft of this manuscript and Michael Robinson and Robert Eaton for developing the animated version of the Pinus taeda karyotype.

References

Amarasinghe V, Carlson JE (1998) Physical mapping and characterization of 5S rRNA genes in Douglas-Fir. J Hered 89: 495–500.

Biessmann H, Mason JM (1994) Telomere repeat sequences. Chromosoma 103: 154–161.

Bogunić F, Siljak-Yakovlev S, Muratović E, Ballian D (2010) Different karyotype patterns among allopatric *Pinus nigra* (Pinaceae) populations revealed by molecular cytogenetics. Plant Biol no. doi: 10.1111/j.1438-8677.2010.00326.x: 1–7.

Borzan Z, Papes D (1978) Karyotype analysis in *Pinus*: A contribution to the standardization of the karyotype analysis and review of some applied techniques. Silvae Genet 27: 144–150.

Brown GR, Carlson JE (1997) Molecular cytogenetics of the genes encoding 18S-5.8S-26S rRNA and the 5S rRNA in two species of spruce (*Picea*). Theor Appl Genet 95: 1–9.

Burley J (1965) Karyotype analysis of Sitka Spruce, *Picea sitchensis* Bong. Carr Silvae Genet 14: 127–132.

Cai Q, Zhang D, Liu ZL, Wang XR (2006) Chromosomal localization of 5S and 18S rDNA in five species of subgenus *Strobus* and their implications for genome evolution of *Pinus*. Ann Bot 97: 715–722.

Cerbah M, Coulaud J, Siljak-Yakovlev S (1998) rDNA organization and evolutionary relationships in the genus *Hypochaeris* (Asteraceae). J Hered 89: 312–318.

Doudrick RL, Heslop-Harrison JS, Nelson CD, Schmidt T, Nance WL, Schwarzacher T (1995) Karyotype of slash pine (*Pinus elliottii* var. *elliottii*) using patterns of fluorescence *in situ* hybridization and fluorochrome banding. J Hered 86: 289–296.

Farjon A (1990) Pinaceae: Drawings and descriptions of the genera. Rengum Vegetabile 121: 1–330.

Fuchs J, Brandes A, Schubert I (1995) Telomere sequence localization and karyotype evolution in higher plants. Plant Syst Evol 196: 227–241.

Ganal MW, Lapitan NL, Tanksley SD (1991) Macrostructure of the tomato telomeres. Plant Cell 3: 87–94.

Gymnosperm Database (2010) *http://www.conifers.org/zz/pinales.htm* (accessed 22 Sept 2010).

Hanson RE, Islam-Faridi MN, Percival EA, Crane CF, McKnight TD, Stelly DM, Price HJ (1996) Distribution of 5S and 18S-28S rDNA loci in a tetraploid cotton (*Gossypyum hirsutum* L.) and its putative diploid ancestors. Chromosoma 105: 55–61.

Heslop-Harrison JS (1991) The molecular cytogenetics of plants. J Cell Sci 100: 15–21.

Hizume M, Ohgiku A, Tanaka A (1983) Chromosome banding in the genus *Pinus* I. Identification of chromosomes in *P. nigra* by fluorescent banding method. Bot Mag Tokyo 96: 273–276.

Hizume M, Ohgiku A, Tanaka A (1989) Chromosome banding in the genus *Pinus* II. Interspecific variation of fluorescent banding patterns in *P. densiflora* and *P. thunbergii*. Bot Mag Tokyo 102: 25–36.

Hizume M, Arai M, Tanaka A (1990) Chromosome banding in the genus *Pinus* III. Fluorescent banding pattern of *P. luchuensis* and its relationships among Japanese diploxylon pines. Bot Mag Tokyo 103: 103–111.

Hizume M, Ishida F, Murata M (1992) Multiple locations of ribosomal RNA genes in chromosomes of pines, *Pinus densiflora* and *P. thunbergii*. Jap J Genet 67: 389–396.

Hizume M, Kuzukawa Y, Kondo TT (1996) Physical mapping of 5S rDNA locus on chromosomes in *Pseudotsuga menziesii*, Pinaceae. Kromosomo 2(83–84): 2901–2908.

Hizume M, Shibata F, Maruyama Y, Kondo T (2001) Cloning of DNA sequences localized on proximal fluorescent chromosome bands by microdissection *in Pinus densiflora* Sieb & Zucc. Chromosoma 110: 345–351.

Hizume M, Shibata F, Matsusaki Y, Garajova Z (2002a) Chromosome identification and comparative karyotypic analyses of four *Pinus* species. Theor Appl Genet 105: 491–497.

Hizume M, Shibata F, Matsumoto A, Maruyama Y, Hayashi E (2002b) Tandem repeat DNA localization of the proximal DAPI bands of chromosomes in *Larix*, Pinaceae. Genome 45: 777–783.

Islam-Faridi MN, Childs KL, Klein PE, Hodnett G, Menz MA, Klein RR, Rooney WL, Mullet JE, Stelly DM, Price HJ (2002) A molecular cytogenetic map of sorghum chromosome 1: Fluorescence *in situ* hybridization analysis with mapped bacterial artificial chromosomes. Genetics 161: 345–353.

Islam-Faridi MN, Nelson CD, Kubisiak TL (2007) Reference karyotype and cyto-molecular map for loblolly pine (*Pinus taeda* L.). Genome 50: 241–251.

Kim J-S, Islam-Faridi MN, Klein PE, Stelly DM, Price HJ, Klein RR, Mullet JE (2005) Comprehensive molecular cytogenetics analysis of sorghum genome architecture: Distribution of euchromatin, heterochromatin, genes and recombination in comparison to rice. Genetics 171: 1963–1976.

Kozubov GM, Muratova EN (1986) Contemporary gymnosperms (morphology systematic review and karyology). Nauka, Leningrad (in Russian) P192.

Kulak S, Hasterok R, Maluszynska J (2002) Karyotype of *Brassica* ampliploids using 5S and 25S rDNA as chromosome markers. Hereditas 136: 144–150.

Leitch AR, Heslop-Harrison JS (1992) Physical mapping of the 18S-5.8S-26S rRNA genes in barley by *in situ* hybridization. Genome 35: 1013–1018.

Leitch AR, Mosgoller W, Shi M, Heslop-Harrison JS (1992) Different patterns of rDNA organization at interphase in nuclei of wheat and rye. J Cell Sci 101: 751–757.

Li L, Arumuganathan K, Gill KS, Song Y (2004) Flow sorting and microcloning of maize chromosome 1. Hereditas 141: 55–60.

Liu B, Zhang S-G, Zhang Y, Lan T-Y, Qi L-W, Song W-Q (2006) Molecular cytogenetic analysis of four *Larix* species by bicolor fluorescence *in situ* hybridization and DAPI banding. Int J Plant Sci 167: 367–372.

Liu Z, Zhang D, Hong D, Wang X (2003) Chromosomal localization of 5S and 18S-5.8S-25S ribosomal DNA sites in five Asian pines using fluorescence *in situ* hybridization. Theor Appl Genet 106: 198–204.

Lubaretz O, Fuchs J, Ahne R, Meister A, Schubert I (1996) Kryotyping of three Pinaceae species via flurescent *in situ* hybridization and computer-aided chromosome analysis. Theor Appl Genet 92: 411–416.

MacPherson P, Filion WG (1981) Karyotype analysis and the distribution of constructive heterochromatin in five species of *Pinus*. J Hered 72: 193–198.

Maluszynska J, Heslop-Harrison JS (1993) Physical mapping of rDNA loci in *Brassica* species. Genome 36: 774–781.

Mergen F, Burley J (1964) *Abies* karyotype analysis. Silvae Genet 13: 63–68.

Meyne J, Baker RJ, Hobart HH, Hsu TC, Ryder OA et al. (1990) Distribution of non-telomeric sites of the $(TTAGGG)_n$ telomeric sequence in vertebrate chromosomes. Chromosoma 99: 3–11.

Morgenstern EK (1962) Note on chromosome morphology in *Picea rubens* Sarc and *Picea mariana* Mill. BSP. Silvae Genet 11: 163–164.

Puizina J, Sviben T, Krajačič-Sokol I, Zoldoš-Pećnik V, Siljak-Yakovlev S, Papeš D, Besendrofer V (2008) Cytogenetic and molecular characterization of the *Abies alba* genome and its relationship with other members of the Pinaceae. Plant Biol 10: 256–267.

Richards EJ, Ausubel FM (1988) Isolation of a higher eukaryotic telomere from *Arabidopsis thaliana*. Cell 53: 127–136.

Šafář J, Bartoš J, Janda J, Bellec A, Kubaláková M et al. (2004) Dissecting large and complex genomes: flow sorting and BAC cloning of individual chromosomes from bread wheat. Plant J 39: 960–968.

Sarri V, Minelli S, Panara F, Morgante M, Jurman I, Zuccolo A, PG Cionini PG (2008) Characterization and chromosomal organization of satellite DNA sequences in *Picea abies*. Genome 51: 705–713.

Saylor LC (1961) A karyotype analysis of selected species of *Pinus*. Silvae Genet 10: 77–84.

Saylor LC (1983) Karyotype analysis of the genus *Pinus*-subgenus *Strobus*. Silvae Genet 32: 119–121.

Sax K, Sax HJ (1933) Chromosome number and morphology in the conifers. J Arnold Arbor 14: 356–375.

Schubert I, Rieger R, Fuchs J (1995) Alteration of basic chromosome number by fusion-fission cycles. Genome 38: 1289–1292.

Schweizer D, (1976) Reverse fluorescent chromosome banding with chromamycin and DAPI. Chromosoma 58: 307–324.

Silen RR, (1978) Genetics of Douglas-fir. USDA Forest Service Research Paper, WO-35.

Siljak-Yakovlev S, Cerbah M, Coulaud J, Stoian V, Brown SC (2002) Nuclear DNA content, base composition, heterochromatin and rDNA in *Picea omorika* and *Picea abies*. Theor Appl Genet 104: 505–512.

Stiff ML (1952) The geographical distribution and cytology of the Coniferales. PhD Thesis, Univ of Virginia, Charlottesville, VA, USA.

Taketa S, Ando H, Takeda K, Bothmer RV (2001) Physical locations of 5S and 18S-25S rDNA in Asian and American diploid *Hordeum* species with the I genome. Heredity 86: 522–530.

Tanksley SD, Ganal MW, Martin GB (1995) Chromosome landing: A paradigm for map based gene cloning in plants with large genomes. Trends Genet 11: 63–68.

Vischi M, Jurma I, Bianchi G, Morgante M (2003) Karyotype of Norway spruce by multicolor FISH. Theo Appl Genet 107: 591–597.

Wang X-Q, Tank DC, Sang T (2000) Phylogeny and divergence times in Pinaceae: Evidence from three genomes. Mol Biol Evol 17: 773–781.

Whetten RW (2005) Conifers. Encyclopedia of Life Sciences. John Wiley, Hoboken, New Jersey, USA, pp 1–3.

Zhang SG, Yang WH, Han SY, Han BT, Li MX, Qi LW (2010) Cytogenetic analysis of reciprocal hybrids and their parents between *Larix leptolepis* and *Larix gmelinii*: implications for identifying hybrids. Tree Genet Genomes 6: 405–412.

Zhou R-N, Hu Z-M (2007) The development of chromosome microdissection and microcloning technique and its applications in genomic research. Curr Genom 8: 67–72.

Zoldoš V, Papeš D, Cerbah M, Besendorfer V, Siljak-Yakovlev S (1999) Molecular-cytogenetic studies of ribosomal genes and heterochromatin reveal conserved genome organization among 11 *Quercus* species. Theor Appl Genet 99: 969–977.

4

Neutral Patterns of Genetic Variation and Applications to Conservation in Conifer Species

Francesca Bagnoli,[1,a] *Bruno Fady,*[2,c] *Silvia Fineschi,*[1,b] *Sylvie Oddou-Muratorio,*[2,d] *Andrea Piotti,*[3] *Federico Sebastiani*[4,e] and *Giovanni G. Vendramin*[4,f,*]

ABSTRACT

This chapter describes how neutral genetic markers can be used for the study of population and conservation genetics, phylogeography and gene flow in conifers. It includes a comprehensive review of the studies performed in these research fields. The chapter starts with a review of the different kinds of neutral genetic markers most frequently used in conifers in the recent literature. In the second part, it describes how variation is organized within and among natural populations at the three conifer genomes (chloroplast, mitochondrial and nuclear). In the third part, it highlights how stochastic processes have shaped this organization focusing on two large areas of investigation in population genetics: phylogeography and gene flow. Finally, it demonstrates

[1]Plant Protection Institute, CNR, Via Madonna del Piano 10, 50019 Sesto Fiorentino (FI), Italy;
[a]e-mail: *bagnoli@ipp.cnr.it*
[b]e-mail: *fineschi@ipp.cnr.it*
[2]INRA, UR629, Ecologie des Forêts Méditerranéennes, Domaine Saint Paul, Site Agroparc, 84914 Avignon, France;
[c]e-mail: *bruno.fady @ avignon.inra.fr*
[d]e-mail: *sylvie.oddou@avignon.inra.fr*
[3]Department of Environmental Sciences, University of Parma, Viale Usberti 11 / A, 43100 Parma, Italy; e-mail: *andrea.piotti@nemo.unipr.it*
[4]Plant Genetics Institute, CNR, Via Madonna del Piano 10, 50019 Sesto Fiorentino (FI), Italy;
[e]e-mail: *federico.sebastiani@unifi.it*
[f]e-mail: *giovanni.vendramin@igv.cnr.it*
*Corresponding author

that neutral genetic markers and the information they generate are fundamental for the conservation and management of genetic resources. This chapter is addressed to plant molecular geneticists as well as plant breeders in the public and private sectors.

Keywords: neutral markers; diversity; differentiation; phylogeography; gene flow; conservation; conifers.

4.1 Introduction

Conifer forests are important reservoirs of biological diversity (at gene, individual, and community level) as a consequence of their complex history and environmental variation at local and regional scales. Conifers are keystone organisms in European ecosystems; they directly support rich plant and animal communities that rely on them and mediate nutrient and water ecological cycles. Impacts of global change on forests are expected to be acute, resulting in notable changes in species range, ecosystem functioning and interactions among species. Because they are sessile but long-lived, trees will either disappear, have to disperse to other places via their seeds and pollen or be able to adapt *in situ* over a reduced number of generations. To adapt, trees will rely more on standing genetic variation and recombination than on new mutations (Aitken et al. 2008).

The genomics revolution of the last 10 years has improved our understanding of the genetic make-up of living organisms. Together with the achievements represented by complete genome sequences for an increasing number of species, high throughput and parallel approaches are available for the analysis of transcripts, proteins, insertional and chemically induced mutants. All this information facilitates the understanding of the function of genes in terms of their relationship to the phenotype. Despite its great relevance, such an understanding could be of little value to population and conservation genetics if it only elucidates the relationship between genetic variant and a mutant phenotype but fails to elucidate the relationship between genetic variation in gene sequences and phenotypic variation in traits. The relationships between complex trait variation and molecular diversity of genes can be studied based on a genomic approach. However, the identification of genes responsible for trait variation is still a difficult task, especially in long lived organisms such as forest trees. Work in model plant species has started to unveil an ever-increasing number of genes involved in the determination of traits of adaptive significance such as phenology and abiotic stress tolerance/resistance. These progresses will hopefully allow ecological and conservation genetics to analyze directly variation in genes involved in adaptive processes rather than in neutral markers as was traditionally done in the past.

The increased availability of genomic tools, however, does not hinder the role of traditional neutral genetic markers in population and ecological genetics. Patterns of genetic diversity, that is the variation in allelic frequencies within and among populations, is the intricate result of mutation, selection, drift, migration and the mating system, i.e., demographic factors and long term historical processes. A major challenge of population genetics is to disentangle the "functional" fraction of genetic diversity that is contributed by variants causing changes in metabolic or phenotypic traits and affecting individual fitness, from the "neutral" fraction of genetic diversity, contributed by variants that are not subjected to positive, negative or balancing selection (Marsjan and Oldenbroek 2007). This challenge cannot be addressed without neutral genetic markers, because (1) neutral genetic markers are the most efficient tools for making inferences on stochastic processes affecting natural population evolution, such as migration and bottlenecks, and (2) because the comparison of genetic diversity of populations at neutral and adaptive loci will make it possible to identify key ecological factors responsible for the observed structures and to finally elucidate selective processes (Gebremedihin et al. 2009).

In this chapter, we focus on neutral genetic markers that are the most common tools in studies on phylogeny, gene flow, spatial structures of populations, and conservation genetics (Hoffmann and Willi 2008). We first review the different kinds of neutral genetic markers most frequently used in conifers in the recent literature. Then, we describe how variation at the three conifer genomes (chloroplast, mitochondrial and nuclear), with their different modes of transmission, is organized within and among natural populations. Finally, we emphasize how stochastic processes have shaped this organization focusing on two large areas of investigation in population genetics, phylogeography and gene flow.

Plants in general and conifers in particular offer excellent models to investigate how past and present ecological and demographic processes have shaped genetic diversity. In this chapter, we demonstrate that neutral genetic markers and the information they generate are fundamental for the conservation and management of genetic resources, for example for identifying species and populations under threat and key regions deserving priority for conservation (Petit et al. 2003).

4.2 Neutral Genetic Diversity and Genetic Markers

Genetic markers are fragments of DNA (or some direct expression of these) that are simple enough to be clearly described, that are heritable (transmitted from one generation to the next), that can be characterized at low cost for a great number of individuals, and that show differences among individuals in a sample. Genetic markers differ in their DNA sequences because of mutations caused by insertion, deletion, duplication or inversion of nucleotides. Genetic markers (or loci) can be inherited bi-parentally at fecundation. They are then found in the nucleus and contain genetic information both from the father and the mother of the individual studied. This double information is either identical (homozygous individual) or different (heterozygous individual) at any given locus in the genome. A set of different loci constitute a genotype. Genetic markers can also be inherited uni-parentally, either from the mother (usually mitochondrial DNA or chloroplast DNA in most angiosperms) or from the father (chloroplast DNA in most conifers, and in the Pinaceae in particular). Variants at different locations on the mitochondrial or chloroplast genomes constitute haplotypes.

The advent of DNA-based molecular markers has represented an important improvement in population genetic studies. A large number of such markers are now available, in particular those based on amplification via PCR (polymerase chain reaction) of genomic DNA fragments (Table 4-1). Molecular marker technology has developed rapidly over the last decades and two forms of sequence based marker, simple sequence repeats (SSRs) and single nucleotide polymorphisms (SNPs), now predominate applications in modern genetic analysis (Duran et al. 2009).

SSRs (also called microsatellites) are made of tandem repeats of simple DNA base-pair motifs, typically 1–6 base-pair long, each repeated from 10 to 30 times on average (Jarne and Lagoda 1996). They can be selectively amplified from nuclear (nSSR) and organelle (both chloroplast, cpSSR and mitochondrial, mtSSR) genomes. Nuclear SSRs are codominant whereas cpSSRs and mtSSRs are haploid. SSRs have a wide range of applications such as population genetic studies, paternity analysis, genotyping and genetic mapping, systematic taxonomy, molecular evolution, hybrid selection (Morgante and Olivieri 1993). SSRs are abundant, dispersed throughout the genome, highly reproducible and highly polymorphic in most organisms, including plants, and thus serve as a universal source of highly informative genetic markers. Although SSRs generally behave as neutral markers (Awadalla and Ritland 1997), it was demonstrated that they can also be involved in gene expression, regulation and function (Gupta et al. 1994; Kashi et al. 1997).

The most recently developed genetic markers are SNPs. They correspond to a mutation at a single nucleotide and can therefore be considered the finest resolution of a DNA sequence. SNPs are generally abundant in populations and have a low mutation rate (Duran et al. 2008). They are widely distributed throughout genomes (Halushka et al. 1999), although their occurrence and distribution varies among species. SNPs are evolutionary stable and their low mutation rate makes them good markers for understanding genome evolution (Duran et al. 2009). SNPs are suitable

Table 4-1 Methods most commonly used to assess neutral genetic diversity in natural populations. All traditional methods rely on electrophoresis although new high-throughput sequencing methods do not.

	Allozymes	RFLP	AFLP	PCR-RFLP	SSR	SNP	Sequencing
Principle	Differential mobility of amino acids	Presence of restriction sites	PCR of random DNA fragments after restriction by endonucleases	Restriction by endonuclease of specific PCR fragments	PCR of tandemly repeated DNA regions	PCR of specific single nucleotide polymorphic sites	Linear sequence of specific PCR products
Type of polymorphism	Substitution of electrically charged non synonymous amino acids	DNA Fragment length polymorphism due to substitutions, indels, inversions of nucleotides	DNA Fragment length polymorphism due to substitutions, indels, inversions of nucleotides	DNA Fragment length polymorphism due to substitutions, indels, inversions of nucleotides	DNA Fragment length polymorphism due to modifications of number of repeats (indels)	DNA sequence changes due to single point substitutions of nucleotides	DNA sequence changes due to substitutions of nucleotides, indels, etc.
Frequency of sites in the genome	Low	High	High	High	Medium to high depending on the genome	Very high	Low to very high depending on the genome
Level of polymorphism	Low	Medium	Low	Medium	High	Low	Low to very high depending on the genome
Quantity of material needed	A few mg of fresh tissue	10 mg of DNA	1–2 mg of DNA	100 ng of DNA	50 ng of DNA	25 ng of DNA	25 ng of DNA

Table 4-1 contd....

Table 4-1 contd....

	Allozymes	RFLP	AFLP	PCR-RFLP	SSR	SNP	Sequencing
Type of dominance on the nuclear genome	Co-dominant (mostly nuclear)	Co-dominant (mostly nuclear)	Dominant (can be carried by all genomes)	Co-dominant (but often used on organelle DNA)	Co-dominant (but often used on organelle DNA)	Co-dominant (possibly found on organelle DNA)	Co-dominant (can be used for organelle DNA)
Possibility to automatize	No or very limited	Limited	Somewhat	Limited	Yes	Yes	Yes
Replicability	High	High	Medium	High	High	High	High
Cost	Low	Medium to high	Low	Medium to high	Medium to high	High	Medium to high

markers for several purposes, such as for population genetic studies and construction of ultra high-density genetic maps. In most organisms studied to date, SNPs are more prevalent in the non-coding than in the coding regions of the genome (Soleimani et al. 2003).

As both SSR and SNP markers often require extensive and costly molecular procedures for their identification and characterization, the more versatile AFLP (amplified fragment length polymorphism, Vos et al. 1995) markers, for which no knowledge of the genes are required prior to their use as population genetic markers, are in some cases used instead in ecological studies (genome scan approach, Alvarez et al. 2009).

The choice of the most appropriate markers for a given study depends on its objectives (Table 4-2) as well as many other factors, among which species' life history traits and availability of information and genome origin (cpDNA, mtDNA and nuclear DNA) are the most important ones. Having the possibility to choose among three types of genomes, either bi-parentally (nuclear genome) or uni-parentally (organelles) inherited, is a key feature in plants, unavailable in other eukaryotes. In angiosperms, both mitochondria and chloroplasts are maternally inherited, although exceptions are also known (Petit and Vendramin 2007). In gymnosperms, however, mitochondria are generally maternally inherited (therefore dispersed through seed), whereas chloroplasts are paternally inherited (therefore dispersed through pollen and then through seeds), although exceptions are known: for example, both chloroplasts and mitochondria are paternally inherited in *Cupressus*, *Araucaria*, *Podocarpus*, *Taxodium* and *Metasequoia* (Whittle and Johnston 2002).

4.2.1 Genetic Markers from the Chloroplast Genome

The chloroplast (also called plastid) genome derives from a cyanobacterial ancestor that was captured early-on in the evolution of the eukaryotic cell. As the site in the cell where photosynthesis takes place, chloroplasts are responsible for much of the world's primary production, making chloroplasts essential to the life of plants as well as all other organisms. Because of its small size and limited number of repeated elements, the chloroplast genome was the first plant genome to be characterized.

The conifer plastid genome ranges an average size of 120–160 kb, and contains about 130 genes. Chloroplast genomes are sufficiently large and complex to include structural and point mutations (Cronn et al. 2008) and consists of a single, circular chromosome typically organized into three regions: a large region of single copy genes (LSC), a small region of single copy genes (SSC), and two copies of an inverted repeat (IR A and IR B) that separate the two regions of single copy genes. Chloroplast genomes have a highly conserved gene content and organization (Cronn et al. 2008). Because

Table 4-2 Which genetic method for which purpose? A ranking of methods from not appropriate (–) to most appropriate (+++) according to the assessment needed.

Type of assessment needed/method used	Allozymes	RFLP/PCR-RFLP of nDNA	RFLP/PCR-RFLP of organelle DNA	SNP	AFLP	SSR of nDNA	SSR of organelle DNA	Sequencing of nDNA	Sequencing of organelle DNA
Species identification	+	–	–	–	–	–	+	++	+++
Hybridization	++	−++	++	++	++	+	++	++	++
Phylogeography	–	–	++	–	–	+	+	–	+++
Differentiation among populations	+++	++	++	++	++	++	++	++	+++
Gene flow	++	++	–	++	–	+++	++	+++	+
Within population genetic diversity	++	−++	+	+++	++	++	++	+++	++
Individual identification	–	++	–	+++	+++	+++	++	+++	–

the only real difference among plastid genomes is related to the repeated sequences (IR), the plastid genomes are classified as: a) Group I genomes, which lack the large (20–25 kb) inverted repeat that characterizes most land plants (certain legumes and conifers, see the pioneering paper of Strauss et al. (1988) on the chloroplast genome structure of two conifers, *Pseudotsuga menziesii* and *Pinus radiata*); b) Group II genomes, which contain inverted repeats (almost all plants); c) Group III sort of oddball genomes, which have tandem repeats (*Euglena*, a photosynthetic protist).

In the 1980s completely sequenced chloroplast genomes became available, thus originating the development of consensus (if not universal) primers of interest for intraspecific studies. The first conifer chloroplast genomes completely sequenced were that of *P. thunbergii* (Wakasugi et al. 1994) (119,707 bp) and *P. koraiensis* (116,866 bp); both genomes are significantly smaller than those of most angiosperms (Steane 2005). Recently, the improvements in second generation sequencing, made it possible to assess genetic diversity at the genome scale and to sequence at a fraction of the time and cost of traditional approaches (Duran et al. 2009). In this way, the chloroplast genome of one spruce species (*P. stichensis*) and seven pine species (*P. contorta, P. lambertiana, P. gerardiana, P. krempfii, P. longaeva, P. monophylla,* and *P. nelsonii*) were sequenced by Cronn et al. (2008). The genome sizes were very similar to the known genome sizes for *P. thunbergii* and *P. koraiensis*.

The availability of full chloroplast genome sequences allowed designing primers in conserved (generally coding) regions separated by more variable regions (Petit and Vendramin 2007). The region amplified are either large and may be used in combination with restriction enzymes (usually 4-bp cutters) (Demesure et al. 1995; Dumolin-Lapègue et al. 1997) as is often the case in angiosperms, or are very small (< 200 bp) but contain potentially variable mononucleotide single strand repeats (cpSSRs) (Vendramin et al. 1996; Weising and Gardner 1999) as is often the case in conifers.

The occurrence of mononucleotide repeats within the chloroplast genome of seed plants, bryophytes and algae was firstly documented by Powell et al. (1995). Furthermore, Powell et al. (1995) also demonstrated that simple mononucleotide repeats in the chloroplast genome of conifers exhibit length variation, and that polymorphism within these regions may be used to study both intra- and interspecific variability. Conifer cpSSR markers have a high degree of transferability between species and primers designed in one species can often be used in closely related species (Vendramin et al. 1996). The haploid state and uni-parental transmission gives chloroplast genes and genomes an effective population size approximately one-half of a nuclear locus. This has the effect of making chloroplast genes more responsive than nuclear genes to stochastic processes like drift and founder events, a property that has been exploited for testing hypotheses of seed (and less commonly pollen) dispersal, migration/colonization routes, intra-specific differentiation and inter-specific introgression (Cronn et al. 2008).

4.2.2 Genetic Markers from the Mitochondrial Genome

Mitochondria are the site of energy metabolism and therefore play a fundamental role within the eukaryotic cell, although in plants, chloroplasts represent a second energy-generating system. Mitochondria are integrated endosymbionts, originating from a large group of eubacteria (Gray 1993). This origin, also common to chloroplasts, confers a limited self sustenance to the organellar genetic systems from the nucleus. Mitochondria, as chloroplasts, have their own set of unique genetic rules, including uni-parental inheritance, somatic recombination, vegetative segregation, gene expression and genome organization. Mitochondrial and chloroplast gene functions complement those, but are not independent, of nuclear genes. The mitochondrial genome encodes only a fraction (estimated at 20–30 proteins) of the gene products required for its function, whereas their vast majority is encoded by the nucleus. The presence of genetic information within the mitochondria and chloroplasts requires that some form of coordinate gene expression must occur with the nucleus. In comparison to the chloroplast genome, the size of the mitochondrial genome is quite variable. Even within one plant family a 10-fold difference in mitochondrial genome size can be observed.

Studies on mtDNA variation are less numerous than studies on cpDNA. However, since the mitochondrial genome represents the only possibility to trace seed flow for most conifers, several studies on the phylogeography of Pinaceae have been carried out using mtDNA markers. Jeandroz et al. (2002) constructed a set of mtDNA primers designed for Norway spruce that can be used also for other *Picea, Abies* and *Pinus* species. Most conifer studies to date have dealt with Northern Hemisphere species, e.g., spruces (Gugerli et al. 2001; Sperisen et al. 2001; Jaramillo-Correa et al. 2003, 2004, 2006), pines (Sinclair et al. 1998; Godbout et al. 2005, 2008) and firs (Tsumura and Suyama 1998; Jaramillo-Correa et al. 2008).

4.2.3 Genetic Markers from the Nuclear Genome

The nuclear genome of gymnosperms, and particularly that of conifers, is very large. Estimates of genome size are available for 12 gymnosperm families, with the most numerous estimates found for the Pinaceae, followed by the Cupressaceae and Podocarpaceae. The Pinaceae include the conifer species with the largest genome size, *Pinus lambertiana* (Murray 1998).

Despite its large size (> 20,000 Mbp for pine species; Wakamiya et al. 1993), little is known about the genomic structure and composition of the nuclear genome of conifer species. With the exception of chromosome numbers, which are highly conserved (2n = 24) (Khoshoo 1961), there are few investigations of the undoubtedly high proportion of non-coding DNA in the genomes of gymnosperms. Re-association kinetics data revealed 75% of the genome to be repetitive DNA (Dhillon 1987). For slash pine (*Pinus elliottii* Engelm. var. *elliottii*; 2n = 2x = 24) only 18S-5.8S-25S rRNA genes and the *Ty*1-*copia*-retrotransposon *TPE1* have been characterized as major classes of repetitive DNA (Doudrick et al. 1995; Kamm et al. 1996) and a *Ty*3- *gypsy* retrotransposon element was characterized as highly amplified in *Pinus radiata* (Kossack and Kinlaw 1999). A satellite DNA family has been cloned from *Picea* species and localized along the chromosomes (Brown et al. 1998).

The high complexity of the conifer genome and the high proportion of repetitive DNA have represented a serious limitation for the development and optimization of SSR markers. Moreover, the transferability rate of nSSR markers from one species to closely related species is in general not very high. Recently, with the advent of new generation high-throughput sequencing methods, a large number of sequences, in particular ESTs (expressed sequence tags) have been generated, allowing the detection of a high number of more easily transferable SSRs in conifers.

For additional and complementary details about nuclear, chloroplast and mitochondrial molecular markers see Gernandt et al. (Chapter 1), Ritland et al. (Chapter 5) and Burdon and Wilcox (Chapter 7, Table 7.1).

4.3 Distribution of Genetic Diversity in Conifers

Genetic diversity is traditionally estimated at two levels of organization, within and among populations. Populations are the smallest group of organisms from a single species which exchange genes using sexual reproduction. Populations constitute a mating unit, where the probability of gene exchange depends on geographic distance and pollen and seed mode of dispersal. Delineating populations that actually exchange genes is difficult, and often a population structure is assumed for sampling, based on the location of its individuals and (vague) prior knowledge of gene flow (pollen and seed dispersal). Individuals found in the vicinity of one another are considered as members of the same population. Alternatively, within the newly defined field of landscape genetics (Manel et al. 2003), methods with no (or less) assumptions have been proposed where groups of populations are defined a posteriori by maximizing the proportion of total genetic variance due to differences among user-defined groups of populations (Dupanloup et al. 2002), or where populations are constituted a posteriori using a Bayesian approach that minimizes departures from Hardy-Weinberg equilibrium (Pritchard et al. 2000; Corander et al. 2003) or using assignment tests and the probability of finding an individual's multilocus genotype across the landscape (Manel et al. 2007).

4.3.1 Genetic Diversity within Populations of Conifers

In trees in general, and in conifers in particular, most of the diversity at bi-parentally (nuclear loci) and paternally (chloroplast DNA loci) inherited neutral loci resides within populations. A typical distribution of variation at these markers is a few frequent and a majority of rare alleles (nuclear loci) or haplotypes (chloroplast loci). In sharp contrast, maternally inherited loci (mitochondrial DNA loci) often show low (or sometimes no) diversity. This situation is well exemplified by genotypic data collected for the two European species of spruce, *Picea abies* and *Picea omorika*.

In *Picea abies* (a widespread European conifer, Skrøppa 2003), within population gene diversity is high. Tollefsrud et al. (2009) found that for seven nSSRs in 1,715 individuals, the mean number of alleles per locus was 22 and the mean gene diversity over loci was 0.640. Close to 97% of all variability was found within populations. Using three cpSSR loci in 1,105 individuals, Vendramin et al. (2000) found an average of seven variants per locus, making altogether 41 haplotypes, each population tested having more than four haplotypes on average. Mean within-population haplotypic diversity was 0.635. Using mtDNA sequence variation at 10 loci in 4,876 individuals, Tollefsrud et al. (2008) found only one variable locus and 28 haplotypes. Mean within-population haplotypic diversity was 0.257. By contrast, in the related species *Picea omorika*, a narrow endemic restricted to a few localities in Serbia and Bosnia-and-Herzegovina, gene diversity was low; Nasri et al. (2008) analyzed 94 individuals at five cpSSR loci and showed that only two were polymorphic with an average of 1.6 variants per locus, making altogether four haplotypes, each population tested having 1.9 haplotypes on average. Mean within-population haplotypic diversity was 0.279.

The high levels of diversity generally observed in widespread conifer species are classically considered to result from several life-history traits of trees in general, and conifers in particular. First, conifers are long-lived organisms, with a long juvenile phase and second, they are capable of dispersing pollen (and seeds) over long distances (high levels of gene flow). Newly found conifer populations are thus made of migrants coming from different populations over long periods of time rather than by the progeny of one or few immigrants. Taken together, these features explain why trees are able to maintain high levels of within population genetic diversity at nuclear genes (Austerlitz et al. 2000), although the climatic history of the Quaternary, which led to several range contractions and expansions, should have resulted in loss of within population genetic diversity.

Because of their life history traits, conifers are thus able to maintain large effective population sizes. A decrease in within population genetic diversity is thus the result of random loss of alleles or haplotypes when populations contract, for example as a result of habitat fragmentation (Young et al. 1996). Species with large distribution areas and low within population genetic diversity at nuclear or paternally inherited genes are thus rare. When they occur, they indicate a dramatic demographic bottleneck effect in the more or less recent past, as shown for *Pinus pinea*. This widespread typically Mediterranean thermophilous conifer (Fady et al. 2004) has almost no diversity at all chloroplast (Vendramin et al. 2008) and nuclear (Fallour et al. 1997) loci investigated, most likely as a result of a major contraction of its distribution area during one of the Quaternary glacial periods.

Biogeographic scale current distribution of within-population genetic diversity is a powerful tool for understanding species history. For example, within population genetic diversity of conifers is higher in the Mediterranean than in temperate regions (Fady 2005). The Mediterranean basin was a refugial zone for temperate and Mediterranean-type organisms during the glacial cycles of the Quaternary (Hewitt 2000). There, species distribution areas, and thus their effective population sizes, were much smaller than their current ones, especially those of temperate species which refugia were close to the southernmost extant of the ice cap that covered most of northern Europe. These low diversity refugia were the front runners of recolonization of Europe when climate warmed during the Holocene circa 10,000 years ago. Mediterranean-type species on the contrary, were not as strongly affected by recolonization and were able to maintain higher effective population sizes during the glacial phases of the Quaternary, particularly in the eastern Mediterranean (Fady and Conord 2010).

Lower effective population size can also explain why rare congeners of widespread species (Gitzendanner and Soltis 2000 and see the example of *Picea omorika* presented above) and marginal/rear edge populations of species (Eckert et al. 2008a) tend to have lower levels of genetic diversity. In self-compatible species (such as conifers), a decrease in population size is not necessarily associated with an increase in consanguineous mating (Leimu et al. 2006). However, in predominantly mixed-mating conifers, which are mostly outcrossed in large and dense populations, marginality is correlated with an increase in selfed reproductive events (Restoux et al. 2008). Ultimately, a reduction in within population genetic diversity will have consequences on individual fitness and population persistence (Hughes et al. 2008). In conifers, pinpointing regions where demographic bottlenecks shaped the populations' genetic diversity and where diversity is higher than average remain key issues in conservation planning and genetic resource sampling (Fady and Conord 2010).

4.3.2 Genetic Diversity among Populations of Conifers

Diversity among populations (differentiation) is the main source of overall genetic diversity at maternally inherited loci (mtDNA in conifers), whereas it is much weaker for paternally inherited or bi-parentally inherited loci. In *Pinus sylvestris* for example, out of 1,380 individuals analyzed range-wide, only three mtDNA haplotypes were found at the *nad*1 gene (although 10 mtDNA regions had been surveyed for polymorphism), most populations carrying only a single haplotype. As a result, over 80% of total genetic diversity was found among populations. Together with paleo-ecological data, this high differentiation made it possible to identify three main disjunct glacial refugial areas: the Iberian Peninsula, Italy and series of interconnected refugia in the Balkans and Central Europe (Cheddadi et al. 2006). Comparatively, diversity at the 178 paternally-inherited cpSSR haplotypes detected resided mostly within population and differentiation was much lower (6%), making it possible to only identify the Iberian Peninsula as a potential refugium. Note however that, although weak, among population genetic differentiation at bi-parentally and paternally inherited markers is often highly significant. In the example of Scandinavian *Picea abies*, the 3% of nSSR diversity found among populations was highly significant and made it possible to identify two recolonization routes from a single refugium (Tollefsrud et al. 2009).

The main factor that explains the divergent estimates of differentiation at differently inherited markers is differential gene flow via seed and pollen. Although nuclear and chloroplast genes are both carried by pollen and then by seed, maternally inherited genes only travel via seed at a much reduced pace. Petit et al. (2005) calculated the median ratio of pollen flow to seed flow to be ~17 (considering both conifers and angiosperm species). For a given set of markers, differentiation will increase with fragmentation and habitat contraction as they reduce gene flow among populations and increase random loss of alleles within populations when their effective size decreases (Young et al. 1996). For instance, in *Cedrus libani*, a mountainous eastern Mediterranean conifer, high among population genetic divergence at different spatial scales indicated the presence of at least two zones of glacial refugia range-wide (Lebanon and Turkey) and the recent effect of deforestation in Lebanon as opposed to Turkey (Fady et al. 2008). In Table 4-3 some examples of genetic differentiation estimates obtained with markers of the three genomes are reported.

The distribution of diversity within and among populations of conifer species is the intricate result of different processes related to the mating system (conifers can often be both selfed and outcrossed), aptitude for gene flow, population size and population history. In the following sections, we review recent results highlighting how these different processes can explain the observed patterns of diversity at differently inherited markers.

Table 4-3 Genetic differentiation in conifer species estimated using mitochondrial, chloroplast and nuclear markers.

Species	Distribution	Markers		
		mitochondrial	chloroplast	nuclear
Abies alba	European	Gst = 0.919 (Liepelt et al. 2002, in: Du et al. 2009)	Gst = 0.251 (Liepelt et al. 2002, in: Du et al. 2009)	Fst = 0.044 (allozymes) (Bergmann 1991)
Pinus cembra	European	-	Fst = 0.127 (Höhn et al. 2009)	Fst = 0.074 (allozymes) (Belokon et al. 2005)
Pinus pinaster	European	Gst = 1.00 (Burban and Petit 2003)	Gst = 0.146 (Vendramin et al. 1998, in: Du et al. 2009)	Gst = 0.170 (allozymes) (Petit et al. 1995)
Pinus sylvestris	European	Gst = 0.800 (Cheddadi et al. 2006)	Gst = 0.060 (Cheddadi et al. 2006)	Gst = 0.14 (rDNA) (Kahru et al. 1996)
Picea abies	Eurasian	Gst = 0.676 (Sperisen et al. 2001)	Gst = 0.078 (Vendramin et al. 2000, in: Du et al. 2009)	Gst = 0.052 (allozymes) (Lagercrantz and Ryman 1990)
Picea asperata complex	Asian	Gst = 0.895 (Du et al. 2009)	Gst = 0.563 (Du et al. 2009)	Fst = 0.224 (RAPDs) (Xue et al. 2007)
Picea crassifolia	Asian	Gst = 0.512 (Meng et al. 2007)	Gst = 0.093 (Meng et al. 2007)	-
Picea jezoensis	Asian	Fst =0.921 (Moriguchi et al. 2009)	Fst = 0.056 (Moriguchi et al. 2009)	Fst = 0.102 (SSR) (Moriguchi et al. 2009)
Pinus densata	Asian	Gst = 0.905 (Song et al. 2003)	Gst = 0.533 (Song et al. 2003)	Gst = 0.112 (allozymes) (Yu et al. 2000)
Pinus luchuensis	Asian	Gst = 0.523	Gst = 0.189	-

Table 4-3 contd....

Table 4-3 contd....

Species	Distribution	Markers		
		mitochondrial	chloroplast	nuclear
		(Chiang et al. 2006, in: Du et al. 2009)	(Chiang et al. 2006, in: Du et al. 2009)	
Pinus tabulaeformis	Asian	Gst = 0.7[illegible]8 (Chen et al. 2008)	Gst = 0.188 (Chen et al. 2008)	Gst = 0.149 (RAPDs) (Li et al. 2008)
Picea chihuahuana	American	Gst = 1.0[illegible] (Jaramillo-Correa et al. 2006)	Gst = 0.362 (Jaramillo-Correa et al. 2006)	Fst = 0.248 (allozymes) (Ledig et al. 1997)
Picea mariana	American	Gst = 0.6[illegible]1 (Jaramillo-Correa et al. 2004)	-	Fst = 0.027 (RAPDs) (Isabel et al. 1995)
Pinus banksiana	American	Gst = 0.5[illegible]9 (Godbout et al. 2005)	Gst = 0.016 (Dong and Wagner 1994, in: Du et al. 2009)	Gst = 0.155 (RAPDs) (Ye et al. 2002)
Pinus contorta	American	Gst = 0.3[illegible]5 (Godbout et al. 2008)	Gst = 0.033 (Dong and Wagner 1994, in: Du et al. 2009)	Gst = 0.079 (RAPDs) (Fazekas and Yeh 2006)
Pinus ponederosa	American	Gst = 0.9[illegible]5	Gst = 0.519	Fst = 0.062 (allozymes)
		(Latta and Mitton, 1999, in: Du et al. 2009)	(Latta and Mitton, 1999, in: Du et al. 2009)	(Latta and Mitton, 1999)
Pinus radiata	American	Gst = 0.7[illegible]5 (Strauss et al. 1993)	Gst = 0.073 (Hong et al. 1993, in: Du et al. 2009)	Gst = 0.260 (RAPDs) (Wu et al. 1999)

4.4 Phylogeographical Approach of Historical Processes Shaping Conifer Genetic Diversity

The term "phylogeography" was first introduced in 1987 to describe "the field of study concerned with the principles and processes governing the geographical distribution of genealogical lineages, especially those at the intra-specific level" (Avise et al. 1987). Phylogeography can provide essential background information to disentangle current from past demographic processes and to understand the consequences of crucial events such as colonization in the life and longevity of plant species (Petit et al. 2003). The comprehension of the past dynamics of diversity can then be used to predict future demographic impacts related to the climate changes (Pitelka et al. 1997).

4.4.1 Methods to Infer Phylogeographic Patterns

Phylogeography associates studies of population genetics, phylogenetics and systematics (micro- and macroevolutionary concepts) with a spatio-temporal distribution of genetic variation (Avise et al. 1987; Avise 2000). The same genetic markers can be used both in population genetics and phylogeograpic studies, although the two disciplines differ in the objectives and methods of analysis. The main difference between the two fields is that population genetics considers the differences in the allelic distribution on the basis of recent gene flow, while phylogeography aims at understanding the historical processes that shaped the current distribution of genetic variation. Allele numbers and distribution is determined by demography, therefore different historical events can be deduced from the allelic patterns. Phylogeography integrates the analysis of fossil remains, such as pollen and macrofossils, or provides knowledge on the history of species for which fossil remains are very scarce or indistinguishable from other taxa (Pleines et al. 2009), by studying the reconstructed histories of individual genes (gene trees) sampled from different populations (Knowles and Maddison 2002). Inferences of past events are possible because most mutations arise at single points in time and space. Based on which ancient or recent processes most likely influence the structure of genetic diversity among and within populations (geographical barriers, dispersal events, population size changes, gene flow), the analysis interprets patterns of congruence or incongruence between the current distribution of alleles and their genealogical relationships. The inheritance relationships between alleles are typically represented as a gene genealogy, similar in form to a phylogenetic tree.

Phylogeographic studies must be followed by statistical inferences (e.g., Templeton et al. 1995) and two basic approaches can be identified: the first is based on the coalescent theory and tools from computational statistics (Slatkin 1987; Griffiths and Tavare 1994; Kuhner et al. 1995; Wakeley and Hey 1997; Beerli and Felsenstein 1999; Nielsen and Wakeley 2001; Beaumont et al. 2002), the second is based on the analysis of the estimated gene trees (or networks) in a cladistic framework (Templeton et al. 1987, 1995; Posada et al. 2000, 2005, 2006; Templeton 2004).

The coalescent theory is a retrospective model of population genetics based on the genealogy of gene copies within and among related species (Felsenstein 1971; Griffiths 1980; Tavaré 1984; Hudson 1990, 1998). This theory uses mathematics to describe the characteristics of the joining of lineages (coalescence) back in time to a common ancestor, and provides the basis for estimation of the expected time to coalescence and for establishing the relationships of coalescence times to population size, age of the most recent common ancestor, and other population genetic parameters.

The main coalescent methods rely on maximum-likelihood approaches (e.g., Bayesian inference) which attempt to maximize the probability (or likelihood) to observe the data (Nielsen and Beaumont 2009), i.e., the types of different genetic variants and their frequencies in a sample under a given model characterized by n parameters, to be estimated (e.g., population size and date of coalescence events). The likelihood in population genetics can be calculated by combining computational methods from phylogenetics with coalescence models (Felsenstein 1988, 1992). Phylogenetics developed methods for connecting genetic data with a tree, whereas coalescence theory provided mathematical methods for connecting demographic or ecological models with a tree (Nielsen and Beaumont 2009). These methods are based on Markov Chain Monte Carlo (MCMC) and importance sampling (IS), which are both based on the simulation of a large number of trees (for more details see Nielsen and Beaumont 2009). When complete likelihood methods are not feasible or flexible enough a number of approximation methods have been recently developed, including the likelihood-free inference or Approximate Bayesian Computation (ABC) (Beaumont et al. 2002; Sisson et al. 2007) and the product of approximating conditionals (PAC) (Li and Stephens 2003). Recently evolved lineages are graphically represented by a network based on distance methods that aim to minimize the distances (number of mutations) among haplotypes (Posada and Crandall 2001). The most commonly used method is the statistical parsimony network (Templeton et al. 1992), which links haplotypes through a series of evolutionary steps. The connections between haplotypes throughout the network represent coalescent events. Following some of the principles of coalescent theory, it is possible to recognize the dualism of old vs. young haplotypes from the shape of the network together with haplotype frequencies (Castelloe and Templeton 1994). Tip haplotypes or clades (linked to the network by only one branch) are commonly younger than interior ones (linked to the network with more than one branch), which generally display higher haplotype frequencies. From a geographical point of view, old alleles are expected to distribute broadly, because they have had a long time to disperse, whereas haplotypes with only one connection (singletons) are likely to be connected to haplotypes of the same population because they evolved recently and had no time to disperse (Freeland 2005).

Once the genealogical relationships among haplotypes have been established, the next step is to identify the historical and geographical factors that influenced the current distribution of haplotypes. In recent years a growing number of methods and specific software have been developed for these purposes: the most commonly used are listed in Table 4-4.

The most popular method in phylogeographic studies is the nested clade phylogeographic analysis (NCPA), also known as nested clade analysis (NCA) (Templeton et al. 1995). NCPA is able to distinguish between recurrent gene flow and a variety of historical processes, such as fragmentation, long distance colonization and range expansion (Pleines et al. 2009). The method uses statistical parsimony to construct a statistically supportable haplotype network as the one outlined above. Then, it tests for an association between geography and haplotype distribution, and works through an inference key to identify the processes that could have produced the association. The oldest and the newest haplotypes are located at the center and at the periphery of the network, respectively. As a result, the nested arrangements correspond to evolutionary time, with higher nested levels corresponding to earlier coalescent events (Freeland 2005). The following step is to overlap the clades with geography and to calculate two measures of distance: the mean distance of clade members from the geographical center of the clade (D_c) and the mean distance of nested clade members from the geographical center of the nested clade (D_n). The existence or not of a non-random association between genetic lineages and geographical locations is verified by permutation tests and if the hypothesis of no association (null hypothesis) can be excluded, an a posteriori inference key is used to determine the most likely alternative scenarios to explain the patterns that have been observed (Templeton 2004). Hence, specific hypothesis about the geographical distribution of lineages based on both organellar and nuclear sequence data can be tested using the NCPA method, although this analysis is limited by sampling size, because the network may be inaccurate if too few individuals or populations are considered (Freeland 2005). During the last 10 years, the complex NCPA analysis has been implemented in computer programs (TCS, Clement et al. 2000; GeoDis, Posada et al. 2000) and more recently several approaches have been developed to automate the procedure (Zhang et al. 2006; Panchall 2007).

Table 4-4 Some of the software used in phylogeographic studies.

SOFTWARE	WEBSITE	FEATURES	REFERENCE
msBayes	http://msbayes.sourceforge.net/	The software analyzes data from multiple species/population pairs under a hierarchical model and employs approximate Bayesian computation (ABC) under a hierarchical coalescent model to test for simultaneous divergence (TSD).	Hickerson MJ, Stahl E, Takebayashi N (2007). msBayes: Pipeline for testing comparative phylogeographic histories using hierarchical approximate Bayesian computation. BMC Bionf 8 Article Number: 268
GenGIS	http://kiwi.cs.dal.ca/GenGIS/Download	The software combines digital map data with information about biological sequences collected from the environment	Parks DH, Porter M, Churcher S, Wang SW, Blouin C, Whalley J, Brooks S, Beiko RG (2009). GenGIS: A geospatial information system for genomic data. Genome 19 (10): 1896–1904
TESS	http://membres-timc.imag.fr/Olivier.Francois/tess.html	The software builds a network structure which describes the prior relationships between the individuals, given individual geographical locations. The program implements a Bayesian clustering algorithm for spatial population genetic analyses	Chen C, Durand E, Forbes F, Francois O (2007). Bayesian clustering algorithms ascertaining spatial population structure: a new computer program and a comparison study. Mol Ecol Notes 7 (5): 747–756
Mesquite	http://mesquiteproject.org/mesquite/mesquite.html	The software includes phylogenetic analyses (parsimony, likelihood, comparative method, simulations and randomizations of characters and trees) and population genetics analyses (coalescence)	Maddison, W. P.,.Maddison D.R. (2007). Mesquite: a modular system for evolutionary analysis. Version 2.0 http://mesquiteproject.org

DNA sequences are the markers best suited to infer genealogical lineages. Allele frequencies, however, can provide information on gene flow and the genetic subdivision of populations and therefore offer useful contributions in phylogeographic studies. In principle, phylogeographic studies can be based on information from all variable DNA regions (nuclear, mitochondrial and chloroplast), although, because they exhibit a much higher population subdivision (Petit et al. 2005), maternally inherited markers are more often used.

The highly polymorphic SSRs are also often used in phylogeographic studies. However, their high level of variability can represent a major disadvantage because of homoplasy (Doyle et al. 1998), which increases with age of divergence and genetic distance among taxonomic units (Provan et al. 2001; Jakob et al. 2007). To reduce the possible effect of homoplasy, SNPs have recently been proposed as the marker of choice for multi-locus population analyses (Brumfield et al. 2003; Zhang and Hewitt 2003). SNPs have simple patterns of variation, but more importantly, have low mutation rates (Brumfield et al. 2003) and thus lower level of homoplasy.

During the last few years, an increasing number of phylogeographic studies using nuclear markers were performed. Nuclear DNA, unlike what happens for that of organelles, undergoes recombination events. This may results in data interpretation problems: mosaic sequences might be included in the analysis and can change gene or locus genealogies. If the rate of recombination at a given locus is similar to the rate of nucleotide substitutions, each allele could have more than one ancestor, resulting in different evolutionary histories for the different parts of the same locus (Freeland 2005). Nevertheless, a recent review of nuclear gene phylogeographic studies (Hare 2001) suggested that recombination is not an invincible problem and that it can be identified with specific software (Holmes et al. 1999; Husmeier and Wright 2001). Once a recombination has been identified, the relevant sequence regions might be removed before doing genealogical analyses.

By combining mitochondrial or chloroplast DNA sequences with nuclear markers, demographic processes acting at different time scales will be captured because organelle and nuclear markers have different modes of inheritance, effective population size and mutation rates (see Hewitt 2000; Semerikov and Lascoux 2003). Thus, to fully address the population history of an organism, several distinct genealogies from independent genetic markers are needed (e.g., Ballard and Whitlock 2004).

4.4.2 *Phylogeographic Studies in Conifers*

Since phylogeography was first defined as the process that determines the geographic distribution of genealogical lineages (Avise et al. 1987; Avise 2000), several studies have been carried out on conifer DNA variation using the organelle and the nuclear genomes. The main objectives of phylogeographic studies are the identification of glacial refugia and the characterization of migration and colonization dynamics. During the last decade the literature on phylogeographic studies of European and American conifers has considerably increased, whereas little is known about the role played by climatic oscillations on plant species in Asia, Africa and the Southern Hemisphere (Pleines et al. 2009). As for the first pioneer works on genetic variation of tree populations, *Abies*, *Picea* and some *Pinus* species have been the first object of phylogeographic studies. Phylogeographic studies on plants demonstrated that this analytical approach may be used to address unresolved issues concerning genetic exchange and differentiation within and among conifer species. Very likely, the identification of glacial refugia and the description of post-glacial colonization dynamics represent the most innovative contribution that phylogeography has given to population studies. The first phylogeographic results published on European forest tree species revealed the important role played by southern refugia (Iberian, Italian and Balkan). Later, when cryptic and more northern refugial areas were also described, the complex history of vegetation during the climatic oscillation of the Quaternary age was more clearly described and the accepted hypothesis that temperate areas were exclusively colonized from southern refugia was somewhat modified (Provan and Bennet 2008).

4.4.2.1 Phylogeography of European and Mediterranean Conifers

The literature on phylogeographic studies of widespread conifers, such as *Picea* and *Pinus* species, is particularly rich. The history of Norway spruce (*Picea abies*) is reported above in the text (see "genetic diversity within populations of conifers" and references therein.).

Several studies have been performed in the genus *Abies*. A spatial organization of haplotypes and a positive correlation between genetic and geographical distances was described in silver fir (*Abies alba*; Vendramin and Ziegenhagen 1997; Vendramin et al. 1999). The European *Abies* complex has been studied by Parducci et al. (2001) and more recently by Liepelt et al. (2009). In their study, Parducci et al. (2001) investigated the highly endangered species *Abies nebrodensis* from northern Sicily, together with populations of *A. alba*, *A. cephalonica* and *A. numidica*. Within-population haplotypic diversity was generally high, but somewhat reduced in *A. nebrodensis* compared with the other *Abies* species. Despite the extreme reduction in population size, the few remaining *A. nebrodensis* individuals still retain a considerable amount of their original variation. In their synthesis from palaeo-botanic and genetic data on silver fir, Liepelt et al. (2009) analyzed the postglacial history of *A. alba* and of European

Abies complex. The geographic distribution of genetic lineages and allele frequencies together with fossil records confirmed the presence of multiple refugia.

Several phylogeographic studies were performed on Mediterranean pine species, particularly on maritime pine, *P. pinaster*, whose distribution range is scattered across the western Mediterranean region. Contrasting results on genetic differentiation are reported by Vendramin et al. (1998) and Ribeiro et al. (2001) that can be explained by the mixing of genetic material caused by the human intervention on this species. The existence of separate refugia of *P. pinaster* during the last ice age was identified by Burban and Petit (2003) and contrasting patterns of variation were detected by chloroplast and mitochondrial DNA.

Both *P. pinaster* and *P. halepensis* were the object of a study on the chloroplast diversity and differentiation performed by Gómez et al. (2005), who identified the most likely refugial areas. Results showed that these two species, which occur in the same geographical area, have different levels and patterns of genetic diversity distribution.

Recently, Vendramin et al. (2008) investigated the phylogeography of another Mediterranean pine, *P. pinea*. The umbrella pine is a genetically depauperate species with an unusually low level of genetic diversity at nuclear (Fallour et al. 1997) as well as at chloroplast genes (Vendramin et al. 2008). The geographic pattern of chloroplast variation was characterized by the presence of an almost unique haplotype spread throughout the entire distribution range.

Contrary to *P. pinea*, the European black pine (*P. nigra*) is characterized by high levels of chloroplast DNA diversity in most populations analyzed by Afzal-Raffi and Dodd (2007), who sampled west of the Balkans, thus including most, but not all, taxonomic sub-entities of this complex species. The most likely glacial refugia were identified in southern regions. Alpine populations were clearly distinct from the other western populations studied; however, a comparison with populations from the south-eastern range in the Balkans is still lacking. The regional structure was supported by a biogeographical analysis that detected five barriers, with the two most significant separating the Alps from Corsica and southern Italy, and southern Spain from the Pyrenees.

P. sylvestris is the most widely distributed Eurasian conifer, its range spreading from arid, mountainous areas of Spain and Turkey to subarctic forests of northern Scandinavia and Siberia. Several phylogeographic studies (Sinclair et al. 1999; Soranzo et al. 2000; Chedaddi et al. 2006; Naydenov et al. 2007; Pyhäjärvi et al. 2008) have identified Spain and Italy as the most probable locations of its glacial refugia, as they are for many European species (Taberlet et al. 1998). However, populations from the Italian and Iberian peninsulas have not contributed to the postglacial colonization of central and northern Europe which was done from northern refugia as for *Picea abies* (see above).

Swiss stone pine, *P. cembra*, is a European species considered to be a glacial relict. It occurs in two disjunct regions: the continental parts of the Alps (central Alps), which is considered to be its core natural range, and the Carpathian Mountains, where isolated populations exist. Höhn et al. (2009) have investigated the post glacial history of this pine species. The populations of *P. cembra* within the two parts of the species' range share many cpDNA haplotypes, suggesting a common gene pool conserved from a previously large, continuous distribution range.

The phylogeographic studies of several Eurasian larch species (*L. decidua*, *L. sibirica*, *L. gmelinii*, *L. olgensis*, *L. kaempferi* and *L. sukaczewii*) helped to identify different glacial refugia and illustrated the post glacial migration routes of the species (Semerikov and Lascoux 2003; Araki et al. 2008).

Moroccan populations of *Cedrus atlantica* (Atlas cedar) from the Rif, Middle Atlas, and High Atlas mountains, were analyzed by Terrab et al. (2006) and Cheddadi et al. (2009) using cpDNA markers. The populations are separated by valleys and confront considerable barriers to gene flow and poor geographic structure was revealed among the analyzed populations (Terrab et al. 2006). The analysis of Moroccan and Turkish populations recognized the existence of two *C. libani* taxa, one in Lebanon and one in Turkey; moreover, Turkish populations probably emerged from several refugia (Fady et al. 2008).

A recent study based on nuclear DNA was performed on cypress, *Cupressus sempervirens*, in its distribution range (Bagnoli et al. 2009). This species is supposed to have originated in the eastern Mediterranean area and experienced a strong human impact during the last thousands of years; as a consequence, the present distribution of the species around the Mediterranean appears to be broader than it was originally. The authors emphasized a different history of cypress compared to the current one based entirely on human introduction of cypress in Italy, suggesting that probably a mosaic of recently introduced trees and remnants of ancient, depauperate populations exist today in central Mediterranean cypress range. It is further suggested that, as already demonstrated for cork oak (Magri et al. 2007), the timescale for understanding tree population dynamics, usually starting from the end of the last glaciation, has to be repositioned to more ancient times.

The range-wide population structure and phylogeography of *Juniperus thurifera* L. revealed that the Strait of Gibraltar represented an efficient barrier against gene flow between the Moroccan and European populations for a very long time, and consequently support that the Moroccan populations should be recognized as a distinct subspecies (*J. thurifera* L. subsp. *africana* (Maire) Romo and Boratyńsky) (Terrab et al. 2008).

4.4.2.2 Phylogeography of North American Conifers

Several North American pine species, characterized by either continuous (*Pinus banksiana, P. contorta, P. ponderosa*) or scattered (*P. flexilis, P. balfouriana, P. albicaulis*) distributions have been the object of phylogeographical studies (Mitton et al. 2000; Richardson et al. 2002; Johansen and Latta 2003; Godbout et al. 2005, 2008; Eckert et al. 2008b). Results on organelle and/or nuclear genome led to the identification of glacial refugia and to the description of population genetic structure and biogeographic patterns of genetic variation.

On the North American continent, mitochondrial and nuclear markers revealed that the widespread *Picea mariana* (black spruce) recolonized its current range from widely separated glacial refugia (Gamache et al. 2003; Jaramillo-Correa 2004).

Phylogeography was studied in two Mexican pine species: *P. strobiformis* (Moreno-Letelier and Piñero 2009) and *P. leyophylla* (Rodríguez-Banderas et al. 2009). In both cases chloroplast markers revealed the presence of a phylogeographic structure, that allowed defining separate lineages and geographic groups.

4.4.2.3 Phylogeography of Asian Conifers

Population fragmentation and contraction, as a consequence of Quaternary glacial cycles, is responsible for the disjointed distribution of the Chinese endemic *Abies ziyuanensis* (Tang et al. 2008). After the last glaciation, in Southern China *Abies* populations were replaced by broadleaved evergreen tree species and few *A. ziyuanensis* populations survived in colder habitats.

The extraordinary, and mostly unexplored, richness of conifers in China is the object of an increasing interest for phylogeographical studies on Asian species. The literature includes widespread as well as rare and endangered trees, such as *Cathaya argyrophylla* (Pinaceae) that is restricted to subtropical mountains of China. The genus *Cathaya* was widespread in northern America and Europe until Late Tertiary climatic deterioration and Quaternary glaciation caused its extinction in the American and the European continents (Liu and Basinger 2000), at present *C. argyrophylla* is the only representative of the genus. The phylogeographical study by Wang and Ge (2006) suggests the existence of at least four separate glacial refugia.

Among widespread species, an interesting study by Chen et al. (2008) revealed a significant phylogeographic structure in *Pinus tabulaeformis*, a major component of coniferous forests, endemic to northern China. The spatial distribution of mitochondrial haplotypes suggests the presence of five distinct population groups.

Recently, uni-parentally inherited chloroplast markers have been used to study populations of *Juniperus tibetic* (Opgenoorth et al. 2010). The genetic data strongly suggest that the juniper forest islands and isolated tree stands of the southern Tibetan Plateau are remnants of a former interstadial forest that were fragmented during the last glacial maximum (LGM) and that experienced postglacial local expansions before again experiencing fragmentation and marginalization as a result of anthropogenic influence as well as desiccation.

A phylogeographic structure was identified by the analysis of chloroplast DNA diversity in *Microbiota decussata*, an endangered species of Cupressaceae with disjoint distribution in the Sikhote Alin Mountains in eastern Russia. Results by Artyukova et al. (2009) suggest that the distribution area of *M. decussata* was fragmented a long time ago by extinction of populations in the adjusted territory. Furthermore, *M. decussata* was able to survive throughout the range with population expansion, successive fragmentation and isolation. Opposite to *M. decussata*, the other Cupressaceae formerly present in the area vanished or migrated southwards.

4.5 Gene Flow in Conifers

Gene flow is a major evolutionary force, homogenizing allelic frequencies across populations and reducing effective population size within local neighborhoods (sensu Wright 1943) when it is restricted. Trees in general and conifers in particular experience remarkably higher levels of gene flow than herbaceous plants. Gene flow can counteract changes in gene frequency triggered by selection, imposing a limit to local adaptation (Lenormand 2002). Gene flow occurs through pollen dispersal, seed dispersal, and establishment of fertile adult trees. The recent development of highly variable molecular markers and of new statistical methods to estimate contemporary gene flow has shed light on the important role of migration in evolution, in particular by disentangling the roles of dispersal and post-dispersal processes (fertilization, germination, and competition). In this section, we summarize the main methodological features of the classical (historical approach) and more recent (contemporary approach) genetic methods (see Table 4-5) and review their most prominent results in conifer species.

Table 4-5 Some of the most used methods to infer historical or contemporary gene flow, and software implementing these methods.

Time scale	Spatial scale	Type of gene flow inferred	Estimate provided	Sampling design *	Sampling effort	Typical number of polymorphic markers required	Principle of inference method	Methodological reference	Software
Historical	among pop.	Total gene flow (seed + pollen) with biparentally /paternally inherited markers. Seed flow estimate with maternally inherited markers	$N_e\sigma^2_e$	30 individuals/ population	Low	Minimum 3 SSR or 50 AFLP	The traditional auto-correlation approach: regression of differentiation among population (Fst/1-Fst) againt (logarithm of) distance	Hardy and Vekemans, 1999; Rousset 1997	Spagedi (Hardy and Vekemans 2002), Genepop (Rousset, 2008), GeneAlex (Peakall and Smouse 2006)
Historical	within pop.		$d_e\sigma^2_e$	• Minimum 50–60 individuals • Ideally distributed on a transect (range of σ^2_e - 20 σ^2_e) • Splitting across cohorts advised (i.e., adult, seedlings, saplings)	Low	Minimum 3 SSR or 50 AFLP	The traditional auto-correlation approach: regression of genetic relatdness among individuals againt (logarithm of) distance	Rousset, 2000; Vekemans and Hardy 2004	

Table 4-5 contd....

Table 4-5 contd....

Time scale	Spatial scale	Type of gene flow inferred	Estimate provided	Sampling design *	Sampling effort	Typical number of polymorphic markers required	Principle of inference method	Methodological reference	Software
Contemporary	within pop.	Pollen flow	• Φ_{ft}	• Maternal progenies collected on N_m mother trees (N_s seed/trees)	Medium	≥ 5 SSR	TwoGener and Kindist	Austerlitz and Smouse, 2001; Austerlitz and Smouse, 2002; Smouse et al. 2001; Robledo-Arnuncio et al. 2006	PolDisp (Robledo-Arnuncio et al. 2007)
			• pollen dispersal kernel (shape and range)	• + Some potential fathers (without coordinates)					
			• *Nep*	• For a given N_m*N_s, N_m should increase (and N_s, decrease) with decreasing Φ_{ft}					
Contemporary	within pop.	Pollen flow	• Pollen dispersal kernel (shape, range, asymetry)	• Maternal progenies collected on Nm mother trees (Ns seed/trees)	High	≥ 5 SSR	The neighborhood model, or Spatially explicit mating model	Adams and Birkes, 1991; Burczyk et al. 2002; Oddou-Muratorio et al. 2005	Neighbor (Burczyk, unpublished), Nm+ (Chybicki & Burczyk, unpublished), MEMM (Klein et al. unpublished)

Contemporary	within pop.	Seed and pollen flow	• Selection gradient on phenotypic variables affecting male fertility • d_e/d_{obs}	• + ALL potential fathers within a given neighborhood • + phenotypic traits of potential fathers (size, flowering…)	High	≥5 SSR	The seedlings neighborhood model or Spatially explicit mating model	Burczyk et al., 2006; Oddou-Muratorio and Klein, 2008	Nm+ (Chybicki & Burczyk unpublished)
			• selection gradient on phenotypic variables affecting female/male fertility	• + all potential parents within a given neighbourhood (N_p)					
				• + phenotypic traits of potential parents (size, flowering…);					
				'-Ns should be Np					

*unless specified, both spatial coordinates and genetic material need to be collected.
N_e effective population size; d_e effective population density; σ^2_e is the mean-squared axial dispersal distance; Φ_{ft}; *Nep* effective pollen pool size.

4.5.1 Historical Approaches to Estimate Total Gene Flow

Traditionally, standard methods for estimating gene flow from genetic data rely on measures of genetic differentiation among populations (Wright 1951) or among individuals within populations (Rousset 2000). Assuming that, in a species, population structure follows an infinite island model at evolutionary equilibrium, different variants of Wright (1951)'s F_{ST} (fixation index) have been used to estimate $N_e m_e$, the effective number of migrants per generation (see references in Whitlock and McCauley 1999). In their historical review of isozyme (a nuclear codominant genetic marker) diversity and genetic structure in 213 species, Hamrick et al. (1992) showed that tree species have low levels of differentiation among populations, with an average G_{ST} (amount of differentiation observed over multiple loci) of 0.07 for gymnosperms and 0.11 for angiosperms. In some endemic species however, high values of differentiation were measured as a result of low population size and geographic isolation: for instance, G_{ST} = 1 in *Pinus torreyana* (Ledig and Conkle 1983); G_{ST} = 0.20 in *Pinus muricata* (Millar 1983). But in species with continuous distribution area and high population size, levels of differentiation were generally very low: for instance, G_{ST} = 0.036 in *Pinus contorta* (Wheeler and Guries 1982); G_{ST} = 0.030 in *Pinus banksiana* (Dancik and Yeh 1983). Under the assumption of Wright's island model of population structure, where $G_{ST} = 1/1{+}4Nm$ at equilibrium, low differentiation translates into high estimates of the effective number of migrants among populations per generation (e.g., N_m= 3.32 for G_{ST} = 0.07).

Forest tree populations can deviate in several ways from the assumptions of Wright's island model, and gene flow estimates based on F_{ST} or G_{ST} should be interpreted with caution (Whitlock and McCauley 1999). First, long-lived trees may only rarely reach equilibrium and the distribution of genetic diversity may be mostly influenced by population history and demography, not by current gene flow (Austerlitz et al. 2000). Second, in continuous populations the isolation-by-distance model is more appropriate than the infinite island model because genetic differentiation and geographical distance are positively correlated (Wright 1943).

Under the isolation-by-distance model, pairwise differentiation between subpopulations is used to estimate gene dispersal relative to effective population density by examining the regression of $F_{ST}/(1-F_{ST})$ against geographical distance (Rousset 1997). The increasing realization that the equilibrium hypothesis is critical for the reliability of gene dispersal inferences based on F_{ST} (Whitlock and McCauley 1999) has led to focus the estimation process on a local scale, where this equilibrium is more likely to be quickly reached and occurrence of mutations can be neglected (Leblois et al. 2003). Considering thus a continuous population exhibiting isolation by distance, the decay rate of genetic relatedness between individuals with distance has been shown to be proportional to $1/d_e\sigma^2_e$, where σ^2_e is the mean-squared axial dispersal distance, and d_e the effective density of individuals (Rousset 2000; Vekemans and Hardy 2004). Intuitively, the product $d_e\sigma^2_e$ expresses the degree of overlap between individual "gene shadows" (the spatial distribution of gene dispersal events around each parent). It implies that the intensity of genetic structuring decreases both with increasing dispersal and increasing individual density. In practice, it is notoriously difficult to estimate effective density in natural populations, and thus to get independent estimate of σ^2_e in a given population under study.

The preliminary steps to estimate indirectly gene flow from spatial genetic structure (SGS) are (1) to characterize SGS and to test it so as to show that the spatial distribution of alleles or genotypes within population is non-random; and (2) to disentangle the effects of dispersal from all the different factors that contribute to SGS. Indeed, fine scale spatial patterns of genetic variation result from the complex interplay of several factors: the local and long distance component of the dispersal process, genetic drift and other evolutionary processes such as natural selection. The study of how the within population component of genetic variation is spatially structured is important for understanding the evolutionary consequences of micro-geographical genetic heterogeneity, mating patterns and inbreeding levels, demographic factors such as the extent of effective population size, competition and, in general, density-dependent processes. For a detailed summary of factors influencing SGS, their expected change over time or with increasing density, and its resulting effect on SGS see the Introduction and Table 1 in Troupin et al. (2006).

Historically SGS within population was primarily studied using spatial autocorrelation methods (reviewed in Epperson 2003). The rationale of such techniques is to measure the correlation of allelic or genotypic states between individuals separated by defined distances within the whole population. Genetic correlation is calculated using different statistics: Moran's I (Cliff and Ord 1981) and measures of relatedness (e.g., Loiselle et al. 1995) are the most used in the literature regarding geographical genetics of forest trees. The multivariate method by Smouse and Peakall (1999) implemented in the program GenAlEx (Peakall and Smouse 2005), provides a multilocus estimate of pairwise relatedness between individuals, which minimizes the stochasticity found in single locus or single allele estimates of relatedness. Vekemans and Hardy (2004) recently proposed the Sp statistics as a useful measure of intensity of SGS. Sp is computed as $Sp = b/(F_1-1)$, where b is the regression slope of the kinship (or coancestry) estimator Fij computed among all pairs of individuals I and j against geographical distances, and F_1 is the average kinship coefficient between individuals of the first distance class (< 50 m). Sp. has the desirable characteristic of being comparable among species or sites, allowing quantitative comparison among studies.

4.5.2 *Spatial Genetic Structure (SGS) and Historical Estimates of Gene Flow in Conifers*

In conifers, only a few studies have investigated patterns of SGS within population, and they usually report weak patterns, in particular at the adult stage. In their review and reanalysis of SGS in 47 plants species, Vekemans and Hardy (2004) could only report two studies on conifer species.

Early studies on SGS within population were based on spatial distribution of allozyme genetic variability analyzed by spatial autocorrelation. Back in the 1990s, the first studies following this approach explored some of the most interesting fields of application of geographical genetics. Among these pioneer studies, the null hypothesis of random distribution of genotypes was rejected only in a few cases in continuous undisturbed conifer populations, and in those cases a weak SGS was detected. Epperson and Allard (1989), Knowles (1991) and Leonardi et al. (1996) showed that, in general, SGS is limited to a few allozyme loci and to the shorter distance classes (0–25 m), and that clusters of trees with genotypes more similar than expected were more frequent in younger age classes. As a consequence, these studies rarely provided quantitative, indirect estimates of gene flow. Interestingly however, they highlight the difficulties in disentangling the effect of gene flow from those of other demographic processes in natural populations, and in particular of habitat fragmentation, anthropogenic and natural disturbances, and colonization of new habitats.

At the population level, fragmentation reduces population size and increases isolation, creating genetic bottlenecks. Remnant populations experience increased genetic drift, increased inbreeding and limited gene flow from surrounding populations. This is expected to increase SGS through non-random mating, lower population density and potential aggregation of reproductive individuals. The smaller the population size the greater these effects are on SGS. The degree of genetic connectivity through long-distance dispersal (LDD) among fragments can counterbalance such effects, decreasing SGS. The impact of forest fragmentation on Spanish maritime pine (*Pinus pinaster* Aiton) populations has recently been assessed in two population pairs, each formed by one continuous and one fragmented population. Fragmented populations showed significant SGS up to 20 m, whereas large continuous populations had either weak or no SGS. The results on continuous populations confirmed previous evidence for this species (De-Lucas et al. 2009 and references therein). Interestingly, the authors integrated their experimental data using a simulation approach to elucidate the factors that might have produced the observed pattern. Simulations suggested that under fat-tailed (including a significant long distance component) dispersal, small population size is a stronger determinant of SGS than genetic isolation, while under normal dispersal, genetic isolation has a stronger effect.

Plants are expected to move northward and upward in response to increasing temperatures. This implies a strong effect of fragmentation at the rear edge of their distribution as well as colonization at the leading edge (Jump et al. 2009). Colonization events in forest trees were modelled by Austerlitz et al. (2000). Their results show that, in species with a long juvenile phase and a delayed first reproductive event, as in many conifers, the colonization process is primarily sustained by new migrants from source populations, in numbers large enough to mitigate or avoid founder effects. For several tree species, it has been shown that colonization events did not cause founder effects (e.g., Lefèvre et al. 2004). Such population dynamics is expected to generate weak or no SGS. Doligez et al. (1998) predicted that SGS will be stronger in older populations than in recently established ones due to generation overlap and increased kinship between mates.

What has been observed in expanding populations of conifers is in agreement with such expectation. In a recent study on Eastern white pine (*Pinus strobus* L.), a conifer with winged wind-dispersed seeds and early age of first reproduction (approximately 5–10 years), a weak SGS (up to 10 m) was found in a recently colonized plot. Comparing an old growth Scots pine (*Pinus sylvestris* L.) stand with a recently colonized one from the same continuous forest, Chybicki et al. (2008) obtained similar results. Scotti et al. (2008) tried to disentangle the effect of pollen and seed dispersal in shaping SGS in recently colonized areas. They analyzed SGS in a Norway spruce (*Picea abies* [L.] Karst.) stand close to a mountain pasture using both mitochondrial and chloroplast molecular markers. In the part of the study plot corresponding to the forest-meadow border, high sapling density reflected a recent colonization. Spatial genetic analysis was carried out separately for the colonization area and the dense stand area (the remaining part of the stand that was part of a continuous forest). Chloroplast genetic variability analysis did not show spatial clumping in either area. Mitochondrial haplotypes showed the typical autocorrelogram of patchy spatial structures, with positive and significant autocorrelation at short distances and negative, although not significant, autocorrelation at long distances in the dense area. On the contrary, SGS was entirely lacking in the border area, where neither adults nor saplings showed any significant values in any distance class. This suggests a high gene flow via-pollen and via-seed in the recently colonized area. High gene flow and non-negligible probability of long distance dispersal via seed were recently demonstrated in the upward shift of a *P. abies* tree-line population (Piotti et al. 2009), resulting in an absence of SGS in seedlings (A. Piotti unpubl.). Finally, Troupin et al. (2006) proposed a novel approach to the study of the temporal variation of SGS in an expanding population. They followed the dynamics of successive cohorts of the same population over time, rather than analyzing different age cohorts at the same time. Using aerial photos of the study area from 1944 and individual tree-ring dating, they were able to reconstruct population expansion in detail over 30 years (1944–1973). They found that the population, now composed by more than 2000 individuals, originated from five putative ancestor trees before 1944, reaching a population size of 168 adult individuals in 1973. No SGS developed in the first 20 years, whereas significantly positive SGS of the reproductive population was found beginning in 1966 and its magnitude increased over time. These results confirmed that at the beginning of the colonization process, when established individuals do not yet contribute to regeneration, gene flow from outside represents the only source of new colonizers. Once sexual maturity was reached by the first colonizers, a weak SGS developed due to local reproduction. The higher the delay is in sexual maturity, the lower the probability is to detect SGS in colonization areas.

Natural and anthropogenic disturbances can alter the partitioning of genetic variation within populations. The comprehension of the consequences of disturbances on SGS, such as fires and silvicultural practices, can be crucial to preserve forest resources through appropriate forest management practices. There is no general rule regarding the effects of disturbances on SGS, as it will depend on the spatial and genetic composition of the seed sources contributing to regeneration after disturbance. As an example, Knowles et al. (1992) studied the SGS of two tamarack (*Larix laricina* [Du Roi] K. Koch) populations characterized by different establishment histories. They demonstrated that the successful reproduction of the few local remnant adult trees was sufficient to determine a SGS in newly established seedlings following a disturbance (20 years old clearcut), whereas no SGS was detected in a colonization area of the same age where only external seed sources were available. Similarly, Boyle et al. (1990) compared two contrasting black spruce (*Picea mariana* [Mill.] B.S.P.) populations showing how SGS characterizing an undisturbed site can be lost after fire disturbance because of high and diversified gene flow into a large open area.

Little is known about the changes in SGS determined by the most common forest management practices. Simulations have shown that both density and spatial distribution of adult trees (which can be modified by management practice) strongly affect the emergence of SGS (Sagnard et al. 2011), but experimental work is scarce. Marquardt et al. (2007) studied SGS in six populations of Eastern white pine (*Pinus strobus*) under different management systems: shelterwood, pine release, plantation and old growth. In general, they found a weak genetic clumping at the shortest distances. The strongest SGS was found in the old-growth forest, whereas shelterwood managed stand showed the lowest spatial genetic autocorrelation. These results confirmed the negative impact of some management practices on SGS detected by the same authors in similar studies based on the comparison between old-growth and logged stands (Epperson and Chung 2001; Marquardt and Epperson 2004). However, the lack of comparative experiments with managed and non-managed stands in the literature was underlined by Garcia-Gil et al. (2009), in a paper where a new method to jointly estimate the fine-scale genetic structure and inbreeding coefficient was proposed to overcome spurious results related to deviations from Hardy-Weinberg equilibrium.

4.5.3 *Direct Estimate of Contemporary Pollen Flow*

Recently, contemporary estimates of gene flow have become available through assignment methods, which use individual multilocus genotypes instead of allele frequencies to ascertain population membership or parental origin of individuals (Manel et al. 2005). These approaches have benefited from the development of highly variable molecular markers (e.g., microsatellite loci) that provide unequivocal individual fingerprints, even with 5–10 loci and a large number of individuals analyzed. The classical direct approach to estimate pollen dispersal relies on paternity analyses, which consist in using a set of polymorphic markers to genotype a sample of fruiting plants, a sample of their seeds and all males within a circumscribed area, in order to detect the most likely father of each seed (categorical assignment) or to evaluate the likelihood of each male as the father of the considered seed (fractional assignment; Meagher 1986; Devlin and Ellstrand 1990; Jones and Ardren 2003). Table 4-6 gives contemporary gene flow estimates in some conifer species.

Paternity analyses demonstrated that long-distance pollen flow can be extensive in many tree species (Table 4-6; see also Table 2 in Petit and Hampe 2006). For example, in an isolated *Pinus sylvestris* population in Spain, 4.3% of fertilizing pollen came from a distance of at least 30 km (Robledo-Arnuncio and Gil 2005). In an isolated stand of *Pinus densiflora* in Japan, Lian et al. (2001) reported a pollen immigration rate of 31% although the investigated population was surrounded by a residential area, with only few park and garden pines around. In the subtropical wind-pollinated conifer tree species *Araucaria angustifolia*, Bittencourt and Sebenn (2007) showed that 10% of the pollen fertilizing trees within a forest fragment originated from an isolated group of trees approximately 2 km away.

To get a full picture of pollen dispersal patterns, the most natural approach is to estimate the dispersal kernel, defined as the probability density function of the final position of the pollen grain relative to the position of the adult from which it was issued. The dispersal kernel enables representing the fine variations of gene flow with distance. This is quite

Table 4-6 Mean pollination distance and contemporary gene flow estimate in a sample of conifer tree species.

Species	Method	Plot area/ number of pop	distance to the nearest population (non-genotyped)	#mothers	#seeds	Selfing rate	Migration rate	Average dispersal distance (m)	Nep	Reference
Araucaria angustifolia	Paternity analyses	5.4 ha	> 4km	11	220	-	0.06	164.82	12.6 + –2.3	Bittencourt and Sebbenn 2007
Araucaria angustifolia	Two Gener	1 transect	-	10 (among 56 females)	190	-	-	85–98	6.4 (4.2–10.2)	Bittencourt and Sebbenn 2008
Eurycorymbus cavaleriei	Paternity analyses	3 ha	370 km	8	240	-	0.01	292.60	8.8 (5-10)	Wang et al. 2008
Pinus pinaster	TwoGener	24 pop	-	61	720	0.003	-	111.90		De-Lucas et al. 2008
Pinus sylvestris	Paternity analyses & Spatially explicit mating model	20	30 km	34	813	0.25	0.043	135.5		Robledo-Arnuncio and Gil 2005
Pinus densiflora	Paternity analyses	9.12 ha	-	1	874	0.045	0.31	68		Lian et al. 2001
Pinus attenuata	Spatially explicit mating model	15 ha	-	4	880	0.032 (se 0.01)	0.55 (se 0.031)	5.34	59.2	Burczyk et al. 1996
Pinus flexilis	Paternity analyses	15 ha	2 km	71	518	0.02–0.03	0.065.	133–140 m		Schuster and Mitton 2000
Picea abies	Spatially explicit mating model	0.89 ha (seed orchard)	-	-	2000	0.06	0.83	6.8		Burczyk et al. 2004

important, since the impact of the general shape of the pollen dispersal kernel (leptokurtic vs. platykurtic kernels) and of the specific shape of its tail (fat-tailed vs. thin-tailed kernels) on major processes in population biology has been highlighted by various theoretical and experimental studies. For instance, the pollen dispersal kernel allows one to gauge the risk of contamination of seed crops by other fields (Bateman, 1947), while the seed dispersal kernel affects strongly both the rate of colonization and the diversity in newly-founded populations (e.g., Le Corre et al. 1997; Clark 1998; Nathan and Muller-Landau 2000).

Neighborhood models such as those proposed by Adams and Birkes (1991), Adams (1992), Burczyk et al. (2002) and Oddou-Muratorio et al. (2005) can be used to jointly estimate the pollen dispersal kernel and the heterogeneity in fecundity among phenotypically or environmentally defined groups of males. A great advantage of neighborhood models is that they can decompose the inter-individual variance in male reproductive success into a spatial component due to the positions of father-trees relatively to mother-trees and to the pollen dispersal kernel, and into an inter-individual variance of male fecundity (determining the effective male reproductive density). These approaches have been used in a few angiosperms and conifer species to investigate the shape and the range of the pollen dispersal kernel and the variance of male fecundity due to a few covariates individually measured on the putative fathers (Burczyk et al. 1996, 2002, 2004; Burczyk and Prat 1997; Bacles et al. 2005; Oddou-Muratorio et al. 2005). For instance, in knobcone pine (*Pinus atteanuata*), Burczyk et al. (1996) showed that distance and direction of individuals males from mother trees and the size of males (tree height) played significant roles in determining outcross mating patterns within a neighborhood.

A major drawback of paternity-based approaches is that they rely on an exhaustive sampling of the males found in the vicinity of the sampled females, requiring substantial sampling efforts as pollen can come from males that are far from the sampled site (Smouse and Sork 2004). An alternative strategy is the TWOGENER analysis of Smouse et al. (2001), based on the differentiation among the inferred pollen pools of a sample of females, spread across the landscape, and encapsulated in a synthetic parameter Φ_{ft}, that is analogous to F_{ST}, but which relates only to a single bout of pollination. The virtue of this method is that, unlike paternity analysis, it does not require exhaustive sampling of the adults of the population. The global estimate of Φ_{ft}, computed from the entire collection of sampled mothers, is easily translated into an estimate of the mean pollination distance and the effective number of pollinators (Smouse et al. 2001). As an extension of TWOGENER, we can use the computation of pairwise Φ_{ft} between the pollen pools sampled by all pairs of sampled females to estimate multiple parameters jointly, among them the adult density and the average distance of pollen dispersal (Austerlitz and Smouse, 2002; Austerlitz et al. 2004). The most important assumption underlying TWOGENER analyses is that selfing is negligible.

The TWOGENER approach is increasingly used to estimate pollen flow in situations where exhaustive sampling of the parental population is impossible, such as in tropical forests and in high density species. In *Araucaria angustifolia*, Bittencourt and Sebenn (2008) observed a high and significant level of differentiation among pollen clouds ($\Phi_{ft} = 7.8\%$), and a high level of correlated paternity ($r_p = 0.156$), with an average pollination distance between 85 m and 98 m (using either a Gaussian or exponential kernel). In *Pinus pinaster*, De-Lucas et al. (2008) showed that pollen dispersal kernels were very leptokurtic (exponential power distributions with b <0.5) with mean dispersal distances from 78.4 m to 174.4 m. The correlated paternity was quite low ($r_p = 0.048$), indicating that most seeds on any given seed tree are half-sibs.

In summary, the results of available studies support the importance of long-distance pollen flow in many tree species, as highlighted by the predominance of fat-tailed dispersal kernels (i.e., fatter than the exponential). Gene flow thus extends at least hundreds of meters, but long-range dispersal is poorly quantified. Widely distributed northern conifers and other wind-pollinated trees may spread a part of their pollen even further. These studies also demonstrated highly uneven contributions among reproductive individuals, resulting in low effective numbers of pollen donors (e.g., Burczyk et al. 1996; Burczyk and Prat 1997 and review in Smouse and Sork 2004).

4.5.4 Direct Estimate of Contemporary Seed Flow

In the same way that paternity analyses can be used to investigate pollen dispersal, parentage analyses can be used to investigate seed flow. In the case of plant populations, parentage analysis consists in genotyping a sample of dispersed seeds or established seedlings and all the reproductive plants within a circumscribed area for a set of shared polymorphic markers, in order to detect the parent pair of each seedling (Meagher 1986).

Despite the importance of pollination dynamics, fitness of adult plants depends on seedling establishment (Dow and Ashley 1996). Therefore, the assessment of parentage of established seedlings is the only approach that allows documentation of genetically relevant dispersal events, the so called "effective" dispersal of pollen and seeds (Cain et al. 2000; Hardesty et al. 2006). The resulting recruitment pattern is defined as the result of the interaction between dispersal and survival functions. The theory was first described by Janzen (1970) and Connell (1971) ("J-C recruitment pattern model"), and recently modelled by Nathan and Casagrandi (2004).

The main problem related to parentage analysis on dispersed seeds or established seedlings is to discriminate between male and female parentage of seed/seedlings. One solution is to genotype maternally inherited tissues collected on dispersed seeds (Jones et al. 2005; Jordano et al. 2007). When dealing with established seedlings, where purely maternal tissues are no longer available, average effective pollen/seed dispersal distance can be directly estimated from parent-offspring genotype data using model fitting, such as the Neighborhood model (Burczyk et al. 2006). These approaches have gained a broad acceptance among population geneticists and ecologists because they make it possible first, to characterize the seed and pollen dispersal process and the heterogeneity in male/female fertility at ecological time scale (Morgan and Conner 2001; Burczyk et al. 2006), and second to investigate ecological factors that are likely to influence these patterns, such as parental phenotypic traits (Gonzalez-Martinez et al. 2006a), seed dispersers behavior (Jordano et al. 2007), or spatial environmental heterogeneity (Jones et al. 2005).

While the parentage approach can provide relevant answers, few studies have been conducted on contemporary seed flow in conifers. In two studies carried out on the wind-dispersed conifer *P. pinaster*, Gonzalez-Martinez et al. (2002, 2006a) found mean seed dispersal distances ranging from 26.53 m to 58.16 m. Limited distance of effective seed dispersal and recruitment was also detected by Lian et al. (2008) in *Abies sachalinensis*. Parentage analysis seems to be a promising technique to assess the scale and the quality of long-distance dispersal (LDD) events. The study of LDD is crucial for understanding how plants can respond to global environmental changes (Trakhtenbrot et al. 2005). By means of parentage analysis on established juveniles, Troupin (2005) found some colonization events longer than 500 m in *Pinus halepensis*, and Burczyk et al. (2006) observed high levels of pollen and seed flow in a Scots pine stand. Long distance dispersal and establishment events were also detected in angiosperm tree species using the same approach (e.g., Bacles et al. 2006, Hardesty et al. 2006).

In a recent application of parentage analysis, Piotti et al. (2009) studied for the first time gene flow patterns in a recently colonized treeline area. They sampled all Norway spruce individuals in the ecotonal area between the upper continuous forest limit and the upper altitude isolated trees. The focus of the work was on the relationship between reproductive success of local adults vs. gene flow from the outside. This is particularly interesting at treeline given the high selective pressure from unfavorable abiotic conditions on seedling establishment and growth. They found that only 11.1% of the juveniles had a local parent pair, whereas two-thirds of the gamete pool they sampled was not produced at treeline. Effective seed dispersal distance distribution was characterized by a peak far from the seed source [mean 344.66 m ± 191.02 SD (standard deviation)]. Reproductive success was skewed, with six local adults that generated almost two thirds (62.4%) of juveniles with local parents. They concluded that, although a few local adults play an important role in the colonization process at treeline, large levels of gene flow from outside were maintained, suggesting that the potential advantages of local adults (such as local adaptation, proximity to the colonization area, phenological synchrony) did not prevent a large gamete immigration in such a harsh environment.

4.6 Conclusions and Perspectives for Conservation and Use of Genetic Resources in Conifers

Trees in general, and conifers in particular, are a highly genetically diverse group of organisms (Hamrick et al. 1992; Nybom 2004) and except for genes that are transmitted maternally via seed, most of the genetic diversity is within population. Differentiation is typically less than 10% of total variation in paternally and bi-parentally inherited genes and over 50–80% in maternally inherited genes. This particular pattern of variation, due to a differential gene flow between seed and pollen, has made it possible to draw convincing phylogeographic reconstructions. Within populations, gene flow can also be restricted, resulting in spatial genetic structures at the local scale. Such patterns and their study have made it possible to understand the role of demographic and historical factors in shaping the evolutionary trajectories of conifer species, and to highlight conservation threats and strategies in natural populations.

In the field of phylogeography, sequence-type data, or genetic data for which an unambiguous mutational pathway can be assessed between the different variants, is best suited. Similarly, mtDNA (when maternally inherited as in most conifers) is better suited for phylogeography and understanding long term imprints of history as its signal is mostly among populations and not obscured by recombination following reproduction (Petit et al. 2005). However, with the advent of sophisticated statistical analyses, more diverse types of markers, whether uni-parentally or bi-parentally inherited, are often used for phylogeographic reconstructions. When they are used in conjunction with maternally inherited mtDNA and high resolution paleo-ecological data (such as carbon dated fossil pollen), these methods can provide a clear picture of how species recolonized their current distribution areas from refugia they occupied 20,000 years ago (e.g. *Cedrus atlantica*, Cheddadi et al. 2009; *Abies alba*, Liepelt et al. 2009). Nuclear markers such as nSSRs (which recombine at fecundation) and paternally inherited cpSSRs are well suited for pinpointing shorter term demographic events (and threats) and are thus the focus of the field of conservation genetics (e.g., *Cedrus libani*, Fady et al. 2008). Therefore, allelic and haplotypic richness and uniqueness as well as spatial genetic autocorrelations are the statistics of interest, which indicate particular demographic histories. Discrepancies between estimated allelic frequencies and patterns at equilibrium under a random mating model and actual frequencies and patterns, which might indicate the demographic consequences of restricted pollen flow, restricted seed dispersal, localized seedling mortality and non-random mating, are of prime interest to conservationists.

Phylogeography can provide essential background information to disentangle current from past processes and to understand the consequences of crucial events such as migration and colonization for plant species. The comprehension of the past dynamics of genetic diversity can be extremely useful to predict the possible future migrations related to the expected climate changes. In conservation and management of genetic resources, phylogeography can help in the prioritization of areas of high value for conservation. Phylogeographic analyses can also play an important role in defining evolutionary significant units (ESU, Ryder 1986), a unit of conservation below the species level that is often defined on unique geographic distribution and genetic patterns. In addition, phylogeographic surveys may allow tracing of wood and other plant products, providing tools to combat illegal logging or to label products originating from sustainable managed regions.

However, the assessment of possible current gene flow limitations seems particularly urgent in those ecological contexts where anthropogenic and climatic changes are determining the fragmentation of habitats. The ecological consequences of habitat fragmentation, such as pollen limitation and recruitment failure, have been assessed on forest tree species only in a few studies. In particular, little is known about the impact of fragmentation on reproductive patterns and the mating system of wind-pollinated species. Wind pollinated species were not expected to be as sensitive as insect- and animal-pollinated species to fragmentation (O'Connell et al. 2006), although it is becoming now clear that the mating system is highly influenced by local density (Restoux et al. 2008).

There is a clear gap in the literature that precludes the comprehension of forest tree species' responses to habitat fragmentation: the study of contemporary gene flow among fragmented populations combined with the assessment of its ecological consequences (Koenig and Ashley 2003, Aguilar et al. 2008). Experiments designed for the comparison of genetic and ecological parameters among populations subject to increasing levels of fragmentation can help understand how and on what geographical scale gene flow patterns vary, and eventually, to highlight ecologically relevant size thresholds over which a fragmented population should not have experience reproductive failure. Neutral genetic markers are effective in determining gene flow pattern and genetic structure of populations.

Forest genomics now provides tools to identify the genes controlling adaptive traits and methods to carry out new-generation population genetic studies (Gonzalez-Martinez et al. 2006b). The recent development of single neucleotide polymorphism (SNP) markers from candidate genes makes it possible for the first time to study the distribution of genome-wide potentially adaptive variability at the population level (Gonzalez-Martinez et al. 2007; Eveno et al. 2008, Namroud et al. 2008; Neale and Ingvarsson 2008). New and highly efficient SNP-discovery and SNP-genotyping techniques have provided an almost unlimited source of markers and genotyping capacity (Hirschhorn and Daly 2005; Namroud et al. 2008) laying out the foundations of a new landscape genomic approach that will play an important role in the selection of populations for 100 genetic reserves or for establishing *ex situ* conservation plantations (Gonzalez-Martinez et al. 2006b). The comparative study of how neutral and potentially adaptive genes spread or are lost across a fragmented landscape will shed light on the genetic basis of the ecological consequences of habitat fragmentation, providing the conceptual link between the high potential for gene flow and the adaptive failure at fragmented range margins of forest trees.

Although knowledge of population genetics of conifers has progressed tremendously over the years, making clearer our understanding of the role of past demographic history, gene flow and the mating system on the evolution of genetic diversity, much remains to be done. Studies have generally focused on widespread economically important conifer species. Rarer species and marginal populations have drawn less interest. However, in an ever increasingly changing environment, such species and populations could prove valuable per se, as a source of novel gene resources, but also as models to further understand the consequences of ecological drivers on evolutionary trajectories, and vice versa, the ecological consequences of genetic diversity.

References

Adams WT (1992) Gene dispersal within forest tree populations. New For 6: 217–240.
Adams W, Birkes D (1991) Estimating mating patterns in forest tree populations. In: S Fineschi, M Malvolti, F Cannata, H Hattemer (eds) Biochemical Markers in the Population Genetics of Forest Trees. SPB Academic Publ, Hague, The Netherlands, pp 157–172.
Adams WT, Griffin AR, Moran GF (1992) Using paternity analysis to measure effective pollen dispersal in plant populations. Am Nat 140: 762–780.
Afzal-Raffi Z, Dodd R (2007) Chloroplast DNA supports a hypothesis of glacial refugia over postglacial recolonization in disjunct populations of black pine (*Pinus nigra*) in western Europe. Mol Ecol 16: 723–736.
Aguilar R, Quesada M, Ashworth L, Herrerias-Diego Y, Lobo J (2008) Genetic consequences of habitat fragmentation in plant populations: susceptible signals in plant traits and methodological approaches. Mol Ecol 17: 5177–5188.
Aitken SN, Yeaman S Holliday JA, Wang T, Curtis-McLane S (2008) Adaptation, migration or extirpation: Climate change outcomes for tree populations. Evol Appl 1: 95–111.

Alvarez N, Thiel-Egenter C, Tribsch A, Holderegger R, Manel S, Taberlet P, Kupfer P, Brodbeck S, Gaudeul M, Gielly L, Mansion G, Negrini R, Paun O, Pellecchia M, Rioux D, Schonswetter P, Schupfer F, Van Loo M, Winkler M, Gugerli F, and Intra Bio Div Consortium (2009) History or ecology? Substrate type as a major driver of spatial genetic structure in Alpine plants. Ecol Lett 12: 632–640.

Araki NHT, Khatab IA, Hemamali KKGU, InomataN, Wang XR, Szmidt AE (2008) Phylogeography of *Larix sukaczewii* and *Larix sibirica* inferred from nucleotide variation of nuclear genes. Tree Genet Genomes 4: 611–623.

Artyukova EV, Kozyrenko MM, Gorovoy PG, Zhuravlev YN (2009) Plastid DNA variation in highly fragmented populations of *Microbiota decussata* Kom. (Cupressaceae), an endemic to Sikhote Alin Mountains. Genetica 137: 201–212.

Austerlitz F, Smouse PE (2001) Two-generation analysis of pollen flow across a landscape. II. Relation between Fft, pollen dispersal and interfemale distance. Genetics 157: 851–857.

Austerlitz F, Smouse PE (2002) Two-generation analysis of pollen flow across a landscape IV. Estimating the dispersal parameter. Genetics 161: 355–363.

Austerlitz F, Mariette S, Machon N, Gouyon PH, Godelle B (2000) Effects of colonization processes on genetic diversity: differences between annual plants and tree species. Genetics 154: 1309–1321.

Austerlitz F, Dick CW, Dutech C, Klein EK, Oddou-Muratorio S, Smouse PE, Sork VL (2004) Using genetic markers to estimate the pollen dispersal curve. Mol Ecol 13: 937–954.

Avise JC (2000) Phylogeography: The History and Formation of Species. Harvard Univ Press, Cambridge, UK.

Avise JC, Arnold J, Ball RM, Bermingham E, Lamb T, Neigel JE, Reed CA, Saunders NC (1987) Intraspecific phylogeography: the mitochondrial DNA bridge between population genetics and systematics. Annu Rev Ecol Syst 18: 489–522.

Awadalla P, Ritland K (1997) Microsatellite variation and evolution in the *Mimulus guttatus* species complex with contrasting mating systems. Mol Biol Evol 14: 1023–1034.

Bacles CFE, Burczyk J, Lowe AJ, Ennos RA (2005) Historical and contemporary mating patterns in remnant populations of the forest tree *Fraxinus excelsior*. Evolution 59: 979–990.

Bacles CFE, Lowe AJ, Ennos RA (2006) Effective seed dispersal across a fragmented landscape. Science 311: 628.

Bagnoli F, Vendramin GG, Buonamici A, Doulis AG, González-Martínez SC, La Porta N, Magri D, Sebastiani F, Raddi P, Fineschi S (2009) Is *Cupressus sempervirens* native in Italy? An answer from genetic and palaeobotanical data. Mol Ecol 18: 2276–2286.

Ballard JWO, Whitlock MC (2004) The incomplete natural history of mitochondria. Mol Ecol 13: 729–744.

Bateman AJ (1947) Contamination in seed crops III. Relation with isolation distance. Heredity 1: 303–336.

Beaumont MA, Zhang WY, Balding DJ (2002) Approximate Bayesian computation in population genetics. Genetics 162: 2025–2035.

Beerli P, Felsenstein J (1999) Maximum-likelihood estimation of migration rates and effective population numbers in two populations using a coalescent approach. Genetics 152: 763–773.

Belokon MM, Belokon YS, Politov DV, Altukhov YP (2005) Allozyme Polymorphism of Swiss Stone Pine *Pinus cembra* L. in Mountain Populations of the Alps and the Eastern Carpathians. Rus J Genet 41 (11): 1268–1280.

Bergmann F (1991) Causes and consequences of species-specific genetic variation patterns in European forest tree species: examples with Norway Spruce and Silver Fir. In: G Müller-Starck, M Ziehe (eds) Genetic Variation in European Populations of Forest Trees J.D. Sauerländer's, Frankfurt am Main, Germany.

Bittencourt JVM, Sebbenn AM (2007) Patterns of pollen and seed dispersal in a small, fragmented population of the wind-pollinated tree *Araucaria angustifolia* in southern Brazil. Heredity 99: 580–591.

Bittencourt JVM, Sebbenn AM (2008) Pollen movement within a continuous forest of wind-pollinated *Araucaria angustifolia*, inferred from paternity and Two GENER analysis. Conserv Genet 9: 855–868.

Boyle T, Liengsiri C, Piewluang C (1990) Genetic structure of black spruce on two contrasting sites. Heredity 65: 393–399.

Brown GR, Newton CH, Carlson JE (1998) Organization and distribution of a *Sau*3A tandem repeated DNA sequence in *Picea* (*Pinaceae*) species. Genome 41: 560–565.

Brumfield RT, Beerli P, Nickerson DA, Edwards SV (2003) The utility of single nucleotide polymorphisms in inferences of population history. Trends Ecol Evol 18: 249–256.

Burban C, Petit RJ (2003) Phylogeography of maritime pine inferred with organelle markers having contrasted inheritance. Mol Ecol 12: 1487–1495.

Burczyk J, Prat D (1997) Male reproductive success in *Pseudotsuga menziesii* (Mirb.) Franco: the effects of spatial structure and flowering characteristics. Heredity 79: 638–647.

Burczyk J, Adams WT, Shimizu JY (1996) Mating patterns and pollen dispersal in a natural knobcone pine (*Pinus attenuata* Lemmon) stand. Heredity 77: 251–260.

Burczyk J, Adams WT, Moran GF, Griffin AR (2002) Complex patterns of mating revealed in a *Eucalyptus regnans* seed orchard using allozyme markers and the neighbour hood model. Mol Ecol 11: 2379–2391.

Burczyk J, Lewandowski A, Chalupka W (2004) Local pollen dispersal and distant gene flow in Norway spruce (*Picea abies* [L.] Karst.). For Ecol Manag 197: 39–48.

Burczyk J, Adams WT, Birkes DS, Chybicki IJ (2006) Using genetic markers to directly estimate gene flow and reproductive success parameters in plants on the basis of naturally regenerated seedlings. Genetics 173: 363–372.

Cain ML, Milligan BG, Strand AE (2000) Long-distance seed dispersal in plant populations. Am J Bot 87: 1217–1227.

Castelloe J, Templeton AR (1994) Root probabilities for intraspecific gene trees under neutral coalescent theory. Mol Phyl Evol 3: 102–113.

Cheddadi R, Vendramin GG, Litt T, François L, Kageyama M, Lorentz S, Laurent LM, de Beaulieu JL, Sadori L, Jost A, Lunt D (2006) Imprints of glacial refugia in the modern genetic diversity of *Pinus sylvestris*. Global Ecol Biogeogr 15: 271–282.

Cheddadi R, Fady B, François L, Hajar L, Suc JP, Huang K, Demarteau M, Vendramin GG (2009) Putative glacial refugia of *Cedrus atlantica* from Quaternary pollen records and modern genetic diversity. J Biogeogr 36: 1361–1371.

Chen C, Durand E, Forbes F, Francois O (2007). Bayesian clustering algorithms ascertaining spatial population structure: a new computer program and a comparison study. Mol Ecol Notes 7: 747–756.

Chen KM, Abbott RJ, Milne RI, Tian XM, Liu JQ (2008) Phylogeography of *Pinus tabulaeformis* Carr. (Pinaceae), a dominant species of coniferous forest in northern China. Mol Ecol 17: 4276–4288.

Chiang YC, Hung KH, Schaal BA, Ge XJ, Hsu TW, Chiang TY (2006) Contrasting phylogeographical patterns between mainland and island taxa of the *Pinus luchuensis* complex. Mol Ecol 15: 765–779.

Chybicki IJ, Dzialuk A, Trojankiewicz Slawski M, Burczyk J (2008) Spatial genetic structure within two contrasting stands of Scots pine (*Pinus sylvestris* L.). Silvae Genet 57: 193–202.

Clark JS (1998) Why trees migrate so fast: confronting theory with dispersal biology and the paleorecord. Am Nat 152: 204–224.

Clement M, Posada D, Crandall KA (2000) TCS: a computer program to estimate gene genealogies. Mol Ecol 9: 1657–1659.

Cliff AD, Ord JK (1981) Spatial Processes: Models and Applications. Pion Ltd, London, UK.

Connell JH (1971) On the role of natural enemies in preventing competitive exclusion in some marine animals and in forest trees. In: JD den Boer, GR Gradwell (eds) Dynamics of Populations. Centre for Agricultural Publishing and Documentation, Wageningen, The Netherlands, pp 298–312.

Corander J, Waldmann P, Sillanpa MJ (2003) Bayesian analysis of genetic differentiation between populations. Genetics 163: 367–374.
Cronn R, Liston A, Parks M, Gernandt DS, Shen R, Mockler T (2008) Multiplex sequencing of plant chloroplast genomes using Solexa sequencing-by-synthesis technology. Nucl Acids Res 36: e122doi: 10.1093/nar/gkn502.
Dancik B, Yeh F (1983) Allozyme variability and evolution of lodgepole pine (*Pinus contorta* var latifolia) and jack pine (*Pinus banksiana*) in Alberta. Can J Genet Cytol 25: 57–64.
De-Lucas AI, Robledo-Arnuncio JJ, Hidalgo E, Gonzalez-Martinez SC (2008) Mating system and pollen gene flow in Mediterranean maritime pine. Heredity 100: 390–399.
De-Lucas AI, Gonzalez-Martinez SC, Vendramin GG, Hidalgo E, Heuertz M (2009) Spatial genetic structure in continuous and fragmented populations of *Pinus pinaster* Aiton. Mol Ecol 18: 4564–4576.
Demesure B, Sodzi N, Petit RJ (1995) A set of universal primers for amplification of polymorphic non-coding regions of mitochondrial and chloroplast DNA in plants. Mol Ecol 4: 129–131.
Devlin B, Ellstrand NC (1990) Male and female fertility variation in wild radish, a hermaphrodite. Am Nat 136: 87–107.
Dhillon SS (1987) DNA in tree species. In: JM Bouga, DJ Deuzan (eds) Cell and Tissue Culture in Forestry. Martinus Nijhoff, Dordrecht, The Netherlands, pp 298–313.
Doligez A, Baril C, Joly HI (1998) Fine-scale spatial genetic structure with non uniform distribution of individuals. Genetics 148: 905–919.
Dong J, Wagner DB (1994) Paternally Inherited Chloroplast Polymorphism in Pinus: Estimation of diversity and population subdivision, and tests of disequilibrium with a maternally inherited mitochondrial polymorphism. Genetics 136: 1187–1194.
Doudrick RL, Heslopharrison JS, Nelson CD, Schmidt T, Nance WL, Schwarzacher T (1995) Karyotype of slash pine (*Pinus-elliottii* var *elliottii*) using patterns of fluorescence *in situ* hybridization and fluorochrome banding. J Hered 86: 289–296.
Dow B, Ashley M (1996) Microsatellites analysis of seed dispersal and parentage of saplings in bur oak, *Quercus macrocarpa*. Mol Ecol 5: 615–627.
Doyle JJ, Morgante M, Tingey SV, Powell W (1998) Size homoplasy in chloroplast microsatellites of wild perennial relatives of soybean (*Glycine subgenus* Glycine). Mol Biol Evol 15: 215–218.
Du FK, Petit RJ, Liu JQ (2009) More introgression with less gene flow: chloroplast vs. mitochondrial DNA in the *Picea asperata* complex in China, and comparison with other Conifers. Mol Ecol 18: 1396–1407.
Dumolin-Lapègue S, Pemonge M-H, Petit RJ (1997) An enlarged set of consensus primers for the study of organelle DNA in plants. Mol Ecol 6: 393–397.
Dupanloup I, Schneider S, Excoffier L (2002) A simulated annealing approach to define the genetic structure of populations. Mol Ecol 11: 2571–2581.
Duran C, Appleby N, Clark T, Wood D, Imelfort M, Batley J, Edwards D (2008) AutoSNPdb: an annotated single nucleotide polymorphism database for crop plants. Nucl Acids Res 37: D951–D953.
Duran C, Appleby N, Edwards D, Batley J (2009) Molecular genetic markers: discovery, applications, data storage and visualisation. Curr Bioinformat 4: 16–27.
Eckert CG, Samis KE, Lougheed SC (2008a) Genetic variation across species' geographical ranges: the central-marginal hypothesis and beyond. Mol Ecol 17: 1170–1188.
Eckert AJ, Tearse BR, Hall BND (2008b). A phylogeographical analysis of the range disjunction for foxtail pine (*Pinus balfouriana*, Pinaceae): the role of Pleistocene glaciation. Mol Ecol 17: 1983–1997.
Epperson BK (2003) Geographical Genetics. Princeton Univ Press, Princeton, New Jersey, USA.
Epperson BK, Allard RW (1989) Spatial autocorrelation analysis of the distribution of genotypes within populations of lodgepole pine. Genetics 121: 369–377.

Epperson BK, Chung GM (2001) Spatial genetic structure of allozyme polymorphisms within populations of *Pinus strobus* (Pinaceae). Am J Bot 88: 1006–1010.

Eveno E, Collada M, Angeles Guevara M, Léger V, Soto A, Diaz L, Léger P, Gonzalez-Martinez SC, Cervera MT, Plomion C, Garnier-Géré PH (2008) Contrasting patterns of selection at *Pinus pinaster* Ait. drought stress candidate genes as revealed by genetic differentiation analyses. Mol Biol Evol 25: 417–437.

Fady B (2005) Is there really more biodiversity in Mediterranean forest ecosystems? Taxon 54: 905–910.

Fady B, Conord C (2010) Macroecological patterns of species and genetic diversity in vascular plants of the Mediterranean Basin. Divers Distrib 16: 53–64.

Fady B, Fineschi S, Vendramin GG (2004) EUFORGEN Technical Guidelines for genetic conservation and use for Italian stone pine (*Pinus pinea*). IPGRI, Rome, Italy.

Fady B, Lefèvre F, Vendramin GG, Ambert A, Régnier C, Bariteau M (2008) Genetic consequences of past climate and human impact on eastern Mediterranean *Cedrus libani* forests. Implications for their conservation. Conserv Genet 9: 85–95.

Fallour D, Fady B, Lefèvre F (1997) Study on isozyme variation in *Pinus pinea* L.: evidence for low polymorphism. Silvae Genet 46: 201–207.

Fazekas AJ, Yeh FC (2006). Postglacial colonization and population genetic relationships in the *Pinus contorta* complex. Can J Bot 84: 223–234.

Felsenstein J (1971) The rate of loss of multiple alleles in finite haploid populations. Theor Popul Biol 2: 391–403.

Felsenstein J (1988) Phylogenies and quantitative characters. Annu Rev Ecol Syst 19: 445–471.

Felsenstein J (1992) Estimating effective population-size from samples of sequences—inefficiency of pairwise and segregating sites as compared to phylogenetic estimates. Genet Res 59:139–147.

Freeland JR (2005) Molecular Ecology. John Wiley, Chichester, England, UK.

Gamache I, Jaramillo-Correa JP, Payette S, Bousquet J (2003) Diverging patterns of mitochondrial and nuclear DNA diversity in subarctic black spruce: imprint of a founder effect associated with postglacial colonization. Mol Ecol 12: 891–901.

García-Gil MR, Olivier F, Kamruzzahan S, Waldmann P (2009) Joint analysis of spatial genetic structure and inbreeding in a managed population of Scots pine. Heredity 103: 90–96.

Gebremedhin B, Ficetola GF, Naderi S, Rezaei HR, Maudet C, Rioux D, Luikart G, Flagstad O, Thuiller W, Taberlet P (2009) Frontiers in identifying conservation units: from neutral markers to adaptive genetic variation. Anim Consserv 12: 107–109.

Gitzendanner MA, Soltis PS (2000) Patterns of genetic variation in rare and widespread plant congeners. Am J Bot 87: 783–792.

Godbout J, Jaramillo-Correa JP, Beaulieu J, Bousquet J (2005) A mitochondrial DNA minisatellite reveals the postglacial history of jack pine (*Pinus banksiana*), a broad-range North American conifer. Mol Ecol 14: 3497–3512.

Godbout J, Fazekas A, Newton CH, Yeh FC, Bousquet J (2008) Glacial vicariance in the Pacific Northwest: Evidence from a lodgepole pine mitochondrial DNA minisatellite for multiple genetically distinct and widely separated refugia. Mol Ecol 17: 2463–2475.

Gómez A, Vendramin GG, González-Martínez SC, Alía R (2005) Genetic diversity and differentiation of two Mediterranean pines (*Pinus halepensis* Mill. and *Pinus pinaster* Ait.) along a latitudinal cline using chloroplast microsatellite markers. Divers Distrib 11: 257–263.

Gonzalez-Martinez SC, Gerber S, Cervera MT, Martinez-Zapater JM, Gil L, Alia R (2002). Seed gene flow and fine-scale structure in a Mediterranean pine (*Pinus pinaster* Ait.) using nuclear microsatellite markers. Theor Appl Genet 104: 1290–1297.

Gonzalez-Martinez SC, Burczyk J, Nathan R, Nanos N, Gil L, Alia R (2006a) Effective gene dispersal and female reproductive success in Mediterranean maritime pine (*Pinus pinaster* Aiton). Mol Ecol 15: 4577–4588.

Gonzalez-Martinez SC, Krutovsky KV, Neale DB (2006b) Forest-tree population genomics and adaptive evolution. New Phytol 170: 227–238.

Gonzalez-Martinez SC, Wheeler NC, Ersoz E, Nelson CD, Neale DB (2007) Association genetics in *Pinus taeda* L. I. Wood property traits. Genetics 175: 399–409.

Gray MW (1993) Origin and evolution of organelle genomes. Curr Opin Genet Dev 3: 884–890.

Griffiths RC (1980) Lines of descent in the diffusion approximation of neutral Wright-Fischer models. Theor Popul Biol 17: 40–50.

Griffiths RC, Tavare S (1994) Simulating probability-distributions in the coalescent. Theor Popul Biol 46: 131–159.

Gugerli F, Sperisen C, Buchler U, Magni F, Geburek T Jeandroz S, Senn J (2001) Haplotype variation in a mitochondrial tandem repeat of Norway spruce (*Picea abies*) populations suggests a serious founder effect during postglacial re-colonization of the western Alps. Mol Ecol 10: 1255–1263.

Gupta M, Chyi YS, Romeroseverson J, Owen JL (1994) Amplification of DNA markers from evolutionarily diverse genomes using single primers of simple-sequence repeats. Theor Appl Genet 89: 998–1006.

Halushka MK, Fan JB, Bently K, Hsie L, Shen N, Weder A, Cooper R, Lipshutz R, Chakravarti A (1999) Patterns of single nucleotide polymorphisms in candidate genes for blood-pressure homeostasis. Nat Genet 22: 239–247.

Hamrick JL, Godt MJW, Sherman-Broyles SL (1992) Factors influencing levels of genetic diversity in woody plant species. New For 6: 95–124.

Hardesty BD, Hubbell SP, Bermingham E (2006). Genetic evidence of frequent long-distance recruitment in a vertebrate-dispersed tree. Ecol Lett 9: 516–525.

Hardy O, Vekemans X (1999) Isolation by distance in a continuous population: reconciliation between spatial autocorrelation analysis and population genetics models. Heredity 83: 145–154.

Hardy O, Vekemans X (2002) SPAGeDi: a versatile compute program to analyse spatial genetic structure at the individual or population levels. Mol Ecol Notes 2: 618.

Hare MP (2001) Prospects for nuclear gene phylogeography. Trends Ecol Evol 16: 700–706.

Hewitt G (2000) The genetic legacy of the Quaternary ice ages. Nature 405: 907–913.

Hickerson MJ, Stahl E, Takebayashi N (2007) msBayes: Pipeline for testing comparative phylogeographic histories using hierarchical approximate Bayesian computation. BMC Bionformat 8: 268.

Hirschhorn JN, Daly MJ (2005) Genome-wide association studies for common diseases and complex traits. Nat Rev Genet 6: 95–108.

Hoffmann AA, Willi Y (2008) Detecting genetic responses to environmental change. Nat Rev Genet 9: 421–423.

Höhn M, Gugerli F, Abran P, Bisztray G, Buonamici A, Cseke K, Hufnagel L, Quintela-Sabarís C, Sebastiani F, Vendramin GG (2009) Variation in the chloroplast DNA of Swiss stone pine (*Pinus cembra* L.) reflects contrasting post-glacial history of populations from the Carpathians and the Alps. J Biogeogr 36: 1798–1806.

Holmes EC, Worobey M, Rambaut A (1999) Phylogenetic evidence for recombination in dengue virus. Mol Biol Evol 16: 405–409.

Hong YP, Hipkins VD, Strauss SH (1993) Chloroplast DNA diversity among trees, populations and species in the California closed-cone pines (*Pinus radiata*, *Pinus muricata* and *Pinus attenuata*). Genetics 135: 1187–1196.

Hudson RR (1990) Gene genealogies and the coalescent process. Oxford Sur Evol Biol 7: 1–44.

Hudson RR (1998) Island models and the coalescent process. Mol Evol 7: 413–418.

Hughes AR, Inouye BD, Johnson MTJ, Underwood N, Vellend M (2008) Ecological consequences of genetic diversity. Ecol Lett 11: 609–623.

Husmeier D, Wright F (2001) Probabilistic divergence measures for detecting interspecies recombination. Bioinformatics 17: S123–S131.

Isabel N, Beaulieut J, Bousquet J (1995) Complete congruence between gene diversity estimates derived from genotypic data at enzyme and random amplified polymorphic DNA loci in black spruce. Proc Natl Acad Sci USA 92: 6369–6373.

Jakob SS, Ihlow A, Blattner FR (2007) Combined ecological niche modelling and molecular phylogeography revealed the evolutionary history of *Hordeum marinum* (Poaceae)—niche differentiation, loss of genetic diversity, and speciation in Mediterranean Quaternary refugia. Mol Ecol 16: 1713–1727.

Janzen DH (1970) Herbivores and the number of tree species in tropical forests. Am Nat 104: 501–528.

Jaramillo-Correa JP, Bousquet J, Beaulieu J, Isabel N, Perron M, Bouillé M (2003) Cross-species amplification of mitochondrial DNA sequence-tagged-site markers in conifers: the nature of polymorphism and variation within and among species in Picea. Theor Appl Genet 106: 1353–1367.

Jaramillo-Correa JP, Beaulieu J, Bousquet J (2004) Variation in mitochondrial DNA reveals multiple distant glacial refugia in black spruce (*Picea mariana*), a transcontinental North American conifer. Mol Ecol 13: 2735–2747.

Jaramillo-Correa JP, Beaulieu J, Ledig FT, Bousquet J (2006) Decoupled mitochondrial and chloroplast DNA population structure reveals Holocene collapse and population isolation in a threatened Mexican-endemic conifer. Mol Ecol 15: 2787–2800.

Jaramillo-Correa JP, Aguirre-Planter E, Khasa DP, Eguiarte LE, Piñero D, Furnier GR, Bousquet J (2008) Ancestry and divergence of subtropical montane forest isolates: molecular biogeography of the genus *Abies* (Pinaceae) in southern México and Guatemala. Mol Ecol 17: 2476–2490.

Jarne P, Lagoda PJL (1996) Microsatellites, from molecules to populations and back. Trends Ecol Evol 11: 424–429 .

Jeandroz S, Bastien D, Chandelier A, Du Jardin P, Favre JM (2002) A set of primers for amplification of mitochondrial DNA in *Picea abies* and other conifer species. Mol Ecol Notes 2: 389–392.

Johansen AD, Latta RG (2003) Mitochondrial haplotype distribution, seed dispersal and patterns of postglacial expansion of ponderosa pine. Mol Ecol 12: 293–298.

Jones AG, Ardren WR (2003) Methods of parentage analysis in natural populations. Mol Ecol 12: 2511–2523.

Jones FA, Chen J, Weng GJ, Hubbell SP (2005) A genetic evaluation of seed dispersal in the neotropical tree *Jacaranda copaia* (Bignoniaceae). Am Nat 166: 543–555.

Jordano P, Garcia C, Godoy JA, Garcia-Castano JL (2007). Differential contribution of frugivores to complex seed dispersal patterns. Proc Natl Acad Sci USA 104: 3278–3282.

Jump AS, Mátyás C, Peñuelas J (2009) The altitude-for-latitude disparity in the range retractions of woody species. Trends Ecol Evol 24: 694–701.

Kamm A, Doudrick RL, HeslopHarrison JS, Schmidt T (1996) The genomic and physical organization of Ty1-copia-like sequences as a component of large genomes in *Pinus elliottii* var. *elliottii* and other gymnosperms. Proc Natl Acad Sci USA 93: 2708–2713.

Kashi Y, King D, Soller M (1997) Simple sequence repeats as a source of quantitative genetic variation. Trends Genet 13: 74–78.

Khoshoo, TN (1961) Chromosome numbers in gymnosperms. Silvae Genet l: 1–9.

Knowles LL, Maddison WP (2002) Statistical phylogeography. Mol Ecol 11: 2623–2635.

Knowles P (1991) Spatial genetic structure within two natural stand of black spruce [*Picea mariana* (Mill.) B.S.P.]. Silvae Genet 40: 13–19.

Knowles P, Perry D, Foster HA (1992) Spatial genetic structure in two tamarack (*Larix laricinia* (du roi) K. Koch) populations with differing establishment histories. Evolution 46: 572–576.

Koenig WD, Ashley M (2003) Is pollen limited? The answer is blowin' in the wind. Trends Ecol Evol 18:157–159.

Kossack DS, Kinlaw CS (1999) IFG, a gypsy-like retrotransposon in *Pinus* (*Pinaceae*), has an extensive history in pines. Plant Mol Biol 39: 417–426.

Kuhner MK, Yamato J, Felsenstein J (1995) Estimating effective population-size and mutation-rate from sequence data using Metropolis—Hastings sampling. Genetics 140: 1421–1430.

Lagercrantz U, Ryman N (1990) Genetic structure of Norway spruce (*Picea abies*): concordance of morphological and allozymic variation. Evolution 44: 38–53.

Latta RG, Mitton JB (1999) Historical separation and present gene flow through a zone of secondary contact in ponderosa pine. Evolution 53: 769–776.

Le Corre V, Machon N, Petit RJ, Kremer A (1997) Colonisation with long-distance seed dispersal and genetic structure of maternally inherited genes in forest trees: a simulation study. Genet Res 69: 117–125.

Leblois R, Estoup A, Rousset F (2003) Influence of mutational and sampling factors on the estimation of demographic parameters in a "Continuous" population under isolation by distance. Mol Biol Evol 20: 491–502.

Ledig FT, Conkle MT (1983) Gene diversity and genetic structure in a narrow endemic, Torrey Pine (*Pinus torreyana* Parry ex Carr.). Evolution 3: 79–85.

Ledig FT, Jacob-Cervantes V, Hodgskiss PD, Eguiluz-Piedra T (1997) Recent evolution and divergence among populations of a rare Mexican endemic, Chihuahua spruce, following Holocene climatic warming. Evolution 51: 1815–1827.

Lefevre F, Fady B, Fallour-Rubio D, Ghosn D, Bariteau M (2004) Impact of founder population, drift and selection on the genetic diversity of a recently translocated tree population. Heredity 93: 542–550.

Leimu R, Mutikainen P, Koricheva J, Fischer M (2006) How general are positive relationship between plant population size, fitness, and genetic variation? J Ecol 94: 942–952.

Lenormand T (2002) Gene flow and the limits to natural selection. Trends Ecol Evol 17: 183–189.

Leonardi S, Raddi S, Borghetti M (1996) Spatial autocorrelation of allozyme traits in a Norway spruce (*Picea abies*) population. Can J For Res 26: 63–71.

Li C, Chai B, Wang M (2008). Population Genetic Structure of *Pinus tabulaeformis* in Shanxi Plateau, China. Rus J Ecol 39: 34–40.

Li N, Stephens M (2003) Modelling linkage disequilibrium, and identifying recombination hotspots using SNP data. Genetics 165: 2213–2233.

Lian CL, Miwa M, Hogetsu T (2001) Outcrossing and paternity analysis of *Pinus densiflora* (Japanese red pine) by microsatellite polymorphism. Heredity 87: 88–98.

Lian C, Goto S, Kubo T, Takahashi Y, Nakagawa M, Hogetsut (2008) Nuclear and chloroplast microsatellite analysis of *Abies sachalinensis* regeneration on fallen logs in a sub-boreal forest in Hokkaido, Japan. Mol Ecol 17: 2948–2962.

Liepelt S, Bialozyt R, Ziegenhagen B (2002) Wind-dispersed pollen mediates gene flow among refugia. Proc Natl Acad Sci USA 99: 14590–14594.

Liepelt S, Cheddadi R, de Beaulieu JL, Fady B, Gömöry D, Hussendörfer E, Konnert M, Litt T, Longauer R, Terhürne-Berson R, Ziegenhagen B (2009) Postglacial range expansion and its genetic imprints in *Abies alba* (Mill.). A synthesis from paleobotanic and genetic data. Rev Palaeobot Palyn 153: 139–149.

Liu YS, Basinger LF (2000) Fossil Cathaya (Pinaceae) pollen from the Canadian High Arctic. Int J Plant Sci 161: 829–847.

Loiselle BA, Sork VL, Nason J, Graham C (1995) Spatial genetic structure of a tropical understory shrub, *Psychotria officinalis* (Rubiaceae). Am J Bot 82: 1420–1425.

Maddison WP, Maddison DR (2007) Mesquite: a modular system for evolutionary analysis. Version 2.0: *http://mesquiteproject.org*

Magri D, Fineschi S, Bellarosa R, Buonamici A, Sebastiani F, Schirone B, Simeone MC, Vendramin GG (2007) The distribution of *Quercus suber* chloroplast haplotypes matches the palaeogeographical history of the western Mediterranean. Mol Ecol 16: 5259–5266.

Manel S, Schwartz M, Luikart G, Taberlet P (2003) Landscape genetics: combining landscape ecology and population genetics. Trends Ecol Evol 18: 189–197.

Manel S, Gaggiotti OE, Waples RS (2005) Assignment methods: matching biological questions with appropriate techniques. Trends Ecol Evol 20: 136–142.
Manel S, Berthoud F, Bellemain E, Gaudeul M, Luikart G, Swenson JE, Waits LP, Taberlet P (2007) A new individual-based spatial approach for identifying genetic discontinuities in natural populations. Mol Ecol 16: 2031–2043.
Marquardt PE, Epperson BK (2004) Spatial and population genetic structure of microsatellites in white pine. Mol Ecol 13: 3305–3315.
Marquardt PE, Echt CS, Epperson BK, Pubanz DM (2007) Genetic structure, diversity, and inbreeding of eastern white pine under different management conditions. Can J For Res 37: 2652–2662.
Marsjan PA, Oldenbroek JK (2007) Molecular markers, a tool for exploring genetic diversity. In: B Rischkowsky, D Pilling (eds) The State of the World's Animal Genetic Resources for Food and Agriculture, section C, part 4. FAO, Rome, Italy, pp 359–379.
Meagher TR (1986) Analysis of paternity within a natural population of *Chamaelirium luteum* I. Identification of most-likely male parents. Am Nat 128: 199–215.
Meng L, Yang R, Abbott RJ, Miehe G, Hu T, Liu J (2007) Mitochondrial and chloroplast phylogeography of *Picea crassifolia* Kom. (Pinaceae) in the Qinghai-Tibetan Plateau and adjacent highlands. Mol Ecol 16: 4128–4137.
Millar CI (1983) A steep cline in *Pinus muricata*. Evolution 37: 311–319.
Mitton JB, Kreiser BR, Latta RG (2000) Glacial refugia of limber pine (*Pinus flexilis* James) inferred from the population structure of mitochondrial DNA. Mol Ecol 9: 91–97.
Moreno-Letelier A, Piñero D (2009) Phylogeographic structure of *Pinus strobiformis* Engelm. across the Chihuahuan Desert filter-barrier. J Biogeogr 36: 121–131.
Morgan MT, Conner JK (2001) Using genetic markers to directly estimate male selection gradients. Evolution 55: 272–281.
Morgante M, Olivieri A (1993) PCR-amplified microsatellites as markers in plant genetics. Plant J 3: 175–182.
Moriguchi Y, Kang KS, Lee KY, Lee SW, Kim YY (2009) Genetic variation of *Picea jezoensis* populations in South Korea revealed by chloroplast, mitochondrial and nuclear DNA markers. J Plant Res 122:153–160.
Murray BG (1998) Nuclear DNA amounts in gymnosperms. Ann Bot 82: 3–15.
Namroud M-C, Beaulieu J, Juge N, Laroche J, Bousquet J (2008) Scanning the genome for gene single nucleotide polymorphisms involved in adaptive population differentiation in white spruce. Mol Ecol 17: 3599–3613.
Nasri N, Bojovic S, Vendramin G.G, Fady B (2008). Population genetic structure of the relict Serbian spruce, *Picea omorika* [Panc.] Purk, inferred from plastid DNA. Plant Syst Evol 271: 1–7.
Nathan R, Muller-Landau HC (2000) Spatial patterns of seed dispersal, their determinants and consequences for recruitment. Trends Ecol Evol 15: 278–285.
Nathan R, Casagrandi R (2004). A simple mechanistic model of seed dispersal, predation and plant establishment: Janzen-Connell and beyond. J Ecol 92: 733–746.
Naydenov K, Senneville S, Beaulieu J, Tremblay FM, Bousquet J (2007) Glacial vicariance in Eurasia: mitochondrial DNA evidence from Scots pine for a complex heritage involving genetically distinct refugia at mid-northern latitudes and in Asia Minor. BMC Evol Biol 7: 233.
Neale DB, Ingvarsson PK (2008) Population, quantitative and comparative genomics of adaptation in forest trees. Curr Opin Plant Biol 11: 149–155.
Nielsen R, Wakeley J (2001) Distinguishing migration from isolation: a Markov chain Monte Carlo approach. Genetics 158: 885–896.
Nielsen R, Beaumont MA (2009) Statistical inferences in phylogeography. Mol Ecol 18: 1034–1047.
Nybom H (2004) Comparison of different nuclear DNA markers for estimating intraspecific genetic diversity in plants. Mol Ecol 13: 1143–1155.

O'Connell LM, Mosseler A, Rajora OP (2006) Impacts of forest fragmentation on the reproductive success of white spruce (*Picea glauca*). Can J Bot 84: 956–965.

Oddou-Muratorio S, Klein EK (2008) Comparing direct vs. indirect estimates of gene flow within a population of a scattered tree species. Mol Ecol 17: 2743–2754.

Oddou-Muratorio S, Klein EK, Austerlitz F (2005) Pollen flow in the wild service tree, *Sorbus torminalis* (L.) Crantz II. Pollen dispersal and heterogeneity in mating success inferred from parent-offspring analysis. Mol Ecol 14: 4441–4452.

Opgenoorth L, Vendramin GG, Mao K, Miehe G, Miehe S, Liepelt S, Liu J, Ziegenhagen B (2010) Tree endurance on the Tibetan Plateau marks the world's highest known tree line of the Last Glacial Maximum. New Phytol 185: 332–342.

Panchal M (2007) The automation of nested clade phylogeographic analysis. Bioinformatics 4: 509–510.

Parducci L, Szmidt AE, Madaghiele A, Anzidei M, Vendramin GG (2001) Genetic variation at chloroplast microsatellites (cpSSRs) in *Abies nebrodensis* (Lojac.) Mattei and three neighbouring Abies species. Theor Appl Genet 102: 733–740.

Parks DH, Porter M, Churcher S, Wang SW, Blouin C, Whalley J, Brooks S, Beiko RG (2009) GenGIS: A geospatial information system for genomic data. Genome 19 (10): 1896–1904.

Peakall R, Smouse PE (2005) GenAlEx V6: Genetic analysis in Excel Population genetic software for teaching and research. Australian National Univ, Canberra: *http://www.anu.edu.au/BoZo/GenAlEx*

Peakall R, Smouse PE (2006) GenAlEx 6: genetic analysis in Excel Population genetic software for teaching and research. Mol Ecol Notes 6: 288–295.

Petit RJ, Hampe A (2006) Some evolutionary consequences of being a tree. Annu Rev Ecol Evol Syst 37: 187–214.

Petit RJ, Vendramin GG (2007) Plant phyologeography based on organelle genes: an introduction. In: S Weiss, F Nuno (eds) Phylogeography of Southern European Refugia: Evolutionary Perspectives on the Origins and Conservation of European Biodiversity. Springer, Heidelberg, Germany, pp 23–97.

Petit RJ, Bahrman N, Baradat PH (1995). Comparison of genetic differentiation in maritime pine (*Pinus pinaster* Ait.) estimated using isozyme, total protein and terpenic loci. Heredity 75: 382–389.

Petit RJ, Aguinagalde I, de Beaulieu JL, Bittkau C, Brewer S, Chedaddi R, Ennos R, Fineschi S, Grivet D, Lascoux M, Mohanty A, Muller-Starck G, Demesure-Musch B, Palme A, Martin JP, Rendell S, Vendramin GG (2003) Glacial refugia: hotspot but not melting pots of genetic diversity. Science 300: 1563–1565.

Petit RJ, Duminil J, Fineschi S, Hampe A, Slavini D, Vendramin GG (2005) Comparative organization of chloroplast, mitochondrial and nuclear diversity in plant populations. Mol Ecol 14: 689–701.

Piotti A, Leonardi S, Piovani P, Scalfi M, Menozzi P (2009) Spruce colonization at treeline: where do those seeds come from? Heredity 103: 136–145.

Pitelka LF, Gardner RH, Ash J, Berry S, Gitay H, Noble IR, Saunders A, Bradshaw RHW, Brubaker L, Clark JS, Davis MB, Sugita S, Dyer JM, Hengeveld R, Hope G, Huntley B, King GA, Lavorel S, Mack RN, Malanson GP, McGlone M, Prentice IC, Rejmanek M (1997) Plant migration and climate change. Am Sci 85: 464–473.

Pleines T, Jakob SS, Blattner FR (2009) Application of non-coding DNA regions in intraspecific analyses. Plant Syst Evol 282: 281–294.

Posada D, Crandall KA (2001) Intraspecific gene genealogies: trees grafting into networks. Trends Ecol Evol 16: 37–45.

Posada D, Crandall KA, Templeton AR (2000) GeoDis: a program for the cladistic nested analysis of the geographical distribution of genetic haplotypes. Mol Ecol 9: 487–488.

Posada D, Maxwell TJ, Templeton AR (2005) TreeScan: a bioinformatics application to search for genotype/phenotype associations using haplotype trees. Bioinformatics 21: 2130–2132.

Posada D, Crandall KA, Templeton AR (2006) Nested clade analysis statistics. Mol Ecol Notes 6: 590–593.

Powell W, Morgante M, McDevitt R, Vendramin GG, Rafalski AJ (1995) Polymorphic simple sequence repeat regions in the chloroplast genome: applications to the population genetics of pines. Proc Natl Acad Sci USA 92: 7759–7763.

Pritchard JK, Stephens M, Donnelly P (2000) Inference of population structure using multilocus genotype data. Genetics 155: 945–959.

Provan J, Bennett KD (2008). Phylogeographic insights into cryptic glacial refugia. Trends Ecol Evol 23: 564–571.

Provan J, Powell W, Hollingsworth PM (2001) Chloroplast microsatellites: new tools for studies in plant ecology and evolution. Trends Ecol Evol 16:142–147.

Pyhäjärvi T, Salamela MJ, Savolainen O (2008) Colonization routes of *Pinus sylvestris* inferred from distribution of mitochondrial DNA variation. Tree Genet Genomes 4: 247–254.

Restoux G, Silva ED, Sagnard F, Torre F, Klein E, Fady B (2008) Life at the margin: the mating system of Mediterranean conifers. Web Ecol 8: 94–102.

Ribeiro MM, Plomion C, Petit R, Vendramin GG, Szmidt AE (2001) Variation in chloroplast single-sequence repeats in Portuguese maritime pine (*Pinus pinaster* Ait.). Theor Appl Genet 102: 97–103.

Richardson BA, Brunsfeld SJ, Klopfenstein NB (2002) DNA from bird-dispersed seed and wind-disseminated pollen provides insights into postglacial colonization and population genetic structure of white bark pine (*Pinus albicaulis*). Mol Ecol 11: 215–227.

Robledo-Arnuncio JJ, Gil L (2005) Patterns of pollen dispersal in a small population of *Pinus sylvestris* L. revealed by total-exclusion paternity analysis. Heredity 94: 13–22.

Robledo-Arnuncio JJ, Austerlitz F, Smouse PE (2006) A new method of estimating the pollen dispersal curve independently of effective density. Genetics 173: 1033–1045.

Robledo-Arnuncio JJ, Austerlitz F, Smouse PE (2007) POLDISP: a software package for indirect estimation of contemporary pollen dispersal. Mol Ecol Notes 7: 763–766.

Rodríguez-Banderas A, Vargas-Mendoza CF, Buonamici A, Vendramin GG (2009) Genetic diversity and phylogeographic analysis of *Pinus leiophylla*: a post-glacial range expansion. J Biogeogr 36: 1807–1820.

Rousset F (1997) Genetic differentiation and estimation of gene flow from F-Statistics under isolation by distance. Genetics 145: 1219–1228.

Rousset F (2000) Genetic differentiation between individuals. J Evol Biol 13: 58–62.

Rousset F (2008) Genepop'007: a complete re-implementation of the genepop software for Windows and Linux. Mol Ecol Res 8: 103–106.

Ryder O.A., 1986. Species conservation and systematics: the dilemma of subspecies. Trends Ecol Evol 1: 9–10.

Sagnard F, Oddou-Muratorio S, Pichot C, Vendramin G.G, Fady B (2010) Effect of seed dispersal, adult tree and seedling density on the spatial genetic structure of regeneration at fine temporal and spatial scales. Tree Genetics and Genomes. In press.

Schuster WSF, Mitton JB (2000) Paternity and gene dispersal in limber pine (*Pinus flexii* James). Heredity 84: 348–361.

Scotti I, Gugerli F, Pastorelli R, Sebastiani F, Vendramin GG (2008) Maternally and paternally inherited molecular markers elucidate population patterns and inferred dispersal processes on a small scale within a subalpine stand of Norway spruce (*Picea abies* [L.] Karst.). For Ecol Manag 255: 3806–3812.

Semerikov VL, Lascoux M (2003) Nuclear and citoplasmic variation within and between Eurasian Larix (Pinaceae) species. Am J Bot 90: 1113–1123.

Sinclair WT, Morman JD, Ennos RA (1998) Multiple origins for Scots pine (*Pinus sylvestris* L.) in Scotland: evidence from mitochondrial DNA variation. Heredity 80: 233–240.

Sinclair WT, Morman JD, Ennos RA (1999) The postglacial history of Scots pine (*Pinus sylvestris* L.) in western Europe: evidence from mitochondrial DNA variation. Mol Ecol 8: 83–88.

Sisson SA, Fan Y, Tanaka MM (2007) Sequential Monte Carlo without likelihoods. Proc Natl Acad Sci USA 104: 1760–1765.

Skrøppa T (2003) EUFORGEN Technical Guidelines for genetic conservation and use for Norway spruce (*Picea abies*). IPGRI, Rome, Italy.

Slatkin M (1987) The average number of sites separating DNA sequences drawn from a subdivided population. Theor Popul Biol 32: 42–49.

Smouse PE, Peakall R (1999) Spatial autocorrelation analysis of individual multi allele and multilocus genetic structure. Heredity 82: 561–573.

Smouse PE, Sork VL (2004) Measuring pollen flow in forest trees: an exposition of alternative approaches. For Ecol Man 197: 21–38.

Smouse PE, Dyer RJ, Westfall RD, Sork VL (2001) Two-generation analysis of pollen flow across a landscape I. Male gamete heterogeneity among female. Evolution 55: 260–271.

Soleimani VD, Baum BR, Johnason DA (2003) Efficient validation of single nucleotide polymorphisms in plants by allele-specific PCR, with an example from Barley Plant. Mol Bio Rep 21: 281–288.

Song BH, Wang XQ, Wang XR, Ding KY, Hong DY (2003) Cytoplasmic composition in *Pinus densata* and population establishment of the diploid hybrid pine. Mol Ecol 12: 2995–3001.

Soranzo N, Alia R, Provan J, Powell W (2000) Patterns of variation at a mitochondrial sequence-tagged-site locus provides new insights into the postglacial history of European *Pinus sylvestris* populations. Mol Ecol 9: 1205–1211.

Sperisen C, Büchler U, Gugerli F, Mátyas G, Geburek T, Vendramin GG (2001) Tandem repeats in plant mitochondrial genomes: application to the analysis of population differentiation in the conifer Norway spruce. Mol Ecol 10: 257–263.

Steane DA (2005) Complete nucleotide sequence of the chloroplast genome from the Tasmanian Blue Gum, *Eucalyptus globulus* (Myrtaceae). DNA Res 12: 215–222.

Strauss SH, Palmer JD, Howe GT, Doerksen AH (1988) Chloroplast genomes of two conifers lack a large inverted repeat and are extensively rearranged. Proc Natl Acad Sci USA 85: 3898–3902.

Strauss SH, Hong YP, Hipkins VD (1993) High levels of population differentiation for mitochondrial DNA haplotypes in *Pinus radiata, muricata,* and *attenuata*. Theor Appl Genet 86: 573–578.

Taberlet P, Fumagalli L, Wust-Saucy AG, Cosson JF (1998) Comparative phylogeography and postglacial colonization routes in Europe. Mol Ecol 7: 453–464.

Tang S, Dai W, Li M, Zhang Y, Geng Y, Wang L, Zhong Y (2008) Genetic diversity of relictual and endangered plant *Abies ziyuanensis* (Pinaceae) revealed by AFLP and SSR markers. Genetica 133: 21–30.

Tavaré S (1984) Line-of-descent and genealogical processes, and their applications in population genetic models. Theor Popul Biol 26: 119–164.

Templeton AR (2004) Statistical phylogeography: methods of evaluating and minimizing inference errors. Mol Ecol 13: 789–809.

Templeton AR, Boerwinkle E, Sing CF (1987) A cladistic analysis of phenotypic associations with haplotypes inferred from restriction endonuclease mapping 1. Basic theory and an analysis of alcohol-dehydrogenase activity in *Drosophila*. Genetics 117: 343–351.

Templeton AR, Crandall KA, Sing CF (1992) A cladistic analysis of phenotypic associations with haplotypes inferred from restriction endonuclease mapping and DNA sequence data. III. Cladogram estimation. Genetics 132: 619–633.

Templeton AR, Routman E, Phillips CA (1995) Separating population structure from population history—a cladistic analysis of the geographical distribution of mitochondrial DNA haplotypes in the tiger salamander, *Ambystoma tigrinum*. Genetics 140: 767–782.

Terrab A, Paun O, Talavera S, Tremetsberger K, Arista M, Stuessy TF (2006) Genetic diversity and population structure in natural populations of Moroccan Atlas cedar (*Cedrus atlantica*; Pinaceae) determined with cpSSR markers. Am J Bot 93: 1274–1280.

Terrab A, Schönswetter P, Talavera S, Vela E, Stuessy TF (2008) Range-wide phylogeography of *Juniperus thurifera* L., a presumptive keystone species of western Mediterranean vegetation during cold stages of the Pleistocene. Mol Phylogenet Evol 48: 94–102.

Tollefsrud MM, Kissling R, Gugerli F, Johnsen Ø, Skrøppa T, Cheddadi R, van der knaap WO, Latalowa M, Terhürne-Berson R, Litt T, Geburek T, Brochmann C, Sperisen C (2008) Genetic consequences of glacial survival and postglacial colonization in Norway spruce: combined analysis of mitochondrial DNA and fossil pollen. Mol Ecol 17: 4134–4150.

Tollefsrud MM, Sonstebo JH, Brochmann C, Johnsen Ø, Skrøppa T, Vendramin GG (2009) Combined analysis of nuclear and mitochondrial markers provide new insight into the genetic structure of North European *Picea abies*. Heredity 102: 549–562.

Trakhtenbrot A, Nathan R, Perry G, Richardson DM (2005) The importance of long-distance dispersal in biodiversity conservation. Divers Distrib 11: 173–181.

Troupin D (2005) Genetic structure and seed dispersal in Aleppo pine (*Pinus halepensis*). MSc Thesis, The Hebrew Univ of Jerusalem, Israel.

Troupin D, Nathan R, Vendramin GG (2006) Analysis of spatial genetic structure in an expanding *Pinus halepensis* population reveals development of fine-scale genetic clustering over time. Mol Ecol 15: 3617–3630.

Tsumura Y, Suyama Y (1998) Differentiation of mitochondrial DNA polymophisms in populations of five Japanese *Abies* Species. Evolution 52: 1031–1042.

Vekemans X, Hardy OJ (2004) New insights from fine-scale spatial genetic structure analyses in plant populations. Mol Ecol 13: 921–935.

Vendramin GG, Ziegenhagen B (1997) Characterization and inheritance of polymorphic plastid microsatellites in *Abies*. Genome 40: 857–864.

Vendramin GG, Lelli L, Rossi P, Morgante M (1996) A set of primers for the amplification of 20 chloroplast microsatellites in Pinaceae. Mol Ecol 5: 595–598.

Vendramin GG, Anzidei M, Madaghiele A, Bucci G (1998) Distribution of genetic diversity in *Pinus pinaster* Ait. as revealed by chloroplast microsatellites. Theor Appl Genet 97: 456–463.

Vendramin GG, Degen B, Petit RJ, Anzidei M, Madaghiele A, Ziegenhagen B (1999) High level of variation at *Abies alba* chloroplast microsatellite loci in Europe. Mol Ecol 8: 1117–1126.

Vendramin GG, Anzidei M, Madaghiele A, Sperisen C, Bucci G (2000) Chloroplast microsatellite analysis reveals the presence of population subdivision in Norway spruce (*Picea abies* K.). Genome 43: 68–78.

Vendramin GG, Fady B, Gonzalez-Martinez SC, Hu FS, Scotti I, Sebastiani F, Soto A, Petit RJ (2008) Genetically depauperate but widespread: The case of an emblematic Mediterranean pine. Evolution 62: 680–688.

Vos P, Hagers R, Bleeker M, Reijans M, van de Lee T, Hornes M, Friters A, Pot J, Paleman J, KuiperM, Zabeau M (1995) Aflp: new technique for DNA fingerprinting. Nucl Acids Res 23: 4407–4414.

Wakamiya I, Newton RJ, Johnston JS, Price HJ (1993) Genome size and environmental-factors in the genus *Pinus*. Am J Bot 80: 1235–1241.

Wakasugi T, Tsudzuki J, Ito S, Nakashima K, Tsudzuki T, Sugiura M (1994) Loss of all ndh genes as determined by sequencing the entire chloroplast genome of the black pine *Pinus thunbergii*. Proc Natl Acad Sci USA 91: 9794–9798.

Wakeley J, Hey J (1997) Estimating ancestral population parameters. Genetics 145: 847–855.

Wang HW, Ge S (2006) Phylogeography of the endangered *Cathaya argyrophylla* (Pinaceae) inferred from sequence variation of mithocondrial and nuclear DNA. Mol Ecol 15: 4109–4122.

Wang J, Ye Q, Kang M, Huang H (2008) Novel polymorphic microsatellite loci and patterns of pollen-mediated gene flow in an *ex situ* population of *Eurycorymbus cavaleriei* (Sapindaceae) as revealed by categorical paternity analysis. Conserv Genet 9: 559–567.

Weising K, Gardner RG (1999) A set of conserved PCR primers for the analysis of single sequence repeat polymorphisms in chloroplast genomes of dicotyledonous angiosperms. Genome 42: 9–19.

Wheeler N, Guries R (1982) Population-structure, genic diversity, and morphological variation in *Pinus contorta* Dougl. Can J For Res 12: 595–606.

Whitlock MC, McCauley DE (1999) Indirect measures of gene flow and migration: Fst=1/(4Nm+1). Heredity 82: 117–125.
Whittle CA, Johnston MO (2002) Male-driven evolution of mitochondrial and chloroplastidial DNA sequences in plants. Mol Biol Evol 19: 938–949.
Wright S (1943) Isolation by distance. Genetics 28: 114–138.
Wright S (1951) The genetical structure of populations. Ann Eugen 15: 323–354.
Wu J, Krutovskii KV, Strauss SH (1999) Nuclear DNA diversity, population differentiation, and phylogenetic relationships in the California closed-cone pines based on RAPD and allozyme markers. Genome 42: 893–908.
Xue X, Wang Y, Korpelainen H, Li C (2007) Genetic diversity of *Picea asperata* populations based on RAPDs. Plant Biol 9: 101–108.
Ye TZ, Yang RC, Yeh FC (2002) Population structure of a lodgepole pine (*Pinus contorta*) and jack pine (*P. banksiana*) complex as revealed by random amplified polymorphic DNA. Genome 45: 530–540.
Young A, Boyle T, Brown T (1996) The population genetic consequences of habitat fragmentation for plants. Trends Ecol Evol 11: 413–418.
Yu H, Ge S, Hong D-Y (2000) Allozyme diversity and population genetic structure of *Pinus densata* Masters in northwestern Yunnan, China. Biochem Genet 38: 139–147.
Zhang A-B, Tan S, Sota T (2006) AUTOINFER 1.0: a computer program to infer biogeographical events automatically. Mol Ecol Notes 6: 597–599.
Zhang D-X, Hewitt GM (2003) Nuclear DNA analyses in genetic studies of populations: practice, problems and prospects. Mol Ecol 12: 563–584.

5

Genetic Mapping in Conifers

Kermit Ritland,[1,*] Konstantin V. Krutovsky,[2] Yoshihiko Tsumura,[3] Betty Pelgas,[4,a] Nathalie Isabel[4,b] and Jean Bousquet[5]

ABSTRACT

This chapter summarizes the history and current status of genetic mapping in conifers. We review the development of molecular markers, methods to construct genetic maps, and the resulting conifer genetic maps. Genetic maps are subdivided into (1) linkage maps of genetic markers, (2) quantitative trait loci (QTL) maps, and (3) comparative maps. Comparative maps involve alignment of marker genes and even QTLs between species. Physical mapping is also briefly discussed. Emphasis is placed up problems and approaches unique to conifers, and the involvement of new genomics technologies.

Keywords: genetic markers, genetic mapping, quantitative trait loci mapping, comparative mapping

5.1 Introduction

Genetic mapping is the ordering of specific genes or DNA fragments (genetic markers) along a chromosome, based up observed frequencies of recombination in pedigrees. It provides the approximate locations of these

[1]Department of Forest Sciences, University of British Columbia, Vancouver, British Columbia V6T 1Z4, Canada; e-mail: *kermit.ritland@ubc.ca*

[2]Department of Ecosystem Science and Management, Texas A&M University, College Station, Texas 77843-2138, USA; e-mail: *k-krutovsky@tamu.edu*

[3]Forestry and Forest Products Research Institute, Tsukuba, Ibaraki 305-8687, Japan; e-mail: *ytsumu@ffpri.affrc.go.jp*

[4]Natural Resources Canada, Canadian Forest Service, Laurentian Forestry Centre, 1055 du P.E.P.S., P.O. Box 10380, Stn Sainte-Foy, Québec, Québec G1V 4C7, Canada;

[a]e-mail: *betty.pelgas@RNCan-NRCan.gc.ca*

[b]e-mail: *nisabel@cfl.forestry.ca*

[5]Canada Research Chair in Forest and Environmental Genomics, Centre d'étude de la forêt, Université Laval, Québec, Québec G1V 0A6, Canada; e-mail: *jean.bousquet@sbf.ulaval.ca*

*Corresponding author

entities, which can serve as DNA "landmarks" for further studies (Ott 1999). Physical mapping, in contrast, uses various molecular techniques to reassemble the actual DNA into contiguous stretches, such that numbers of bases separating genes are approximately known, as in for example the chloroplast genome (Tsumura et al. 1993) and the nuclear genome (Amarasinghe and Carlson 1998). Quantitative trait loci (QTL) mapping places the locations of putative genes underlying a quantitative trait onto a genetic map (Lander and Botstein 1989). Conifers have enormous genomes, on the order of tens of billions of nucleotides (Murray 1998). This prohibits physical mapping, and suggests that marker/QTL mapping may continue to dominate conifer genetics research (Neale et al. 1994; White et al. 2007). In addition, the conserved nature of conifer evolution places greater importance on comparing genetic and QTL maps (comparative mapping) in conifers (Krutovsky et al. 2004) and transferring information among these species.

Conifers provide unique opportunities but also problems for genetic mapping. Most notably, the gametophyte allows direct observation of the haploid product of maternal meiosis (Cairney and Pullman 2007). Secondly, conifers are outbred, and issues in data analysis arise from the fact that parents and grandparents are heterozygous for markers and QTL, requiring more complex approaches for data analysis (Liu 1998). Thirdly, the large genome size of conifers, a consequence of repeated DNA elements (Morse et al. 2009), make protocols for marker screening more complex, and the development of markers more difficult compared to most angiosperms (Kinlaw and Neale 1997). Finally, the enormous evolutionary distance between conifers and angiosperms, separated by 300 hundred million years of evolution (Savard et al. 1994), makes gene identification and annotation in conifers very difficult (Kirst et al. 2003; Ralph et al. 2008). Here, we review the current state of conifer genome mapping, with reference to current advances in genomics studies of conifers.

5.2 Types and Properties of Genetic Markers for Conifers

Over the past 20 years, the increasing availability of molecular genetic markers such as restriction fragment length polymorphisms (RFLPs), amplified fragment length polymorphisms (AFLPs), microsatellites or simple sequence repeats (SSRs), single nucleotide polymorphisms (SNPs), and conserved orthologous sets (COS), has resulted in the development of numerous genetic linkage maps in conifers. The important conifer species—many pines and spruces, Sugi and Douglas-fir—have been mapped, though marker density is still low in relation to genome size. For a typical conifer, a map with 1,000 markers would have, on average, 10–40 million nucleotide sites separating adjacent markers.

5.2.1 First Generation Markers

Before molecular markers became popular in the 1980s, isozymes or allozymes were used for molecular population genetic investigations in conifers. Isozymes are enzymes that differ in amino acid sequence but catalyze the same chemical reaction. Thus, they are representative of differences at the DNA level. Isozymes gave the first revelation about DNA variation in conifers, and for a period centering about the 1980s, many conifers were the subject of isozyme investigations. The dawn of the isozyme era was heralded by a seminal 1979 symposia on "Isozymes in Forest Genetics and Forest Insects" (Conkle 1981). The dusk of the isozyme era was after the 1990 IUFRO symposium, published in the journal New Forests, Volume 6, and in book form by Adams (1992), on the more general topic of "Isozymes in Forest Trees". These two symposia bookmark this era. Isozymes have been placed in genetic maps, but they are not numerous enough to show much linkage. Typically, 20–30 loci are the maximum number of loci that can be assayed, so that in a typical genome of 2,000–3,000 centiMorgans, few loci will be linked.

Another first generation marker used is the RFLP, a co-dominant polymorphism for the presence/absence of restriction sites. Bands were visualized via Southern blots, which require a probe or sequence complementary to the region about the polymorphism. The RFLP technique is relatively laborious to develop and implement compared to the more recent polymerase chain reaction (PCR) based methods. Neverthess, a number of conifer linkage maps were constructed using RFLP markers during 1985–1995. This marker is regarded as "first generation", as their numbers were still quite limited.

5.2.2 Second Generation Markers

RAPD markers consist of fragments generated via the PCR using a randomly selected ten base primer (Williams et al. 1990). A number of RAPD maps were constructed during the period 1992–98. "Random" refers to the fact that primers are chosen at random, without prior knowledge of any specific primer sites in the genome. Hence the step of cloning and identification of specific sequences is skipped. However RAPDs exhibit dominance, wherein heterozygotes cannot be distinguished from dominant homozygotes (which are the band phenotype; band-less phenotypes are recessive). Also, the RAPD gel band patterns often lack reproducibility, making this class of marker only reliable for studies involving controlled crosses such as genetic mapping, where segregation ratios can verify proper inheritance. A variant of RAPDs is "inter simple sequence repeats" (ISSRs), which are randomly amplified markers produced by PCR amplification with short primers that contain both a microsatellite motif and a random sequence (Bornet and Branchard 2001). These have seen some applications for mapping.

AFLPs are a new class of dominant markers that avoid many of the pitfalls of RAPDs. Assaying for this marker involves restriction digestion of genomic DNA, then PCR amplification of a subset of these fragments (Vos et al. 1995). These markers share many of the characteristics of RAPDs, including dominance and the appearance of many loci on one gel. But the fragment patterns are more reliable, and many more fragments per gel are scoreable. AFLPs have made dense linkage maps possible. However, DNA fragments generated by this technique differ by as little as a single base, requiring use of vertical acrylamide gels or automatic fragment analyzers for clear separation.

Due to the large genome size of conifers, modifications of the AFLP technique for conifers are needed, as the standard +3/+3 primer combinations used for AFLP result in too many bands. With large genomes, one might think that one can limit the pool of selectively amplified DNA by merely increasing the number of selective nucleotides. Vos et al. (1995) found that primers with 4 or more added nucleotides actually suffered a loss of selectivity. For conifers, with their huge genome size, as a means to select subsets of fragments beyond this limit (and to also increase template concentrations), Remington et al. (1999) introduced an additional step prior to the main amplification, termed the "preamplification". It corresponds to a normal amplification, but with shorter primer combinations, usually +1/+1 or +2/+2. Numerous conifer genetic maps have been constructed with AFLPs since the end of the 1990s.

5.2.3 Third Generation Markers

The last class of markers requires *a-priori* knowledge of the DNA sequence at, or around, the marker of interest. A hybrid between second and third generation marker is characterized-sequence amplified region (SCAR), which is developed by cloning RAPD or AFLP markers, and finding the nucleotide sequence about these markers. SCARs are not suited for mapping, as the procedure is laborious; they are useful for finding candidate genes closely linked to anonymous RAPD and AFLP markers found linked to a trait of interest, but they do find themselves occasionally included in conifer genetic maps.

True third generation markers include SSRs (simple sequence repeats or microsatellites), expressed sequence tag (EST)-SSRs (microsatellites in expressed DNA regions), ESTPs (expressed sequence tag polymorphisms, a marker found in ESTs) and finally, the gold standard, the SNP (single nucleotide polymorphism, which directly reflects nucleotide polymorphism at a specific nucleotide site). For further information about markers in plants, see Ritland and Ritland (2000) and Weising et al. (2005).

The completion of the genome sequences in model species, and the accumulation of numerous EST and genomic sequences in many other species, will provide rich resources for the development of these third generation markers.

5.2.3.1 Simple Sequence Repeats

SSRs are markers that are polymorphic for the numbers of repeats of a simple motif (usually 2 to 4 bases long, for example dinucleotide repeats ATATAT….). SSRs are usually co-dominant, highly variable, and somatically stable (Morgante and Olivieri 1993). The locus is amplified by primers that flank the locus. The flanking primers are usually highly species specific. The cost of finding and designing the primers, which must be done for every species, does limit the use of this technique. Sometimes SSRs can be "transferred" to closely related species but at the risk of high null-allele frequency (when the allele does not amplify due to primer mismatches). One disadvantage of SSRs is that they cannot be multiplexed very well, making high density maps impractical.

A special class of SSRs is "EST-SSRs". These are SSRs found in EST sequences and because they are in or near coding regions, the primer regions are more conserved and better able to amplify across species, but the markers are also less polymorphic (Rungis et al. 2004; Ellis and Burke 2007).

5.2.3.2 Single Nucleotide Polymorphisms

SNPs are clearly becoming the marker of choice for species that have been subject to genomic work (either for ESTs, or for genome sequencing). This is because high-throughput genotyping is possible for SNPs, and the number of SNP loci is virtually unlimited. Common high-throughput genotyping methods include the Illumina GoldenGate and Infinium assays (*www.illumina.com*) (Pavy et al. 2008). However while the genotyping costs are much lower (typically 5–10 cent per genotype, compared to 50–100 cents for other methods), the scale of assay (96 SNPs, 480 sample minimum) makes each experiment a big budget item, on the order of many thousands of dollars.

The large number of SNPs uncovered by high-throughput technologies also presents itself with opportunities for marker transfer between species. Even if few in percentage, valuable anchor loci are provided. In the Arborea genome project, for a subset of 1,964 SNPs successfully genotyped in eastern *P. glauca*, 1,565 (80%) were found polymorphic among 11 western *P. glauca* individuals, 728 (37%) among nine *P. sitchensis*, 386 (20%) among 10 *P. abies*, and 321 (16%) among ten *P. mariana* (J. Bousquet et al. unpubl. data). Also, about 10% of SNPs identified in loblolly pine (*Pinus taeda* L., subgenus *Pinus*, section *Pinus*, subsection *Australes*) amplified in white spruce (D Neale et al. unpubl. data); these are currently being utilized in the Treenomix II project for map synteny comparisons.

5.2.3.3 *Conserved Orthologous Sets (COS) and Orthology of Maps*

Comparative mapping relies on orthologous markers. The concepts of orthology and paralogy are essential to construct comparative maps. Orthologous gene pairs are directly descended from a common ancestor. Paralogous genes are separated by gene duplication events and may reside in different locations, but also be very closely linked necessitating sequencing to ascertain orthology. These concepts are essential for comparing maps between species (Gogarten and Olendzenski 1999; Koonin 2005; Theissen 2005; Pelgas et al. 2006).

COS markers are genes of low copy number within a genome, and also have low rate of evolution among species. COS markers are identified by self-BLASTing ESTs within a species, to identify genes of low copy number, then cross-BLASTing these sequences among taxa; genes of low copy number within taxa, and low divergence between taxa, are identified as COS markers (Fulton et al. 2002).

Krutovsky et al. (2006) identified COS markers for conifers using sequence comparisons between *Arabidopsis*, rice, black cottonwood, loblolly pine, white spruce, Douglas-fir, and sugi. Interestingly, almost half of the single-copy genes in the non-tree species *Arabidopsis* and rice had additional copies and homologs in poplar and conifers. However, laboratory assays indicate that the high level of evolutionary conservation of COS markers also results in lower gene diversity within populations, and less available polymorphism for mapping purposes (Liewlaksaneeyanawin et al. 2009).

In lieu of these difficulties of COS markers, with the larger number of genes available through automated genomic investigations, there is the hope that orthologous markers from the huge library of SNPs for conifers can be identified, to anchor genetic maps (Le Dantec et al. 2004; Pavy et al. 2006; Pavy et al. 2008).

5.2.4 Public Databases for Third Generation Markers

Public databases contain a wealth of *in silico* data for marker development. Expressed sequence tags (ESTs) are segments of genes expressed as messenger RNA. Hence they are most useful for identifying SNPs of putative function. For ESTs, the most intensively surveyed conifer species are pine and spruce. NCBI's Entrez Taxonomy Browser (*ncbi.nlm.nih.gov*), as of September 2010, contained 629,815 ESTs for *Pinus* and 542,939 ESTs for *Picea*. Within *Pinus*, the numbers of ESTs are (in parenthesis) are: *P. taeda* (328,756), *P. contorta* (40,483), *P. banksiana* (36,379), *P. pinaster* (34,044), *and P. radiata* (7,538). Within *Picea*, the numbers are *P. glauca* (313,110), *P. sitchensis* (186,637), *P. engelmannii x P. glauca* (28,174) and *P. abies* (14,224). Smaller EST collections exist for other conifers including the family Cupressacea (cedars), which has 72,146 ESTs deposited, mainly for *Crytomeria japonica*. For in silico SNP development, a large number of ESTs are required, unless the deposited ESTs are used to design primers to amplify a small panel of individuals to find SNPs. For pure in silico marker development, a given gene must have at least four overlapping ESTs, in which case a SNP will be detected if two of four nucleotide sites differ in base composition (this mostly rules out sequencing error).

5.3 Mapping Strategies in Conifers

5.3.1 Detecting Recombination

The techniques of marker mapping date from Mendel's crosses. The pioneer of genetic mapping, Thomas Hunt Morgan, showed that recombination frequency can estimate distance separating genes; the distance over which 1% crossover frequency occurs was named by JBS Haldane as the "Morgan", and map distances are generally labeled in centiMorgans (cM) (Ott 1999).

In the context of conifer genetics, issues arise about determining linkage phase. Because conifers are heterozygous, linkage phase cannot be directly ascertained. For example, in the simplest cross, the "backcross", where for two loci A and B, with alleles A_1, A_2 and B_1, B_2, respectively, a cross of grandparent genotype $A_1A_1B_1B_1$ with $A_2A_2B_2B_2$ results in double heterozygote parent $A_1A_2B_1B_2$. Progeny from a backcross of this genotype with either grandparent genotypes may reveal recombinants $A_1A_2B_1B_1$ or $A_1A_1B_1B_2$. In the "intercross", the double heterozygote parents $A_1A_2B_1B_2$ can be crossed with another double heterozygote. As recombination can be detected in both parents in the "intercross", the data are more informative, up to twice as informative when linkage is tight (Ott 1999). However, this assumes linkage phase is known, and grandparents are homozygous.

If grandparents are not homozygous, and/or the grandparents are not genotyped, either single-heterozygote progeny or double-heterozygote progeny (but not both) can be recombinant. This is analogous to the inference of haplotypes from diploid population samples as originally investigated by Clark (1990), in that the phase is indirectly determined by reference to a population sample. Wu et al. (2002) describe how linkage phase can be inferred for outcrossing species with unknown heterozygosity of grandparents, commonly found in conifers. Margarido et al. (2007) implement this procedure in "OneMap".

5.3.2 Assembling Linkage Maps

Lander et al. (1987) popularized genetic mapping with their widely used software, MAPMAKER followed by MAPMAKER/EXP and its close descent, MAPMAKER/QTL. Since then, dozens of programs for both linkage and quantitative trait loci (QTL) mapping have been made freely available. A comprehensive list of linkage and QTL mapping software can be found at *http://linkage.rockefeller.edu/soft*. MAPMAKER starts with a two-point linkage analysis (recombination estimated between all pairs of loci). It then uses a "greedy" algorithm, which builds up linkage groups by sequentially adding markers. This does not guarantee correct orders, so various permutations of maps are done by "rippling". The most commonly used mapping program is JoinMap (Stam 1993), discussed below.

Multipoint linkage analysis takes into consideration the segregation of many linked markers simultaneously. With this approach, it becomes possible to identify individual chromosomal breakpoints and establish order with great certainty (Lathrop et al. 1985). This will become of increasing importance with the advent of high-resolution mapping of conifer genomes.

5.3.3 The Pseudo-testcross

With dominant markers, if a locus is heterozygous in one parent and null (double recessive) in the other, this mimics a testcross with 1:1 segregation ratios. This was termed a "two-way pseudo-testcross" by Grattapaglia and Sederoff (1994), and this was meant to resolve the problem with dominance of RAPD and AFLP markers. It was named "pseudo-testcross" because while it is a testcross mapping configuration, the mating configuration of the markers is not known a priori. However, in genetic mapping, one ends up with a map for the female and a second map for the male. The maps must be joined in some way.

5.3.4 Joining Maps

The integration of two or more marker genetic maps into a single unified map, named a "composite" or "consensus" map, requires common markers that segregate in two or more of the mapping populations. The "pseudo-testcross strategy" is a simple case of multiple maps (two). With dominant markers, one can infer maps for the male and female parent separately as in *Eucalyptus* (Grattapaglia and Sederoff 1994). These workers recognized that multiallelic co-dominant markers with alleles heterozygous in parents are needed as "locus bridges". Joining of maps is now a common activity in conifers, as much of the pedigree material resides within breeding programs, which include many small pedigrees in progeny tests or diallel crosses.

Stam (1993) developed a computer program "JoinMap" that joins pairs of LGs that share the same marker(s) using either raw genetic data or recombination frequencies. The "JoinMap" algorithm estimates information about recombination in a given cross from LOD values and then combines estimates among crosses assuming a binomial sampling distribution. With more than two pedigrees, joining maps is more complicated. Hu et al. (2004) presented a likelihood approach for joining genetic maps that uses a joint likelihood function that combines information across all crosses. The main advantage of this method is substantially improved accuracy when dominant or a mixture of dominant and co-dominant markers are used.

A new approach to build verified multilocus consensus genetic maps in which shared markers are integrated into stable consensus orders was recently developed by Mester and his colleagues (Mester et al. 2003, 2004, 2006) and implemented into software (*http://www.multiqtl.com/*). The approach is based on (1) combined analysis of initial mapping data rather than manipulating with previously constructed maps, and (2) "synchronized ordering", facilitated by cycles of resampling.

However, several pitfalls exist in joining genetic maps, the most important being differences in recombination rates between pedigrees. Recombination rates can differ between crosses and individuals due to environment particularly in stressful conditions where recombination increases (Agrawal et al. 2005). It can also differ in relation to sex or age, where recombination is lower in males and in older individuals (Rose and Baillie 1979). In several pine species, significantly less recombination was observed for the female gametes than for the male gametes in radiata pine (Moran et al. 1983), loblolly pine (Groover et al. 1995) and maritime pine (Plomion and O'Malley 1996). However, Pelgas et al. (2005) observed no difference in map length between males and females in white spruce, as did Pelgas et al. (2006) and Pavy et al. (2008) for white and black spruce pedigrees. This suggests that sex specific recombination rates may differ between conifer species. Further investigation is needed on this topic.

Another pitfall in joining maps is that markers can vary in abundance and distribution. In Norway spruce, low- and high-copy-number markers tend to occupy separate genome regions (Scotti et al. 2005). Also, microsatellites may be preferentially associated with nonrepetitive DNA (more coding DNA) in plant genomes (Morgante et al. 2002). Both of these situations indicate that joining maps with different classes of markers might be difficult, as common polymorphic markers between these marker-type classes may not be present in many parts of the genome.

5.3.5 *Improving the Resolution of Maps*

To get beyond the resolution of traditional marker mapping, which is of 5–10 cM resolution for mapping populations of size ca. 100, one can use larger mapping populations, or else physical mapping. Physical mapping involves the cloning and mapping (by fingerprinting) of large plasmid inserts, such as bacterial artificial chromosomes (BACs), normally 150 KB in length. In conifers, which harbor a 10–40 gigabase genome, this would require 200,000 BAC clones for a 1× coverage; ideally 2 million BACs would be needed for a 10 × coverage, as this is the typical required for a BAC tiling path (Soderlund et al. 2000). At least, the repetitive nature of the conifer genome would suggest that assembly of BAC fingerprints into a tiling path is difficult. However, suggestive data indicate that the major period of repetitive DNA activity (transposition) occured over 100 million years ago (Mya) (M Morgante et al. unpubl. data). Such a feature would actually increase the feasibility of genome assembly, since members of the same repeat class have diverged since transposition. This is a current area of research in conifer genomics—the nature of low complexity DNA in conifers and its implication for genome assembly (Nelson et al. 2008).

With high-resolution meiotic maps, a problem is that low frequency of genotyping error (1.5% or less) can influence mapping outcomes. Such an error was observed to reduce power to discriminate orders, dramatically inflate map length, and provide significant support for incorrect over correct orders (Buetow 1991). Occasional genotype errors skew estimates of recombination between closely linked loci; a similar situation occurs in paternity analysis, where just one missscored locus can invalidate the correct parent. Various workers have since dealt with this issue (Sobel et al. 2002) and new SNP genotyping methods have shown to be highly accurate, with error rate below 1% (Pavy et al. 2008).

To increase the rate that meiotic events can be detected, Gasbarra and Sillanpaa (2006), proposed pooling haploid tissue, such as conifer megametophytes, to estimate recombination rates between closely linked loci (< 1 cM). Pools of several hundred were simulated but they found that several pools were better than a single pool.

Selective mapping approach can facilitate the production of high-quality, high-density genome-wide linkage maps (Vision et al. 2000). It was demonstrated that, to construct a map with high genome-wide marker density, it is neither necessary nor desirable to genotype all markers in every individual of a large mapping population. Instead, a reduced sample of individuals bearing complementary recombinational or radiation-induced breakpoints may be selected for genotyping subsequent markers from a large, but sparsely genotyped, mapping population.

5.4 Conifer Linkage Maps

5.4.1 Overview

Genetic mapping in conifers, and in all other species for that matter, has progressed through three generations of development, corresponding to the marker categories described above. The first generation maps involved allozyme and RFLP markers, which rarely revealed genetic linkage because of their sparsity. The second generation maps involved anonymous genetic markers such as RAPDs and AFLPs; "anonymous" in the sense that we have no idea of their gene function. Nevertheless, complete genetic maps were inferred, as these markers were so much more numerous. The third generation maps involved markers of known gene function, mainly SNPs derived from genome projects. This last wave now allows incredibly detailed maps of genomes, both with numerous markers, and with markers linked to genes putatively related to adaptation and other desired traits.

As isozymes are limited in number, they did not play a significant role in linkage mapping; occasionally a few isozyme markers were added to other markers in a complete map. Significant effort into developing RFLP markers for linkage mapping has been done only for *Cryptomeria japonica*, *Pinus elliottii*, *P. taeda*, *P. radiata*, and *Pseudotsuga menziesii* (Table 5-1). Usually RFLPs were analyzed in conjunction with other markers. The most significant early generation molecular marker map was developed in loblolly pine. Devey et al. (1994a) reported an RFLP linkage map for loblolly pine based on a three-generation outbred pedigree. Seventy three of 90 loci (including two isozymes) clustered into 20 linkage groups (LGs). Other studies are summarized in Table 5-1.

The first complete linkage maps in conifers, where the number of large LGs equalled to the haploid number of chromosomes, were made possible by the advent of RAPD and AFLP markers. In the first application of RAPD markers for conifer mapping, Tulsieram et al. (1992) mapped 47 of 61 RAPD markers into 12 LGs in white spruce. Subsequent studies are summarized in Table 5-1. Genetic maps have been constructed for ca. 12 pine species (Table 5-1: *Pinus brutia*, *caribaea*, *contorta*, *densiflora*, *edulis*, *elliottii*, *palustris*, *pinaster*, *radiata*, *strobus*, *sylvestris*, and *taeda*). Maps have been constructed for four spruce species (Table 5-1: *Picea abies*, *glauca*, *mariana* and *rubens*). *Cryptomeria* and *Pseudotsuga* are two other conifers that have received much attention, while isolated work has been done with *Abies*, *Cunninghamia*, *Larix* and *Taxus*. It should be noted that many of these studies used open-pollinated seed progeny of an individual tree assayed for haploid megagametophytes (this also avoids the dominance of RAPD markers). We now discuss more detailed maps made in the four most important conifer genera.

Table 5-1 Genetic linkage maps in conifers. Linkage groups include at least 3 markers. Expected coverage is the ratio between observed and predicted map sizes estimated following Hulbert et al. (1988) and Chakravarti et al. (1991); NA = not available. The haploid number of chromosomes in all species is 12 except *Cryptomeria japonica* ($n = 11$), and *Pseudotsuga menziesii* ($n = 13$). Numbers separated by "/" refer to the maternal/paternal parents of the mapping population, respectively. Marker types are defined in Section 5.2.

Species	Marker type	Markers	Linkage groups	Map length, cM	Expected coverage, %	cM per marker	Reference
Abies nordmanniana	AFLP, RAPD	556	19	1977	80	NA	Hudson (2005)
Cunninghamia lanceolata	AFLP	101/94	11	2283/2566	NA	23/27	Tong and Shi (2004)
Cryptomeria japonica	RFLP, RAPD, Isozymes	91	13	887	NA	10	Mukai et al. (1995)
	RAPD	84/119	14/21	1112/1756	40/62	13/15	Kuramoto et al. (2000)
	AFLP, CAPS	91/132	19/23	1266/1992	50/80	16/18	Nikaido et al. (2000)
	CAPS, Isozymes, SNP. RAPD, RFLP, SSR	438	11	1372	96	3	Tani et al. (2003)
Larix decidua	AFLP, ISSR, RAPD	117	17	1152	80	14	Arcade et al. (2000)
L. kaempferi	AFLP, ISSR, RAPD	125	21	1206	81	14	Arcade et al. (2000)
Picea abies	RAPD	165	17	3584	NA	22	Binelli and Bucci (1994)
	RAPD	82	13	1385	NA	24	Skov & Wellendorf (1998)
	AFLP, SSR	413	22	2198	77	9	Paglia et al. (1998)
	AFLP, ESTP, rDNA, SSR	755	12	2035	NA	3	Acheré et al. (2004)
	AFLP, IRAP, S-SAP, ESTP, SSR	203/152	27/23	2316/1669	66/79	13/ 13	Scotti et al.(2005)
P. glauca	RAPD	61	12	873	NA	14	Tulsieram et al. (1992)
	ESTP, RAPD, SCAR	144/165	19	2008/2059	73/87	9/15	Gosselin et al. (2002)
	AFLP, ESTP, SSR	802	12	1934	89	2.4	Pelgas et al. (2006)
	AFLP, ESTP, SNP, SSR	821	12	2304	98	2.8	Pavy et al. (2008)
	AFLP, ESTP, SNP, SSR	1301	12	2087	NA	1.6	Pelgas et al. (2011)
	AFLP, ESTP, SSR, COS	505	12	1835	NA	3.5	Liewlaksaneeyanawin et al. (unpubl. data)
P. mariana × *P. rubens*	AFLP, ESTP, RAPD, SSR	1124	12	1846	92	1.6	Pelgas et al. (2005)

Table 5-1 contd....

Table 5-1 contd....

Species	Marker type	Markers	Linkage groups	Map length, cM	Expected coverage, %	cM per marker	Reference
P. mariana	AFLP, ESTP, SNP, RAPD, SSR	835	12	1850	98	2.2	Pavy et al. (2008)
Pinus brutia	RAPD	13	6	164	NA	NA	Kaya and Neale (1995)
	AFLP, SAMPL, ESTP, SSR	1111	12	1770	97	1.6	Kang et al. (2010)
P. caribaea	AFLP, SSR	109	27	1658	88	16	Shepherd et al. (2003a)
P. contorta	RAPD	225	16	2287	95	15	Li and Yeh (2001)
P. densiflora	AFLP	152	19	2341	82	18	Kim et al. (2005)
P. edulis	AFLP	338	22	2012	85	9	Travis et al. (1998)
P. elliottii	RAPD	73	13	782	64	11	Nelson et al. (1993)
	RAPD	91	13	953	62	16	Kubisiak et al. (1995)
	ESTP, Isozymes, RAPD, RFLP	154	15	1115	NA	7	Brown et al. (2001)
	AFLP, SSR	78	23	1170	82	15	Shepherd et al. (2003a)
P. palustris	RAPD	133	16	1635	85	15	Nelson et al. (1994)
	RAPD	122	18	1368	81	13	Kubisiak et al. (1995)
P. lambertiana	SNP	399	19	1231	NA	3.1	Jermstad et al. (2010)
P. pinaster	RAPD	263	13	1380	90	9-10	Plomion et al. (1995b)
	Isozymes, RAPD	463	12	1860	93	8.3	Plomion et al. (1995a)
	AFLP, Isozymes, RAPD			1873	93		Costa et al. (2000)
	AFLP, ESTP, SSR	1182	12	1994	NA	10	Ritter et al. (2002)
	AFLP	620	12	1441	NA	NA	Chagné et al. (2002)
	AFLP, ESTP	326	12	1639	NA	NA	Chagné et al. (2003)
P. radiata	RAPD, RFLP, SSR	195	14	1382	NA	7	Devey et al. (1996)
	RFLP, SSR	173	14	1223	75	7	Devey et al. (1999)
	RAPD, SSR	172	19	1117	56	NA	Kuang et al. (1999)
	AFLP, RAPD, SSR	194	21	1144	77	12	Wilcox et al. (2001)
P. strobus	RAPD, SSR, STS	101	12	745	58	14	Echt and Nelson (1997)

P. sylvestris	RAPD	261	14	2639	NA	10	Yazdani et al. (1995)
	AFLP	94/155	15	796/1335	77/86	18/17	Lerceteau et al. (2000)
	AFLP	188/245	12/15	1696/1719	86/99	9/7	Yin et al. (2003)
	AFLP, ESTP, SSR	120/112	21/16	929/1452	66/85	9/12	Komulainen et al. (2003)
P. taeda	Isozymes, RFLP	75	10	632	NA	NA	Devey et al. (1994a)
	AFLP	508	12	1528	99	9	Remington et al. (1999)
	Isozymes, RAPD, RFLP	357	20	1359	82	4	Sewell et al. (1999)
	RFLP, SSR	223	20	1281	75	4	Devey et al. (1999)
	ESTP, Isozymes, RFLP	235	12	1227	NA	5	Brown et al. (2001)
	ESTP, Isozymes, RFLP	302	12	1274	NA	NA	Krutovsky et al. (2004)
	ESTP, Isozymes, RAPD, RFLP, SNP, SSR	373	12	1228	NA	4	Eckert et al. (2009)
	ESTP, Isozymes, RAPD, RFLP, SSR	462					Echt et al. (2011)
P. thunbergii	AFLP, RAPD	207	20	2085	77-78	10	Hayashi et al. (2001)
Pseudotsuga menziesii	RFLP, RAPD	141	17	1062	NA	7.5	Jermstad et al. (1998)
	RAPD	210	16	2279	91	10	Krutovsky et al. (1998)
	RAPD	132	13	2143	NA	NA	Carlson et al. (2007)
	ESTP, Isozymes, RAPD, RFLP, SSR, STS	376	22	1859	NA	NA	Krutovsky et al. (2004)
	AFLP	120	19	939	NA	9	Ukrainetz et al. (2008a)
Taxus brevifolia	RAPD	41	17	306			Göçmen et al. (1996)

Figure 5-1, based upon a meta-analysis of Table 5-1, shows how the types of markers used in genetic maps have changed in the past 20 years. In general, the numbers of maps have declined in the past five years. From this graph, it is evident that RAPD and ISSR markers predominated during 1995–2005, but their use has declined, as they cannot be transferred among pedigrees to build additional maps. AFLPs had a big impact during 2000–2005; again, as they are anonymous markers, but their transferability is limited. Isozymes (the hobbit in the corner) and RFLPs have had a constant impact, but their numbers are still limited. SSRs and SNP/ESTP/STS markers are obviously the markers of choice for future mapping. They have seen increasing useage. These are all sequence based markers that can be transferred among pedigrees and species.

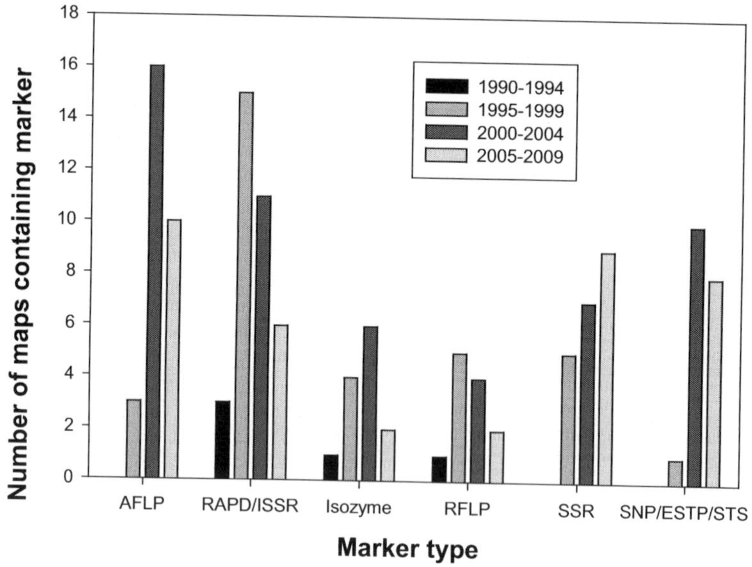

Figure 5-1 Trends in conifer maps over the past 20 years. I. Usage of the various classes of genetic markers for conifer genetic mapping.

Figure 5-2, also based upon Table 5-1, shows how the number of markers used in conifer genetic mapping has increased in the past 20 years. Figure 5-2a shows that the number of markers has obviously increased as expected, but the variance of the number of markers has also increased. This does not include plans from spruce and pine genome projects to radically increase marker number to 5,000 and above. But total map length has remained almost constant (Fig. 5.2b) as markers separated by 30 cM or less (ca. 100 markers total) are sufficient to cover genome length. High density marker maps will be of main use for assembling contigs from genome sequencing projects.

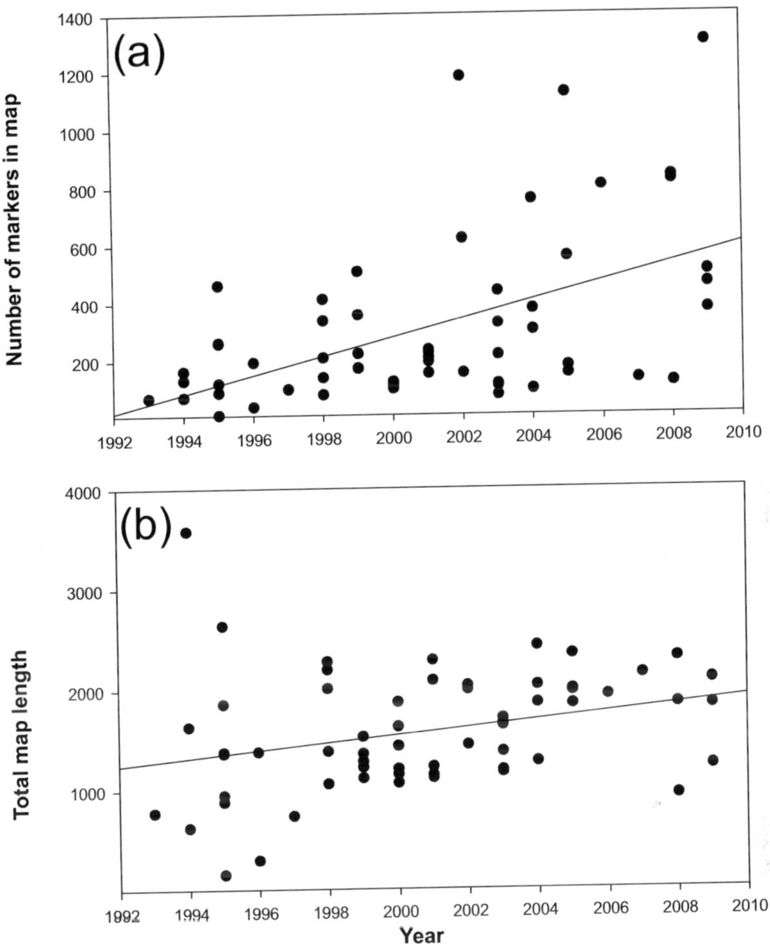

Figure 5-2 Trends in conifer maps over the past 20 years. II. (a) Numbers of markers linked to maps, (b) Total map length explained by markers.

5.4.1.1 Loblolly Pine

Pinus taeda genome maps generally contain ca. 300 loci based mostly on the two standard pedigrees: *base* (Devey et al. 1994b) and *qtl* (Groover et al. 1994). Mapping data for previously reported SSR, RFLP and ESTP markers were combined with new SSR markers to generate a loblolly pine consensus map of 462 markers covering 1,380 cM across 12 LGs, using both the *qtl* pedigree (n = 171) and *base* pedigree (n = 98) (Echt et al. 2011). Of the 234 mapped SSR loci, 171 were newly developed, 81 of which were derived from

EST sequence data. Marker data were obtained for an additional 50 new EST-SSR loci that did not segregate in either mapping population but which were polymorphic in population surveys. One hundred and ninety four mapped loci were given a functional GO assignment; 242 mapped loci were assigned to a NCBI UniGene cluster. Unigene and GO assignments, along with linkage data, aided in identifying duplicated and paralogous marker loci on the map. This species may serve as a reference map in comparative mapping with other pines and even other members of the Pinaceae family such as spruce and Douglas-fir.

5.4.1.2 Spruce

Linkage mapping in spruce (*Picea* spp.) has been directed toward three species of major economic importance: *Picea abies*, a European species, and *P. glauca* and *P. mariana*, both primarily North American species. The first saturated composite map for white spruce was reported by Gosselin et al. (2002), who used 165 RAPD, SCAR and ESTP markers to join maps from two individuals. They noted that co-dominant markers were needed to join the maps. In Norway spruce, Acheré et al. (2004) developed the second map, involving 755 markers. Interestingly, 150 of these markers were tested for their pattern of population differentiation differing from neutral expectations, and nine of these markers were found to be "outliers", or genes that showed excessive population divergence, compared to the majority of markers, suggesting they were linked to QTLs for adaptation (Acheré et al. 2005).

More recently, the Arborea project in Canada has constructed several linkage maps involving both individual and composite maps for white spruce and black spruce. A map for the black spruce × red spruce species complex was constructed (Pelgas et al. 2005), and for white spruce alone (Pelgas et al. 2006). Most notably, Pavy et al. (2008) assembled a white spruce linkage map with markers assayed via the Illumina GoldenGate SNP genotyping platform. The resulting composite map had 821 loci including 461 AFLPs, 12 SSRs, 31 ESTPs and 317 gene SNPs, and map coverage was > 90%. This map also positioned genes with SNPs involved in among-population differentiation of eastern white spruce; 50 outlier SNPs were identified (Namroud et al. 2008); these genes are putatively involved in adaptive differentiation. An expanded white spruce composite map containing 836 gene loci has recently been published (Pelgas et al. 2011).

The most recent white spruce gene composite map emerging from the Arborea project integrates two pedigrees of 500 progeny and has an increased resolution of 0.9 cM with 2,255 positioned loci including 455 AFLPs, 12 SSRs and 1,788 gene SNPs. The map covers 2,065.4 cM over 12

LGs. The average gene density is 1.16 cM. The current published spruce map has 826 genes; the largest number of mapped genes in a conifer species.

5.4.1.3 *Douglas-fir*

In *Pseudotsuga menziesii*, the most recent marker development has focused on ESTP and SNP markers (Krutovsky et al. 2004), which together with SSR markers, have added to the RFLP and RAPD linkage maps (Jermstad et al. 1998). The most recently published genetic map of Douglas-fir consists of 376 markers, including 172 RFLP, 77 RAPD, 2 isozyme, 20 SSR, 4 sequence tagged site (STS), and 101 expressed sequence tag (EST) markers (Krutovsky et al. 2004). This map is organized into 22 LGs that have three or more linked markers and spans 1,859 cM. Several hundred SNP markers were developed recently (Eckert et al. 2009), and their mapping is under way. When enough markers are mapped, the number of LGs should coalesce into 13, corresponding to the 13 chromosome pairs in Douglas-fir. It would be valuable to map additional ESTP, EST-SSR and SNP markers to create a high-density map that can be used for QTL, candidate gene and physical mapping to facilitate eventual complete Douglas-fir genome sequencing.

5.4.1.4 *Sugi*

Sugi (*Cryptomeria japonica*) has been planted widely throughout Japan over an area of 4.5 million ha, accounting for 44% of all the Japanese artificial forest. A second generation linkage map for Sugi was constructed by integrating linkage data from two unrelated third-generation pedigrees. The progeny segregation data of the first pedigree, which involved a cross between half-sibs, were derived from cleaved amplified polymorphic sequences (CAPS), SSRs, RFLPs, and SNPs (Tsumura et al. 1997; Iwata et al. 2001). The data of the second pedigree, which involved a self-pollinated individual, were derived from CAPS, isozyme markers, morphological traits, RAPDs, and RFLPs. The co-dominant DNA markers such as CAPS, RFLP and SNP were developed from ESTs and cDNA clones from several kinds of cDNA libraries (Ujino-Ihara et al. 2000; Ujino-Ihara et al. 2005). More than 95 % of the markers were gene-based markers.

Using JoinMap, linkage analyses were done for the first pedigree assuming cross-pollination, and for the second pedigree assuming selfing. Four hundred and thirty eight markers were assigned to 11 large LGs (corresponding to the 11 chromosomes of *C. japonica*), 1 small LG, and 1 non-integrated LG from the second pedigree; their total length was

1,372.2 cM (Tani et al. 2003). On average, the consensus map showed one marker every 3.0 cM. PCR-based co-dominant DNA marker such as CAPS, microsatellite and SNP were distributed over all LGs and represented about a half of mapped loci.

5.4.2 Genome Sizes

Besides providing a linear map of markers along a genome, mapping experiments can also provide estimates of genome size, in terms of map units. Hulbert et al. (1988) gave the first estimate of genome size based upon observed recombination between randomly selected pairs of markers. Chakravarti et al. (1991) improved this with a maximum likelihood method for estimating genome size. Many conifer mapping studies have provided estimate of genome size from either method; estimates range from ca. 2,000 to 3,000 map units. Relatively few numbers of markers can estimate genome size, as long as some are linked.

Genome size can also be estimated by flow cytometry, in terms of picograms (pg) of DNA per nucleus, which can be translated into millions of base pairs using the relationship 1 pg = 978 million base pairs. This gives an idea of how many nucleotides separate linked markers. Genome size in the Pinaceae ranges from 5.8 to 32.2 pg with 20 pg (20 billion base pairs) a rough average (Murray 1998); this is 100 times larger than *Arabidopsis thaliana* (0.18 pg).

Genome evolution in the gymnosperm lineage of seed plants has given rise to many of the most complex and largest plant genomes; however the elements involved are poorly understood. Most of the enormous genome complexity of pines can be explained by divergence of retrotransposons (Morse et al. 2009); however the elements responsible for genome size variation are yet to be identified. This is currently a very active area of research in conifer genomics.

5.4.3 Physical Mapping Opportunities

Physical mapping complements genetic mapping. Unfortunately the large physical genome size of conifers as just described prohibits most of these approaches. Approaches that are free from constraints from large genome size involve hybridization of certain genes to chromosomes. Earlier works used fluorescence *in situ* hybridization (FISH) experiments to identify location and distribution of ribosomal RNA. In Sitka spruce, 5s rDNA was found to be restricted to one chromosome, whereas 18S-5.8S-26S rDNA repeats occurred on five chromosomes (Brown and Carlson 1997). Both distribution and location of large tandem repeats on the genome of white spruce and Sitka spruce were comparable (Brown et al. 1998). A reference karyotype was presented recently for loblolly pine based on FISH and using 18S–28S rDNA, 5S rDNA, and an *Arabidopsis*-type telomere repeat sequence, *A*-type TRS signals (Islam-Faridi et al. 2007). Statistically, only seven of the 12 loblolly pine chromosomes could be distinguished by their relative lengths. However, the position and relative strength of the rDNA and telomeric sites made it possible to uniquely identify all of the chromosomes, providing a reference karyotype for use in comparative genome analyses. A dichotomous key was developed to aid in the identification of loblolly pine chromosomes and their comparison to chromosomes of other *Pinus* spp. A cytomolecular map was developed using the interstitial 18S–28S rDNA and *A*-type TRS signals. A total of 54 bins were assigned, ranging from three to five bins per chromosome. This is the first report of a chromosome-anchored physical map for a conifer that includes a dichotomous key for accurate and consistent identification of the loblolly pine chromosomes.

Recently, bacterial artificial chromosome (BAC) hybridization has been developed as an alternative to rDNA hybridization, which allows very specific identification of chromosomes, and such methods would be fruitful to apply to conifers, particularly the Pinaceae, as chromosomal morphology is hardly distinguishable among the dozen or so chromosomes. This method has been used in many plant species (Zhang et al. 2004) but not in conifer.

The normal activity of physical mapping is to construct a library of inserts, then to construct "tiling paths" to obtain an ordered set of clonal inserts that span the entire genome. For coverage of a conifer genome (5–10×), about two million BAC clones are needed, too large for practical work. Nevertheless, BACs are useful for conifers, and there are currently BAC libraries available for white spruce and loblolly pine. The spruce library is unarrayed and about 5× coverage, while the loblolly pine library is arrayed and about 8× coverage (Liu et al. 2009). Currently, both random BACs and targeted BACs (BAC identified as having a gene of interest) are being sequenced from both libraries (J MacKay et al. unpubl. data; DG Peterson et al. unpubl. data; K Ritland et al. unpubl. data).

5.5 Conifer Comparative Mapping

Alignments of genetic or QTL maps among species demonstrate the evolutionary conservation of gene linkages among species. An early paradigm was set by work with the grass family (Gale and Devos 1998). Conserved chromosomal number in the pines family (Pinaceae) suggested that similar comparisons could be made in pine family members. The "Conifer Comparative Genomics Project" organized by David Neale and his colleagues at UC Davis has verified that such approaches can be used in conifers (e.g., Krutovsky et al. 2004). The end goal is to transfer information between species about co-localization of QTL and candidate genes among species. In genome sequencing projects, it also predicts the reliability that related conifer genomes can be "resequenced", once a reference genome is sequenced.

To facilitate the identification of orthologous markers for comparative mapping, sequence-based gene markers such as ESTPs and SNPs are best because they are usually orthologous across congeneric species, and more reliable than anonymous markers. Hidden paralogy is the ghost of map construction (Huynen and Bork 1998; Remm et al. 2001; Pelgas et al. 2006). To reduce the risk of paralogous amplification, primer pairs should be designed with a primer matching in the 3′ UTR gene region (e.g., Perry and Bousquet 1998; Brown et al. 2001; Chagné et al. 2003; Pavy et al. 2008). In conifers, resequencing from megagametophyte DNA indicates paralogous polymorphisms by the presence of double peaks on sequence chromatograms (Pelgas et al. 2004).

Until recently, limited numbers of orthologous markers were available for useful map comparisons. SNPs are virtually in unlimited number. Because they can be annotated and are dense along linkage maps, SNPs can better determine gene orthology, and serve as anchor markers for intra- and interspecific map comparisons (Pelgas et al. 2006; Pavy et al. 2008).

5.5.1 *Pine Species Comparisons*

Historically, the most extensive genetic maps have involved loblolly pine. Detailed comparative maps are needed to study conifer genome evolution and to leverage genomic information of adaptive and economic traits from the relatively well-studied species, such as loblolly pine, to other conifers. Most comparative maps among *Pinus* species are within the subgenus *Pinus* and based on comparisons of ESTP markers. They contain 41 common loci between *P. taeda* and *P. sylvestris* (Komulainen et al. 2003) and 32 common loci between *P. taeda* and *P. pinaster* (Chagné et al. 2004). Both of these studies used prior published *P. taeda* maps (Krutovsky et al. 2004). Recently, maps from the two subgenera of *Strobus* and *Pinus* could be compared, based on neaerly 400 gene SNPs (Jermstad et al. 2010). All 19 linkage groups of *P. lambertiana* co-aligned with the 12 linkage groups of *P. taeda*, providing a basis for integrated structural genomics approaches across pine subgenera.

5.5.2 *Spruce Species Comparisons*

The first comparative map of white spruce (Pelgas et al. 2006) revealed remarkable synteny with black spruce (*P. mariana*) and Norway spruce (*P. abies*); identical LGs and conservation of gene content and gene order was found. One breakdown of synteny between *P. glauca* and the other taxa involved an inter-chromosomal rearrangement of an insertional translocation. Analysis of marker colinearity also revealed a putative segmental duplication. This three-species comparison showed that genome comparisons among *Picea* species can provide a platform for transfer of genomic information across species of spruce.

More recently, a detailed analysis of synteny and macro-colinearity between *P. glauca* and *P. mariana*, using 215 anchor markers, consisting mainly SNPs, found that 98% of the anchor genes were in synteny (Pavy et al. 2008). Translocations were validated in the case of previously reported *PgMyb4*, and three new translocations involving three genes were indicated. However, the sequencing of haploid megagametophytes for these genes indicated that these new cases were likely false positives, involving paralogous variation. Macro-colinearity was also well conserved among homologous LGs between species, with 82% of syntenic anchor markers positioned in the same order. Exceptions to colinearity involved small inversions also observed between individual maps within species, indicating that that most of these inversions were artefacts.

Figure 5-3 shows a relatively high density genetic map for both white and black spruce (LGs III-VI only), with the maps also aligned between the two species. Map distances in centiMorgan are indicated with a scale on the left side. The composite map of each species was obtained by first assembling two parental datasets for each species, using JoinMap (Stam 1993); then maps were aligned between species using common markers. There are five types of markers in these maps: SNPs (bold), ESTPs (bold and underlined), SSRs (bold and italics), RAPDs (italics and underlined) and AFLPs (others). Typically AFLPs are the most in such maps with several types of markers, but they are not useful for joining maps between species (the loci are named after the primer combination used and the band migration distance). Syntenic marker loci between spruce species are indicated in black, and these are typically gene-based markers. These syntenic markers are identified with a red solid line (colinear markers) or a red dashed line (non-colinear markers). Orthologous markers positioned onto non-homologous LGs are indicated in white with red background and paralogous markers are identified in white with blue background. Overall, there is a remarkable preservation of gene order between white and black spruce, and the exceptions may be mistaken cases of orthology and merit further investigation.

5.5.3 *Pine Family Species Comparisons*

The first intergeneric comparative map in conifers was constructed between loblolly pine and Douglas-fir with ESTP and RFLP markers (Krutovsky et al. 2004). Comparison of Douglas-fir and loblolly pine maps revealed 10 LGs (LG1–LG10) in loblolly pine that shared 2–10 orthologous markers with 12 apparently syntenic LGs in Douglas-fir based on 46 orthologous markers. The comparisons revealed extensive synteny and colinearity of

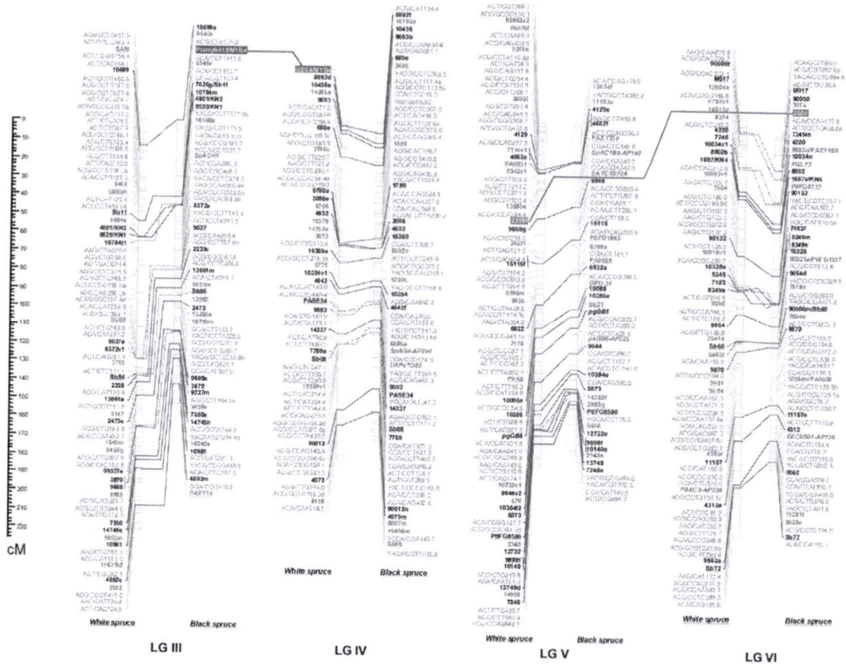

Figure 5-3 Comparison of homologous linkage groups between white spruce (*Picea glauca*) and black spruce (species complex *Picea mariana* × *P. rubens*).

Color image of this figure appears in the color plate section at the end of the book.

gene order between the two genomes, consistent with the hypothesis of conservative chromosomal evolution among even distantly related species in the Pinaceae family. This study established a working framework that the Pinaceae can be viewed as a single genetic system.

Homology of Pinaceae LGs was more recently extended to three spruce species (Pelgas et al. 2006). Between spruce and loblolly pine, 26 of 29 anchor markers were in synteny, identifying 11 homologous LGs. In this study, orthology of anchor gene markers was checked by extensive resequencing of single haploid megagametophytes in the various species. For the three exceptions to synteny, sequencing of megagametophytes indicated at least two cases of paralogy, while the third case remained dubious, implicating a conserved gene family. Between spruce and Douglas-fir, synteny could be assessed with 20 anchor markers, of which just one proved to be paralogous after megagametophyte resequencing. Of the remaining markers, three were not in synteny, including two markers on LG13 of Douglas-fir, confirming that the supernumerary chromosome of Douglas-fir is the result of fission (Krutovsky et al. 2004; Pelgas et al. 2005). The remaining marker, in synteny between spruce and lodgepole pine, was translocated to a different LG in

Douglas-fir, thus indicating that chromosome rearrangements occurred in the lineage leading to Douglas-fir. This study established rigorous criteria for determining orthology of genetic markers among species, and only after this criteria is met, can we make reliable inferences about chromosomal rearrangements among species.

Figure 5-4 shows a recent syntenic map for Douglas-fir, loblolly pine and Norway spruce. This was identified as LG6 of loblolly pine, as the high level of synteny and conservation of gene order allows homologous LGs among pine species to be identified (Neale and Krutovsky 2004). Orthologous comparative mapping markers are underlined and shown in bold (this is based upon unpublished data kindly provided by Craig Echt, USDA Forest Service, Southern Institute of Forest Genetics, Saucier, Mississippi, USA [for pine] and by Michela Troggio, IASMA Research and Innovation Centre, San Michele, Italy [for spruce]). Overall, the alignment of maps between species separated by over 100 million years of evolution is remarkable

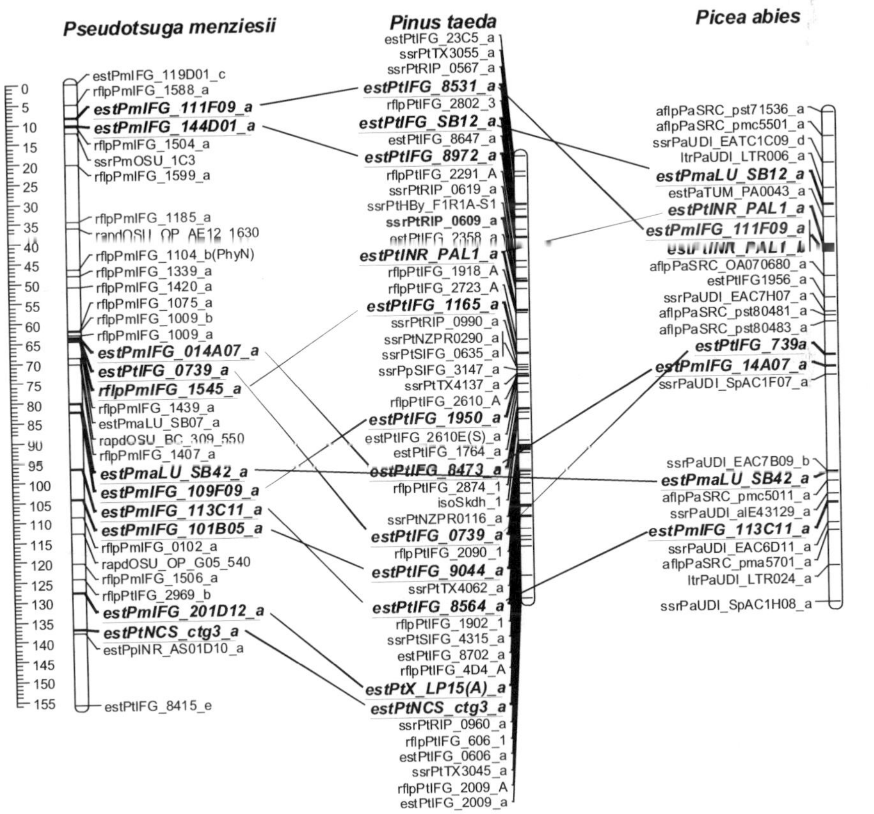

Figure 5-4 Comparison of homologous linkage groups between Douglas-fir (*Pseudotsuga menziesii*), loblolly pine (*Pinus taeda*) and Norway spruce (*Picea abies*).

for the plant kingdom, and suggests that the pine family (Pinaceae) can be viewed as one genetic system, allowing genomic information to be readily transferred, in contrast to angiosperm species with even one-tenth evolutionary separation.

However, on these maps, there are three instances of apparent segmental inversions, two between Douglas-fir and pine, and one between pine and spruce. A case where a pair of markers is reversed is likely due to mistaken orthology. However, between Douglas-fir and pine, four markers are involved with an apparent rearrangement (involving orthologous markers 3–6 in pine, which are linked to Douglas-fir). To have four, instead of two, markers involved in an apparent inversion provide much stronger evidence of true orthology. This suggests that the genetic system is less homologous in Douglas-fir, as indeed its time since evolutionary divergence is greater than between pine and spruce, and that there are limits to the transfer of genomic information between conifer taxa.

5.6 Quantitative Trait Loci Mapping in Conifers

The last aspect of mapping in conifers involves identifying genes underlying quantitative traits along the marker maps. The co-segregation of genetic markers with phenotypes within pedigrees can reveal individual genes underlying quantitative traits. The ultimate objective of QTL mapping is to infer the "genetic architecture" of the quantitative trait, e.g., the numbers of gene loci controlling the trait, the magnitudes of their effects, and their location in LGs, epistatic interactions, and gene-by-environment interactions. While the idea of using markers to study quantitative traits dates from Sax (1923), who used single-locus morphological markers as categories for continuous traits, the landmark paper that provided the modern paradigm is Lander and Botstein (1989), who considered the multiple marker mapping of QTL mapping.

QTL mapping involves associating alternative marker alleles with phenotypes in segregating progenies. The major issue in conifers is that parents should be heterozygous for both genetic markers and QTLs. Separate QTL maps (but not marker maps) need to be constructed for each parent. However, if a given marker is heterozygous in both parents, the QTL cannot be assigned to a parent, unless there is a priori knowledge about linkage. Issues about QTL mapping in outbred pedigrees are discussed in Williams (1998).

Candidate genes can also be used as marker loci in QTL mapping. For example, Wheeler et al. (2005) used 29 putative cold-hardiness candidate genes for mapping cold-hardiness related traits in Douglas-fir, and Pot et al. (2006) used 10 candidate genes involved in the biosynthesis and deposition of the secondary cell wall in maritime pine. Recently, Pelgas et al. (2011) used 836 candidate genes as marker loci for QTL mapping of various adaptive traits. However, location of the candidate gene within a QTL interval is not proof of causality; further testing using functional or association genetic approaches are required for proof that such a candidate gene underlies the quantitative trait.

5.6.1 Crossing Designs for QTL Analysis

There are several possible experimental designs for QTL detection. In conifers, breeding programs offer genetic material for QTL analyses. The most common test is a progeny test, where a small number (8–20) of open-pollinated progeny are grown, which can estimate the genetic value of the female parent (White et al. 2007). This number is too small to estimate QTL effect in any single family, and variation among parents for QTL content adds complexity. In outbred conifers, each parent will have different QTLs. Ideally, large (> 100) full sib families are needed for reliable inference of QTL, in order to avoid the bias of inference of QTL effect due to small family size (Beavis 1998). However this ignores variation of QTL among individuals in the larger population.

For QTL mapping, the two major designs are the "inter-specific F_1" design, and the "three-generation full-sib pedigree" design. Interspecific F_1 designs are rare if non-existent in conifers as they are based upon hybridization between subspecies that are usually fixed for alternative QTL and alternative markers. The three-generation design has been employed for Douglas-fir and loblolly pine. An intermediate situation is often encountered: factorial crossing designs with 10–50 progeny per family (a complete factorial design is where *N* males are individually crossed with *M* females, resulting in *NM* families). This design is used to estimate general and specific combining abilities on both the male and female side (Verhoeven et al. 2005).

QTLs found in one pedigree may not exist in other pedigrees. "Validation" of QTLs is the replication of the finding on a second population. In association genetic studies, validation in other populations is a requirement. In QTL studies, this is a difficult task as emphasized by Williams et al. (2007). They point out that a given QTL may not be polymorphic in the second pedigree, and that other segregating QTLs can cause gene interactions that obscure the QTL in another pedigree. In conifers, replicate pedigrees are few due to the long generation times.

The density of markers needed for QTL mapping need not be that high. Darvasi et al. (1993) found that QTL detection probability for a map with 10 cM spacing of markers was virtually the same as that for a map with an infinite number of markers. Since SNPs usually have just two alleles per locus, a larger number of SNPs are needed to obtain the ideal 10 cM marker spacing. SSR markers are usually highly heterozygous and if on the order of 100 markers are used; their distribution is sufficiently dense such that a given individual is usually heterozygous for at least one locus over a small (10 cM) genome interval.

5.6.2 QTL Traits of Interest

As in the choice of markers for genetic mapping, the phenotypic traits of interest need to be identified. In conifers, the two main phenotypic traits targeted in breeding programs are growth characteristics and wood quality. Total volume, height and ring width are usually used as growth measures. Wood quality is defined in terms of end-uses, and often involves several traits related to wood density, chemical composition and fiber properties. In the area of tree adaptation, phenological traits (timing of bud set and bud burst), as well as cold-hardiness, are traits of interest.

New technologies are increasing the types and numbers of quantitative traits that can be examined, and thus studied for their QTL architecture. At the wood quality level, traits such as stem straightness, stiffness, wood specific gravity, fiber coarseness, and microfibril angle can be measured with x-ray diffraction, the SilvaScan technology, or near-infrared technology (Byram et al. 2005). At the gene level, microarray technologies allow monitoring of a vast number of gene transcripts, whose expression levels are regarded as quantitative traits. Genes involved with the lignin biosynthetic pathway are often of interest, as these genes are putatively involved with wood quality and perhaps phenology. Wood cellulose carbon isotope composition, $\delta^{13}C$, is another important trait of interest, as it is regarded as a time integrated estimate of water use efficiency. A vast number of metabolites can also be assayed via gas chromatography, especially when interfaced with mass spectrometry or high performance liquid chromatography. Like gene expression, metabolite levels can also be considered a quantitative trait; however, they are not directly tied to a gene locus like gene expression levels are. Considering global climate change it becomes very important to study genetic control of adaptive traits such as phenology, cold-hardiness and drought resistance related traits.

5.6.3 QTL Maps

5.6.3.1 Loblolly Pine

In the first QTL map for a conifer, Groover et al. (1994) inferred male and female QTL maps in loblolly pine from a full-sib family of 177 progeny assayed for RFLPs. Five genome regions contained one or more RFLP loci for wood specific gravity. In an analysis of male-female QTL homology, they inferred that the male can have a different QTL segregating at the same locus than the female, and that these alleles can have epistatic interactions.

Following this original work, Knott et al. (1997) analyzed the same data for evidence of multiple QTL in the same linkage interval, finding discordant results with Groover et al. (1994). Kaya et al. (1999) used the pedigree of Groover et al. (1994), termed "*qtl*", as well as second pedigree, "*base*", used previously by Devey et al. (1994b). Thirteen height and eight diameter QTLs were detected, suggesting control by few genes of large effect. However, a given QTL was rarely expressed in multiple years or multiple genetic backgrounds.

A series of works then ensued with the "*qtl*" pedigree. Sewell et al. (2000) used the "*qtl*" pedigree to infer physical traits of wood: wood specific gravity (*wsg*), volume percentage of latewood (*vol%*) and microfibril angle (*mfa*), in both earlywood and latewood. Nine unique QTLs were detected for wood specific gravity, five for volume percentage of latewood, and five for microfibril angle (*mfa*). Most QTL for specific gravity were specific to either earlywood or latewood, whereas each *mfa* QTL occurred in both earlywood and latewood. Sewell et al. (2002) found eight unique chemical wood property QTLs, with differences among populations for QTL. Brown et al. (2003) stressed that verification of QTL is necessary, comparing inferred QTL among populations and within populations for different years. They found that QTL expressed within pedigrees were more stable than QTL expressed among pedigrees.

An unusual approach to QTL mapping, which takes advantage of the conifer megagametophyte, was undertaken by Gwaze et al. (2003). As megagametophytes are haploid, QTL haplotypes can be traced from the offspring back to individual founders in outbred pedigrees by combining founder-origin probabilities with fully informative flanking markers. A large QTL accounting for 11.3 % of the phenotypic variance in the growth rate was detected in a loblolly pine pedigree; the QTL haplotype was traced from offspring to its founder, GP_3.

5.6.3.2 Maritime Pine

Some of the earliest conifer QTL studies also occurred in *Pinus pinaster*. Plomion et al. (1996a) assayed 126 F_2 progeny for RAPD markers, including assay of megagametophytes to determine the linkage phase of the parents. Height growth components related to the initiation (controlled by the apical meristem) and elongation of shoot cycles (controlled by the subapical meristem) were mapped to different chromosomes, suggesting that the activity of these meristems is controlled by separate genetic mechanisms. Plomion et al. (1996b) further studied this cross to find a major QTL for delta 3-carene, a monoterpene, which is a constituent of turpentine. In addition, a qualitative approach found that the "C" locus that controls the relative quantity of delta 3-carene was associated with RAPD markers near the major QTL. This was the first study of co-localization of QTL.

Markussen et al. (2003) found 10 QTLs for height or diameter and 40 QTLs for seven wood parameters in *P. pinaster*. They found that two SSR markers linked to QTL also were linked in a QTL mapped for *P. taeda* (Devey et al, 1999); such markers could be used for comparative QTL studies. Using a second *P. pinaster* three-generation pedigree, Brendel et al. (2002) found four QTLs for $\delta^{13}C$ (the first time found in a tree) and two QTLs for ring width, but they did not co-locate with the $\delta^{13}C$ QTL. On the same pedigree, Pot et al. (2006) detected 54 QTLs. QTL for different traits in the same map position also showed genetic correlations as estimated by traditional quantitative genetic analyses. Chagné et al. (2003) compared QTL maps of Maritime pine and loblolly pine, using 32 common mapped ESTP markers. The positions of two QTLs controlling wood density and cell wall components were conserved between the two species. This was the first ever comparison of QTL maps between conifer species.

5.6.3.3 Radiata Pine

In *Pinus radiata*, efforts for QTL mapping were directed towards eventual use for marker-assisted selection (MAS; the use of specific allelic variants detected in mapping population for tree improvement in unrelated populations). In the first investigation (Emebiri et al. 1998a), haploid megagametophytes were assayed, then progeny of the same individuals grown up to evaluate traits for QTL analysis. This is not a pseudo-testcross design, but rather it evaluates QTLs from the female parent only. From 222 RAPD markers, stem diameter, volume and height were compared at 5 months, and at 1, 2 and 3 years of age. In a second study, four QTLs for stem growth efficiency were found, which accounted for 8.5–36.4% of the population variance (Emebiri et al. 1998b). In a third study, wood density was evaluated at three stages (Kumar et al. 2000). The results suggested that early selection can be used in order to increase juvenile wood density, although the putative QTLs detected in this study need to be verified in an independent population.

Devey et al. (2004) mapped QTL for juvenile wood density (JWD) and diameter at breast height (DBH) using a large full-sib family. The percent variance accounted for by several QTL ranged from 0.78% to 3.58%, suggesting a genomic architecture of many genes with small effect. Two unrelated "bridging" families were chosen to identify markers for MAS. Four markers showed consistent association with JWD, providing the first basis for MAS in a conifer.

5.6.3.4 Scots Pine

In *Pinus sylvestris*, Lerceteau et al. (2000) generated both male and females using the two-way pseudo-testcross strategy. On the female size, 12 QTLs were detected, the largest for frost hardiness. A cluster of QTLs for tree height, trunk diameter and volume was located on one LG. On the male map, four QTLs for trunk diameter and volume were detected. Yazdani et al. (2003) also adopted the pseudo-testcross method, and found QTLs for shoot elongation; growth cessation and cold acclimation were found on both maps. Their study concluded that major QTLs control growth rhythm and autumn cold acclimation.

5.6.3.5 Pine hybrids

In the only QTL study of a conifer hybrid (*Slash pine x Caribbean Pine*), a pseudo-testcross QTL detection strategy was used to identify QTLs for wood density, secondary growth, and dry wood mass in a pedigree of size 133 (Shepherd et al. 2003b). Twelve QTLs were identified that clustered into four LGs in the slash pine parent and in only one group in the Caribbean pine parent. QTLs that influenced density and ring width did not co-locate, suggesting independent inheritance of these characters. Two other pedigrees were more recently mapped for QTLs for adventitious rooting (Shepherd et al. 2006). Most small to moderate effect QTL were congruent between the two pedigrees, while a large effect QTL was found only in one pedigree, and was postulated to be a between-species effect. Targeting between-species effects for improvement in synthetic hybrid populations may increase the efficacy and predictability of hybrid breeding.

5.6.3.6 Douglas-fir

A series of studies used a three-generation pedigree to examine various classes of traits for QTL in Douglas-fir (*Pseudotsuga menziesii*). Jermstad et al. (2001a) genotyped 192 progeny for 74 evenly distributed RFLP markers found by Jermstad et al. (1998). Thirty three QTL for timing of spring bud flush were found, and measurements for each of 3 years and 2 test sites showed that several QTLs influence the timing of bud flush over multiple years within sites but not between sites, indicating major QTL of consistent effect within sites but interactions with environment between sites. Using the same material, Jermstad et al. (2001b) found 11 and 15 QTLs affecting fall and spring cold-hardiness, respectively. Three different shoot tissues phenotyped for spring hardiness showed similar QTL, while different tissues phenotyped for fall hardiness showed little QTL similarity, supporting previous reports that spring tissues are more synchronized than fall tissues.

Jermstad et al. (2003) again used the same pedigree and markers, but for additional individuals totaling 460, to investigate QTL interactions of many of the above traits with photoperiod, moisture stress, winter chilling, and spring temperature. In the first investigation of QTL interaction with environment, they found two QTL-by-treatment interactions for growth initiation traits, and several QTL-by-treatment interactions for growth cessation traits. Finally, Wheeler et al. (2005) evaluated QTL for cold-hardiness via artificial freezing and various cold injury assessment methods in two pedigrees of size 170 and 383. Six QTL were found in the first pedigree, eight in the second, of which four were shared between the pedigrees; 17 of 29 putative cold-hardiness candidate genes identified from ESTs were located within the QTL intervals, thus identifying them as high priority for association studies. These works with Douglas-fir demonstrate a unique opportunity of working with trees: long-lived species allow "*immortal*" pedigrees that can be repeatedly phenotyped for different traits after genotyping.

Finally, QTL analyses are normally conducted in single pedigrees. In contrast, Ukrainetz et al. (2008b) examined eight full-sib families, each of size 40 progeny, for wood-related QTLs, using the software "QTL Express" (Seaton et al. 2002). They found that wood fiber and density traits both showed the lowest number of QTLs (3) with relatively small effects; wood chemistry traits showed more QTLs (7), while ring density traits large numbers of QTLs (78) and interesting patterns of temporal variation. Growth traits gave just five QTLs but of major effect. These wood quality traits are the widest suite of traits yet examined for QTL analysis in a conifer. Moreover, examination of multiple families for QTL gives a population perspective of the true extent of QTL variation.

5.6.3.7 Norway spruce

Markussen et al. (2004) employed bulked segregant analysis and AFLP markers to compare Norway spruce (*Picea abies*) individuals with high and low wood density. Of 107 polymorphic AFLP markers, 15 markers showed significant linkage to wood density, and two of these were found to predict wood density in unrelated full-sib families. Markussen et al. (2005) extended this strategy to compare individuals with high and low extractives content. Of 14 polymorphic AFLP markers were detected between the pools, one marker was linked to low extractives content and subsequently verified as above. Recently, a full-sib family of size 250 has been assayed for *Heterobasidion* (root rot), with the objective of mapping QTLs and identifying candidate genes conferring reduced susceptibility to *Heterobasidion* spp. (Jenny Arnerup and Jan Stenlid, Univ of Uppsala, pers. comm.).

5.6.3.8 *North American Spruce Species*

No QTL mapping studies have been conducted in spruce until recently. In the Quebec Arborea genome project, two pedigrees of white spruce, of size 395 and 740, have been established and genotyped for 768 and 1,536 gene SNPs, respectively, using the Illumina GoldenGate assay. Experiments on different sites involving clonal propagation of root cuttings have been used to evaluate genotype-by-environment interactions for growth and adaptive traits (Pelgas et al. 2011). About 34 QTL clusters each explaining generally below 15% of phenotypic variance were found for bud flush, bud set and height growth, with about 20% of these replicated between mapping populations and 50% of them with spatial or temporal stability. At least three occurences of overlapping QTLs were noted, indicative of potential pleiotropic effects. On a smaller scale, a black spruce pedigree of size 283 is being studied for wood quality and phenology traits (J Prunier et al. unpubl. data). As the genes have already been mapped in both this pedigree and in the white spruce pedigrees, this will offer an excellent opportunity to assess QTL homology across species.

The British Columbia Treenomix genome project has worked with two factorial crosses from the spruce weevil resistance breeding program (see Alfaro et al. 2004). In the first, involving Interior spruce, 369 progeny in 3 × 2 factorial were genotyped for 253 informative SNP markers using the Illumina GoldenGate assay (I Porth et al. unpubl. data). Over 300 metabolites were also assayed (R Dauwe et al. unpubl. data). The second cross, involving Sitka spruce, is currently being assayed. An approach called "genetical genomics" may also identify previously unidentified networks of genes unique to conifers.

5.6.3.9 *Sugi*

Yoshimaru et al. (1998) mapped QTLs for growth, flowering and rooting ability in Sugi (*Crypomeria. japonica*). Growth is one of the most important traits for timber-producing woody species and also for carbon dioxide fixation to mitigate global warming. QTLs for juvenile growth, including height and diameter of basal area, were mapped. Flowering is essential for reproduction, but is not necessary for timber production. If the expression of flowering could be controlled, it would be useful not only for breeding but also for forestry and the environment. QTLs for male and female flowers have been mapped at two locations each, respectively. The rooting ability of this species is very important for clonal forestry in the southwestern part of Japan, especially in Kyushu Island. QTLs for rooting ability were found but there were not highly significant in the family used in the study (Yoshimaru et al. 1998). Wood quality QTLs, specifically modulus of elasticity (an important indicator of wood strength), have also been mapped in Sugi (Kuramoto et al. 2000).

Recently, pollinosis (human allergies to pollen) has become a serious social problem; 10 to 20% of Japanese have pollinosis to pollen from Sugi because of a large plantation, which now has matured to flowering. As a countermeasure, the male-sterile lines of *C. japonica* are planned to be used for reforestation. Some male-steriles seem to be controlled by a single recessive locus (Taira et al. 1993). To determine the location of the locus on the linkage map, co-dominant DNA markers have been used for mapping of the gene, using SSRs (Moriguchi et al. 2003; Tani et al. 2004), EST-SSRs (Y Moriguchi et al. unpubl. data), and SNPs (T Ujino-Ihara et al. unpubl. data). After the genome location of this male-sterile gene is found, a selective marker will be developed and used for selection of the male-sterile individuals from the plantation forests and plus trees as breeding materials.

5.7 Prospects

In a seminal review, Remington and Purugganan (2003) stated that future research in plants should expand the number of traits that are intensively studied and make greater use of QTL mapping in wild plant taxa, especially those undergoing adaptive radiations, while continuing to draw on insights from model plants. Conifers are inherently non-domesticated (e.g., wild plant taxa) and the resources provided by breeding programs and genome projects will provide rich resources for testing of candidate gene-trait associations in wild populations, genetic mapping in hybrid zones, and microarray analyses of gene expression.

In conifers, comparative analyses of genetic maps will continue to be a fertile ground for future studies. In sunflower species, a comparative study showed that in the face of extensive hybridization and gene flow, species integrity is maintained (Strasburg et al. 2009). There are many examples of hybrid zones in conifer species, such as the hybridization between Englemann spruce and white spruce in British Columbia. There have been no such studies in conifers that compare patterns of genetic divergence and diversity along chromosomal segments, which can reveal divergent selection for speciation. In conifers, few studies involving "genome scans" have been done (but see Namroud et al. 2008).

Another approach possible for conifers is to use "hitchhiking mapping" to identify regions of recent selective sweeps, due to adaptive divergence. This method starts from a genome scan using a randomly spaced set of molecular markers followed by a fine-scale analysis in the flanking regions of the candidate regions under selection. In fish, the hitchhiking approach identified a selective sweep around candidate locus *Stn90* (Makinen et al. 2008). Fine scale genome maps will help identify candidate loci for adaptation in conifers, particularly those involved with strong ecological gradients, such as that found Sitka spruce from coastal California to coastal Alaska (see Mimura and Aitken 2007).

Yet another new avenue for using QTL maps is "genetical genomics", which combines genetic mapping with gene expression analysis. It uses variation of gene expression induced by segregation within mapping populations to infer interactions among expressed genes or metabolites. Gene networks, and even directed gene networks, can be inferred by the joint analysis of marker genotypes and gene expression and metabolite levels (Rockman 2008). In the Treenomix II project, two genetical genomic studies are nearing completion. These involve a 22K member cDNA microarray, hundreds of assayed metabolites, and scores for weevil resistance in both white spruce pedigree (I Porth et al. unpubl. data; R Dauwe et al. unpubl. data) and a Sitka spruce pedigree (S Verne et al. unpubl. data).

Recently a number of "next-generation" sequencing technologies have been invented, which can sequence fragments of DNA at astoundingly higher rates compared to Sanger sequencing. These include the Illumina/ Solexa, ABI/SOLiD, 454/Roche, Pacific Biosciences/SMRT and Helicos (Morozova and Marra 2008). To date, these technologies have been applied mostly in non-marker contexts, such as whole-genome sequencing (Bentley et al. 2008), targeted resequencing (Gnirke et al. 2009), discovery of transcription factor binding sites, transcript and non-coding RNA expression profiling, and other functional genomic studies (Eveland et al. 2008). These technologies should greatly facilitate genotyping of mapping populations for mapping through direct and parallel sequencing of multiple individuals.

Finally, and last but not least, for the past several years, there has been an initiative to sequence a conifer genome, starting with the seminal paper of Neale et al. (1994). There are several initiatives such as the Pine Genome Initiative (*http://pinegenomeinitiative.org/*) and the International Conifer Genome Initiative (*http://www.pinegenome.org*). It is not clear what strategy is the best, and current initiatives are exploring alternatives. Fine scale genetic mapping will clearly enable the assembly of contigs based upon shotgun sequencing (for example, in the monkeyflower genome project, John Willis pers. comm.). A current goal of the Arborea project is to map 10,000 genes in white spruce (J Bousquet pers. comm.). Other workers in the USA, Canada and Spain have embarked upon exploratory BAC sequencing and gene enrichment of the repetitive genome to discover the structure of conifer genomes, using "gene space" explorations developed such as for maize (Liu et al. 2007). These approaches will interface with genetic mapping to help assemble the first conifer genome.

References

Acheré V, Faivre-Rampant P, Jeandroz S, Besnard G, Markussen T, Aragones A, Fladung M, Ritter E, Favre J-M (2004) A full saturated linkage map of *Picea abies* including AFLP, SSR, ESTP, 5S rDNA, and morphological markers. Theor Appl Genet 108: 1602–1613.

Acheré V, Favre J-M, Besnard G, Jeandroz S (2005) Genomic organization of molecular differentiation in Norway spruce (*Picea abies*). Mol Ecol 14: 3191–3201.

Adams WS, Strauss SH, Copes DL, Griffin AR (1992) Population Genetics of Forest Trees. Klewer, Dordrecht, The Netherlands.

Agrawal AF, Hadany L, Otto SP (2005) The evolution of plastic recombination. Genetics 171: 803–812.

Alfaro RI, vanAkker L, Jaquish B, King J (2004) Weevil resistance of progeny derived from putatively resistant and susceptible interior spruce parents. For Ecol Manag 202: 369–377.

Amarasinghe V, Carlson JE (1998) Physical mapping and characterization of 5S rRNA genes in Douglas-fir. J Hered 89: 495–500.

Arcade A, Anselin F, Faivre Rampant P, Lesage MC, Pâques LE, Prat D (2000) Application of AFLP, RAPD and ISSR markers to genetic mapping of European and Japanese larch. Theor Appl Genet 100: 299–307.

Beavis WD (1998) QTL analyses: power, precision, and accuracy. In: AH Paterson (ed) Molecular Dissection of Complex Traits. CRC Press, Boca Raton, Florida, USA, pp 145–162.

Bentley DR, Balasubramanian S, Swerdlow HP, et al. (2008) Accurate whole human genome sequencing using reversible terminator chemistry. Nature 456: 53–59.

Binelli G, Bucci G (1994) A genetic linkage map of *Picea abies* Karst., based on RAPD markers, as a tool in population genetics. Theor Appl Genet 88: 283–288.

Bornet B, Branchard M (2001) Nonanchored Inter Simple Sequence Repeat (ISSR) markers: reproducible and specific tools for genome fingerprinting. Plant Mol Biol Rep 19: 209–215.

Brendel O, Pot D, Plomion C, Rozenberg P, Guehl J-M (2002) Genetic parameters and QTL analysis of delta C-13 and ring width in maritime pine. Plant Cell Environ 25: 945–953.

Brown GR, Carlson JE (1997) Molecular cytogenetics of the genes encoding 18*s*-5.8*s*-26*s* rRNA and 5s rRNA in two species of spruce (*Picea*). Theor Appl Genet 95: 1–9.

Brown GR, Newton CH, Carlson JE (1998) Organization and distribution of a *Sau*3A tandem repeated DNA sequence in *Picea* (Pinaceae) species. Genome 41: 560–565.

Brown GR, Kadel III EE, Bassoni DL, Kiehne KL, Temesgen B, van Buijtenen JP, Sewell MM, Marshall KA, Neale DB (2001) Anchored reference loci in loblolly pine (*Pinus taeda* L.) for integrating pine genomics. Genetics 159: 799–809.

Brown GR, Bassoni DL, Gill GP, Fontana JR, Wheeler NC, Megraw RA, Davis MF, Sewell MM, Tuskan GA, Neale DB (2003) Identification of quantitative trait loci influencing wood property traits in loblolly pine (*Pinus taeda* L.) III. QTL verification and candidate gene mapping. Genetics 164: 1537–1546.

Buetow KH (1991) Influence of aberrant observations on high-resolution linkage analysis outcomes. Am J Hum Genet 49: 985–994.

Byram TD, Myszewski JH, Gwaze DP, Lowe WJ (2005) Improving wood quality in the western gulf forest tree improvement program: the problem of multiple breeding objectives. Tree Genet Genomes 1: 85–92.

Cairney J, Pullman GS (2007) The cellular and molecular biology of conifer embryogenesis. New Phytol 176: 511–536.

Carlson JE, Traore A, Agrama HA, Krutovsky KV (2007) Douglas-fir. In: C Kole (ed) Genome Mapping and Molecular Breeding in Plants, vol 7: Forest Trees. Springer, Berlin, Tokyo, pp 199–210.

Chagné D, Lalanne C, Madur D, Kumar S, Frigério J-M, Krier C, Decroocq S, Savouré A, Bou-Dagher-Kharrat M, Bertocchi E, Brach J, Plomion C (2002) A high density genetic map of maritime pine based on AFLPs. Ann For Sci 59: 627–636.

Chagné D, Brown G, Lalanne C, Madur D, Pot D, Neale D, Plomion C (2003) Comparative genome and QTL mapping between maritime and loblolly pines. Mol Breed 12: 185–195.

Chagné D, Chaumeil P, Ramboer A, Collada C, Guevara A, Cervera MT, Vendramin GG, Garcia V, Frigerio J-M, Echt C, Richardson T, Plomion C (2004) Cross-species transferability and mapping of genomic and cDNA SSRs in pines. Theor Appl Genet 109: 1204–1214.

Chakravarti A, Lasher LK, Reefer JE (1991) A maximum likelihood method for estimating genome length using genetic linkage data. Genetics 128: 175–182.

Clark AG (1990) Inference of haplotypes from PCR-amplified samples of diploid populations. Mol Biol Evol 7: 111–122.

Conkle MT (1981) Proceedings of the Symposium on Isozymes of North American Forest Trees and Forest Insects, 27 July, 1979, Berkeley, California. Gen Tech Rep PSW-GTR-48: Pacific Southwest Forest and Range Exp Stn, Forest Service, US Department of Agriculture, Berkeley, CA, USA.

Costa P, Pot D, Dubos C, Frigerio JM, Pionneau C, Bodenes C, Bertocchi E, Cervera M-T, Remington DL, Plomion C (2000) A genetic map of Maritime pine based on AFLP, RAPD and protein markers. Theor Appl Genet 100: 39–48.

Darvasi A, Weinreb A, Minke V, Weller JI, Soller M (1993) Detecting marker-QTL linkage and estimating QTL gene effect and map location using a saturated genetic map. Genetics 134: 943–951.

Devey ME, Fiddler TA, Liu BH, Knapp SJ, Neale DB (1994a) An RFLP linkage map for loblolly pine based on a three-generation outbred pedigree. Theor Appl Genet 88: 273–278.

Devey ME, Fiddler TA, Liu BH, Knapp SJ, Neale DB (1994b) An RFLP linkage map for loblolly pine based on a 3-generation outbred pedigree. Theor Appl Genet 88: 273–278.

Devey ME, Bell JC, Smith DN, Neale DB, Moran GF (1996) A genetic linkage map for *Pinus radiata* based on RFLP, RAPD, and microsatellite markers. Theor Appl Genet 92: 673–679.

Devey ME, Sewell MM, Uren TL, Neale DB (1999) Comparative mapping in loblolly and radiata pine using RFLP and microsatellite markers. Theor Appl Genet 99: 656–662.

Devey ME, Carson SD, Nolan MF, Matheson AC, Te Riini C, Hohepa J (2004) QTL associations for density and diameter in *Pinus radiata* and the potential for marker-aided selection. Theor Appl Genet 108: 516–524.

Echt CS, Nelson CD (1997) Linkage mapping and genome length in eastern white pine (*Pinus strobus* L.). Theor Appl Genet 94: 1031–1037.

Echt CS, Saha S, Krutovsky KV, Wimalanathan K, Erpelding JE, Liang C, Nelson CD (2011) An Annotated genetic map of loblolly pine based on microsatellite and cDNA markers. BMC Genetics (in press).

Eckert AJ, Pande B, Ersoz ES, Wright MH, Rashbrook VK, Nicolet CM, Neale DB (2009) High-throughput genotyping and mapping of single nucleotide polymorphisms in loblolly pine (*Pinus taeda* L.). Tree Genet Genomes 5: 225–234.

Ellis JR, Burke JM (2007) EST-SSRs as a resource for population genetic analyses. Heredity 99: 125–132.

Emebiri LC, Devey ME, Matheson AC, Slee MU (1998a) Age-related changes in the expression of QTLs for growth in radiata pine seedlings. Theor Appl Genet 97: 1053–1061.

Emebiri LC, Devey ME, Matheson AC, Slee MU (1998b) Interval mapping of quantitative trait loci affecting NESTUR, a stem growth efficiency index of radiata pine seedlings. Theor Appl Genet 97: 1062–1068.

Ercht CS, Saha S, Krutovsky KV, Wimalanathan K, Erpelding JE, Liang C, Nelson CD (2011) An annotated genetic map of loblolly pine based on microsatellite and cDNA markers. BMC Genetics (in press).

Eveland AL, McCarty DR, Koch KE (2008) Transcript profiling by 3′-untranslated region sequencing resolves expression of gene families. Plant Physiol 146: 32–44.

Fulton TM, Van der Hoeven R, Eannetta NT, Tanksley SD (2002) Identification, analysis, and utilization of conserved ortholog set markers for comparative genomics in higher plants. Plant Cell 14: 1457–1467.

Gale MD, Devos KM (1998) Comparative genetics in the grasses. Proc Natl Acad Sci USA 95: 1971–1974.

Gasbarra D, Sillanpää MJ (2006) Constructing the parental linkage phase and the genetic map over distances < 1 cM using pooled haploid DNA. Genetics 172: 1325–1335.

Gnirke A, Melnikov A, Maguire J, Rogov P, LeProust E, Brockman W, Fennell T, Giannoukos G, Fisher S, Russ C, Gabriel S, Jaffe DB, Lander ES, Nusbaum C (2009) Solution hybrid selection with ultra-long oligonucleotides for massively parallel targeted sequencing. Nat Biotechnol 27: 182–189.

Göçmen B, Jermstad KD, Neale DB, Kaya Z (1996) Development of random amplified polymorphic DNA markers for genetic mapping in Pacific yew (*Taxus brevifolia*). Can J For Res 26: 497–503.

Gogarten JP, Olendzenski L (1999) Orthologs paralogs and genome comparisons. Curr Opin Genet Dev 9: 630–636.

Gosselin I, Zhou Y, Bousquet J, Isabel N (2002) Megagametophyte-derived linkage maps of white spruce (*Picea glauca*) based on RAPD, SCAR and ESTP markers. Theor Appl Genet 104: 987–997.

Grattapaglia D, Sederoff R (1994) Genetic linkage maps of *Eucalyptus grandis* and *Eucalyptus urophylla* using a pseudo-testcross: mapping strategy and RAPD markers. Genetics 137: 1121–1137.

Groover A, Devey M, Fiddler T, Lee J, Megraw R, Mitchel-Olds T, Sherman B, Vujcic S, Williams C, Neale D (1994) Identification of quantitative trait loci influencing wood specific gravity in an outbred pedigree of loblolly pine. Genetics 138: 1293–1300.

Groover AT, Williams CG, Devey ME, Lee JM, Neale DB (1995) Sex-related differences in meiotic recombination frequency in *Pinus taeda*. J Hered 86: 157–158.

Gwaze DP, Zhou Y, Reyes-Valdes MH, Al-Rababah MA, Williams CG (2003) Haplotypic QTL mapping in an outbred pedigree. Genet Res 81: 43–50.

Hayashi E, Kondo T, Terada K, Kuramoto N, Goto Y, Okamura M, Kawasaki H (2001) Linkage map of Japanese black pine based on AFLP and RAPD markers including markers linked to resistance against the pine needle gall midge. Theor Appl Genet 102: 871–875.

Hu X-S, Goodwillie C, Ritland KM (2004) Joining genetic linkage maps using a joint likelihood function. Theor Appl Genet 109: 996–1004.

Hudson EE (2005) Development of a genetic linkage map in *Abies nordmanniana* accession 9M and assessment of disease resistance to *Phytophthora cinnamomi*. North Carolina State Univ, Raleigh, North Carolina, USA.

Hulbert SH, Ilott TW, Legg EJ, Lincoln SE, Lander ES, Michelmore RW (1988) Genetic analysis of the fungus *Bremia lactucae*, using restriction fragment length polymorphisms. Genetics 120: 947–958.

Huynen MA, Bork P (1998) Measuring genome evolution. Proc Natl Acad Sci USA 95: 5849–5856.

Iwata H, Ujino-Ihara T, Yoshimura K, Nagasaka K, Mukai Y, Tsumura Y (2001) Cleaved amplified polymorphic sequence markers in sugi, *Cryptomeria japonica* D. Don, and their locations on a linkage map. Theor Appl Genet 103: 881–895.

Jermstad KD, Bassoni DL, Wheeler NC, Neale DB (1998) A sex-averaged genetic linkage map in coastal Douglas-fir (*Pseudotsuga menziesii* [Mirb.] Franco var '*menziesii*') based on RFLP and RAPD markers. Theor Appl Genet 97: 762–770.

Jermstad KD, Bassoni DL, Jech KS, Wheeler NC, Neale DB (2001a) Mapping of quantitative trait loci controlling adaptive traits in coastal Douglas-fir. I. Timing of vegetative bud flush. Theor Appl Genet 102: 1142–1151.

Jermstad KD, Bassoni DL, Wheeler NC, Anekonda TS, Aitken SN, Adams WT, Neale DB (2001b) Mapping of quantitative trait loci controlling adaptive traits in coastal Douglas-fir. II. Spring and fall cold-hardiness. Theor Appl Genet 102: 1152–1158.

Jermstad KD, Bassoni DL, Jech KS, Ritchie GA, Wheeler NC, Neale DB (2003) Mapping of quantitative trait loci controlling adaptive traits in coastal Douglas fir. III. Quantitative trait loci-by-environment interactions. Genetics 165: 1489–1506.

Jermstad KD, Eckert AF, Wegrzyn JL, Delfino-Mix A, Davis DA, Burton DC, Neale DB (2010) Comparative mapping in Pinus: sugar pine (*Pinus lambertiana Dougl.*) and loblolly pine (*Pinus taeda* L.). Tree Genet Genomes (in press).

Kang B-Y, Mann IK, Major JE, Rajora OP (2010) Near-saturated and complete genetic linkage map of black spruce (*Picea mariana*). BMC Genomics 11: 515.

Kaya Z, Neale DB (1995) Utility of random amplified polymorphic DNA (RAPD) markers for linkage mapping in Turkish red pine (*Pinus brutia* Ten.). Silvae Genet 44: 110–116.

Kaya Z, Sewell MM, Neale DB (1999) Identification of quantitative trait loci influencing annual height- and diameter-increment growth in loblolly pine (*Pinus taeda* L.). Theor Appl Genet 98: 586–592.

Kim Y-Y, Choi HS, Kang B-Y (2005) An AFLP-based linkage map of Japanese red pine (*Pinus densiflora*) using haploid DNA samples of megagametophytes from a single maternal tree. Mol Cells 20: 201–209.

Kinlaw CS, Neale DB (1997) Complex gene families in pine genomes. Trends Plant Sci 2: 356–359.

Kirst M, Johnson AF, Baucom C, Ulrich E, Hubbard K, Staggs R, Paule C, Retzel E, Whetten R, Sederoff R (2003) Apparent homology of expressed genes from wood-forming tissues of loblolly pine (*Pinus taeda* L.) with *Arabidopsis thaliana*. Proc Nat Acad Sci USA 100: 7383–7388.

Knott SA, Neale DB, Sewell MM, Haley CS (1997) Multiple marker mapping of quantitative trait loci in an outbred pedigree of loblolly pine. Theor Appl Genet 94: 810–820.

Komulainen P, Brown GR, Mikkonen M, Karhu A, Garcia-Gil MR, O'Malley D, Lee B, Neale DB, Savolainen O (2003) Comparing EST-based genetic maps between *Pinus sylvestris* and *Pinus taeda*. Theor Appl Genet 107: 667–678.

Koonin EV (2005) Orthologs, paralogs, and evolutionary genomics. Annu Rev Genet 39: 309–338.

Krutovsky KV, Vollmer SS, Sorensen FC, Adams WT, Knapp SJ, Strauss SH (1998) RAPD genome maps of Douglas-fir. J Hered 89: 197–205.

Krutovsky KV, Troggio M, Brown GR, Jermstad KD, Neale DB (2004) Comparative mapping in the Pinaceae. Genetics 168: 447–461.

Krutovsky KV, Elsik CG, Matvienko M, Kozik A, Neale DB (2006) Conserved ortholog sets in forest trees. Tree Genet Genomes 3: 61–70.

Kuang H, Richardson T, Carson S, Wilcox P, Bongarten B (1999) Genetic analysis of inbreeding depression in plus tree 850.55 of *Pinus radiata* D. Don. I. Genetic map with distorted markers. Theor Appl Genet 98: 697–703.

Kubisiak TL, Nelson CD, Nance WL, Stine M (1995) RAPD linkage mapping in a longleaf pine × slash pine F_1 family. Theor Appl Genet 90: 1119–1127.

Kumar S, Spelman RJ, Garrick DJ, Richardson TE, Lausberg M, Wilcox PL (2000) Multiple-marker mapping of wood density loci in an outbred pedigree of radiata pine. Theor Appl Genet 100: 926–933.

Kuramoto N, Kondo T, Fujisawa Y, Nakata R, Hayashi E, Goto Y (2000) Detection of quantitative trait loci for wood strength in *Cryptomeria japonica*. Can J For Res 30: 1525–1533.

Lander ES, Botstein D (1989) Mapping mendelian factors underlying quantitative traits using RFLP linkage maps. Genetics 121: 185–199.

Lander ES, Green P, Abrahamson J, Barlow A, Daly MJ, Lincoln SE, Newburg L (1987) MAPMAKER: An interactive computer package for constructing primary genetic linkage maps of experimental and natural populations. Genomics 1: 174–181.

Lathrop GM, Lalouel JM, Julier C, Ott J (1985) Multilocus linkage analysis in humans: detection of linkage and estimation of recombination. Am J Hum Genet 37: 482–498.

Le Dantec L, Chagné D, Pot D, Cantin O, Garnier-Géré P, Bedon F, Frigerio J-M, Chaumeil P, Léger P, Garcia V, Laigret F, de Daruvar A, Plomion C (2004) Automated SNP detection in expressed sequence tags: statistical considerations and application to maritime pine sequences. Plant Mol Biol 54: 461–470.

Lerceteau E, Plomion C, Andersson B (2000) AFLP mapping and detection of quantitative trait loci (QTLs) for economically important traits in *Pinus sylvestris*: a preliminary study. Mol Breed 6: 451–458.

Li C, Yeh FC (2001) Construction of a framework map in *Pinus contorta* subsp. *latifolia* using random amplified polymorphic DNA markers. Genome 44: 147–153.

Liewlaksaneeyanawin C, Zhuang J, Tang M, Farzaneh N, Lueng G, Cullis C, Findlay S, Ritland CE, Bohlmann J, Ritland K (2009) Identification of COS markers in the Pinaceae. Tree Genet Genomes 5: 247–255.

Liu BH (1998) Statistical Genomics: Linkage, Mapping, and QTL Analysis. CRC Press, Boca Raton, Florida, USA.

Liu R, Vitte C, Ma J, Mahama AA, Dhliwayo T, Lee M, Bennetzen JL (2007) A GeneTrek analysis of the maize genome. Proc Nat Acad Sci USA 104: 11844–11849.

Liu W, Magbanua ZV, Ozkan S, Chouvarine P, Bartlett BD, Peterson DG (2009) BAC libraries for two distantly related conifers, loblolly pine and bald cypress. In: Plant Anim Genome XVII Conf, San Diego, CA, USA.

Mäkinen HS, Shikano T, Cano JM, Merilä J (2008) Hitchhiking mapping reveals a candidate genomic region for natural selection in three-spined stickleback chromosome VIII. Genetics 178: 453–465.

Margarido GRA, Souza AP, Garcia AAF (2007) OneMap: software for genetic mapping in outcrossing species. Hereditas 144: 78–79.

Markussen T, Fladung M, Achere V, Favre JM, Faivre-Rampant P, Aragones A, Silva Perez DD, Harvengt L, Espinel S, Ritter E (2003) Identification of QTLs controlling growth, chemical and physical wood property traits in *Pinus pinaster* (Ait.). Silvae Genet 52: 8–15.

Markussen T, Tusch A, Stephan BR, Fladung M (2004) Identification of molecular markers for selected wood properties of Norway spruce *Picea abies* L. (Karst.) I. Wood density. Silvae Genet 53: 45–50.

Markussen T, Tusch A, Stephan BR, Fladung M (2005) Identification of molecular markers for selected wood properties of Norway spruce *Picea abies* L. (Karst.) II. Extractives content. Silvae Genet 54: 145–152.

Mester D, Ronin Y, Minkov D, Nevo E, Korol A (2003) Constructing large-scale genetic maps using an evolutionary strategy algorithm. Genetics 165: 2269–2282.

Mester DI, Ronin YI, Nevo E, Korol AB (2004) Fast and high precision algorithms for optimization in large-scale genomic problems. Comp Biol Chem 28: 281–290.

Mester DI, Ronin YI, Korostishevsky MA, Pikus VL, Glazman AE, Korol AB (2006) Multilocus consensus genetic maps (MCGM): Formulation, algorithms, and results. Comp Biol Chem 30: 12–20.

Mimura M, Aitken SN (2007) Adaptive gradients and isolation-by-distance with postglacial migration in *Picea sitchensis*. Heredity 99: 224–232.

Moran GF, Bell JC, Hilliker AJ (1983) Greater meiotic recombination in male vs. female gametes in *Pinus radiata*. J Hered 74: 62.

Morgante M, Olivieri AM (1993) PCR-amplified microsatellites as markers in plant genetics. Plant J 3: 175–182.

Morgante M, Hanafey M, Powell W (2002) Microsatellites are preferentially associated with nonrepetitive DNA in plant genomes. Nat Genet 30: 194–200.

Moriguchi Y, Iwata H, Ujino-Ihara T, Yoshimura K, Taira H, Tsumura Y (2003) Development and characterization of microsatellite markers for *Cryptomeria japonica* D. Don. Theor Appl Genet 106: 751–758.

Morozova O, Marra MA (2008) Applications of next-generation sequencing technologies in functional genomics. Genomics 92: 255–264.

Morse AM, Peterson DG, Islam-Faridi MN, Smith KE, Magbanua Z, Garcia SA, Kubisiak TL, Amerson HV, Carlson JE, Nelson CD, Davis JM (2009) Evolution of genome size and complexity in *Pinus*. PLoS One 4: e4332.

Mukai Y, Suyama Y, Tsumura Y, Kawahara T, Yoshimaru H, Kondo T, Tomaru N, Kuramoto N, Murai M (1995) A linkage map for sugi (*Cryptomeria japonica*) based on RFLP, RAPD, and isozyme loci. Theor Appl Genet 90: 835–840.

Murray B (1998) Nuclear DNA amounts in gymnosperms. Ann Bot 82: 3–15.

Namroud M-C, Beaulieu J, Juge N, Laroche J, Bousquet J (2008) Scanning the genome for gene single nucleotide polymorphisms involved in adaptive population differentiation in white spruce. Mol Ecol 17: 3599–3613.

Neale DB, Krutovsky KV (2004) Comparative genetic mapping in trees: The group of conifers. In: H Lörz, G Wenzel (ed) Biotechnology in Agriculture and Forestry: Molecular Marker Systems in Plant Breeding and Crop Improvement. Springer, Berlin, Germany, pp 267–277 .

Neale DB, Kinlaw CS, Sewell MM (1994) Genetic mapping and DNA sequencing of the loblolly pine genome. For Genet 1: 197–206.

Nelson CD, Nance WL, Doudrick RL (1993) A partial genetic linkage map of slash pine (*Pinus elliottii* Engelm. var. *elliottii*) based on random amplified polymorphic DNAs. Theor Appl Genet 87: 145–151.

Nelson CD, Kubisiak TL, Stine M, Nance WL (1994) A genetic linkage map of longleaf pine (*Pinus palustris* Mill.) based on random amplified polymorphic DNAs. J Hered 85: 433–439.

Nelson CD, Peterson DG, Echt CS, Whetten R, Krutovsky KV, Yuceer C, Dean JF (2008) Pine physical mapping and genome sequencing. In: Plant Anim Genome XVI Conf, San Diego, CA, USA.

Nikaido AM, Ujino T, Iwata H, Yoshimura K, Yoshimura H, Suyama Y, Murai M, Nagasaka K, Tsumura Y (2000) AFLP and CAPS linkage maps of *Cryptomeria japonica*. Theor Appl Genet 100: 825–831.

Nurul Islam-Faridi M, Dana Nelson C, Kubisiak T (2007) Reference karyotype and cytomolecular map for loblolly pine (*Pinus taeda* L.) Genome 50: 241–251

Ott J (1999) Analysis of human genetic linkage, 3rd edn. Johns Hopkins Univ Press, Baltimore, Maryland, USA.

Paglia GP, Olivieri AM, Morgante M (1998) Towards second generation STS (sequence-tagged sites) linkage maps in conifers: a genetic map of Norway spruce (*Picea abies* K.). Mol Gen Genet 258: 466–478.

Pavy N, Parsons LS, Paule C, MacKay J, Bousquet J (2006) Automated SNP detection from a large collection of white spruce expressed sequences: contributing factors and approaches for the categorization of SNPs. BMC Genom 7: 174.

Pavy N, Pelgas B, Beauseigle S, Blais S, Gagnon F, Gosselin I, Lamothe M, Isabel N, Bousquet J (2008) Enhancing genetic mapping of complex genomes through the design of highly-multiplexed SNP arrays: application to the large and unsequenced genomes of white spruce and black spruce. BMC Genom 9: 21.

Pelgas B, Isabel N, Bousquet J (2004) Efficient screening for expressed sequence tag polymorphisms (ESTPs) by DNA pool sequencing and denaturing gradient gel electrophoresis (DGGE) in spruces. Mol Breed 13: 263–279.

Pelgas B, Bousquet J, Beauseigle S, Isabel N (2005) A composite linkage map from two crosses for the species complex *Picea mariana* × *P. rubens* and analysis of synteny with other Pinaceae. Theor Appl Genet 111: 1466–1488.

Pelgas B, Beauseigle S, Acheré V, Jeandroz S, Bousquet J, Isabel N (2006) Comparative genome mapping among *Picea glauca, P. mariana x P. rubens* and *P. abies*, and correspondence with other Pinaceae. Theor Appl Genet 113: 1371–1393.

Pelgas B, Bousquet J, Meirmans P, Ritland K, Isabel N (2011) QTL mapping in white spruce: gene maps and genomic regions underlying adaptive traits across pedigrees, years and environments. BMC Genomics (in press).

Perry DJ, Bousquet J (1998) Sequence-tagged-site (STS) markers of arbitrary genes: development, characterization and analysis of linkage in black spruce. Genetics 149: 1089–1098.

Plomion C, O'Malley DM (1996) Recombination rate differences for pollen parents and seed parents in *Pinus pinaster*. Heredity 77: 341–350.

Plomion C, Bahrman N, Durel C-E, O'Malley DM (1995a) Genomic mapping in *Pinus pinaster* (maritime pine) using RAPD and protein markers. Heredity 74: 661–668.

Plomion C, O'Malley DM, Durel CE (1995b) Genomic analysis in maritime pine (*Pinus pinaster*). Comparison of two RAPD maps using selfed seeds and open-pollinated of the same individual. Theor Appl Genet 90: 1028–1034.

Plomion C, Durel C-E, OMalley DM (1996a) Genetic dissection of height in maritime pine seedlings raised under accelerated growth conditions. Theor Appl Genet 93: 849–858.

Plomion C, Yani A, Marpeau A (1996b) Genetic determinism of delta 3-carene in maritime pine using RAPD markers. Genome 39: 1123–1127.

Pot D, Rodrigues J-C, Rozenberg P, Chantre G, Tibbits J, Cahalan C, Pichavant F, Plomion C (2006) QTLs and candidate genes for wood properties in maritime pine (*Pinus pinaster* Ait.). Tree Genet Genomes 2: 10–24.

Ralph SG, Chun HJE, Kolosova N, Cooper D, Oddy C, Ritland CE, Kirkpatrick R, Moore R, Barber S, Holt RA, Jones SJM, Marra MA, Douglas CJ, Ritland K, Bohlmann J (2008) A conifer genomics resource of 200,000 spruce (*Picea* spp.) ESTs and 6,464 high-quality, sequence-finished full-length cDNAs for Sitka spruce (*Picea sitchensis*). BMC Genom 9: 484.

Ran J-H, Wei X-X, Wang X-Q (2006) Molecular phylogeny and biogeography of *Picea* (Pinaceae): Implications for phylogeographical studies using cytoplasmic haplotypes. Mol Phylogenet Evol 41: 405–419.

Remington DL, Purugganan MD (2003) Candidate genes, quantitative trait loci, and functional trait evolution in plants. Int J Plant Sci 164: S7–S20.

Remington DL, Whetten RW, Liu B-H, O'Malley DM (1999) Construction of an AFLP genetic map with nearly complete genome coverage in *Pinus taeda*. Theor Appl Genet 98: 1279–1292.

Remm M, Storm CEV, Sonnhammer ELL (2001) Automatic clustering of orthologs and in-paralogs from pairwise species comparisons. J Mol Biol 314: 1041–1052.

Ritland C, Ritland K (2000) DNA-fragment markers in plants. In: AJ Baker (ed) Molecular Methods in Ecology. Blackwell Publ, Oxford, UK, pp 208–234.

Ritter E, Aragonés A, Markussen T, Acheré V, Espinel S, Fladung M, Wrobel S, Faivre-Rampant P, Favre J-M (2002) Towards construction of an ultra high density linkage map for *Pinus pinaster*. Ann For Sci 596: 37–643.

Rockman MV (2008) Reverse engineering the genotype-phenotype map with natural genetic variation. Nature 456: 738–744.

Rose AM, Baillie DL (1979) The effect of temperature and parental age on recombination and nondisjunction in *Caenorhabditis elegans*. Genetics 92: 409–418.

Rungis D, Bérubé Y, Zhang J, Ralph S, Ritland CE, Ellis BE, Douglas C, Bohlmann J, Ritland K (2004) Robust simple sequence repeat markers for spruce (*Picea* spp.) from expressed sequence tags. Theor Appl Genet 109: 1283–1294.

Savard L, Li P, Strauss SH, Chase MW, Michaud M, Bousquet J (1994) Chloroplast and nuclear gene sequences indicate Late Pennsylvanian time for the last common ancestor of extant seed plants. Proc Nat Acad Sci USA 91: 5163–5167.

Sax K (1923) The association of size differences with seed-coat pattern and pigmentation in *Phaseolus vulgaris*. Genetics 8: 552–560.

Scotti I, Burelli A, Cattonaro F, Chagné D, Fuller J, Hedley PE, Jansson G, Lalanne C, Madur D, Neale D, Plomion C, Powell W, Troggio M, Morgante M (2005) Analysis of the distribution of marker classes in a genetic linkage map: a case study in Norway spruce (*Picea abies* karst). Tree Genet Genomes 1: 93–102.

Seaton G, Haley CS, Knott SA, Kearsey M, Visscher PM (2002) QTL Express: mapping quantitative trait loci in simple and complex pedigrees. Bioinformatics 18: 339–340.
Sewell MM, Sherman BK, Neale DB (1999) A consensus map for loblolly pine (*Pinus taeda* L.) I. Construction and integration of individual linkage maps from two outbred three-generation pedigrees. Genetics 151: 321–330.
Sewell MM, Bassoni DL, Megraw RA, Wheeler NC, Neale DB (2000) Identification of QTLs influencing wood property traits in loblolly pine (*Pinus taeda* L.). I. Physical wood properties. Theor Appl Genet 101: 1273–1281.
Sewell MM, Davis MF, Tuskan GA, Wheeler NC, Elam CC, Bassoni DL, Neale DB (2002) Identification of QTLs influencing wood property traits in loblolly pine (*Pinus taeda* L.) II. Chemical wood properties. Theor Appl Genet 104: 214–222.
Shepherd M, Cross M, Dieters MJ, Henry R (2003a) Genetic maps for *Pinus elliottii* var. *elliottii* and *P. caribaea* var. *hondurensis* using AFLP and microsatellite markers. Theor Appl Genet 106: 1409–1419.
Shepherd M, Cross M, Dieters MJ, Harding K, Kain D, Henry R (2003b) Genetics of physical wood properties and early growth in a tropical pine hybrid. Can J For Res 33: 1923–1932.
Shepherd M, Huang S, Eggler P, Cross M, Dale G, Dieters M, Henry R (2006) Congruence in QTL for adventitious rooting in *Pinus elliottii X Pinus caribaea* hybrids resolves between and within-species effects. Mol Breed 18: 11–28.
Skov E, Wellendorf H (1998) A partial linkage map of *Picea abies* clone V6470 based on recombination of RAPD-markers in haploid megagametophytes. Silvae Genet 47: 273–282.
Sobel E, Papp JC, Lange K (2002) Detection and integration of genotyping errors in statistical genetics. Am J Hum Genet 70: 496–508.
Soderlund C, Humphray S, Dunham A, French L (2000) Contigs built with fingerprints, markers, and FPC V4.7. Genome Res 10: 1772–1787.
Stam P (1993) Construction of integrated genetic linkage maps by means of a new computer package: JoinMap. Plant J 3: 739–744.
Strasburg JL, Scotti-Saintagne C, Scotti I, Lai Z, Rieseberg LH (2009) Genomic patterns of adaptive divergence between chromosomally differentiated sunflower species. Mol Biol Evol 26: 1341–1355.
Taira H, Teranishi H, Kenda S (1993) A case study of male sterility in sugi (*Cryptomeria japonica*) (in Japanese with English summary). J Jap For Soc 75: 377–379.
Tani N, Takahashi T, Iwata H, Mukai Y, Ujino-Ihara T, Matsumoto A, Yoshimura K, Yoshimaru H, Murai M, Nagasaka K, Tsumura Y (2003) A consensus linkage map for sugi (*Cryptomeria japonica*) from two pedigrees, based on microsatellites and expressed sequence tags. Genetics 165: 1551–1568.
Tani N, Takahashi T, Ujino-Ihara T, Iwata H, Yoshimura K, Tsumura Y (2004) Development and characteristics of microsatellite markers for sugi (*Cryptomeria japonica* D. Don) derived from microsatellite-enriched libraries. Ann For Sci 61: 569–575.
Theissen G (2005) Birth, life and death of developmental control genes: New challenges for the homology concept. Theory Biosci 124: 199–212.
Tong C, Shi J (2004) Constructing genetic linkage maps in Chinese-fir using F1 progeny. Acta Genet Sin 31: 1149–1156.
Travis SE, Ritland K, Whitham TG, Keim P (1998) A genetic linkage map of pinyon pine (*Pinus edulis*) based on amplified fragment length polymorphisms. Theor Appl Genet 97: 871–880.
Tsumura Y, Ogihara Y, Sasakuma T, Ohba K (1993) Physical map of chloroplast DNA in sugi, *Cryptomeria japonica*. Theor Appl Genet 86: 166–172.
Tsumura Y, Suyama Y, Yoshimura K, Shirato N, Mukai Y (1997) Sequence-tagged-sites (STSs) of cDNA clones in *Cryptomeria japonica* and their evaluation as molecular markers in conifers. Theor Appl Genet 94: 764–772.

Tulsieram LK, Glaubitz JC, Kiss G, Carlson JE (1992) Single tree genetic linkage mapping in conifers using haploid DNA from megagametophytes. Bio/Technology 10: 686–690.

Ujino-Ihara T, Yoshimura K, Ugawa Y, Yoshimaru H, Nagasaka K, Tsumura Y (2000) Expression analysis of ESTs derived from the inner bark of *Cryptomeria japonica*. Plant Mol Biol 43: 451–457.

Ujino-Ihara T, Kanamori H, Yamane H, Taguchi Y, Namiki N, Mukai Y, Yoshimura K, Tsumura Y (2005) Comparative analysis of expressed sequence tags of conifers and angiosperms reveals sequences specifically conserved in conifers. Plant Mol Biol 59: 895–907.

Ukrainetz NK, Ritland K, Mansfield SD (2008a) An AFLP linkage map for Douglas-fir based upon multiple full-sib families. Tree Genet Genomes 4: 181–191.

Ukrainetz NK, Ritland K, Mansfield SD (2008b) Identification of quantitative trait loci for wood quality and growth across eight full-sib coastal Douglas-fir families. Tree Genet Genomes 4: 159–170.

Verhoeven KJF, Jannink J-L, McIntyre LM (2005) Using mating designs to uncover QTL and the genetic architecture of complex traits. Heredity 96: 139–149.

Vision TJ, Brown DG, Shmoys DB, Durrett RT, Tanksley SD (2000) Selective mapping: a strategy for optimizing the construction of high-density linkage maps. Genetics 155: 407–420.

Vos P, Hogers R, Bleeker M, Reijans M, van de Lee T, Hornes M, Frijters A, Pot J, Peleman J, Kuiper M, Zabeau M (1995) AFLP: a new technique for DNA fingerprinting. Nucl Acids Res 23: 4407–4414.

Weising K, Nybom H, Wolff K, Kahl G (2005) DNA Fingerprinting in Plants: Principles, Methods, and Applications, 2nd edn. CRC Press, Boca Raton, Florida, USA.

Wheeler NC, Jermstad KD, Krutovsky K, Aitken SN, Howe GT, Krakowski J, Neale DB (2005) Mapping of quantitative trait loci controlling adaptive traits in coastal Douglas-fir. IV. Cold-hardiness QTL verification and candidate gene mapping. Mol Breed 15: 145–156.

White TL, Adams WT, Neale DB (2007) Forest Genetics. CABI Pub, Wallingford, Oxfordshire, UK; Cambridge, Massachusetts, USA.

Wilcox PL, Richardson TE, Corbett GE, Ball RD, Lee JR, Djorovic A, Carson SD (2001) Framework linkage maps of *Pinus radiata* D. Don based on pseudotestcross markers. For Genet 8: 109–117.

Williams CG (1998) QTL Mapping in outbred pedigrees. In: AH Paterson (ed) Molecular Dissection of Complex Traits. CRC Press, Boca Raton, Florida, USA, pp 81–94.

Williams CG, Reyes-Valdes MH, Huber DA (2007) Validating a QTL region characterized by multiple haplotypes. Theor Appl Genet 116: 87–94.

Williams JGK, Kubelik AR, Livak KJ, Rafalski JA, Tingey SV (1990) DNA polymorphisms amplified by arbitrary primers are useful as genetic markers. Nucl Acids Res 18: 6531–6535.

Wu R, Ma C-X, Painter I, Zeng Z-B (2002) Simultaneous maximum likelihood estimation of linkage and linkage phases in outcrossing species. Theor Pop Biol 61: 349–363.

Yazdani R, Yeh FC, Rimsha J (1995) Genomic mapping of *Pinus sylvestris* (L.) using random amplified polymorphic DNA markers. For Genet 2: 109–116.

Yazdani R, Nilsson JE, Plomion C, Mathur G (2003) Marker trait association for autumn cold acclimation and growth rhythm in *Pinus sylvestris*. Scand J For Res 18: 29–38.

Yin T-M, Wang X-R, Andersson B, Lerceteau-Köhler E (2003) Nearly complete genetic maps of *Pinus sylvestris* L. (Scots pine) constructed by AFLP marker analysis in a full-sib family. Theor Appl Genet 106: 1075–1083 .

Yoshimaru H, Ohba K, Tsurumi K, Tomaru N, Murai M, Mukai Y, Suyama Y, Tsumura Y, Kawahara T, Sakamaki Y (1998) Detection of quantitative trait loci for juvenile growth, flower bearing and rooting ability based on a linkage map in sugi (*Cryptomeria japonica* D. Don). Theor Appl Genet 97: 45–50.

Zhang P, Li W, Fellers J, Friebe B, Gill BS (2004) BAC-FISH in wheat identifies chromosome landmarks consisting of different types of transposable elements. Chromosoma 112: 288–299.

6

Patterns of Nucleotide Diversity and Association Mapping

González-Martínez S.C.,[1,a,*,†] Dillon S.,[2,†] Garnier-Géré P.H.,[3,†] Krutovsky K.V.,[4,†] Alía R.,[1,b] Burgarella C.,[1,c] Eckert A.J.,[5] García-Gil M.R.,[6] Grivet D.,[1,d] Heuertz M.,[1,7] Jaramillo-Correa J.P.,[1,8] Lascoux M.,[9] Neale D.B.,[10,11] Savolainen O.,[12] Tsumura Y.[13] and Vendramin G.G.[14]

ABSTRACT

Understanding the molecular basis of adaptive traits is a major interest in conservation and population genetics. In commercial species, such as several conifers, it is also interesting for operational breeding. In this chapter, we provide a state-of-the-art view on candidate gene research, from general estimates of nucleotide diversity and recombination to new-generation neutrality tests and association genetics methodologies. Levels of nucleotide diversity in conifers are substantial, although lower than expected given their life-history traits. In addition, linkage disequilibrium seems to decay rapidly in this group of species, at least within genes that are not submitted to natural selection. These two facts makes genetic association studies appealing in conifers, as significant associations may correspond to the actual causal polymorphisms. Population genomic methods also seem appropriate in conifers, in particular for those species with accused population genetic structure and strong response to environmental gradients. New-generation neutrality tests, outlier loci detection methods and genotype/phenotype association studies have revealed various candidate genes and single nucleotide polymorphisms underlying different adaptive phenotypes, despite potential confounding effects of demographical and historical processes. Finally, perspectives about future genomic research in conifers are provided, including its application for conservation and breeding.

For affiliations see at the end of this chapter on page 275.

Keywords: candidate genes; nucleotide diversity; recombination; natural selection; neutrality tests; single nucleotide polymorphism; genetic association

6.1 Introduction

Conifers are long-lived, sessile organisms that occupy extensive landscapes. They are important forest components in many areas of the world, and members of the pine family are especially abundant in cool to cold-temperate and mountainous areas of the Northern Hemisphere.

Conifer forests are the key in terrestrial ecosystems and a major source of biodiversity. They are also economically important, as they provide a full suite of ecosystem services and resources for human use. Conifers are important as a source of timber and other wood products, and are also widely planted as ornamental trees and shrubs. The most important source of softwood timber in the world is trees in the Pinaceae (pine family), which are widely used for building and boat construction. Several species of conifers are tapped or cut and steam-distilled for stem resins, which are used as commercial sources of turpentine, tar oils, rosin, and pitch. Many species of conifers are grown as ornamentals and a wide variety of cultivated shrub forms have been selected for garden use. Recently, bark and leaves of several species of yews have become important as the source of taxol and related alkaloids, which disrupt the process of cell division and are used in cancer therapy.

It should be stressed that despite the ancient use of forests by humans, there is still abundant genetic variation present in natural populations of conifers. Based on molecular marker studies, conifers display higher genetic variation than other plant species (Hamrick and Godt 1996; Nybom and Bartish 2000). Recent studies based on DNA-sequence data for several loci (see below) also showed a considerable amount of genetic variation still present in conifers, even in intensively-managed commercial species.

Conifers, however, differ in their life history traits from most other species where extensive candidate gene studies are available. Most importantly, they are long-lived and in many cases have large effective population sizes, with highly efficient pollen flow between populations (see Savolainen and Pyhäjärvi 2007). Despite the lack of self-incompatibility system, most species produce predominantly outcrossed seed, as most selfed embryos are eliminated by a system of embryonic lethals (Koelewijn et al. 1999). Extensive pollen flow also homogenizes allelic frequencies, such that, for example, Swedish and Chinese populations of Scots pine (*Pinus sylvestris* L.) are only little differentiated at the isozyme level (Wang et al. 1991). In strong contrast, many species show steep clinal variation in adaptive traits over environmental gradients, obviously maintained by natural selection (Morgenstern 1996; Howe et al. 2003; Savolainen et al. 2007). Nucleotide diversity of conifers is apparently affected by their longevity. Conifers have relatively slow nucleotide substitution rate per year compared to other species. Even those conifer species that separated more than 100 Mya have only moderate divergence at synonymous sites and low divergence at non-synonymous sites. Some recently diverged species show a surprisingly high level of shared polymorphism (Bouillé and Bousquet 2005; Chen et al. 2010). Thus, species such as the Mediterranean *P. pinaster* and the boreal *P. sylvestris* have adapted to rather different environments using very similar genomic resources. This provides many interesting possibilities for conifer comparative studies.

In the past few years there has been a tremendous progress in studying patterns of polymorphism at genes within plant genomes, including conifers. However, the large genome size (e.g., about 25.5 pg/C for *P. pinaster*, Chagné et al. 2002, i.e., about 170 times the size of the *Arabidopsis* genome) and high frequency of repetitive sequences in conifers (up to 75% in *Picea abies*, de-Paoli 2005), together with insufficient genomics resources in non-commercial species, still hold back research progress. In this chapter, we describe some of the processes that are shaping nucleotide diversity within conifer species and across their genomes, and how some of this nucleotide variation can be related to phenotypic variation.

6.2 Nucleotide Diversity

Levels of nucleotide diversity are influenced by many factors at individual gene and whole genome levels (see Table 1 in Buckler and Thornsberry 2002) and reflect a history of population size changes, mutation, selection, migration and recombination. Diversity at the gene level is the source of most of the adaptive phenotypic variation. Distribution and levels of nucleotide diversity provide direct genetic data to study molecular evolution and infer selection and demography.

Nucleotide diversity has been estimated in many conifers, but current data are based mostly, if not exclusively, on the functional gene space, i.e., on sequences representing exons, introns and short untranscribed regions (UTRs). These data have been obtained from resequencing studies using relatively small but mostly wide-range population samples. Whole genome data is currently not available in conifers, thus composition and structure of most of the conifer genome still remains unravelled. With this lack of information, it is quite likely that gene-based estimates of nucleotide diversity do not reflect the whole genome estimates. However, currently available information allows us to make comparisons across different species (although it focuses mainly on commercial tree species), at least for nucleotide variation in the gene space.

The average within-species total nucleotide diversity, π, is 0.0043 ± 0.0022 for all conifers, but varies from 0.0016 in *Cryptomeria japonica* to 0.0086 in *Pinus densata* (Table 6-1).

For those studies with 10 and more genes, the mean silent variation (0.0075 ± 0.0027) is two times higher than for all regions and five times higher than non-synonymous variation (0.0016 ± 0.0011). The silent variation is lowest in *Cryptomeria japonica* and *Picea abies* (0.0038 and 0.0040, respectively) and highest in *Pinus pinaster* and *Pinus taeda* (0.0085). One single nucleotide polymorphism (SNP) occurs approximately every 91 bp on average for all conifers. However, nucleotide diversity estimates are highly heterogeneous both across genes within species and between species. For instance, in a study based on a large number of genes, a likelihood ratio test indicated that nucleotide diversity significantly varied across 115 polymorphic genes (Eckert et al. 2009b) in *Pseudotsuga menziesii* when considering all sites, as well as for different categories of sites, such as silent and non-synonymous. Nucleotide diversity can also vary in the different parts of the range of a species. For instance, Grivet et al. (2009) found that Greek populations of Aleppo pine (*Pinus halepensis* Mill.) had about three times more nucleotide diversity than their western counterparts, a likely result of past long-range colonization of its western Mediterranean current distribution.

Nevertheless, nucleotide diversity at silent sites was about three to five times higher than at non-synonymous sites in all species, likely due to purifying selection. The average pairwise divergence (D_{xy}, Nei 1987) at synonymous (K_s) and silent (K_{sil}) sites was greater than at non-synonymous (K_a) sites in several studies, with most genes exhibiting a K_a/K_s ratios of less than one, which is inconsistent with neutral expectations. In general, θ_w was larger (although only slightly) than π in most genes, illustrating an excess of rare variants relative to expectations under neutrality. Both π and θ_w estimate nucleotide diversity, which is expected to equal $4N_e\mu$ for autosomal loci in populations at neutral equilibrium according to the neutral theory of molecular evolution, where N_e is the effective population size and μ is the neutral mutation rate per locus (normally divided by locus length to obtain an estimate per nucleotide site) per generation (Kimura 1983). This parameter summarizes the equilibrium between processes of mutation that generate variation, and random genetic drift, which is assumed to play a more important role for observed genetic variation than natural selection. Violations of the assumptions of the standard neutral model will cause discrepancy between π and θ_w estimates.

Average nucleotide diversity in conifers is higher than in humans and in some cultivated crops, such as *Glycine max* (soybean), but lower than in *Zea mays* ssp. *mays* (maize), and similar to those in *Drosophila* (Table 6-1).

Table 6-1 Nucleotide diversity π (Nei 1987) per site[a] across different gene regions and species. π range in parenthesis.

Species	Loci	All	Coding	Non-coding[b]	Silent[c]	Synonymous	Non-synonymous	bp per SNP	Reference
Cryptomeria japonica (sugi)	7	25 (0.4–52)			38 (2–81)	42 (0–86)	7 (0–20)	118	Kado et al. 2003[d]
	10	16 (2–55)				38 (0–100)	5 (0–25)	188	Kado et al. 2008
	5	36 (10–76)			46 (17–95)	44 (10–85)	12 (0–35)	50	Fujimoto et al. 2008[e]
Chamaecyparis obtusa (hinoki)	10	24 (0–83)				69 (0–234)	11 (0–36)	119	Kado et al. 2008
Chamaecyparis pisifera (sawara)	8	29 (7–89)				68 (0–148)	12 (0–38)	194	Kado et al. 2008
Taxodium distichum (bald cypress)	4	26 (17–36)			58 (46–73)			57	Kado et al. 2006[e]
Pinus pinaster (maritime pine)	8	24 (2–70)	4 (2–12)	24 (8–66)		3 (2–8)	2 (0–10)	164	Pot et al. 2005
	11	55[f] (15–95)			85 (13–142)		28 (11–58)	28	Eveno et al. 2008
Pinus radiata (Monterey pine)	8	19 (5–20)	8 (0–38)	7 (0–14)		8 (0–38)	1 (0–2)	365	Pot et al. 2005
Pinus taeda (loblolly pine)	19	40 (3–173)			64 (4–198)		11 (0–246)	63	Brown et al. 2004
	18	51 (1–118)		63 (0–174)	85 (2–205)	91 (0–296)	17 (0–52)	50	González-Martínez et al. 2006a
Pinus tabuliformis (Chinese pine)	7	85 (39–124)			119 (52–352)		30 (0–54)	22	Ma et al. 2006
Pinus yunnanensis (Yunnan pine)	7	67 (21–132)			95 (15–251)		23 (0–44)	48	Ma et al. 2006
Pinus densata (Sikang pine)	7	86 (19–112)			122 (37–281)		28 (0–59)	22	Ma et al. 2006

Table 6-1 contd....

Table 6-1 contd....

Species	Loci	All	Coding	Non-coding[b]	Silent[c]	Synonymous	Non-synonymous	bp per SNP	Reference
Pinus sylvestris (Scots pine)	16				53 (0–265)			87	Pyhäjärvi et al. 2007[e]
	10	47 (2–253)			61 (0–315)	43 (0–213)	4 (0–17)	34	Palmé et al. 2008
	14	62 (6–209)			77 (8–291)		39 (0–155)	34	Wachowiak et al. 2009
Pseudotsuga menziesii (Douglas–fir)	18	66 (24–138)	46 (16–74)	100 (48–181)	106 (26–184)	128 (4–292)	21 (0–60)	46	Krutovsky and Neale 2005
	121 [g]	44 (0–170)			76 (0–551)	76 (0–551)	20 (0–130)	63	Eckert et al. 2009b
Picea abies (Norway spruce)	22	21 (2–48)			40 (0–156)		9 (0–29)	69	Heuertz et al. 2006
Mean ± SD	17 ± 25	43 ± 22	19 ± 23	49 ± 42	75 ± 27	55 ± 36	16 ± 11	91 ± 83	
Populus tremula (European aspen)	5	111 (59–147)			160 (94–229)	220 (130–303)	59 (28–112)	60	Ingvarsson 2005
	77	42 (3–351)		48[h]		120	17	53	Ingvarsson 2008
Quercus crispula (Japanese oak)	3	69 (67–72)			78 (71–78)			25	Quang et al. 2008
Persea americana (wild avocado)	4	66 (35–123)			102 (50–179)		20 (7–50)	34	Chen et al. 2008
Arabidopsis lyrata ssp. *petraea*	6				240 (13–625)			14 [i]	Wright et al. 2003
	8	116 (35–324)			230 (61–633)				Ramos–Onsins et al. 2004
	77				108[j]				Ross-Ibarra et al. 2008

Arabidopsis lyrata ssp. *lyrata*	6				37 (19–87)			73 [i]	Wright et al. 2003
	3	17 (6–24)			31 (14–64)				Ramos-Onsins et al. 2004
	26	176 (0–447)						85	Wright et al. 2006 [i]
	77				58[k]				Ross-Ibarra et al. 2008
Arabidopsis halleri	8	81 (24–327)			150 (49–569)				Ramos-Onsins et al. 2004
Arabidopsis thaliana (self-pollinated)	6				50 (12–302)			54 [l]	Wright et al. 2003
	3	39 (19–55)			64 (45–84)	79 (91–122)	17 (4–26)		Ramos-Onsins et al. 2004
	357[m]	71		80	90	100	10		Schmid et al. 2005
Boechera stricta (self-pollinated)	86	30 (1–280)			35	41	17	77	Song et al. 2009
Helianthus annuus (wild sunflower)	9	128 (23–356)			234 (44–585)	315 (33–898)	34 (8–116)	19	Liu and Burke 2006
Solanum peruvianum (wild tomato)	8	135 (56–270)			250			14	Arunyawat et al. 2007
Solanum chilense (wild tomato)	8	116 (55–166)			212			18	Arunyawat et al. 2007
Hordeum vulgare ssp. *spontaneum* (wild barley)	9	68 (11–219)	47 (9–179)	103 (4–257) [h]		128 (36–488)	23 (0–88)		Morrell et al. 2003
	18	75 (2–224)						36	Morrell et al. 2005
Glycine max (soybean)	143	13	5	15		8	4	273	Zhu et al. 2003

Table 6-1 contd....

Table 6-1 contd....

Species	Loci	All	Coding	Non-coding[b]	Silent[c]	Synonymous	Non-synonymous	bp per SNP	Reference
Glycine soja (wild soybean)	102	22	11	28		47	10	121	Hyten et al. 2006
Zea mays ssp. *mays* L. (maize)	21 [m]	96	72		111	173	39	28	Tenaillon et al. 2001
	774	64 (0–370)						54	Wright et al. 2005
Zea mays ssp. *parviglumis* (teosinte)	774	95 (0–501)						30	Wright et al. 2005
	5	134 (91–214)							Moeller et al. 2007
Triticum monococcum ssp. *monococcum* (domesticated einkorn)	18	30			54				Kilian et al. 2007
Triticum monococcum ssp. *boeoticum* (wild einkorn)	18	47			85				Kilian et al. 2007
Oryza rufipogon (wild rice)	10	64 (22–183)			72 (13–203)			31	Zhu et al. 2007
Oryza nivara (wild rice)	10	62 (23–164)			63 (2–170)			41	Zhu et al. 2007
Secale cereale (rye)	14	203 (59–530)						58	Varshney et al. 2007
Caenorhabditis elegans (nematode)	6				22 (1–64)	63	0	141	Cutter 2006
Drosophila melanogaster	24	40 (0–[illegible]8)		108 (0–265)	135 (0–335)			90	Moriyama and Powell 1996

Human	75 [m, n]	8.3	8.0	8.5		15.1	5.7	217	Halushka et al. 1999
	106[o]	5.1	5.0	5.2		3.1	2.0	350	Cargill et al. 1999

[a]all π values are multiplied by 10^4.
[b]Introns and untranscribed regions.
[c]Non-coding and synonymous sites.
[d]Unweighed average π calculated from Table 3.
[e]Unweighed average π calculated from Table 2.
[f]0.0046 when two monomorphic loci were included.
[g]Including also 18 genes from Krutovsky and Neale 2005.
[h]Introns only.
[i]Calculated from supplemental Table S2.
[j]Average for four populations, German, Icelandic, Swedish and Russian.
[k]Average for two populations, US and Canadian.
[l]Silent sites.
[m]Calculated as θ.
[n]Europeans and Africans (for Europeans $\pi = 0.0005$, the same as in Cargill et al. 1999).
[o]Europeans only.

Such similarity to *Drosophila* is not completely surprising given that both conifers and *Drosophila* have large population sizes and high outcrossing rates (see also section on LD and recombination rates). In general, nucleotide polymorphism in conifers is low when compared to wild angiosperms (see Table 6-1 and Fig. 6-1) , but heterogeneity between different species is almost as high as heterogeneity between different genes within species. This fact highlights the problems of comparing variation among species based on a few loci. Indeed, comparisons among species should be based on many loci and, ideally, be restricted to orthologous loci. For example, an early study in *Populus tremula* (European aspen) based on only five loci suggested that aspen had higher nucleotide variation than any conifer (Ingvarsson 2005). However, when a larger number of genes were sequenced, the nucleotide variation found was of the same order of magnitude as that of conifers (Ingvarsson 2008).

The currently available estimates of nucleotide diversity in conifers are lower than expected considering their life-history traits and the high heterozygosity levels observed at allozyme and microsatellite loci for these species (Hamrick and Godt 1996; Nybom and Bartish 2000). The level of

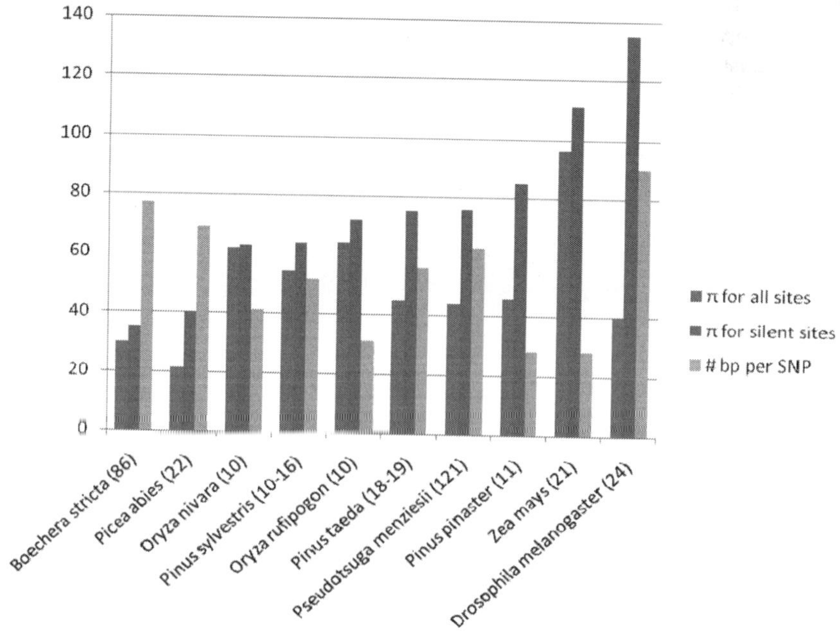

Figure 6-1 Nucleotide diversity estimates for all and silent (synonymous and non-coding) sites, and number of base pairs per SNP for studies where 10 or more loci were studied. The number of loci is in parentheses. Estimates for *Pinus taeda* and *Pinus sylvestris* were averaged for two or more studies. See Table 6.1 for references.

Color image of this figure appears in the color plate section at the end of the book.

genetic polymorphism is determined by several factors, such as effective population size, selection, mutation rate, and demography. For instance, Brown et al. (2004) suggested that the low nucleotide diversity in *Pinus taeda* could be the result of a low mutation rate, estimated at $\mu \approx 0.17 \times 10^{-9}$/bp/year, which is more than one order of magnitude lower than in angiosperms (e.g., 1.5×10^{-8}/bp/year in *Arabidopsis*; Koch et al. 2000). However, later research has shown that these estimates are probably biased downwards due to problems with time calibration. More reliable mutation rates were computed using 11 nuclear genes in four pine species representing pine major lineages ($0.70–1.31 \times 10^{-9}$/bp/year; Willyard et al. 2007). Estimates of molecular divergence between *Pinus* and *Picea* based on expressed sequence tags (ESTs) also gave a higher mutation rate estimate ($\sim 1 \times 10^{-9}$/bp/year; Savolainen and Wright 2004) than suggested by Brown et al. (2004). Although these divergence rates are still approximately 4- to 20-fold slower than in angiosperms, they are more consistent with the high per-generation deleterious mutation rates observed in pines. Moreover, if we express these substitution or mutation rates per generation rather than per year (i.e. by multiplying them by a generation-time of about 15–20 years), they become more similar to the rates estimated for annual plant species.

The effective population size (N_e) of loblolly pine calculated from the formula of neutral molecular evolution $\theta = 4N_e\mu$ turned out to be also relatively low (from 0.9×10^5 to 5.6×10^5 depending on mutation rate estimates; Brown et al. 2004; Willyard et al. 2007) with respect to current census sizes. One possible explanation for this discrepancy could be significant population fluctuations and effective size reductions during the late Pleistocene and the Holocene in this species (Brown et al. 2004 and references therein). An alternative explanation is related to the presence of repeated selective sweeps or the effect of background selection operating at linked genomic areas (which would reduce nucleotide diversity). However, considering that linkage disequilibrium (LD) in conifers decays very rapidly and often does not extend beyond a few hundred or thousand base pairs (see next section), the potential for an overall effect of selective sweeps in conifers may be low (Savolainen and Pyhäjärvi 2007). However, it must be noted that the current estimates of LD in conifers are based only on a relatively small number of loci and in a few species, and LD can vary extensively across an entire genome. Indeed, Fujimoto et al. (2008) extended the estimated value of the recombination rate they had obtained from a sample of five long loci in *C. japonica* to the total genome and obtained a total map distance of 8,500–17,000 cM, a value much higher than the 2,000 cM obtained previously in a pseudo-test cross experiment (Nikaido et al. 2000). It therefore seems that the regions surveyed tend to have higher recombination rates than other parts of the genome. Preliminary data suggest that the same might be also true in Norway spruce (N Gyllenstrand unpubl. data). In addition, recent work has suggested that there is evidence for continuously ongoing sweeps at some loci, which would have required substantial LD. A thorough examination of the potential for background selection and selective sweeps remains to be done in conifers.

Direct estimates of mutation rate at the nucleotide level in conifers are not available yet. Moreover, indirect estimates inferred from observed nucleotide differentiation between closely related pine species are often based on poorly characterized divergence times and on the assumption that silent sites are neutral. Paleobotanical data are also incomplete and inconclusive, although recent work suggests large effective population sizes of conifers during the Holocene or the Pleistocene (Birks and Willis 2008).

6.3 Recombination and Extent of Linkage Disequilibrium

Linkage disequilibrium (LD) is defined as the non-random association of alleles at different loci. Several LD statistics are available in the literature. The square of the correlation coefficient between two loci (r^2) and D' are the most commonly used ones (see, for instance, Flint-García et al. 2003 for a full description of the statistics). A major difference between r^2 and D' is that r^2 considers both recombinational and mutational history while D' is not affected by mutation. However, D' is highly affected by sample size and its use is not recommended when sample size is low. The significance of LD is normally assessed using Fisher exact tests and Bonferroni corrections for multiple testing. Besides recombination and mutation, several factors may affect LD, including species and population attributes such as the mating system, admixture level, population subdivision and population size (see, for instance, Table 1 in Rafalski and Morgante 2004). Of particular interest are the effects of natural and artificial selection on LD. A selective sweep is expected to increase LD locally, in the genome regions surrounding the selected polymorphism (Sabeti 2002; Kim and Nielsen 2004; Voight et al. 2006). This property has been useful to detect selection acting on candidate genes in different plant species (e.g., in *Arabidopsis*, Olsen et al. 2002; but see Nordborg and Tavaré 2002 for caveats in the use of LD to identify positive selection).

Recent reports in conifers have shown low levels of LD and a rapid decay with physical distance. For example, LD decayed about 50% in *P. taeda* (from ~0.5 to ~0.25 r^2; Brown et al. 2004) and *Pseudotsuga menziesii* (from ~0.25 to ~0.10 r^2; Krutovsky and Neale 2005) in about 2,000 bp, and was lower than 0.2 after only ~100 bp in *Picea abies* (Heuertz et al. 2006). A similar fast decay of LD with physical distance has been observed in other outcrossing species with large population size, such as *Drosophila* (Long et al. 1998). These results contrast with those obtained from *Arabidopsis* (in particular at the regional or population levels) or humans, where large haploblocks (> 50 kb) can be found (Rafalski and Morgante 2004). Nevertheless, patterns of LD can be very variable depending on the gene and population assessed (Fig. 6-2). For example, in *Pinus taeda* and *P. pinaster*, some genes, such as *ccoaomt-1* (González-Martínez et al. 2006a; Eveno et al. 2008), have high levels of LD, being arranged in only a few haplotypic lineages, whereas others show little evidence of LD (e.g., *aqua-MIP* in *P. taeda*, *GRP3* in *Pinus pinaster*). In addition, natural selection can substantially increase local LD, as it has been recently shown for the *Y1* gene in maize. Similar to conifers, maize normally shows a rapid decay of LD (in approximately 200–1,500 bp); however, the selective sweep of the *Y1* gene resulted in LD extending over 800 kb (Palaisa et al. 2003). LD can also vary depending on the population or region of origin, as it has been shown in *Picea abies* (Heuertz et al. 2006) and *Pinus sylvestris* (Pyhäjärvi et al. 2007). Finally, it is important to note that most of the results on LD currently available in the literature deal with genic regions, which are generally considered recombinational hotspots, and a different pattern may arise when the intergenic and repetitive regions from the large conifer genomes are screened.

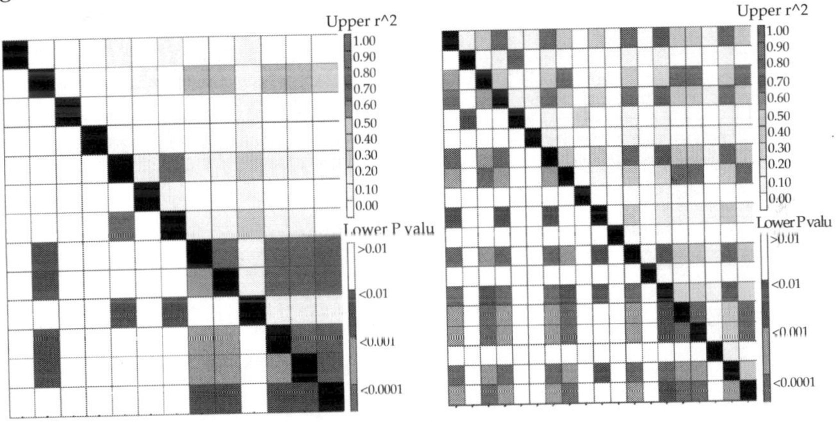

Figure 6-2 LD plots for *dhn1* (left) and *sod-chl* (right) candidate genes for drought in loblolly pine. A LD block is apparent in the lower right part of *dhn1* while LD is distributed more regularly in *sod-chl*.

Color image of this figure appears in the color plate section at the end of the book.

The extent of LD in a given species is important for the design of association studies (see below). In species with rapid decay of LD (as seems the case in conifers), an exhaustive sampling of the genome will be required to be able to identify polymorphisms associated with phenotypic variation. However, once a significant association is found, the functional causative mutation or mutations would likely be in the surrounding genomic region as linkage is expected to disappear rapidly with physical distance along the chromosome. Technical constraints of current SNP genotyping platforms (e.g., Illumina Golden Gate assay) may also affect the development of association studies in species with rapid decay of LD with physical distance, as often only one or two SNPs per amplicon are included in a genotyping assay. Such amount of SNPs would normally not be enough to tag all the relevant haplotypes of a given gene (but see Krutovsky and Neale 2005 in Douglas-fir where only 2–3 haploblocks per gene are suggested to contain most relevant variation).

LD is broken by recombination. Several methods have been developed to estimate recombination, the most popular being those based on composite likelihood (Hudson 2001; McVean et al. 2002) and summary statistics (Haddrill et al. 2005). Accurate recombination estimates are difficult to obtain because they require a high number of sampled chromosomes, long gene fragments, and high levels of polymorphism, as well as stationary populations and absence of population structure (or estimates have to be given by population reducing considerably sample size in most cases). Even rarer are estimates of gene conversion, a form of recombination that involves a nonreciprocal transfer of genetic information and that may be more frequent in wild plants than previously thought (see Morrell et al. 2006 for a comparative study and Brown et al. 2004 for an estimate in pine).

Given the reported low levels of LD within conifer genes and their large genomes (see Lynch 2006), it would be expected that conifers had higher levels of recombination than other organisms. High levels of recombination in conifers would also explain the striking simultaneous findings of relatively low nucleotide diversity (see Savolainen and Pyhäjärvi 2007) and high heterozygosity for other molecular markers assessed in this group of species (allozymes, microsatellites, etc.). However, this is not the case. In several conifers, recombination rate estimates are very low. For example, some Asian conifers, such as sugi (*Cryptomeria japonica*) and hinoki (*Chamaecyparis obtusa*) had low levels of recombination (Kado et al. 2008; Fujimoto et al. 2008). In *C. japonica*, the population recombination rate, $4Nr$, is one-sixth of that in maize and one-half of that in loblolly pine. Conifer genome sizes are around 20 billion bp, i.e., an order of magnitude larger than, for instance, the human genome (3 billion bp). Yet genetic maps are not much longer than those of species with smaller genomes. This therefore seems to imply either that some areas of the genome have very low recombination rates or a substantially low recombination rate overall. A recent meta-analysis including 81 plant taxa has shown that conifers indeed have lower recombination levels than other plants, including other forest trees such as poplar or oak (Jaramillo-Correa et al. 2010). As in the case of LD estimation, better insights on recombination levels at the whole genome scale will be available when intergenic and repetitive regions are sequenced in population samples. New initiatives on high-throughput sequencing of bacterial artificial chromosome (BAC) libraries in pines are, thus, very encouraging.

6.4 Disentangling Demographical and Selective Effects: Overview of Statistical Methods

In the past 20 years, the development of a large body of coalescent-based methods has allowed to analyze and compare nucleotide diversity patterns with the aim of better understanding the genetic basis of adaptation (Ford 2002; Luikart et al. 2003; Wright and Gaut 2005). The debate has progressively shifted from examining general patterns of neutral molecular evolution to a focus on making inferences and providing evidence about natural selection in natural populations (Nielsen 2001, 2005).

Positive selection on advantageous mutations alters nucleotide diversity patterns relative to neutral expectations, and can produce different types of signatures that include reduced levels of genetic variation, skews in the site frequency spectrum (SFS) (i.e., an excess of low- and/or high-frequency derived variants), and stronger haplotype structure due to increased levels of LD (Biswas and Akey 2006). The relative magnitude of these signals depends on many parameters such as the type of advantageous mutation (new or standing variation), the time since the advantageous mutation appeared (and whether it has reached fixation or not), the strength of selection, and local rates of mutation and recombination. Available statistical methods differ in their assumptions and the type of possible signals that they explore, so that they are more or less sensitive to the variation in the parameters mentioned above, with direct consequences on their power to detect some signals over others.

However, population demographic history can produce similar signatures on diversity patterns and cause deviations from neutral expectations. For example, increased population differentiation (or reduced nucleotide diversity) and an excess of rare mutations due to selection for local adaptation can be similar to effects of population isolation, bottlenecks or expansion (Simonsen et al. 1995; Ramos-Onsins and Rozas 2002; Ramirez-Soriano et al. 2008). For these reasons, tests based on the SFS within species have to be carefully examined in the context of different demographic scenarios, as they rely on strong assumptions regarding the demography of populations, while methods comparing data in different species are more robust to demography. In this context, genome-wide control plays an important role in discriminating selective factors from demographic events because unlike selection that usually affects particular genes, demographic events (such as, for instance, bottlenecks and fast expansion) affect all or most genes and have genome-wide effects.

Within-species methods include traditional Tajima's *D*, Fu and Li's *D* and *F*, Fu's *Fs*, and Fay and Wu *H* tests, which assess different properties of the SFS, often by comparing different estimates of θ (Tajima 1989; Fu and Li 1993; Fu 1997; Fay and Wu 2000; reviewed in Biswas and Akey 2006). Among traditional methods that use data from different species are the Hudson-Kreitman-Aguadé (HKA) and the McDonald Kreitman (MK) tests. The HKA compares levels of nucleotide polymorphisms within species and divergence between species across different loci, which should be positively correlated under neutral expectations (Hudson et al. 1987), and allows detecting loci with either increased or reduced levels of polymorphisms or divergence compared to others. A recent version of the test uses a maximum likelihood analysis of multilocus polymorphism and divergence (Wright and Charlesworth 2004). The MK test also uses divergence and polymorphism data. It compares the ratio of non-synonymous to synonymous mutations for sites that are polymorphic within species, and for sites that are fixed between species, the neutral prediction being that both ratios are the same. New tests that explore the joint distribution of different statistics, i.e., "compound" tests, have also been proposed (Zeng et al. 2007). The underlying rationale is that the combined test would altogether be more robust to demography because the distinct statistics differ in their sensitivity to particular demographic assumptions.

Other types of neutrality tests using multiple species are based on K_d/K_s ratios (non-synonymous/synonymous substitution rates ratios, as described above). Those genes that are affected by functional constraints are expected to have K_d/K_s ratios less than 1, unlike positive selection that is expected to lead to $K_d/K_s > 1$, being the neutral expectation around 1. The basic test has been extended to more complex methods accounting for variation in ratios along lineages and across genomic regions (Bielawski et al. 2000; Yang and Nielsen 2000). These methods are free of demographic assumptions and thus more powerful for detecting longer-time scale selection events, but they are a priori less powerful for detecting more recent selection events and their diverse signals (Nielsen 2005).

Population genetic signatures of selection can also be detected among populations as local adaptation can increase their degree of differentiation (Charlesworth et al. 1997; Slatkin and Wiehe 1998). Building from the original Lewontin and Krakauer (1973) test based on the genetic differentiation variance (F_{ST}) among loci, various coalescent-based approaches have been developed to detect loci showing outlier patterns (i.e., that deviate from the simulated neutral distribution) for diversity, differentiation or other summary statistics (e.g., Bowcock et al. 1991; Beaumont and Nichols 1996; Schlötterer 2003; Beaumont and Balding 2004; Nielsen 2005; reviewed in

Storz 2005). The more sophisticated method of Beaumont and Balding (2004), which uses a separation of time scales model for the structured coalescent (Wakeley 1999), allows to test for locus, population, and locus by population effects on the differentiation parameter in a Bayesian framework. This method thus integrates heterogeneity among population migration rates, and detects both outlier loci and outlier populations that might have been affected by particular demographic events. However, although robust to interlocus variation in mutation rates and supposedly able to integrate a large part of the species structure and demographic histories (Beaumont 2005), model-based outlier approaches are not likely to solve all problems related to complex demographic histories and their possible effects on diversity patterns (Nielsen 2005).

In theory, the availability of genome-wide data in one or different populations make empirical scans possible without assuming any (normally simplified or unrealistic) model, as locus-specific effects could be easily distinguished from genome-wide patterns using non-parametric analyses, if the number of loci assessed were large enough. However, empirical scans are not possible yet in conifers due to their large genome size and rapid LD decay (see above). Also, recent studies have shown the limits of genome scans for detecting particular types of selection (Teshima et al. 2006; Kelley et al. 2007). Since there is no explicit reference model for inference in empirical approaches, tail values in the distributions will always lead to the conclusion that some loci are under selection (Tenaillon and Tiffin 2008). This problem will be enhanced in the case of complex demographic history (such as recurrent bottlenecks) that can increase the variance of most statistics and thus the rate of false positives (Nielsen 2005).

In conclusion, as different methods do not assess the same time scale and patterns in the data and do not have the same power in detecting particular selection signatures, stronger evidence may come from the application of a variety of methods, depending on the data available and on method applicability.

6.5 Detection of Selection Signatures in Conifers

Tests for molecular signatures of selection have only been performed in conifers since less than a decade ago, and included only a limited number of candidate genes (e.g., Brown et al. 2004; Krutovsky and Neale 2005; Pot et al. 2005; González-Martínez et al. 2006a; Wachowiak et al. 2009). Candidate genes have generally been selected from expression or functional studies in model and target species, and are often involved in stress tolerance or wood formation. Altogether, only a few genes showing departures from the standard neutral model have been detected, normally using classical test statistics such as Tajima's *D* (Tajima 1989) or Fay and Wu's *H* (Fay and Wu

2000). More recent methods, for instance those based on the joint distribution of different test statistics (e.g., Zeng et al. 2007) or using Approximate Bayesian Computation (ABC) for adjusting the null neutral model, have been rarely applied in conifer species as yet. In *Picea abies*, however, a recent study used the joint distribution of Tajima's *D* and Fay and Wu's *H* after adjustment to demography through Approximate Bayesian Computation demonstrated that the circadian clock gene *PRR3* departs significantly from neutrality (Källman 2009). A similar approach has also been used in Douglas-fir by Eckert et al. (2009b).

Earlier studies were often limited in terms of sampling and unknown demographic history. In contrast, recent studies are based on larger samples and readily incorporated demography in test statistics and interpretation of results. Specific models of demographic history suitable for neutrality testing have been developed for a few conifers. For instance, Heuertz et al. (2006) and Pyhäjärvi et al. (2007) showed that ancient bottlenecks followed by expansion could explain the polymorphism patterns observed at 22 loci in *Picea abies* and 16 loci in *Pinus sylvestris*, both cold-tolerant species, and their deviations from the neutral model at equilibrium. Grivet et al. (2009) found recent bottlenecks in a Mediterranean, drought-adapted conifer (*Pinus halepensis*) using the same approach. Current work also suggests that, in *P. pinaster*, we observe the effect of an ancient and possibly recurrent bottleneck of medium severity, but not followed by expansion (Lepoittevin 2009). Tested against the best fitting demographical models, patterns of polymorphism for some candidate genes have proved to be indeed caused by natural selection. For example, a multilocus HKA test pointed to the supposed action of selection on an abiotic stress-related dehydrin gene in Scots pine (Wachowiak et al. 2009). The application of several different neutrality tests, including those that incorporated explicit demographic models, revealed a suite of six genes consistent with selective sweeps in a large study of 121 cold-hardiness related candidate genes in coastal Douglas-fir (*Pseudotsuga menziesii* var. *menziesii*) (Eckert et al. 2009b).

Ideally designed for genome-wide scans and detection of local selection, F_{ST}-based outlier approaches have been applied to conifers either on a limited number of candidate genes or on a wider (but still very limited) scale with markers derived from expressed sequence tag (EST) databases [cleaved amplified polymorphisms (CAPS) or SNPs]. Tsumura et al. (2007a) identified four outliers out of 37 CAPS markers using the Beaumont and Nichols (1996) approach in one cypress species (*Chamaecyparis obtusa*). Using a similar approach, Krutovsky et al. (2009) found that among 25 allozyme loci in coastal Douglas-fir only the *PGM-1* locus demonstrated an unusually high level of differentiation as an apparent outlier. In *Pinus taeda*, 7% of 55 SNPs showed a level of population differentiation that was seven-fold the species average, possibly because of diversifying selection

(González-Martínez et al. 2006b). Tsumura et al. (2007b) detected four outlier CAPS loci (out of 139 markers) in *Cryptomeria japonica* using a combination of F_{ST}-based outlier detection and the coalescence simulation approach developed by Vitalis et al. (2001). Two of the CAPS markers were associated with differentiation between the two varieties of the species and are strong candidates for genes that have been subject to selection. The first large scale population genomics study in conifers (534 SNPs representing 345 expressed genes) aimed at exploring *Picea glauca* differentiation across six ecological regions in eastern Canada (Namroud et al. 2008). SNP by population effects were suggested at 49 SNPs, which was interpreted as evidence for local adaptation (Namroud et al. 2008). A similar study in *Pinus pinaster*, along with one designed specifically to account for the higher variance of F_{ST} at biallelic loci (Eveno et al. 2008), showed outlier patterns at the haplotype level for five out of 11 drought stress candidate genes. Polymorphism patterns at these genes can be interpreted as the consequence of diversifying selection (*PR-AGP4* and *erd3*), or homogenizing and balancing selection across populations (*dhn1*, *dhn2* and *lp3-1*).

In all the cited studies, the stronger candidates for adaptive loci are those showing different lines of evidence that could be interpreted as selection signals, on top of being strong expressional or functional candidates. For example, a dehydrin (*dhn1*) showed a significant excess of fixed non-synonymous substitutions compared to the number of fixed synonymous substitutions and polymorphic sites in *P. sylvestris*, suggesting the past action of selection (Wachowiak et al. 2009). Further, in a conifer phylogeny, there was evidence of positive selection on *dhn1* (Palmé et al. 2009). Similarly, a high proportion of non-synonymous polymorphic sites was observed at another dehydrin, *dhn1* from *P. pinaster* (homolog of *dhn9* in *P. sylvestris*), which is a strong candidate for balancing selection (Eveno et al. 2008). Most interesting are those loci that show similar patterns of selection in different and non-hybridizing species. This is the case of the *ccoaomt1* locus, involved in the lignification pathway, which showed low levels of differentiation consistent with an excess of intermediate frequency variants across *P. pinaster* populations in the Atlantic coast in France due to the existence of two divergent haplotypes. This same pattern is also observed in native populations of *Pinus taeda* from its southeastern distribution (González-Martínez et al. 2006a). These patterns were interpreted as the possible effect of balancing selection in both species. Signatures of balancing selection can be maintained over very long periods (Bouillé and Bousquet 2005).

One interesting recent approach combined the robustness of K_a/K_s ratio tests to demographic history with SFS methods, which thus allowed searching for genes under positive selection using different time scale signals since the selection events. Orthologous EST data in four conifer species (*Pinus pinaster, Pinus taeda, Picea glauca* and *Pseudotsuga menziesii*) were either obtained from publicly available EST libraries or resequenced to complete the sample (Palmé et al. 2008). A K_a/K_s ratio above 1 supposedly indicates positive selection. Branch-specific K_a/K_s ratios across all genes did not show any estimates above 1, but comparisons with polymorphism patterns revealed that genes with higher K_a/K_s (0.20–0.52) had more negative values of Tajima's D than the genes with lower K_a/K_s values. In this high K_a/K_s ratio group, the HKA test was also significant, suggesting that moderate K_a/K_s values could be indicative of selection in the EST data set.

The scale at which detection of natural selection is done in conifers is likely to change dramatically in the near future, as new sequencing technologies develop. In addition, future genomic scans will certainly integrate explicit demographic scenarios and/or ecological variables (Foll and Gaggiotti 2006; Excoffier et al. 2009). These methods will also benefit from the large resequencing efforts that are currently underway in conifer species. New sequencing technologies will greatly increase sample sizes both within and between populations, thus allowing tests to be applied in a more powerful context. These developments will surely trigger powerful comparative studies among conifer species.

6.6 Validation of Important Polymorphisms in Association Studies: Genotype-Phenotype Correlations

Variation in quantitative traits, such as many traits of interest in conifer species (i.e., mechanical properties of the wood, cold hardiness, drought acclimation, pulp content), is likely to result from allelic variation within multiple genes regulating these traits (Neale and Savolainen 2004; Oraguzie and Wilcox 2007). Association LD mapping aims to identify (using statistical inference) co-segregation of the specific allelic variation with phenotypic variation and, through this, to reveal the functional variant, or a marker in close linkage with it, responsible for phenotypic differences. The use of multiple unrelated populations and high resolution of marker trait associations differentiates this approach from traditional quantitative trait loci (QTL) mapping. Because of their abundance in the genome, SNPs have become the marker of choice to capture allelic variation for association mapping, although other markers such as indels are also of interest (Kruskopf-Österberg et al. 2002; Koornneef et al. 2004). Once SNP association has been identified, and independently validated, this marker is referred to as the quantitative trait nucleotide (QTN). Different QTNs have been suggested to date in plants (e.g., Thornsberry et al. 2001; Olsen et al. 2004; Aranzana et al. 2005; Thumma et al. 2005; González-Martínez et al. 2007; Beló et al. 2008; González-Martínez et al. 2008; Kuittinen et al. 2008; Eckert et al. 2009c), and, once they are validated, their application with conventional quantitative strategies to assist conifer breeding will be of significant value.

6.6.1 Factors to Consider in Association Studies

The power to detect an association is the probability that a statistical test will detect a true effect of a given size, or reject a false null hypothesis. The probability of accepting a false null hypothesis, i.e., of making a type II error, is also referred to as the false negative rate. Whether applying a Bayesian or a frequentist approach, numerous factors have been identified that will influence the power of association tests to detect a true effect (Gordon and Finch 2005; Ball 2007; Zhao et al. 2007). Furthermore, even when the power is optimal, statistical analysis should be used in conjunction with, but not replace, independent evidence of associations such as validation in alternative populations or other functional evidence for the association. Caution should also be applied since many associations could be population-specific, as many SNP effects are likely to depend on the genetic background and interactions with other SNPs involved in the trait variation.

6.6.1.1 Sample Size

The number of individuals sampled will affect experimental power. Small sample sizes can result in insufficient power to detect minor contributions of alleles, or imprecise estimates of the magnitude of the allele effect (Chanock et al. 2007). However the size required to detect a QTN for a given *p*-value (type I error rate) and power is influenced by multiple factors, including allele frequency, the extent of LD, gene and population by environment interaction and number and size of the QTN effects. A few methods have been developed which attempt to estimate optimal sample sizes for association studies from prior information, including Bayesian LD design (Ball 2005), calculation of power for quantitative transmission disequilibrium testing (TDT) design (Purcell et al. 2003) and case control studies (PAWE-3D software, Gordon et al. 2002). In the absence of prior information, or a suitable test for sample size, larger populations are preferable. In the literature, association studies in plants have reported sample sizes between ~75 to ~1,000 individuals (Thornsberry et al. 2001; Kruskopf-Österberg et al. 2002; Kumar et al. 2004; Olsen et al. 2004; Aranzana et al. 2005; Thumma et al. 2005; Ingvarsson et al. 2006; González-Martínez et al. 2007; Zhao et al. 2007; Beló et al. 2008; González-Martínez et al. 2008; Eckert et al. 2009c), but studies at the lower end of this range reveal low power to detect associations (Zhao et al. 2007), and much larger samples sizes (n = several thousand) have been recommended depending on the prior information at hand (Ball 2007).

6.6.1.2 Allele Frequency

The frequency of a SNP under examination can influence the power to detect association. The minor allele frequency (MAF) will affect the number of individuals (n) in each genotype class, and for low frequency SNPs this can be small and the standard errors for that class subsequently large. However, where low MAF at a locus (< 0.1) has been driven by purifying selection there may be improved power in association tests due to a larger effect size (Gorlov et al. 2008).

6.6.1.3 Effect Size

The absence of strong artificial positive or negative selection associated with domestication in the history of most conifers implies that their quantitative traits are likely to be under the control of many loci, each with small effect (Neale and Savolainen 2004). This characteristic will be somewhat limiting for association studies in conifers due to the direct relationship between power and effect size (Gordon and Finch 2005; Zhao et al. 2007; González-Martínez et al. 2008), where generally speaking the larger the effect size, the more easily an individual's genotype can be distinguished based on its phenotype.

6.6.1.4 Genotype/Phenotype Errors

Often difficult to detect, they can significantly reduce power to detect association (Gordon and Finch 2005). Replication of both types of data, although not always feasible, would allow estimation of error rates and adjustment of association tests to account for this. More feasible is the validation of methods for SNP genotyping by directly sequencing some individuals.

6.6.1.5 Linkage Disequilibrium (LD)

The extent of LD is a critical factor in determining the power to detect associations. Humans and many crop species exhibit extensive LD (Neale and Savolainen 2004), and SNPs or other markers associated with a phenotype may be linked to, but be at very long distances from, the QTN. Although this increases the chance of identifying an association in genome-wide scans, the association may be lost in the future if LD in this region is disrupted by recombination. In contrast, the rapid decay of LD shown for many conifers makes them ideal models to perform high-resolution association mapping (see above), but it also implies that a very high SNP density will be required. In addition, estimates of large scale LD across genomic regions are still largely unknown, and LD observed so far in the gene space of conifers is very variable.

6.6.1.6 Population Structure

Population subdivision, stratification or structure occurs when populations are divided into smaller subpopulations within which mating takes place. The use of natural populations or pedigrees for association studies can introduce confounding genetic structure, as has been observed in human and plant studies (The Wellcome Trust Case Control Consortium 2007; Zhao et al. 2007). This is of particular concern given that population genetic structure can create false LD between markers and QTLs (Neale and Savolainen 2004; Ball 2007). Population structure is likely to challenge association studies for some conifer species, in particular if such divergence is driven by fragmentation of natural populations. Additionally, the use of progeny trials will introduce pedigree structure, but may also harbour significant population structure if it was present among the pedigree founders. Evidence of genetic structure has been produced to varying degrees in different conifer species from surveys of genetic diversity. In some commercial species, such as *Pinus pinaster*, high levels of population genetic structure are observed. Other species are arrayed along clines (e.g., *Pinus sylvestris*). However, even low levels of genetic structure can have serious implications for association testing. In *Pinus taeda*, genetic differentiation across the species distribution is low (F_{st} = 0.0083), and significant population structure has not been detected using model-based clustering analyses (González-Martínez et al. 2007). Similar observations were obtained in coastal Douglas-fir based on allozyme and SSR markers (Krutovsky et al. 2009). Nevertheless, a significant interaction has been observed between a SNP marker and population structure delineated west and east of the Mississippi Valley (González-Martínez et al. 2007). In *Pinus radiata*, low to moderate genetic differentiation has been detected among the mainland populations using microsatellite and SNP markers ($F_{st} \approx 0.05$; Karhu et al. 2006). In association studies employing these populations, a large proportion of spurious associations were obtained (34%).

Several methods for the quantification of genetic structure have been developed to date. They include Bayesian cluster analysis (Pritchard et al. 2000; Pritchard and Przeworski 2001) and principal component analysis (Price et al. 2006) to take into account large scale population structure, and kinship estimation (Ritland 1996) to account for family relationships within clusters. These methods permit estimation of genetic structure from molecular data, where no prior knowledge of provenances or pedigrees is available. The structure estimates can subsequently be included in association models (Yu et al. 2006). However, corrections based on population structure estimates have several limitations. Complex structure patterns may not always be adequately captured using a single method (Zhao et al. 2007). Furthermore, adjustments for structure may be overly conservative, resulting in false negatives when divergent allele frequencies have arisen due to population-specific selection for the trait of interest. Nevertheless, as population structure from geographical sampling is often well known in conifers, the most promising approaches might be to directly integrate this structure in mixed models used for testing SNP effects overall, and to separately test those effects within groups or sub-populations.

6.6.2 Populations and Statistical Tests

The structure of a population will determine the appropriate design for association tests. Unrelated populations, where no family or pedigree structure is present, are readily available for many conifers by sampling from natural populations. Typically natural populations will exhibit low LD and some population genetic structure. Analysis of association in such populations can be achieved by standard ANOVA or *F*-testing, which detect correlations between quantitative variation and bi-allelic genotypes. General linear models (GLMs) incorporating terms for population genetic structure and other covariates as fixed effects can adjust for the confounding effect of terms and are already implemented in user-friendly software (e.g., Tassel, see Bradbury et al. 2007).

Particularly for commercially valuable conifer species, it is possible to access populations with pedigree structure, where garden experiments or progeny trials have been established. The utility of pedigree populations to detect associations is limited by those factors already noted, but also by the number of founders they were established from. Altogether, the number of founders will influence the effective population size and the extent of LD (Ball 2007). To account for biases due to fine-scale family structure, mixed models (MLMs) incorporating a term for random effects (of known pedigree relationships or pairwise kinship estimates from independent molecular data) can be applied to test association (Yu et al. 2006). However, biases may still arise in pedigree association tests if a significant population structure was sampled among the pedigree founders, in which case modelling both fixed (population structure) and random (kinship) effects may be beneficial (Zhao et al. 2007).

Transmission disequilibrium testing (TDT) is a pedigree-based method to test for association, which is robust to population structure (Gordon and Finch 2005). This approach studies the transmission of alleles from heterozygous parents to one or more affected offspring using a non-parametric approach on nominal data applied to 2 × 2 contingency tables (Spielman et al. 1993). The method has been extended to include quantitative traits (Xiong et al. 1998; Abecasis et al. 2000a, b), and is therefore suitable for many traits of interest in conifers. This test is also appropriate when large numbers of small families are available (with one or more sibs), but not always appropriate in trees due to their long generation times and anonymity or unavailability of both parent genotypes in existing populations.

6.6.3 *Current Association Studies in Conifers*

International activity on association studies currently targets at least nine major conifer species, although it concentrates on commercial pines and spruces (*Pinus taeda, P. pinaster, P. sylvestris, P. radiata, Picea abies, P. glauca, P. sitchensis, Pseudotsuga menziesii* and *Cryptomeria japonica*). The advancement of association genetics in conifers has spawned from the improved knowledge of conifer genomes generated in surveys of nucleotide diversity, LD and population structure (see sections above). Because of the large size and complexity of most conifer genomes (Ahuja and Neale 2005), diversity studies have focused on sequences within candidate genes and their flanking regions as said before. This approach has been complementary to the low level of LD observed in conifers, which has made genome wide association studies impractical to date (Neale and Savolainen 2004; but see "Conclusions and Perspectives" section below).

Several processes have been applied to select candidate SNPs for association testing in conifers. Manual random selection and selection of SNPs based on their informed position in the gene sequence (i.e., coding vs. non-coding regions, silent vs. non-synonymous) are practical when small numbers of genes and SNPs are considered. However, as DNA sequencing technologies advance and larger numbers of SNPs become available, approaches such as haplotype block partitioning, which minimize the number of representative SNPs required to account for most of the common haplotypes in each block (Ding et al. 2005; Zhang et al. 2005), or automated SNP selection approaches, which consider sequence quality, allele frequency or other user-defined parameters (Pavy et al. 2006; PineSAP, Wegrzyn et al. 2009), have greater utility. Furthermore, SNPs are being selected for association studies based on signatures of selection found in genome-wide scans of natural populations, which may target genes with ecological and adaptive significance (e.g., Namroud et al. 2008; Eckert et al. 2009b, c). High-throughput genotyping of SNP markers in large populations have recently been reported in spruce and pine (Pavy et al. 2008; Eckert et al. 2009a), with a number of unpublished studies currently underway.

To date, only a few association studies have been published in conifers, in loblolly pine (González-Martínez et al. 2007, 2008) and Douglas-fir (Eckert et al. 2009c), and are among the first in this field to identify individual genes regulating quantitative traits in trees. Statistical correlation between SNPs from several candidate genes with quantitative wood properties and carbon isotope discrimination (CID, related to water use efficiency in plants) was examined in the loblolly pine studies using mixed linear models (MLMs) and quantitative TDT tests, respectively. After applying corrections for multiple testing, González-Martínez et al. (2007) reported six significant genotype-phenotype associations for three wood traits (earlywood density, latewood proportion and earlywood microfibril angle, MFA) and SNPs in five genes (*cad, 4cl, sams-2, α-tubulin* and *lp3-1*). The strongest genetic association identified in this study occurred between a SNP in a *α-tubulin* gene and MFA. The lack of LD between this SNP and other polymorphisms in the region suggests this SNP may be the QTN causing phenotypic variation, a view that is supported by the proposed role of cortical microtubules in orienting cellulose deposition (Lindeboom et al. 2008). In González-Martínez et al. (2008), genetic associations with CID were found for SNPs in four genes (*dhn1*, *lp5* homolog, *sod-chl* and *wrky*-like transcription factor). However, after accounting for multiple testing none of these associations were significant. Genetic associations between 384 SNPs from 117 candidate genes and 21 cold-hardiness related traits were studied in Douglas-fir (Eckert et al. 2009c). A generalized linear model approach, including population structure estimates as covariates, was implemented for each marker-trait pair. Thirty highly significant genetic associations across 12 candidate genes and 14 of the 21 traits were discovered. A set of seven markers that had elevated levels of differentiation between sampling sites was also detected. Marker effects were small ($r^2 < 0.05$) and within the range of those published previously for forest trees. The majority of markers were characterized as having largely non-additive modes of gene action, especially under dominance in the case of cold-tolerance related phenotypes. These results were considered in the context of trade-offs between the ability to grow longer and to avoid an early fall cold damage. These associations provide insight into the genetic components of complex traits in coastal Douglas-fir, and highlight the need for landscape genetic approaches to the detection of adaptive genetic diversity.

Individual SNPs are expected to account for a small proportion (between 1–10%) of the overall phenotypic variance of quantitative traits (Neale and Savolainen 2004; Neale and Ingvarsson 2008), depending on the trait. In the *P. taeda* CID study, non-significance of associations after multiple testing most likely resulted from the low level of phenotypic variance accounted for by each SNP (< 1%). Numerical simulations suggested that allelic effects of 2.5% or greater would be necessary to detect association with this trait and experimental design (González-Martínez et al. 2008). Cumulatively, the two genes (*cad* and *sams-2*) affecting earlywood specific gravity, accounted for up to 10% of the phenotypic variance (González-Martínez et al. 2007), whereas the single SNP from *α-tubulin* associating with earlywood MFA accounted for up to 4% of the phenotypic variance. These estimates are in line with expectations, and implore the screening of large numbers of genes and SNPs to dissect a significant proportion of the genetics underlying quantitative traits. Still, these effects are larger than those reported in other organisms such as humans (where the largest effects, with sample sizes of tens of thousands, were less than 0.5%; Weedon et al. 2008). An obvious outcome of association genetics in conifers, and other species, is the application of multiple QTN for gene-assisted breeding (GAS). Not only are QTNs with larger effects more easily detected in association tests, they will also be more appealing to breeders wishing to improve a trait of interest as an equivalent gain can be achieved using fewer loci. Furthermore, interactions among QTNs from different genes might explain more variation and be useful in a particular genetic background.

6.7 Conclusions and Perspectives

"Now is the time to concentrate on experimental design, so that the deluge of genotype data can be fully exploited when it arrives in the future" (Myles et al. 2009).

Obviously, genome-wide association studies are not yet in sight in conifers as genomes are huge, highly repetitive and still poorly characterized. Yet the situation might not be as bleak as it seems, and the success of future association studies in conifers will heavily depend on our ability to exploit the strengths of conifer characteristics and circumvent their inherent difficulties.

First, recently developed methods of high-throughput (HT) sequencing and genotyping will allow reasonable access to the conifer transcriptome. Accessing regulatory areas far from the coding sequence will be much more difficult, although there too conifer research will benefit from advances in other difficult genomes such as maize. For example, maize geneticists have developed library construction methods, such as hypomethylated partial restriction (HMPR), to target gene-rich regions in large, highly repetitive genomes. These methods are currently combined with novel sequencing techniques to identify large number of SNPs (Gore et al. 2009) in maize and initial results on the development of HMPRs in Norway spruce are encouraging.

Second, many conifers species also benefit from a long and strong breeding tradition (see Chapter 2, this book). In those species, a large number of progeny trials are therefore available, as well as good estimates of major quantitative genetic parameters for an array of economically and ecologically important traits. QTL for some of these traits have also been mapped on dense genetic maps (see Chapter 5, this book). Using these resources and developing new ones, as was recently done in maize (Myles et al. 2009), will be instrumental to the success of association mapping efforts in conifers. In particular, this information can be combined with information arising from the candidate gene and surveys of nucleotide polymorphisms. As we have shown above, even if numbers are still modest, both approaches have yielded interesting genes. Co-locating those with QTLs should help confirm their status. Similarly, when possible, expression studies and transformation will bring additional information on those genes.

Third, conifer research is being conducted on many different species, some of which are very closely related at the sequence level, and at the genome structure level (in terms of map positions of homologous loci, e.g., Brown et al. 2001, Chapter 5, this book). It will be important to leverage this in developing association tests and in using information from other species on candidate genes and genome structure. Such studies will also provide very interesting information on genome evolution in conifers.

Fourth, conifers are key organisms in terrestrial ecosystems and the study of landscape patterns of ecologically important traits/genetic variation is relevant for local adaptation of conifers in a changing world. Comparative analysis in this framework is developing rapidly in conifers and they are likely to become some of the best model systems available for landscape genetics.

Finally, there are many lessons for conifer researchers in association studies in humans and crops. Conifers and humans share some biological features –both are random mating and long lived organisms, both have gone through recent exponential growth– and a lot will be gained by carefully following the successes and failures of human geneticists in their quest to understand the genetic architecture of complex traits. At the same time, as recent studies in maize illustrate beautifully (Buckler et al. 2009; McMullen et al. 2009), plant biologists also benefit from more degrees of freedom in this quest than human geneticists.

References

Abecasis GR, Cardon LR, Cookson WO (2000a) A general test of association for quantitative traits in nuclear families. Am J Hum Genet 66: 279–292.

Abecasis GR, Cookson WO, Cardon LR (2000b) Pedigree tests of transmission disequilibrium. Eur J Hum Genet 8: 545–551.

Ahuja MR, Neale DB (2005) Evolution of genome size in conifers. Silvae Genet 54: 126–137.

Aranzana MJ, Kim S, Zhao K, Bakker E, Horton M, Jakob K, Lister C, Molitor J, Shindo C, Tang C, Toomajian C, Traw B, Zheng H, Bergelson J, Dean C, Marjoram P, Nordborg M (2005) Genome-wide association mapping in Arabidopsis identifies previously known flowering time and pathogen resistance genes. PLoS Genet 1: 531–539.

Arunyawat U, Stephan W, Städler T (2007) Using multilocus sequence data to assess population structure, natural selection, and linkage disequilibrium in wild tomatoes. Mol Biol Evol 24: 2310–2322.

Ball RD (2005) Experimental designs for reliable detection of linkage disequilibrium in unstructured random population association studies. Genetics 170: 859–873.
Ball RD (2007) Statistical Analysis and Experimental Design. Springer Science, New York, USA.
Beaumont MA (2005) Adaptation and speciation: what can F_{ST} tell us? Trends Ecol Evol 20: 435–440.
Beaumont MA, Nichols RA (1996) Evaluating loci for use in the genetic analysis of population structure. Proc Roy Soc Lond Ser B 263: 1619–1626.
Beaumont MA, Balding DJ (2004) Identifying adaptive genetic divergence among populations from genome scans. Mol Ecol 13: 969–980.
Beló A, Zheng P, Luck S, Shen B, Meyer DJ, Li B, Tingey S, Rafalski A (2008) Whole genome scan detects an allelic variant of fad2 associated with increased oleic acid levels in maize. Mol Genet Genom 279: 1–10.
Bielawski JP, Dunn KA, Yang Z (2000) Rates of nucleotide substitution and mammalian nuclear gene evolution: approximate and maximum-likelihood methods lead to different conclusions. Genetics 156: 1299–1308.
Birks JHB, Willis KJ (2008) Alpines, trees, and refugia in Europe. Plant Ecol Divers 1: 147–160.
Biswas S, Akey JM (2006) Genomic insights into positive selection. Trends Genet 22: 437–446.
Bouillé M, Bousquet J (2005) Trans-species shared polymorphisms at orthologous nuclear gene loci among distant species in the conifer Picea (Pinaceae): implications for the long-term maintenance of genetic diversity in trees. Am J Bot 92: 63–73.
Bowcock AM, Kidd JR, Mountain JL, Hebert JM, Carotenuto L, Kidd KK, Cavallisforza LL (1991) Drift, admixture, and selection in human evolution: a study with DNA polymorphisms. Proc Natl Acad Sci USA 88: 839–843.
Bradbury PJ, Zhang Z, Kroon DE, Casstevens TM, Ramdoss Y, Buckler ES (2007) Tassel: Software for association mapping of complex traits in diverse samples. Bioinformatics 23: 2633–2635.
Brown GR, Kadel EE 3rd, Bassoni DL, Kiehne KL, Temesgen B, van Buijtenen JP, Sewell MM, Marshall KA, Neale DB (2001) Anchored reference loci in loblolly pine (*Pinus taeda* L.) for integrating pine genomics. Genetics 159: 799–809.
Brown GR, Gill GP, Kuntz RJ, Langley CH, Neale DB (2004) Nucleotide diversity and linkage disequilibrium in loblolly pine. Proc Natl Acad Sci USA 101: 15255–15260.
Buckler ES IV, Thornsberry JM (2002) Plant molecular diversity and applications to genomics. Curr Opin Plant Biol 5: 107–111.
Buckler ES IV, Holland JB, Bradbury PJ, Acharya CB, Brown PJ, Browne C, Ersoz E, Flint-Garcia S, Garcia A, Glaubitz JC, Goodman MM, Harjes C, Guill K, Kroon DE, Larsson S, Lepak NK, Li H, Mitchell SE, Pressoir G, Peiffer JA, Rosas MO, Rocheford TR, Romay MC, Romero S, Salvo S, Sanchez Villeda H, da Silva HS, Sun Q, Tian F, Upadyayula N, Ware D, Yates H, Yu J, Zhang Z, Kresovich S, McMullen MD (2009) The genetic architecture of maize flowering time. Science 325: 714–718.
Cargill M, Altshuler D, Ireland J, Sklar P, Ardlie K, Patil N, Shaw N, Lane CR, Lim EP, Kalyanaraman N, Nemesh J, Ziaugra L, Friedland L, Rolfe A, Warrington J, Lipshutz R, Daley GQ, Lander ES (1999) Characterization of single-nucleotide polymorphisms in coding regions of human genes. Nat Genet 22: 231–238.
Chagné D, Lalanne C, Madur D, Kumar S, Frigério JM, Krier C, Decroocq S, Savouré A, Bou-Dagher-Kharrat M, Bertocchi E, Brach J, Plomion C (2002) A high density genetic map of maritime pine based on AFLPs. Ann For Sci 59: 627–636.
Chanock SJ, Manolio T, Boehnke M, Boerwinkle E, Hunter DJ, Thomas G, Hirschhorn JN, Abecasis G, Altshuler D, Bailey-Wilson JE, Brooks LD, Cardon LR, Daly M, Donnelly P, Fraumeni JF Jr, Freimer NB, Gerhard DS, Gunter C, Guttmacher AE, Guyer MS, Harris EL, Hoh J, Hoover R, Kong CA, Merikangas KR, Morton CC, Palmer LJ, Phimister EG,

Rice JP, Roberts J, Rotimi C, Tucker MA, Vogan KJ, Wacholder S, Wijsman EM, Winn DM, Collins FS (2007) Replicating genotype-phenotype associations. Nature 447: 655–660.

Charlesworth B, Nordborg M, Charlesworth D (1997) The effects of local selection, balanced polymorphism, and background selection on equilibrium patterns of genetic diversity in subdivided populations. Genet Res 70: 155–174.

Chen H, Morrell PL, de la Cruz M, Clegg MT (2008) Nucleotide diversity and linkage disequilibrium in wild avocado (*Persea americana* Mill.). J Hered 99: 382–389.

Chen J, Källman T, Gyllenstrand N, Lascoux M (2010) New insights on the speciation history and nucleotide diversity of three boreal spruce species and a Tertiary relict. Heredity 104: 3–14.

Cutter AD (2006) Nucleotide polymorphism and linkage disequilibrium in wild populations of the partial selfer *Caenorhabditis elegans*. Genetics 172: 171–184.

de-Paoli E (2005) Diversità genetica, linkage disequilibrium e componente ripetitiva del genoma in abete rosso [*Picea abies* (L.) Karst.]. Ph.D. Dissert, Univ of Udine, Italy.

Ding KY, Zhang J, Zhou KX, Shen Y, Zhang XG (2005) htSNPer1.0: Software for haplotype block partition and htsnps selection. BMC Bioinformat 6: 38.

Eckert AJ, Pande B, Ersoz ES, Wright MH, Rashbrook VK, Nicolet CM, Neale DB (2009a) High-throughput genotyping and mapping of single nucleotide polymorphisms in loblolly pine (*Pinus taeda* L.). Tree Genet Genomes 5: 225–234.

Eckert AJ, Wegrzyn JL, Pande B, Jermstad KD, Lee JM, Liechty JD, Tearse BR, Krutovsky KV, Neale DB (2009b) Multilocus patterns of nucleotide diversity and divergence reveal positive selection at candidate genes related to cold hardiness in coastal Douglas-fir (*Pseudotsuga menziesii* var. *menziesii*). Genetics 183: 289–298.

Eckert AJ, Bower AD, Wegrzyn JL, Pande B, Jermstad KD, Krutovsky KV, St Clair JB, Neale DB (2009c) Association genetics of coastal Douglas-fir (*Pseudotsuga menziesii* var. *menziesii*, Pinaceae) I. Cold-hardiness related traits. Genetics 182: 1289–1302.

Eveno E, Collada C, Guevara MA, Léger V, Soto A, Díaz L, Léger P, González-Martínez SC, Cervera MT, Plomion C, Garnier-Géré PH (2008) Contrasting patterns of selection at *Pinus pinaster* Ait. drought stress candidate genes as revealed by genetic differentiation analyses. Mol Biol Evol 25: 417–437.

Excoffier L, Hofer T, Foll M (2009) Detecting loci under selection in a hierarchically structured population. Heredity 103: 285–298.

Fay JC, Wu C-I (2000) Hitchhiking under positive Darwinian selection. Genetics 155: 1405–1413.

Flint-García SA, Thornsberry JM, Buckler ES IV (2003) Structure of linkage disequilibrium in plants. Annu Rev Plant Biol 54: 357–374.

Foll M, Gaggiotti OE (2006) Identifying the environmental factors that determine the genetic structure of populations. Genetics 174: 875–891.

Ford MJ (2002) Applications of selective neutrality tests to molecular ecology. Mol Ecol 11: 1245–1262.

Fu YX (1997) Statistical tests of neutrality of mutations against population growth, hitch-hiking and background selection. Genetics 147: 915–925.

Fu YX, Li WH (1993) Statistical tests of neutrality of mutations. Genetics 133: 693–709.

Fujimoto A, Kado T, Yoshimaru H, Tsumura Y, Tachida H (2008) Adaptive and slightly deleterious evolution in a conifer *Cryptomeria japonica*. J Mol Evol 67: 201–210.

González-Martínez SC, Ersoz E, Brown GR, Wheeler NC, Neale DB (2006a) DNA sequence variation and selection of tag single-nucleotide polymorphisms at candidate genes for drought-stress response in *Pinus taeda* L. Genetics 172: 1915–1926.

González-Martínez SC, Krutovsky KV, Neale DB (2006b) Forest-tree population genomics and adaptive evolution. New Phytol 170: 227–238.

González-Martínez SC, Wheeler NC, Ersoz E, Nelson CD, Neale DB (2007) Association genetics in *Pinus taeda* L. I. Wood property traits. Genetics 175: 399–409.

González-Martínez SC, Huber D, Ersoz E, Davis JM, Neale DB (2008) Association genetics in *Pinus taeda* L. II. Carbon isotope discrimination. Heredity 101: 19–26.

Gordon D, Finch SJ (2005) Factors affecting statistical power in the detection of genetic association. J Clin Invest 115: 1408–1418.

Gordon D, Finch SJ, Nothnagel M, Ott J (2002) Power and sample size calculations for case-control genetic association tests when errors present: Application to single nucleotide polymorphisms. Hum Hered 54: 22–33.

Gore MA, Wright MH, Ersoz ES, Bouffard P, Szekeres ES, Jarvie TP, Hurwitz BL, Narechania A, Harkins TT, Grills GS, Ware DH, Buckler IV ES (2009) Large-scale discovery of gene-enriched SNPs. Plant Genet 2: 121–133.

Gorlov IP, Gorlova OY, Sunyaev SR, Spitz MR, Amos CI (2008) Shifting paradigm of association studies: Value of rare single-nucleotide polymorphisms. Am J Hum Genet 82: 100–112.

Grivet D, Sebastiani F, González-Martínez SC, Vendramin GG (2009) Patterns of polymorphism resulting from long-range colonization in the Mediterranean conifer Aleppo pine. New Phytol 184: 1016–1028.

Haddrill PR, Thornton KR, Charlesworth B, Andolfatto P (2005) Multilocus patterns of nucleotide variability and the demographic and selection history of *Drosophila melanogaster* populations. Genome Res 15: 790–799.

Halushka MK, Fan JB, Bentley K, Hsie L, Shen N, Weder A, Cooper R, Lipshutz R, Chakravarti A (1999) Patterns of single-nucleotide polymorphisms in candidate genes for blood-pressure homeostasis. Nat Genet 22: 239–247.

Hamrick JL, Godt MJW (1996) Effects of life history traits on genetic diversity in plant species. Phil Trans Roy Soc Lond Ser B Biol Sci 351: 1291–1298.

Heuertz M, de Paoli E, Källman T, Larsson H, Jurman I, Morgante M, Lascoux M, Gyllenstrand N (2006) Multilocus patterns of nucleotide diversity, linkage disequilibrium and demographic history of Norway spruce [*Picea abies* (L.) karst]. Genetics 174: 2095–2105.

Howe GT, Aitken SN, Neale DB, Jermstad KD, Wheeler NC, Chen THH (2003) From genotype to phenotype: unraveling the complexities of cold adaptation in forest trees. Can J Bot 81: 1247–1266.

Hudson RR (2001) Two-locus sampling distributions and their application. Genetics 159: 1805–1817.

Hudson RR, Kreitman M, Aguadé M (1987) A test of neutral molecular evolution based on nucleotide data. Genetics 116: 153–159.

Hyten DL, Song Q, Zhu Y, Choi IY, Nelson RL, Costa JM, Specht JE, Shoemaker RC, Cregan PB (2006) Impacts of genetic bottlenecks on soybean genome diversity. Proc Natl Acad Sci USA 103: 16666–16671.

Ingvarsson PK (2005) Nucleotide polymorphism and linkage disequilibrium within and among natural populations of European aspen (*Populus tremula* L., Salicaceae). Genetics 169: 945–953.

Ingvarsson PK (2008) Multilocus patterns of nucleotide polymorphism and the demographic history of *Populus tremula*. Genetics 180: 329–340.

Ingvarsson PK, Garcia MV, Hall D, Luquez V, Jansson S (2006) Clinal variation in phyb2, a candidate gene for day-length-induced growth cessation and bud set, across a latitudinal gradient in European aspen (*Populus tremula*). Genetics 172: 1845–1853.

Jaramillo-Correa JP, Verdú M, González-Martínez SC (2010) The contribution of recombination to heterozygosity differs among plant evolutionary lineages and life-forms. BMC Evol Biol 10: 22.

Kado T, Yoshimaru H, Tsumura Y, Tachida H (2003) DNA variation in a conifer, *Cryptomeria japonica* (Cupressaceae sensu lato). Genetics 164: 1547–1559.

Kado T, Ushio Y, H. Yoshimaru H, Tsumura Y, Tachida H (2006) Contrasting patterns of DNA variation in natural populations of two related conifers, *Cryptomeria japonica* and *Taxodium distichum* (*Cupressaceae sensu lato*). Genes Genet Syst 81: 103–113.

Kado T, Matsumoto A, Ujino-Ihara T, Tsumura Y (2008) Amounts and patterns of nucleotide variation within and between two Japanese conifers, sugi (*Cryptomeria japonica*) and hinoki (*Chamaecyparis obtusa*) (Cupressaceae sensu lato). Tree Genet Genomes 4: 133–141.

Källman T (2009) Adaptive evolution and demographic history of Norway Spruce (*Picea abies*). Ph.D. Dissert, Univ of Uppsala, Sweden.

Karhu A, Vogl C, Moran GF, Bell JC, Savolainen O (2006) Analysis of microsatellite variation in *Pinus radiata* reveals effects of genetic drift but no recent bottlenecks. J Evol Biol 19: 167–175.

Kelley JL, Madeoy J, Calhoun JC, Swanson W, Akey JM (2007) Genomic signatures of positive selection in humans and the limits of outlier approaches. Genome Res 16: 980–989.

Kilian B, Ozkan H, Walther A, Kohl J, Dagan T, Salamini F, Martin W (2007) Molecular diversity at 18 loci in 321 wild and 92 domesticate lines reveal no reduction of nucleotide diversity during *Triticum monococcum* (einkorn) domestication: Implications for the origin of agriculture. Mol Biol Evol 24: 2657–2668.

Kim Y, Nielsen R (2004) Linkage disequilibrium as a signature of selective sweeps. Genetics 167: 1513–1524.

Kimura M (1983) The neutral theory of molecular evolution. Cambridge Univ. Press, Cambridge, UK.

Koch MA, Haubold B, Mitchell-Olds T (2000) Comparative evolutionary analysis of chalcone synthase and alcohol dehydrogenase loci in *Arabidopsis*, *Arabis*, and related genera (Brassicaceae). Mol Biol Evol 17: 1483–1498.

Koelewijn HP, Koski V, Savolainen O (1999) Magnitude and timing of inbreeding depression in Scots pine (*Pinus sylvestris* L.). Evolution 53: 758–768.

Koornneef M, Alonso-Blanco C, Vreugdenhil D (2004) Naturally occurring genetic variation in *Arabidopsis thaliana*. Annu Rev Plant Biol 55: 141–172.

Kruskopf-Österberg M, Shavorskaya O, Lascoux M, Lagercrantz U (2002) Naturally occurring indel variation in the *B. nigra* COL1 gene is associated with variation in flowering time. Genetics 161: 299–306.

Krutovsky KV, Neale DB (2005) Nucleotide diversity and linkage disequilibrium in cold-hardiness- and wood quality-related candidate genes in Douglas fir. Genetics 171: 2029–2041.

Krutovsky KV, St. Clair JB, Saich R, Hipkins VD, Neale DB (2009) Estimation of population structure in coastal Douglas-fir [*Pseudotsuga menziesii* (Mirb.) Franco var. *menziesii*] using allozyme and microsatellite markers. Tree Genet Genomes 5: 641–658.

Kuittinen H, Niittyvuopio A, Rinne P, Savolainen O (2008) Natural variation in *Arabidopsis lyrata* vernalization requirement conferred by a FRIGIDA indel polymorphism. Mol Biol Evol 25: 319–329.

Kumar S, Echt C, Wilcox PL, Richardson TE (2004) Testing for linkage disequilibrium in the New Zealand radiata pine breeding population. Theor Appl Genet 108: 292–298.

Lepoittevin C (2009) Description de la diversité nucléotidique dans des gènes candidats impliqués dans la formation du bois chez le pin maritime et association avec les composants de la qualité du bois. PhD Dissert, Univ of Bordeaux, France.

Lewontin RC, Krakauer J (1973) Distribution of gene frequency as a test of theory of selective neutrality of polymorphisms. Genetics 74: 175–195.

Lindeboom J, Mulder BM, Vos JW, Ketelaar T, Emons AMC (2008) Cellulose microfibril deposition: Coordinated activity at the plant plasma membrane. J Microsc 231: 192–200.

Liu A, Burke JM (2006) Patterns of nucleotide diversity in wild and cultivated sunflower. Genetics 173: 321–330.

Long AD, Lyman RF, Langley CH, Mackay TFC (1998) Two sites in the *Delta* gene region contribute to naturally occurring variation in bristle number in *Drosophila melanogaster*. Genetics 149: 999–1017.

Luikart G, England PR, Tallmon D, Jordan S, Taberlet P (2003) The power and promise of population genomics: from genotyping to genome typing. Nat Rev Genet 4: 981–994.

Lynch M (2006) The origins of eukaryotic gene structure. Mol Biol Evol 23: 450–468.

Ma X-F, Szmidt AE, Wang X-R (2006) Genetic structure and evolutionary history of a diploid hybrid pine *Pinus densata* inferred from the nucleotide variation at seven gene loci. Mol Biol Evol 23: 807–816.

McMullen MD, Kresovich S, Villeda HS, Bradbury P, Li H, Sun Q, Flint-Garcia S, Thornsberry J, Acharya C, Bottoms C, Brown P, Browne C, Eller M, Guill K, Harjes C, Kroon D, Lepak N, Mitchell SE, Peterson B, Pressoir G, Romero S, Oropeza Rosas M, Salvo S, Yates H, Hanson M, Jones E, Smith S, Glaubitz JC, Goodman M, Ware D, Holland JB, Buckler ES IV (2009) Genetic properties of the maize nested association mapping population. Science 325: 737–740.

McVean G, Awadalla P, Fearnhead P (2002) A coalescent-based method for detecting and estimating recombination from gene sequences. Genetics 160:1231–1241.

Moeller DA, Tenaillon MI, Tiffin P (2007) Population structure and its effects on patterns of nucleotide polymorphism in Teosinte (*Zea mays* ssp. *parviglumis*). Genetics 176: 1799–1809.

Morgenstern EK (1996) Geographic variation in forest trees: genetic basis and application of knowledge in silviculture. UBC Press, Vancouver, BC, Canada.

Morrell PL, Lundy KE, Clegg MT (2003) Distinct geographic patterns of genetic diversity are maintained in wild barley (*Hordeum vulgare* ssp. *spontaneum*) despite migration. Proc Natl Acad Sci USA 100: 10812–10817.

Morrell PL, Toleno DM, Lundy KE, Clegg MT (2005) Low levels of linkage disequilibrium in wild barley (*Hordeum vulgare* ssp. *spontaneum*) despite high rates of self-fertilization. Proc Natl Acad Sci USA 102: 2442–2447.

Morrell PL, Toleno DM, Lundy KE, Clegg MT (2006) Estimating the contribution of mutation, recombination and gene conversion in the generation of haplotypic diversity. Genetics 173: 1705–1723.

Moriyama EN, Powell JR (1996) Intraspecific nuclear DNA variation in Drosophila. Mol Biol Evol 13: 261–277.

Myles S, Peiffer J, Brown PJ, Ersoz ES, Zhang Z, Costich DF, Buckler ES IV (2009) Association mapping: critical considerations shift from genotyping to experimental design. Plant Cell 21: 2194–2202.

Namroud MC, Beaulieu J, Juge N, Laroche J, Bousquet J (2008) Scanning the genome for gene single nucleotide polymorphisms involved in adaptive population differentiation in white spruce. Mol Ecol 17: 3599–3613.

Neale DB, Savolainen O (2004) Association genetics of complex traits in conifers. Trends Plant Sci 9: 325–330.

Neale DB, Ingvarsson PK (2008) Population, quantitative and comparative genomics of adaptation in forest trees. Curr Opin Plant Biol 11: 149–155.

Nei M (1987) Molecular Evolutionary Genetics. Columbia Univ Press, New York, USA.

Nielsen R (2001) Statistical tests of selective neutrality in the age of genomics. Heredity 86: 641–647.

Nielsen R (2005) Molecular signatures of natural selection. Annu Rev Genet 39: 197–218.

Nikaido AM, Ujino T, Iwata H, Yoshimua K, Yoshimaru H, Suyama Y, Murai M, Nagasaka K, Tsumura T (2000) AFLP and CAPS linkage maps of *Cryptomeria japonica*. Theor Appl Genet.100: 825–831.

Nordborg M, Tavaré S (2002) Linkage disequilibrium: what history has to tell us? Trends Genet 18: 83–90.

Nybom H, Bartish IV (2000) Effects of life history traits and sampling strategies on genetic diversity estimates obtained with RAPD markers in plants. Perspect Plant Ecol Evol Syst 3: 93–114.

Olsen KM, Womack A, Garrett AR, Suddith JI, Purugganan MD (2002) Contrasting evolutionary forces in the *Arabidopsis thaliana* floral developmental pathway. Genetics 160: 1641–1650.

Olsen KM, Halldorsdottir SS, Stinchcombe JR, Weinig C, Schmitt J, Purugganan MD (2004) Linkage disequilibrium mapping of Arabidopsis cry2 flowering time alleles. Genetics 167: 1361–1369.

Oraguzie NC, Wilcox PL (2007) An overview of association mapping. Springer Science, New York, USA.

Palaisa KA, Morgante M, Williams M, Rafalski A (2003) Contrasting effects of selection on sequence diversity and linkage disequilibrium at two phytoene synthase loci. Plant Cell 15: 1795–1806.

Palmé AE, Wright M, Savolainen O (2008) Patterns of divergence among conifer ESTs and polymorphism in *Pinus sylvestris* identify putative selective sweeps. Mol Biol Evol 25: 2567–2577.

Palmé AE, Pyhäjärvi T, Wachowiak W, Savolainen O (2009) Selection on nuclear genes in a *Pinus* phylogeny. Mol Biol Evol 26: 893–905.

Pavy N, Parsons LS, Paule C, MacKay J, Bousquet J (2006) Automated SNP detection from a large collection of white spruce expressed sequences: Contributing factors and approaches for the categorization of SNPs. BMC Genom 7: 174.

Pavy N, Pelgas B, Beauseigle S, Blais S, Gagnon F, Gosselin I, Lamothe M, Isabel N, Bousquet J (2008) Enhancing genetic mapping of complex genomes through the design of highly-multiplexed SNP arrays: Application to the large and unsequenced genomes of white spruce and black spruce. BMC Genom 9: 21.

Pot D, McMillan L, Echt C, Le Provost G, Garnier-Géré P, Cato S, Plomion C (2005) Nucleotide variation in genes involved in wood formation in two pine species. New Phytol 167: 101–112.

Price AL, Patterson NJ, Plenge RM, Weinblatt ME, Shadick NA, Reich D (2006) Principal components analysis corrects for stratification in genome-wide association studies. Nat Genet 38: 904–909.

Pritchard JK, Przeworski M (2001) Linkage disequilibrium in humans: Models and data. Am J Hum Genet 69: 1–14.

Pritchard JK, Stephens M, Donnelly P (2000) Inference of population structure using multilocus genotype data. Genetics 155: 945–959.

Purcell S, Cherny SS, Sham PC (2003) Genetic power calculator: Design of linkage and association genetic mapping studies of complex traits. Bioinformatics 19: 149–150.

Pyhäjärvi T, García-Gil MR, Knürr T, Mikkonen M, Wachowiak W, Savolainen O (2007) Demographic history has influenced nucleotide diversity in European *Pinus sylvestris* populations. Genetics 177: 1713–1724.

Quang ND, Ikeda S, Harada K (2008) Nucleotide variation in *Quercus crispula* Blume. Heredity 101: 166–174.

Rafalski A, Morgante M (2004) Corn and humans: recombination and linkage disequilibrium in two genomes of similar size. Trends Genet 20: 103–111.

Ramírez-Soriano A, Ramos-Onsins SE, Rozas J, Calafell F, Navarro A (2008) Statistical power analysis of neutrality tests under demographic expansions, contractions and bottlenecks with recombination. Genetics 179: 555–567.

Ramos-Onsins SE, Rozas J (2002) Statistical properties of new neutrality tests against population growth. Mol Biol Evol 19: 2092–2100.

Ramos-Onsins SE, Stranger BE, Mitchell-Olds T, Aguade M (2004) Multilocus analysis of variation and speciation in the closely related species *Arabidopsis halleri* and *A. lyrata*. Genetics 166: 373–388.

Ritland K (1996) Estimators for pairwise relatedness and individual inbreeding coefficients. Genet Res 67: 175–185.

Ross-Ibarra J, Wright SI, Foxe JP, Kawabe A, DeRose-Wilson L, Gos G, Charlesworth D, Gaut BS (2008) Patterns of polymorphism and demographic history in natural populations of *Arabidopsis lyrata*. PLoS ONE 3: e2411.

Sabeti PC (2002) Detecting recent positive selection in the human genome from haplotype structure. Nature 419: 832–837.

Savolainen O, Wright M (2004) Estimating divergence rates of conifers based on EST sequences conifer EST sequences. In: Population, Evolutionary and Ecological Genomics of Forest Trees. IUFRO Sections Population Genetics and Genomics, 13–17 Sept, Pacific Grove, CA, USA, p 7.

Savolainen O, Pyhäjärvi T (2007) Genomic diversity in forest trees. Curr Opin Plant Biol 10: 162–167.

Savolainen O, Pyhäjärvi T, Knürr T (2007) Gene flow and local adaptation in forest trees. Annu Rev Ecol Evol Syst 37: 595–619.

Schlötterer C (2003) Hitchhiking mapping—functional genomics from the population genetics perspective. Trends Genet 19: 32–38.

Schmid KJ, Ramos-Onsins S, Ringys-Beckstein H, Weisshaar B, Mitchell-Olds T (2005) A multilocus sequence survey in *Arabidopsis thaliana* reveals a genome-wide departure from a neutral model of DNA sequence polymorphism. Genetics 169: 1601–1615.

Simonsen KL, Churchill GA, Aquadro CF (1995) Properties of statistical tests of neutrality for DNA polymorphism data. Genetics 141: 413–429.

Slatkin M, Wiehe T (1998) Genetic hitchhiking in a subdivided population. Genet Res 71: 155–160.

Song BH, Windsor AJ, Schmid KJ, Ramos-Onsins S, Schranz ME, Heidel AJ, Mitchell-Olds T (2009) Multilocus patterns of nucleotide diversity, population structure and linkage disequilibrium in *Boechera stricta*, a wild relative of *Arabidopsis*. Genetics 181: 1021–1033.

Spielman RS, McGinnis RE, Ewens WJ (1993) Transmission test for linkage disequilibrium: The insulin gene region and insulin-dependent diabetes mellitus (iddm). Am J Hum Genet 52: 506–516.

Storz JF (2005) Using genome scans of DNA polymorphism to infer adaptive population divergence. Mol Ecol 14: 671–688.

Tajima F (1989) Statistical method for testing the neutral mutation hypothesis by DNA polymorphism. Genetics 123: 585–595.

Tenaillon MI, Tiffin PL (2008) The quest of adaptive evolution: a theoretical challenge in a maze of data. Curr Opin Plant Biol 11: 110–115.

Tenaillon MI, Sawkins MC, Long AD, Gaut RL, Doebley JF, Gaut BS (2001) Patterns of DNA sequence polymorphism along chromosome 1 of maize (*Zea mays* ssp. *mays* L.). Proc Natl Acad Sci USA 98: 9161–9166.

Teshima KM, Coop G, Przeworski M (2006) How reliable are empirical genome scans for selective sweeps? Genome Res 16: 702–712.

The Wellcome Trust Case Control Consortium (2007) Genome-wide association study of 14,000 cases of seven common diseases and 3,000 shared controls. Nature 447: 661–678.

Thornsberry JM, Goodman MM, Doebley J, Kresovich S, Nielsen D, Buckler ES IV (2001) Dwarf8 polymorphisms associate with variation in flowering time. Nat Genet 28: 286–289.

Thumma BR, Nolan MR, Evans R, Moran GF (2005) Polymorphisms in Cinnamoyl CoA Reductase (CCR) are associated with variation in microfibril angle in Eucalyptus spp. Genetics 171: 1257–1265.

Tsumura Y, Matsumoto A, Tani N, Ujino-Ihara T, Kado T, Iwata H, Uchida K (2007a) Genetic diversity and the genetic structure of natural populations of *Chamaecyparis obtusa*: implications for management and conservation. Heredity 99: 161–172.

Tsumura Y, Kado T, Takahashi T, Tani N, Ujino-Ihara T, Iwata H (2007b) Genome-scan to detect genetic structure and adaptive genes of natural populations of *Cryptomeria japonica*. Genetics 176: 2393–2403.

Varshney RK, Beier U, Khlestkina EK, Kota R, Korzun V, Graner A, Börner A (2007) Single nucleotide polymorphisms in rye (*Secale cereale* L.): discovery, frequency, and applications for genome mapping and diversity studies. Theor Appl Genet 114: 1105–1116.

Vitalis R, Dawson K, Boursot P (2001) Interpretation of variation across marker loci as evidence of selection. Genetics 158: 1811–1823.

Voight BF, Kudaravalli S, Wen X, Pritchard JK (2006) A map of recent positive selection in the human genome. PLoS Biol 4: e72.

Wachowiak W, Balk PA, Savolainen O (2009) Search for nucleotide diversity patterns of local adaptation in dehydrins and other cold-related candidate genes in Scots pine (*Pinus sylvestris* L.). Tree Genet Genomes 5: 117–132.

Wakeley J (1999) Non-equilibrium migration in human history. Genetics 153: 1836–1871.

Wang XR, Szmidt AE, Lindgren D (1991) Allozyme differentiation among populations of *Pinus sylvestris* (L.) from Sweden and China. Hereditas 114: 219–226.

Weedon MN, Lango H, Lindgren CM, Wallace C, Evans DM, Mangino M, Freathy RM, Perry JR, Stevens S, Hall AS, Samani NJ, Shields B, Prokopenko I, Farrall M, Dominiczak A; Diabetes Genetics Initiative; Wellcome Trust Case Control Consortium, Johnson T, Bergmann S, Beckmann JS, Vollenweider P, Waterworth DM, Mooser V, Palmer CN, Morris AD, Ouwehand WH; Cambridge GEM Consortium, Zhao JH, Li S, Loos RJ, Barroso I, Deloukas P, Sandhu MS, Wheeler E, Soranzo N, Inouye M, Wareham NJ, Caulfield M, Munroe PB, Hattersley AT, McCarthy MI, Frayling TM (2008) Genome-wide association analysis identifies 20 loci that influence adult height. Nat Genet 40: 575–583.

Wegrzyn JL, Lee JM, Liechty J, Neale DB (2009) PineSAP-sequence alignment and SNP identification pipeline. Bioinformatics 25: 2609–2610.

Willyard A, Syring J, Gernandt DS, Liston A, Cronn R (2007) Fossil calibration of molecular divergence infers a moderate mutation rate and recent radiations for Pinus. Mol Biol Evol 24: 90–101.

Wright SI, Charlesworth B (2004) The HKA test revisited: a maximum-likelihood-ratio test of the standard neutral model. Genetics 168: 1071–1076.

Wright SI, Gaut B (2005) Molecular population genetics and the search for adaptive evolution in plants. Mol Biol Evol 22: 506–519.

Wright SI, Lauga B, Charlesworth D (2003) Subdivision and haplotype structure in natural populations of *Arabidopsis lyrata*. Mol Ecol 12: 1247–1263.

Wright SI, Bi IV, Schroeder SG, Yamasaki M, Doebley JF, McMullen MD, Gaut BS (2005) The effects of artificial selection on the maize genome. Science 308: 1310–1314.

Wright SI, Foxe JP, DeRose-Wilson L, Kawabe A, Looseley M, Gaut BS, Charlesworth D (2006) Testing for effects of recombination rate on nucleotide diversity in natural populations of *Arabidopsis lyrata*. Genetics 174: 1421–1430.

Xiong MM, Krushkal J, Boerwinkle E (1998) TDT statistics for mapping quantitative trait loci. Ann Hum Genet 62: 431–452.

Yang Z, Nielsen R (2000) Estimating synonymous and nonsynonymous substitution rates under realistic evolutionary models. Mol Biol Evol 17: 32–43.

Yu J, Pressoir G, Briggs WH, Vroh Bi I, Yamasaki M, Doebley JF, McMullen MD, Gaut BS, Nielsen DM, Holland JB, Kresovich S, Buckler ES IV (2006) A unified mixed model method for association mapping that accounts for multiple levels of relatedness. Nat Genet 38: 203–208.

Zeng K, Shi S, Wu CI (2007) Compound tests for the detection of hitchhiking under positive selection. Mol Biol Evol 24: 1898–1908.

Zhang K, Qin Z, Chen T, Liu JS, Waterman MS, Sun F (2005) HapBlock: Haplotype block partitioning and tag snp selection software using a set of dynamic programming algorithms. Bioinformatics 21: 131–134.

Zhao K, Aranzana MJ, Kim S, Lister C, Shindo C, Tang C, Toomajian C, Zheng H, Dean C, Marjoram P, Nordborg M (2007) An Arabidopsis example of association mapping in structured samples. PLoS Genet 3: e4.

Zhu Q, Zheng X, Luo J, Gaut BS, Ge S (2007) Multilocus analysis of nucleotide variation of *Oryza sativa* and its wild relatives: severe bottleneck during domestication of rice. Mol Biol Evol 24: 875–888.

Zhu YL, Song QJ, Hyten DL, Van Tassell CP, Matukumalli LK, Grimm DR, Hyatt SM, Fickus EW, Young ND, Cregan PB (2003) Single-nucleotide polymorphisms in soybean. Genetics 163: 1123–1134.

[1]Department of Forest Ecology and Genetics, Center of Forest Research, INIA, 28040 Madrid, Spain;
[a]e-mail: *santiago@inia.es*
[b]e-mail: *alia@inia.es*
[c]e-mail: *concettaburgarella@hotmail.com*
[d]e-mail: *dgrivet@inia.es*
[2]CSIRO Plant Industry, GPO Box 1600, Canberra, ACT 2601, Australia;
e-mail: *shannon.dillon@csiro.au*
[3]INRA, UMR1202 Biodiversity Genes & Communities, 69 route d'Arcachon, 33612 Cestas Cedex, France; e-mail: *pauline@pierroton.inra.fr*
[4]Department of Ecosystem Science & Management, Texas A&M University, College Station TX77843-2138, USA; e-mail: *k-krutovsky@tamu.edu*
[5]Section of Evolution and Ecology and Center for Population Biology, University of California at Davis, Davis, CA 95616, USA; e-mail: *ajeckert@ucdavis.edu*
[6]Umeå Plant Science Center, Swedish University of Agricultural Science, SE 901 83 Umeå, Sweden; e-mail: *m.rosario.garcia@genfys.slu.se*
[7]Université Libre de Bruxelles, Faculté des Sciences, Behavioural and Evolutionary Ecology cp160/12, av. F.D. Roosevelt 50, 1050 Brussels, Belgium; e-mail: *mheuertz@ulb.ac.be*
[8]Department of Evolutionary Ecology, Ecology Institute, Universidad Nacional Autónoma de México, Ciudad Universitaria, Tercer circuito Exterior, Apartado Postal 70-275, México, D.F.; e-mail: *jaramillo@miranda.ecologia.unam.mx*
[9]Program in Evolutionary Functional Genetics, Evolutionary Biology Centre, Uppsala University, 75326 Uppsala, Sweden; e-mail: *Martin.Lascoux@ebc.uu.se*
[10]Department of Plant Sciences, University of California at Davis, Davis, CA 95616, USA; e-mail: *dbneale@ucdavis.edu*
[11]Institute of Forest Genetics, Pacific Southwest Research Station, U.S. Department of Agriculture Forest Service, Placerville, CA 95667, USA.
[12]Department of Biology, University of Oulu, 90014 Oulu, Finland;
e-mail: *outi.savolainen@oulu.fi*
[13]Department of Forest Genetics, Forestry and Forest Products Research Institute, Tsukuba, Ibaraki 305-8687, Japan; e-mail: *ytsumu@ffpri.affrc.go.jp*
[14]Consiglio Nazionale delle Ricerche, Istituto di Genetica Vegetale, Via Madonna del Piano 10, 50019 Sesto Fiorentino (Firenze), Italy; e-mail: *giovanni.vendramin@igv.cnr.it*
[†]These four authors have contributed equally to this paper.
Authors in alphabetical order, except leading authors.
[*]Corresponding author

7

Integration of Molecular Markers in Breeding

Rowland D. Burdon[1,a] and *Phillip L. Wilcox*[1,b,*]

ABSTRACT

Applications of genetic markers in conifer breeding fall mostly in two main areas: population management and selection.

Conifers pose problems, for selection in particular, on account of their size and typical generation intervals, their typically outbreeding behavior, the essentially wild state of their genetic systems, and their huge genomes.

Applications for population management are various. For obtaining background information, they include studying breeding systems, and characterizing populations among and within taxa, doing the latter wherever possible in conjunction with common-garden field trials. More specifically, use of markers (1) can be used to inform structuring of breeders' metapopulations, in terms of sizes and subdivisions, and (2) offers the prospect of retaining the benefits of full pedigree without having to incur the costs and effort of exhaustive controlled crossing. Full practical application, however, is being widely achieved in using markers to verify genotype and parentage, thereby safeguarding the capture of genetic gain.

Use of markers for selection in conifers is in principle attractive, because of the time frames of conventional conifer breeding and the problems of obtaining good phenotypic expression. Yet developing marker-trait associations for marker-assisted selection (MAS) is greatly hampered by the same features, plus extremely limited population-wide linkage disequilibrium and an evident paucity of major genes. Scope for selection using linkage between quantitative trait loci (QTL) and

[1]Scion: New Zealand Forest Research Institute Ltd;
[a]e-mail: *rowland.burdon@scionresearch.com*
[b]e-mail: *phillip.wilcox@scionresearch.com*
*Corresponding author

neutral markers is therefore very limited, pointing to a need to identify DNA polymorphisms of direct functional significance for selection, i.e., using gene-assisted selection (GAS). For GAS, gene discovery is indicated, using a combination of association genetics and candidate genes, employing as needed "-omics" technologies to track pathways to phenotypic expression. Association genetics will remain hampered by paucity of major genes, and gene discovery hampered by the size of conifer genomes. Using candidate genes, however, should be helped by widespread gene homologies, or even orthologies, in the plant kingdom, and remarkably close synteny among various conifer species. Breeding for disease resistance is seen as having special potential for using DNA polymorphisms for selection.

Joint use of DNA markers for population management and selection will depend not only on inherent suitability of marker types for different purposes but also on historical factors, such as existing development of makers and genotyping already done. Nonetheless there is potential to combine objectives to reduce genotyping costs.

Brief comparisons are drawn with selected marker-assisted selection scenarios with some other agronomically important organisms.

Applications of genetic markers pose not only technical problems, but also institutional ones, relating to distinct cultures and allocation of resources between conventional breeding and molecular biology disciplines. Yet successful application will depend on the infrastructure of conventional breeding.

Keywords: Marker-assisted selection, Gene-assisted selection, genomic selection, DNA markers, population management, pedigree reconstruction, single nucleotide polymorphisms, microsatellites

7.1 Introduction

The practical applications of molecular markers lie essentially in two broad areas (cf Burdon and Wilcox 2007):

- Providing knowledge whereby the breeder can structure the metapopulation as appropriately as possible. This structuring includes choice and representation of base populations, the stratification of the metapopulation and the appropriate sizes of the population strata, and the crossing schemes.
- Providing knowledge to better target and accelerate and/or enhance genetic gain. Applications involving faster and more efficient selection are crucial here. However, the same knowledge can also be important in eventually using genetic engineering to supplement the breeder's efforts in order to produce material for commercial deployment.

Successful use of genetic markers in these areas depends not only on obtaining the knowledge, but integrating the research and resulting knowledge with the breeding operations. To address this, we consider the following:

- The distinctive features of conifers in this connection
- Overview of properties of different classes of markers
- Population-management applications:
  - Taxonomy and base-population studies
  - Characterizing breeding systems
  - Genetic fingerprinting and applications
  - Pedigree reconstruction and implications
  - Role in estimating genetic parameters
  - Setting population sizes, and
  - Continued monitoring of genetic diversity of metapopulations.
- Selection-oriented applications:
  - Detection and characterizing chromosome regions governing quantitative variation (quantitative trait loci, or QTL), for selection. We also discuss the use of markers for integrating oligogenic disease resistance, and the use of genome-wide scans for selection.
  - Identifying specific genes or other DNA sequences governing quantitative variation (quantitative trait nucleotides, or QTN), for selection. Specific applications considered here are gene-assisted selection (GAS) and marker-assisted recovery of genotype (MARG).
  - Reviewing existing selection applications in other agronomically important species.
- Integrating population-management and selection issues
- Prospective roles of epigenetic markers
- Institutional challenges
- Concluding remarks.

7.2 Distinctive Features of Conifers for Use of Genetic Markers

It is appropriate to recapitulate on several features of conifers that influence the scope for use of genetic markers and ease of using them: These features are reviewed, or at least strongly implied in this book by Gernandt et al. (2011); Mullin et al. (2011); Bagnoli et al. 2011; Ritland et al. 2011; González-Martínez et al. (2011); and Dauwe et al. (2011).

- Their economic importance
- Almost all strictly diploid
- Genetic systems generally in essentially wild state

- Wind pollination and breeding systems
- Huge genomes
- High levels of synteny and colinearity
- The haploid megagametophyte
- Phenotypic information generally slow and costly to obtain
- Evident scarcity of large-effect alleles (major genes), and
- Interaction of these factors with general problems of using genetic markers.

7.2.1 Economic Importance

The economic importance of conifers is enormous (Mullin et al. 2011). Indeed, it is out of all proportion to the number of species—around 650 (Gernandt et al. 2011), with many of them extremely localized. This reflects the areas that conifers occupy and their typical status as keystone species of ecosystems. Many conifers, especially pines, are not very demanding of soil fertility, and produce wood that is suited to a wide range of high-volume processes and end-uses. Because of these factors, together with pressures on high-quality land for producing food and even certain hardwoods, conifers will surely remain of great commercial importance. The economic importance has already driven intensive breeding programs, and will continue to do so. This, in turn, is conducive to use of highly sophisticated breeding technology.

7.2.2 Ploidy

Almost all conifers that are appreciably domesticated are strictly diploid (Williams 2009; Gernandt et al. 2011), with a haploid chromosome complement of around 12. This feature facilitates genomic mapping and the detection of quantitative trait loci (QTL) and quantitative trait nucleotides (QTN), which will be considered later. Indeed, even minor departures from perfect diploidy, such as trisomics, typically have catastrophic effects on field fitness (Mergen 1963). The classic exception is the hexaploid coast redwood (*Sequoia sempervirens* [D. Don] Endl.) which, while economically significant, is not subject to large-scale domestication, let alone to intensive breeding. Another, less-known exception is the tetraploid *Fitzroya cupressoides* (Molina) I.M. Johnstone which, while it has also been commercially significant, appears to have almost no prospects for significant domestication. The Podocarpaceae show more variable chromosome numbers than other conifer families (Gernandt et al. 2011), but they are essentially undomesticated.

7.2.3 State of Genetic Systems

As forest trees, conifers almost all have their genetic systems still in essentially a wild state. They have had only very short histories of intensive artificial breeding (White et al. 2007; Genrandt et al. 2011). This contrasts with the long history of artificial breeding, deliberate or otherwise, that has characterized highly domesticated crop plants, often converting those plants to inbreeding systems. There is good news in that the natural genetic diversity is basically intact. There is bad news, however, in very little of the linkage disequilibrium (LD) whereby selectively neutral marker alleles can be consistently associated with certain QTLs, so LD between neutral markers and the alleles that govern functional variation is confined almost entirely to families. Therefore, any population-wide association between an allele in a DNA polymorphism and a quantitative trait is likely to reflect a QTN, unlike the associations that can easily arise just within families. However, finding associations can be more like the proverbial hunt for a needle in a haystack.

7.2.4 Wind Pollination and Breeding Systems

Being wind-pollinated, conifers have a greater potential for medium- to long-distance pollen contamination (Williams 2009), which can complicate pedigree reconstruction. The typical outbreeding, in conjunction with wind pollination and the almost complete lack of historical domestication, accounts well for the almost complete lack of population-wide linkage disequilibrium (González-Martínez et al. 2011).

7.2.5 The Size of Conifer Genomes

Conifers have huge genomes, ~1-4 × 10^{10} base pairs, (Ahuja 2001; Ahuja and Neale 2005; Gernandt et al. 2011; Morgante and De Paoli 2011), despite being almost entirely diploid. This size reflects an extremely high proportion of non-coding DNA. Why this proportion is so much higher than in most angiosperms is a mystery, but recent findings (e.g., Mattick 2004) increasingly argue against the "junk-DNA" interpretation.

Despite the genome sizes, which are reflected in visibly large chromosomes, centiMorgan map lengths of conifers (see Plomion et al. 2007 and references therein; Ritland et al. 2011) appear not to be correspondingly greater than those of diploid angiosperm species (e.g., Yin et al. 2004; Brondani et al. 2006).

Anyway, the sheer size of conifer genomes has to date made the task of complete genomic sequencing prohibitive. Even with much higher-throughput technology the task of sequencing and interpreting the sequence data will remain very challenging. Along with no general LD, the genome size creates a major obstacle to direct gene discovery, given the limitations of foreseeable bioinformatics searches. Meanwhile, we will be forced to leverage genomic information from angiosperm species, with heavy use of candidate genes for gene discovery.

The very high proportion of non-coding DNA might be expected to favor finding high-quality selectively-neutral markers. However, reservations exist concerning neutrality or genetic stability of the markers. There are sometimes problems with null alleles and hypermutability at some loci. Moreover, additional problems may well arise with duplication of loci.

An important complication is that functional genetic diversity can arise in various ways. Classically, the study of functional genetic diversity has focused on coding-region polymorphisms, which involve amino-acid substitutions in protein chains. The pathways of expression for such polymorphisms can be followed by studying proteomics, involving activities of either enzyme variants or the impacts of variants of "structural" proteins. Yet it is becoming increasingly clear in biology that at least some functional variation can be governed by polymorphisms outside the coding regions (e.g., Morgante and Salamini 2003; Paran and Zamir 2003). These can certainly involve regulatory sequences, but the effects of such variants lack even the predictability of amino-acid substitutions. Moreover, study of regulatory-sequence polymorphisms can be complicated by many such sequences not even adjoining the coding regions, although the prevalence of such sequences remains to be quantified. And there is the still-obscure role of various RNA sequences that are transcribed from non-coding DNA but somehow help orchestrate developmental processes. The nature and importance in forest trees of functionally significant polymorphisms in such sequences is still conjectural, but such a polymorphism evidently governs a radical difference between the human and chimpanzee brains (Pollard et al. 2006).

It is unknown whether the very high proportion of non-coding DNA, which makes the conifer genomes so large, means a comparatively high proportion of functional variation being governed by polymorphisms in non-coding regions. This could have a significant impact on the exact nature of selection applications.

7.2.6 High Levels of Synteny and Colinearity

Despite apparently being comparatively susceptible to mutation (Gernandt et al. 2011), conifer genomes have shown great long-term stability (Bagnoli et al. 2011; Ritland et al. 2011). Chromosome configurations are very uniform among taxa that are not even very closely related (e.g., Krutovsky et al. 2004). Moreover, among conifer taxa, the same genes typically appear to perform the same roles, showing close synteny; and the locations of particular genes on specific chromosomes are very similar among species, representing high levels of colinearity. Furthermore, there are evidently widespread orthologies, extending to angiosperm species (Krutovsky et al. 2006). Because of these features, obtaining genomic information for individual conifer species should eventually allow a rapid spread of genomic knowledge among other species.

7.2.7 The Haploid Megagametophyte

Historically, without present-day DNA assays, the relatively large haploid megagametophyte allowed easy genomic mapping without recourse to codominant markers (e.g., Devey et al. 1995; White et al. 2007). This, however, is no longer the crucial advantage that it was, given that easily genotyped codominant markers, including microsatellites or simple sequence repeats (SSRs) and single nucleotide polymorphisms (SNPs), are now widely available. It does, however, facilitate studies of genetic architecture for individual genotypes in a range of contexts (e.g., Wilcox et al. 1996; Plomion et al. 2007 and references therein).

7.2.8 Delays and Costs of Obtaining Phenotypic Information

Any use of markers as an aid to selection, or even as a basis for it, depends on linking DNA polymorphisms to expression of phenotype in association genetics. Yet phenotypic expression for breeding-goal traits, and even various selection traits, is often unavoidably slow and costly to obtain (e.g., Wilcox et al. 2007), creating a major hurdle for such use of markers. In conifers, the delays are accentuated by several factors. Conifers lack the high early relative growth rates that allow the quick acquisition of reliable phenotypic information that is often possible with eucalypts, for example. Moreover, conifers tend to be grown on sites that do not favor super-fast growth, and are generally not grown on very short pulpwood rotations. For some traits, notably wood properties, even measuring phenotypes can be expensive. Use of proxies for economic traits is almost universal, given that the economic traits represent harvest-age product-performance traits, in contrast to the early-rotation tree phenotypes. Some proxies, such as laboratory determinations of wood chemistry or detailed wood structure, need not only show good juvenile-mature correlations of this sort, but also reflect properties without adverse side-effects on field fitness, which will be very slow to confirm.

In some instances, phenotypic data generation can be circumvented by using legacy records and derived measures such as breeding values. In association genetics research, this is particularly useful, not only to avoid additional cost and time delay, but also to build a repository of genotypic information regarding breeding-population genotypes that could be used for other purposes.

7.2.9 *Evident Scarcity of Large-Effect Alleles (Major Genes)*

The search for specific genetic factors, be they chromosomal regions or QTN, that definitely have quite large phenotypic effects (e.g., controlling ≥ 20% of phenotypic variance), has generally been disappointing (Wilcox et al. 1997). Indeed, well-quantified QTL effects have generally been much less (Ball 2001; Brown et al. 2004). This is despite finding such factors in a range of domesticated taxa (Dekkers and Hospital 2002). Without large-effect alleles, very large numbers of individuals are needed to detect loci where phenotypic effects are being governed and quantify those effects. Given the outlays and time lags for obtaining good phenotypic information, dealing with the requisite numbers of individuals tends to become extremely costly. This creates a special call for efficient partitioning of the environmental effects that contribute to phenotypes.

Possible exceptions to the dearth of major-gene effects in forest trees exist for disease resistance in some conifers (Devey et al. 1995; Wilcox et al. 1996), which we discuss later.

7.2.10 *Interactions among Features of Conifers*

Overall, the problems posed by large genomes, lack of linkage disequilibrium, apparent rarity of large-effect QTL or QTN, and costs and delays of obtaining good phenotypic information, are strongly accentuated by the co-occurrence of these features.

7.3 Overview of Different Marker Types

Table 7-1 lists various marker categories. In anticipation of the following material, it summarizes the key properties of such markers, and their suitability for different breeding purposes. The detailed list of marker categories has expanded during recent years (Jones et al. 2009), and will presumably continue to do so. Application of new types of marker to conifers, however, is likely to be subject to some time lags, as marker systems are typically developed primarily for smaller less complex angiosperm genomes.

Of note is that certain marker types can be evaluated using different assay platforms. For example, a wide range of assay technologies are available for single nucleotide polymorphisms (SNPs), and vary considerably in terms of cost. The general rule with SNP technologies is that

Table 7-1 Summary of properties and suitability of different marker types for different classes of application.

A. Properties	Marker System[1]						
	Isozymes	RFLP	AFLP & RAPD	SSR	SNP	Indels	DArT
Dominant (D) or codominant (C)	C	C	D	C	C	C	D
Single (S) or multiple (M) loci	S	S	M	S	S	S	M
Number of loci available	<30	10-200	1000+	10-300	1,000+	1,000+	1,000+
Number of alleles	Mostly low	Low	Biallelic	High	Usually 2	Biallelic	Biallelic
Prevalence of null alleles	Some	Zero?	N/A[1]	Significant	Low	N/A	N/A
Mutability	V. low	V. low	Variable	Variable	V. low	Low	Low
Selective significance (0 = neutral)	>0	Generally low	Generally low	Generally low or 0	Variable	Variable	Generally low or 0
Prior sequence data needed?	No	No	No	Yes	Yes	Yes	No
Assay costs (per locus)	High	High	Low	High	Variable – technology-dependent	Low-moderate	Low-moderate
Potential for spurious associations	Low	Low	Moderate	Low	Low	Low	Low
Amenability to high-throughput assays	Low	Low	Low	Moderate	High	Moderate-high	High
Rate of technology improvement	Almost none	Almost none	Low	Low	High	Moderate	Moderate
Development costs	Moderate	Moderate-high	Low-moderate	High	Moderate	Moderate	Low-moderate
B. Applications (suitability[2])							
Characterizing breeding system	••••	••	••	••••	•••	••	••
Taxonomic studies	•••	••	••	••••	•••	••	••
Clonal/parental verification	•••	••	•••••	•••••	••••	••	••••
Pedigree reconstruction	••	••	••	•••••	•••	••	•••
Selection	•	•	•	••	•••••	••••	•••

[1]Acronyms: RFLP = restriction fragment length polymorphisms, AFLP = amplified fragment length polymorphisms, RAPD = random amplified polymorphic DNA, SSR = simple sequence repeat (also referred to as 'microsatellite'), SNP = single nucleotide polymorphism, Indels= sequence insertion/deletion, DArT = diversity array.

[2]Taking into account cost-efficiency of information, i.e., value in relation to cost of obtaining it.

N/A denotes not applicable.

overall experimental cost is usually inversely proportional to per-data-point costs. Technologies such as the Illumina Infinium platform (Gunderson et al. 2005; Ritchie et al. 2009) involve tens to hundreds of thousands of loci (cf Barbazuk et al. 2007; Edwards et al. 2008; Pindo et al. 2008), and are very cheap on a per-data-point basis, but are limited in that the platform cannot be cost-effectively down-scaled to few loci, necessitating other technologies that are more expensive on a per- data-point basis. Currently, most conifer experiments have to date involved hundreds to a few thousands of SNP markers, thus technologies such as Illimina Gold Gate assays, ABI TaqMan, and Sequenome iPlex platforms (see Ragoussis 2006 and references therein; Dean 2011) are typically more suited to the smaller-scale applications typical in forestry-related research. However, the scale will most likely increase over time.

The rate of technology development for the various marker systems differs considerably (Table 7-1), which impacts choice of marker system. Older technologies such as isozymes and restriction fragment length polymorphisms (RFLPs) are virtually obsolete. SNP technologies have been subject to considerable development in the recent five years, resulting in a variety of SNP platforms (Ragoussis 2006). In comparison, little development of SSR technologies has been undertaken in the same period. Therefore, despite being less informative on a per-locus basis, SNPs could replace SSRs in the medium term for applications such as pedigree reconstruction and parentage analysis. Thus using a single technology that is broadly applicable may be preferable to a mix of marker technologies that serve each single application optimally. In the longer term, DNA sequencing is likely to be increasingly used for genotyping, as sequencing technologies improve in throughput and reduce in cost.

7.4 Population-Management Applications

The potential impacts of advances in genomics on population management for forest trees in general have been reviewed by Burdon and Wilcox (2007). Here, the focus is more specifically on conifers, with some updating in the light of subsequent publications.

7.4.1 *Taxonomy and Population Structures*

The breeder can operate with much greater confidence from good information on the taxonomy of populations and the nature of taxonomic and intra-taxon differentiation among populations. Classical, herbarium-based taxonomy has long been supplemented in forest trees with common-garden experiments which, while often slow and expensive, can give crucial information on the functional genetic variation involving the traits that interest the breeder. Genetic markers are now a powerful adjunct, which can shed much additional light on species hierarchies (Gernandt et al. 2005) and population differences (Burdon and Wilcox 2007; Bagnoli et al. 2011; González-Martínez et al. 2011 and references therein). Often involving selectively neutral alleles, and purely cryptic variation, they need to be used warily, in conjunction with other measures of variation. However, they can be a powerful tool for inferring the evolutionary history of species and populations thereof. Genetic divergence revealed by markers, for instance, can give pointers to the prospects of achieving hybrid vigor, or heterosis, in various interpopulation crosses; this application is called Diversity Index Breeding (White et al. 2007). For this purpose, neutral markers may give an early guide. Moreover, when such information is to be used, neutral-marker data may be the only available marker information to go on.

In principle, much can be inferred from DNA makers concerning the genetic history of populations and what DNA sequences have been subject to selection (Gernandt et al. 2011 and references therein; González-Martínez et al. 2011 and references therein). Aspects of the genetic history include the timing and sizes of past population bottlenecks and population coalescences, which tend to be manifested in patterns of linkage disequlibrium. Influences of selection pressures will be manifested in various ways, such as the ratios of synonymous to non-synonymous polymorphisms and differences among chromosome regions in nature and levels of polymorphism. Such information, however, typically allows only very tentative inferences on population history, given the simultaneous uncertainties concerning the rates and nature of mutations. All this allows the breeder to make only slightly more educated guesses in various areas of population management. Also, evidence of selection having operated on certain DNA sequences is subject to the reservation that the selective forces operating in domesticated populations may be quite different from those operating in the wild. Furthermore, levels of conservation of DNA sequences, while often thought to be a guide to significance of the sequences for evolutionary fitness, can be unreliable in this respect (Monroe 2009).

7.4.2 Characterizing Breeding Systems

Various types of markers can be used for studying breeding systems, which are characterized by the levels and patterns of outcrossing. Isozymes serve the purpose well and for many years represented the only widely convenient marker technology for this purpose. Nowadays, other types of markers can be used (Table 7-1), and the choice may even depend on what marker type is preferred for some other purpose(s). The key benefits of markers, compared with phenotypic data, are that they not reliant upon common-garden studies, can provide data faster, and/or are applicable to natural populations *in situ*. On the other hand markers can be expensive and time consuming to develop—particularly SSR markers—or in some cases are not completely informative due to dominance (Table 7-1).

Breeding systems are of interest for several reasons. First, knowledge of a breeding system gives some indication of the potential and likely avenues for genetic improvement. Where a species is fully self-fertile, as in *Pinus resinosa*, within-population functional genetic diversity may be too small to warrant selective breeding (Fowler and Lester 1970; Mullin et al. 2011), although useful genetic variation may exist among populations. On the other hand, in western red cedar (*Thuja plicata* D. Don), which is largely inbreeding, there is appreciable tree-to-tree genetic variation (Russell et al. 2003) which, along with other factors (addressed later), has allowed a breeding program to proceed (Russell and Ferguson 2008; Mullin et al. 2011).

Second, knowledge of the breeding system is important for studying heritability and testing of progenies. For instance, with a mixed mating system, with significant but variable rates of selfing, open-pollinated progenies can give seriously biased estimates of heritability (Borralho 1994) and parental breeding values. In these cases, effects of inbreeding depression, which typically does occur but with variable severity, can actually be much greater than the effects of inbreeding inflating coefficients of relationship. Indeed, open-pollinated progeny performance can even be poorly correlated in eucalypts with breeding-value estimates from controlled outcrossing (Hodge et al. 1996). While such situations will almost certainly be the exception in conifers, they cannot always be ruled out.

Third, care may have to be taken to minimize (e.g., by rigorous nursery culling) the contribution of self- (or other inbred) seedlings to offspring performance.

Fourth, strong outbreeding can lead to population-wide linkage disequilibrium (LD) being negligible, typically ≤ 2 kb in conifers (e.g., Brown et al. 2004; Wilcox et al. 2007; González-Martínez et al. 2011). This means that any method of QTL detection or gene discovery based on LD between marker loci and loci governing functional variation requires information specific to the various individual pedigrees.

Fifth, markers can differentiate nuclear and organellar inheritance. In most conifers, chloroplast DNA is paternally inherited (e.g., Neale and Sederoff 1989; Cato and Richardson 1996; Ahuja 2001), allowing direct evaluation of paternity and pollen flow. This has also facilitated paternity assignment in seed orchards, and associated development of paternity-specific markers (e.g., Cato and Richardson 1996).

7.4.3 Genetic Fingerprinting-Verifying Genetic Identity of Material

The earliest, simplest and most successful use of genetic markers has been verification of genetic identity of material. This involves both parents and offspring. Verifying identity of parent clones for breeding populations or seed orchards is a general issue (e.g., Bell et al. 2004; Kumar and Richardson 2005; Burdon and Wilcox 2007), and has been effectively addressed by numerous marker systems. Wind pollination, however, with its greater potential for pollen contamination, makes identity of offspring parentage more of an issue than with insect pollination. Various types of markers can be used in this area (Table 7-1), but SSRs have to date tended to be favored because they are both codominant and multiallelic.

7.4.4 Pedigree Reconstruction

Maintaining full pedigree is preferred in tree breeding programs (Burdon and Wilcox 2007), if it is practical. It has been viewed as allowing maximum control of coancestry, thus minimizing both inbreeding and loss of genetic diversity. However, it entails the effort and logistics of 100% controlled pair-crossing, which can be extremely difficult and is almost always expensive. Indeed, the costs and difficulty can make such crossing counterproductive in terms of maintaining genetic diversity, because they can severely limit the effective size of fully pedigreed populations (Burdon 1997; Burdon and Wilcox 2007). Therefore, the attractions of being able to reconstruct pedigrees, after avoiding expensive and time-consuming controlled crossing, are great. Modern DNA marker technology makes this essentially feasible.

There are several possible applications:

1. Inferring pollen parentage after polycrossing;
2. Dispensing totally with controlled pollination, in favor of using open-pollinated families;
3. Reconstructing pedigrees in totally unpedigreed commercial plantings, e.g., in seedlots from open-pollinated seed orchards;
4. Identifying pollen contaminants in unpedigreed gene-resource plantings.

These are discussed below.

1. Inferring pollen parentage after polycrossing was explicitly proposed and explored by Lambeth et al. (2001). This has since been pursued by Kumar et al. (2006, 2007) for *Pinus radiata* D. Don. In principle, it is not very difficult. While the maternal parent is nominally known in each case, maternal identity can be readily verified via means such as genotyping haploid megagametophytes. Provision for unequal representation of pollen parents in the seed produced will, however, require a safety margin in increased family size. Unless pollen mixes are very large, pollen-contaminants should also be readily identifiable.

2. Dispensing totally with controlled pollination can be effectively forced upon the breeder in species that are simply not amenable to routine controlled pollination, e.g., various eucalypts. In such cases, insect pollination is likely to make pedigree reconstruction easier, since it will tend to reduce the effective pool of potential pollen parents. This has been addressed by Gea et al. (2007) for *Eucalyptus nitens* (H.Deane et Maiden) Maiden (cf Grattapaglia 2007), and it has been vigorously advocated, as "breeding without breeding" (BWB) for appropriate cases by El-Kassaby and Lstibůrek (2009). Wang et al. (2010) have simulated genetic gains from BWB for a Scots pine (*Pinus sylvestris* L.) scenario in Sweden, after initial seed-orchard establishment. They compared expected genetic gains in forwards-selected seed orchards from BWB, in relation to assumed inputs ("man-years"), compared with expected gains from controlled pair-crossing and forwards selection using field family-plus-individual progeny-test data. BWB scenarios, which featured phenotypic selection in open-pollinated progeny trials, revised on the basis of reconstructed pedigree information, gave much earlier and greater expected gains per unit of input.

Indeed, what amounts to BWB is being adopted for part of the breeding population in the recently revised breeding strategy for the New Zealand Radiata Pine Breeding Company (Dungey et al. 2009). With a wind-pollinated species, however, a more powerful set of genetic markers is likely to be needed, i.e., many more marker loci (preferably SSR); even then, some individuals are likely to be identified as having unknown, contaminant pollen parentage.

3. Recourse to selection in totally unpedigreed plantings, such as from seed orchards, has been proposed as a possible response to a biotic crisis whereby acceptable genetic resistance to a new pest or pathogen is so rare as to make an existing breeding population inadequate (Burdon 1997; Burdon and Wilcox 2007). This is contrary to a past notion, based largely on a lack of technology to reconstruct pedigrees, that such plantings are a genetic dead-end for breeding. Where vegetative multiplication has been practiced there is a potential bonus of being able to identify some candidate selections as members of single seedling clones, which could afford much stronger evidence of true genotypic superiority. On the other hand, being able to infer unrelatedness among resistant selections would help to minimize the narrowing of the genetic base that would occur. If, however, some seed-orchard clones are no longer extant, and have not been genotyped, their offspring will not be identifiable. Moreover, this general application is also liable to point to unknown, contaminant pollen parentage, because resistance inherited from any orchard clone should not be truly rare.

4. Identifying pollen contaminants in unpedigreed gene resources has been reviewed as an issue by Burdon and Kumar (2002). Such pollen contamination can severely compromise the value of such resources for their intended purpose. This is especially likely when some selection of parents is done for commercial acceptability; while this is not a purist approach, the loss of genetic diversity through the selection per se is likely to be greatly outweighed by the consequent genetic improvement and the reduced opportunity costs of maintaining such material. For managing such resources, it could suffice to check candidate seed parents at the end of each generation for contaminant status, which both reduces the number of individuals to be checked and allows more time for the marker technology to advance. Again, this is likely to be a more important issue with wind-pollinated species than with insect-pollinated ones.

7.4.4.1 Choice of Marker Systems

For these purposes, selectively neutral markers suffice in principle, especially if they are codominant, multi-allelic and cheap to assay. If neutral however, occasional markers may show excessive duplication, deletions, or general hypermutability.

For purposes 1–3, high levels of within-population polymorphism are clearly preferable. Typically, SSR markers would tend to be favored, with the caveats mentioned above. However, for purpose 4, where the breeder wants to avoid cross-contamination among populations, strong differentiation of populations is needed, preferably in the form of private or semi-private alleles. Opinion is evidently moving in favor of SNPs rather than SSRs (BS Weir pers. comm.). Large numbers of SNPs are being generated via large-scale genome (re)sequencing for other purposes, in addition to large-scale genotyping platforms, effectively reducing SNP development cost and time relative to that of SSRs.

7.4.5 Estimating Genetic Parameters

Estimating genotypic variances and heritabilities from family information depends on knowledge of coefficients of relationship/co-ancestry. For instance, true half-sib families, produced by totally random mating in a very large population, have an additive genetic coefficient of relationship of 0.25. For random full-sib families the coefficient is 0.5, subject to greater contamination with non-additive gene effects. Estimation of the genetic parameters from open-pollinated families, however, can be much quicker and cheaper than from controlled crosses, if the mating system and coefficient of relationship are known. Using markers, it is possible to infer the coefficient of relationship (e.g., Kumar and Richardson 2005). Not only that, departures from random mating can be partitioned into components of inbreeding and of effects of finite effective numbers of unrelated pollen parents. The latter need to be major before serious bias is expected (Namkoong 1966). Appreciable inbreeding, however, can cause serious bias if there is inbreeding depression; levels of inbreeding can be family-specific (Williams and Savolainen 1996), as can susceptibility to inbreeding depression (Kumar 2004).

Methods for estimating co-ancestry parameters (e.g., Weir 1996) have been widely used for many years, from the heyday of isozymes. Fortunately, open-pollinated conifer families often approximate well to half-sibs, but it is desirable to be able to confirm this for the material concerned. However, good estimates of co-ancestry parameters require adequate sets of markers. In principle, one can estimate coefficients of relationship for individual families for improved estimates of the genetic parameters, and especially parental values, but good estimates of the coefficients evidently require quite a powerful set of genetic markers; Kumar and Richardson (2005), using eight SSR markers, obtained a poor correlation between pedigree-based and marker-based measures of coefficients of relationship.

In principle, this use of markers can be applied to estimating genetic parameters in natural populations without common-garden testing. If the set of markers is good enough, family structures and coefficients of relationship could be inferred with no prior knowledge of genetic relationships. Moreover, spatial analysis (White et al. 2007 and references therein) can be used to substitute for block layouts in order to partition off major environmental effects. Such methods have the potential benefit of allowing one to obtain information from existing forests (e.g., natural forests or commercial plantations), thereby obviating the need for genetic testing involving progenies, thus speeding up the generation of information, particularly for species where little a priori information is available. For example, Andrew et al. (2005) reported estimates of heritability for some foliar defence-related chemicals in *Eucalyptus melliodora* A.Cunn., some of which were consistent with independent estimates based on phenotype. The conditions for reliability of such estimates, however, remain to be ascertained. Simulation studies may be useful in determining the requisite number and characteristics of markers, and where and how spatial analysis of phenotypes can help.

7.4.6 Setting Population Sizes

Appropriate sizes of breeding populations and/or of underpinning gene resources, in terms of numbers of parents and family sizes, is an important issue. Subject to costs and resources, appropriate choices will depend on the nature and frequencies of alleles of interest, which can be very uncertain. If there is truly polygenic control, with large numbers of loci involved, and largely intermediate allele frequencies, quite small populations (30 parents or even less) can be expected to allow near-optimal long-term genetic gain (e.g., Rosvall et al. 1998). Indeed, there have been selection experiments with annual plants that have given continued response to selection over many generations (e.g., Dudley 1977; Hill 2005). Where, however, there are valuable or potentially valuable alleles that are of large effects and rare, the appropriate population sizes can be much greater, the probabilities of retaining alleles at give frequencies for given population sizes having been modeled by Yanchuk (2001). Yet just retaining such alleles may not suffice, in that the breeder would want to capture them in a number of independent pedigrees (Burdon 1997). If that cannot be achieved in naturally occurring material, there may well be a call for incorporating or multiplying such alleles by genetic engineering, which would likely be used in commercially deployed clonal material, i.e., in clonal forestry (Burdon and Lstibůrek 2010).

Detecting QTL or SNP alleles of large effect and ascertaining their frequencies, is not only relevant for appropriate population sizes. Indeed, it is a crucial application of DNA sequences for selection, which is reviewed later.

7.4.7 Continued Monitoring of Genetic Diversity of Metapopulations

In the context of a breeding program, metapopulation diversity represents the diversity contained in the assemblage of populations used by the breeder. As such, it represents the aggregate diversity among and within the population units. It is governed by the diversity of the founder material, loss of diversity as generations are turned over by the breeder, and any countervailing accretion of diversity by mutation (sensu lato), breaking of tight linkages, or infusion of fresh germplasm. The diversity of the founder material has itself among- and within-population components, the former reflecting both variation among founder populations and the proportions in which they are represented. Moreover, as a breeding program progresses, variation among population sub-units can be imposed by the breeder.

Use of DNA markers to measure diversity is not straightforward, because real interest lies in functional diversity. As with characterizing founder material (earlier), DNA marker data typically serves only as a proxy for functional variation, and cross-calibration can be difficult, and even uninformative (e.g., Karhu et al. 1996 with regard to population differences), unless DNA sequences are identified that govern functional variation (Burdon and Wilcox 2007).

Where pedigree information is available, or can be effectively retrieved by use of markers, measures of effective population size can be used as indicators of decline in diversity. Most measures relate to size of an idealized population that would generate an equivalent rate of increase in level of inbreeding (Caballero 1994). However, a measure using explicit representation of individual founders in the ancestry of a population at a specified point in time is considered to reflect more appropriately the current breadth of genetic base. Such a measure is the Status number (N_s) (e.g., Lindgren et al. 1996, 1997) which is given by;

$$N_s = 1/\Sigma(p_i^2) = 1/2\Theta$$

where p_i is the total proportion of ancestry contributed to the population by the ith founder parent ($\Sigma p_i = 1$), typically calculated from pedigrees;

Θ = average coefficient of co-ancestry in population (including self co-ancestry; $\theta_i = 0.5$ for an outbred individual, *i*).

Where $N_s = N$ (census number), all individuals are fully outbred and unrelated, and insofar as $N_s < N$ there is some identity by descent. Hence N_s implies an infinite-alleles model, and has the property of declining very rapidly in a typical closed breeding population if the initial state is assumed to be $N_s = N$. In that many loci are not polymorphic, and there will be some accretion of genetic variation through mutation or "quasi-mutational events" (e.g., breakage of tight repulsion linkages, transposon insertions or tandem duplication of functional sequences), the rapid decline in N_s will give a very exaggerated picture of the decline in functional diversity. That, however, is not seen as negating the advantage of reflecting the expected contributions of the various population founders.

As mentioned earlier, explicit measures of diversity can be based on observed DNA polymorphism data, but the interpretation of such measures can be highly problematic. Within-population heterozygosity statistics, expected and observed (H_e and H_o respectively), can be calculated from the data, as can among-population diversity statistics (G_{st}) (Weir 1996). Such statistics can be calculated for individual loci or pooled across loci. However, species or even populations may vary in the degree to which their functional diversity is reflected in different types of DNA marker. Moreover, for small populations and limited numbers of loci population bottlenecks can mean that low-frequency alleles, if they are not lost, can remain in elevated frequencies, thus boosting *H* statistics.

Functional diversity is governed by various factors: number of polymorphic loci, magnitude of phenotypic effects of allelic differences at those loci, and allele frequencies since alleles make maximum contributions to diversity at intermediate frequencies. The information available to the breeder, however, is typically very incomplete, because very few conifer loci, let alone SNPs, have had quantitative phenotypic effects reliably attributed to them. This may change with association genetics, albeit rather slowly.

Moreover, possible accretion of genetic variation is still conjectural, and as such, it is often not addressed, even though much of the early theoretical work on population genetics assumed mutation rates in formulating the expected course of allele frequencies under selection (Falconer and Mackay 1996).

A general advantage of measurable diversity is indicated by the change of H_o during the life cycle; typically, it increases steadily from the stage of ungerminated seed through to a mature stand. Yet diversity is not an infallible measure of fitness. If the measure of diversity involves markers that are not selectively neutral, H_o may reflect the presence of genetic load, in the form of alleles of suboptimal fitness. This has been observed with isozyme loci (Strauss and Libby 1987; Bush and Smouse 1992).

Use of heterozygosity statistics from DNA data may, therefore, be just an ancillary tool for monitoring the time trajectory of genetic diversity in breeding populations. Losses of low-frequency alleles, which are simpler statistics, may often be as informative. For the most part, such DNA statistics seem likely to give clear indications of a decline in diversity only when the genetic base has been severely constricted. That is not a case for abandoning or ignoring such statistics, but rather for using them very cautiously.

7.4.8 Overview of Population-Management Applications

Use of DNA markers is having outstanding success in verifying clonal identity and parentage, thereby averting immediate loss of potential genetic gain, without major direct costs and opportunity costs. The benefits will become ever greater with time, as the genetic gains at issue increase. Markers have long been used to characterize breeding systems, and represent a powerful ancillary tool for characterizing population diversity. Use of genetic markers for reconstructing pedigrees would save much labor in conventional breeding. Controlled crossing between individuals, to retain pedigree and guarantee prescribed representation of parentage, is usually costly and laborious, and in some species is prohibitive. Pedigree reconstruction promises major cost savings through using open pollination without losing pedigree information. The cost savings can be immediate, meaning a big economic advantage, but challenges will remain in reducing the costs of genotyping and achieving reliable reconstruction, especially with wind-pollination. In this general area, there is thus already some effective integration of use of molecular markers into breeding operations.

7.5 Selection-Oriented Applications

Given the typical delays and costs of obtaining phenotypic expression in conifers, and the environmental "noise" effects that can largely mask the contribution of genotype to phenotype, the attractions of being able to use DNA markers for selection are great. Use of markers for selection depends on the development of association genetics, namely establishing associations between marker polymorphisms and polymorphisms governing functional variation. Association genetics can be applied in two ways for selection:

(1) Applying purely statistical associations between particular marker alleles and chromosome regions that govern desirable phenotypic effects. This is marker-assisted (or marker-aided) Selection, or marker-based selection if used without also using phenotypic on the candidates, the two cases being collectively termed MAS. The default assumption is that chromosomal linkage is involved between some essentially neutral marker(s) and desirable "alleles" of polymorphic chromosome segments. This does not, however, absolutely preclude the possibility of the marker allele representing the determinative sequence, e.g., if it is a single-nucleotide polymorphism (SNP) that represents a quantitative trait nucleotide (QTN).

(2) Selecting for DNA alleles (functional markers) that have been shown to govern desired phenotypic effects, either in isolation or in conjunction with phenotypic data on candidates. We refer to this as gene-assisted selection (GAS) (Wilcox et al. 2007). While such polymorphisms may typically involve QTNs, they could also involve insertions or deletions (collectively, indels), which could include transposons.

7.5.1 *Marker-Assisted Selection*

Reliable detection and quantification of QTL is a crucial first step. In practice, this can be very difficult to achieve. It is based on relationships between phenotypes and marker alleles, which in turn depends on chromosomal linkage between QTL and the marker alleles. While this almost certainly requires sufficiently dense genomic maps, achieving map density is no longer a major challenge. In principle, MAS is economically attractive (Johnson et al. 2000; Wilcox et al. 2001). In practice, there are major problems in using it with conifers (Ball 2007; Butcher and Southerton 2007; Wilcox et al. 2007):

- Large study populations will be needed, unless there are QTLs of very large effect.
- Associations detected will relate to quite broad chromosome regions rather than to single-nucleotide polymorphisms (SNPs).
- Where QTN exerting individual effects are in tight repulsion linkage, there may be no detectable QTL.
- With conifers, the extent of linkage disequilibrium tends to be extremely limited, typically in the order of ≤ 2 kb (see earlier), meaning that almost all QTL-marker associations will be pedigree-specific.
- Genetic segregation will also mean that certain loci containing important population-level polymorphisms will be uninformative in pedigrees that are not polymorphic for both the marker- and QTL alleles.
- The number of loci involved can lead to large numbers of random spurious associations (false positives), for which statistical correction needs to be attempted.
- This also means that errors of estimating QTL effects will tend to be distributed normally about zero, creating "selection bias".
- Where epistatic effects, which represent interactions between phenotypic effects of alleles at different loci, are involved the main genetic effects of alleles at individual loci will tend to be diluted. However, the number of potential inter-locus combinations gives huge scope for false positives. Very stringent genome-wide thresholds for detection statistics are needed to control the rate of false positives, and power of valid detection can be very low unless individual epistatic effects are large (Wei et al. 2010).
- The maximum theoretical efficiency of MAS being greatest with large-effect QTL combined with low heritability represents a difficult case for achieving proof of concept in terms of efficacy.
- A lack of QTL of genuinely large effect, which appears to be the rule for conifers, exacerbates these various problems.

These considerations largely militate against success of QTL detection and MAS, so it is not surprising that there has been very little use of MAS in conifers. Putative QTLs have often not been confirmed upon resampling study populations (e.g., Wilcox et al. 1997), although there are reports of independently verified QTLs (e.g., Brown et al. 2003). Nonetheless, independent verification in unrelated genotypes is expensive but necessary (see Wilcox et al. 2001). Moreover, Bayesian analysis, which tests more rigorously for genuine QTL, has tended to overturn positive findings (Ball 2001, 2007). On the other hand, results of the classic simulation study of Meuwissen et al. (2001) indicate that "false negatives", i.e., genuine QTL that are not detected or else discounted for want of robust evidence for their existence, can cause greater loss of selection efficiency.

Because population-wide LD is typically almost non-existent in conifers, use of neutral markers in MAS will generally be confined to within-family selection. While within-family selection is in principle good for maintaining effective population size doing such selection by MAS would only be practicable for a small subset of families and, as such, would really be confined to top-ranking strata in a breeder's metapopulation. By contrast with most conifers, the domestic apple (*Malus* × *domestica* Borkh.) has given much more positive results (e.g., Bus et al. 2000; Stoeckli et al. 2008). Very large and repeatable QTLs have been detected for a range of traits. Of potential interest for us, however, is that disease resistance was among the traits showing large QTL.

There remains the worrying paradox that, while the greatest theoretical advantage of MAS exists with large QTL but low heritability (Lande and Thompson 1990), high heritability is far more conducive to both proving the concept and detecting and reliably quantifying any QTL effects. A low heritability, by contrast, militates against powerful and reliable detection of such effects, although this can be mitigated by improving phenotypic information, through better correction for local environmental effects and/or clonal replication of individual genotypes.

For conifers, the best prospects for MAS would be in small subsets of pedigrees of especial interest, possibly for shortlisting clones being screened for clonal forestry and/or mass vegetative propagation of preferred genotypes within top-ranking full-sib families. Such a focus on only a small subset of pedigrees, however, may not be in the best interests of sound, long-term population management. Family-specific LD does not sit well with the emphasis on within-family selection that is indicated for maintaining effective population size, unless applied to a sufficient number of families, substantially increasing cost. And the development of general within-population LD will tend to reflect a very undesirable reduction of effective population size.

7.5.1.1 Genome-Wide Scans for Selection (GWS)

This approach, originally proposed for cattle breeding by Meuwissen et al. (2001), involves screening individuals using moderate-densely spaced markers spanning most or all of the genome, and utilizes extant LD that has accumulated in closed breeding populations over multiple generations. If the background LD is sufficient, at least some markers will be in LD with QTN of interest, and will account for at least some of the heritable variation, and thus can be used as a selection tool. GWS usually involves (a) establishing statistical relationships between a suite of markers and trait(s) of interest—usually estimated breeding values (EBVs) that have been previously determined from phenotypic data on a "training population" of genotypes, and (b) genotyping breeding populations with all markers for the purposes of forwards selection. The correlation between EBVs and "genomic breeding values" (GBVs) depends on the size of the training population used; genetic architecture(s) (defined here as the sizes, locations and modes of action of QTL) of traits of interest; extent of LD in the training set; and marker density.

GWS is being increasingly used in animal breeding, particularly for sex-limited traits such as milk production (Hayes et al. 2009, Strauss 2010), where bulls can be selected based on GBVs rather than waiting an additional generation of phenotypic records from daughters. It is also being used in plant breeding for species such as wheat, maize, and oil palm (Remington et al. 2001; Bernardo and Yu 2007; Wong and Bernardo 2008 and references therein). Habier et al. (2009) proposed that moderate genetic gains were possible using low-density marker panels of up to 10 cM spacing, although in reality most panels involve somewhat more densely spaced markers.

Advantages of GWS include:

(a) By using sufficiently many markers, GWS avoids identification of actual QTN and stringent confirmation of individual QTL effects. This in turn avoids the need for extensive candidate-gene screening technologies;
(b) For polygenic traits, training sets do not need to involve the large sample sizes needed for association tests for GAS, because GWS involves multiple markers in a single analysis, not individual markers or haplotypes in multiple analyses for a single trait;
(c) Additional traits can be easily accommodated at relatively little extra cost. Compared to association tests for GAS, which require independent candidate-gene screening for independent traits, GWS can easily incorporate new traits without additional genotyping;
(d) Gains will be made from selecting individuals with highest BVs according to relative grandparental contributions (see the following section describing Marker Assisted Estimation of Breeding Value), thus enabling both among- and within-family selection; and
(e) Much earlier selection. While GWS is unlikely to be as efficient as phenotypic selection (Hayes et al. 2009), the opportunity to significantly reduce generation length, and/or to short-list selection candidates for testing, should more than compensate.

However, there are disadvantages:

(a) GWS is entirely dependent upon extant LD. Such LD may not occur in natural and semi-domesticated outbred species such as conifers, except in small breeding populations (such as 10–50 parent elites, for example). However, the accretion of co-ancestry in such populations over generations could offset long-term gains unless concomitant efforts are made to purge deleterious alleles that contribute to genetic load and inbreeding depression;

(b) Need for near-complete genomic (re)sequence which, while not absolutely necessary, is particularly advantageous when designing marker assays. However, in conifers this is currently too expensive and technically demanding with current sequencing technologies because of the size and repetitive nature of the conifer genome;

(c) High genotyping costs—because hundreds to tens-of-thousands of markers are needed, genotyping costs per sample can be prohibitive, particularly if large numbers of individuals need to be screened. Thus there may be relatively limited opportunity to substantively increase selection intensity; and

(d) The rapid breakdown of disequilibrium means that the efficacy of marker sets reduces with time and necessitates re-evaluation of training sets with new markers, although for species with long generation intervals this may not be prohibitive.

Overall, if GS/GWS is used, it is most likely to be for early multi-trait forwards selection in small closed breeding populations such as elites, rather than for mainline advancement or evaluation of new introductions from wild germplasm. Indications from unpublished analyses (Grattapaglia and Resende 2011) suggest that in these situations GWS is likely to yield significant improvements in gain per unit time, although implementation cost needs to be determined. In general, meeting the conditions for GWS to be efficient (Rafalski 2010) will typically be very challenging with forest trees.

Among species a possible exception is western redcedar (*Thuja plicata*). It has several distinctive features (Russell and Ferguson 2008), which appear to favor use of DNA markers for selection, including GWS. It is largely inbreeding, with a history of rapid post-Pleistocene recolonization. Early attempts to find molecular diversity failed, leading some to conclude that the species lacked the genetic diversity needed for any successful breeding program. Since then, however, worthwhile functional genetic diversity has been detected, as have some DNA polymorphisms. Thus, with the breeding system and the migration history, there are good prospects of much more linkage disequilibrium than in typical conifers. Moreover, it is possible to self-fertilize the species without significant loss of viability or vigor, although individuals that survive to function as seed parents show much lower inbreeding statistics than the rest of their cohorts early in the life cycle. Furthermore, it is possible to reduce the generation interval to as little as two years. Self-fertilization is the procedure of choice for producing the next generation, and use of gibberellin can induce extremely precocious flowering (as in various other members of the Cupressaceae sensu stricto). Thus, while detectable DNA polymorphisms may remain sparse, there appears to be real potential for use of DNA makers as a selection tool. Should that succeed, operational use of DNA polymorphisms for selection would be combined with use of *Thuja* as a model species, almost as a "conifer *Arabidopsis*" (JH Russell pers. comm.). An impediment, however, is the absence of a commitment to intensive, large-scale breeding.

7.5.1.2 Marker-Assisted Estimation of Breeding Value (MAEBV)

In segregating generations (advanced filials or backcrosses), there will be random variation among individuals in the amount of nuclear DNA coming from grandparents. There is the prospect of exploiting such variation, by use of markers to quantify the respective contributions of the various grandparents to individuals' DNA. Here, full-sib families would be genotyped by a suite of markers spanning the entire genome, and the relative grandparental contributions calculated, and used to estimate breeding values of entire progeny arrays, from which selections can be made at seedling stage. It would not depend on any QTL information.

This approach, while attractive in principle, needs evaluation to determine appropriate marker densities, gain efficiencies and cost-effectiveness in specific breeding programs. It also requires breeding values to be available for grandparents, which may not be the case in some breeding programs.

7.5.1.3 Introgressive Genome-Wide Selection

This approach is similar to the above (MEABV), with the addition that specific genomic regions, representing known QTL, would be selected upon to more quickly introgress new attributes into target populations. The most promising context would be introgression by backcrossing of some specific attribute(s) into genetically improved, domesticated populations. Backcrossing would be to the improved "recipient" material, conferring the desired attribute(s) with minimum unwanted DNA from the otherwise inferior "donor" material. If the desired attribute(s) are governed by small numbers of known and sequenced genes of large effect, early screening of candidate offspring can be done for presence of those desired genes combined with a minimal overall DNA contribution from the otherwise inferior donor. Minimizing contributions of donor DNA would depend on (1) the genomes of individual ancestors being characterized well enough to permit identification of the DNA sequences that they contribute to their descendants, and (2) prior knowledge of what genomic region(s) are desired from the donor.

Even without specific genes being identified in the donor, it may be possible to achieve effective very early phenotypic screening of segregant candidates for the desired attributes, followed by culling for otherwise minimal donor DNA. However, where the introgressed attribute is subject to epistatic gene effects whereby loss of desired alleles at some loci has no immediate phenotypic effect, there could be some loss of such alleles under phenotypic screening.

Careful management of population size and inbreeding would be indicated for any such introgressive program. To assure this, backcrossing would be done to different recipient parents each generation. Using a number of donor parents would help, especially where desired donor alleles are not known to be fixed within the donor population.

Among specific attributes to be conferred by such introgression, disease resistance is seen as having special promise, and is discussed in the next section.

7.5.1.4 Disease Resistance

Breeding for disease resistance appears to be an area of special interest, in which MAS and possibly GAS may be important (Burdon and Wilcox 2007; Plomion et al. 2007). Kinloch et al. (1970) detected a large-effect gene for disease resistance in *Pinus lambertiana* Douglas. Since then, resistance genes of large effect are being found in forest trees through careful dissection of pathosystems (e.g., Li et al. 2006; Burdon and Wilcox 2007 and references therein; Nelson et al. 2010), often without obvious biometric signatures. Apart from issues of selection, frequencies of such genes have major implications for appropriate population management (Burdon and Wilcox 2007; Nelson et al. 2010). In selecting for disease resistance, care may be needed to ensure that the resistance is durable against genetic shifts in pathogens (pathotype shifts). This can be ensured, in at least some situations, by "pyramiding" resistance genes (Burdon 2001), which involves ensuring that multiple mechanisms of genetic resistance are present in deployed material (cf Wang et al. 1994).

Pyramiding resistance genes is effectively designed to exploit a form of epistasis, which can be large, and intensely dependent on the often labile genetic composition of pathogen populations. In the absence of a "virulent" (i.e., host-compatible) pathotype strain, "up-front" resistance that effectively prevents pathogen entry will mask the presence or absence of "downstream" resistance genes that slow the spread or sporulation of the pathogen that infected successfully, but downstream resistance genes will exert phenotypic effects against a virulent pathotype strain (cf Burdon 2001). Such genes may not be evident as QTL, and therefore not amenable to MAS even if effects are potentially large. Reliance on phenotypic screening can therefore be conducive to inadvertent loss of such alleles, with a potential to compromise durability of resistance. Therefore, these alleles may have to be detected by gene discovery based on use of candidate genes and selected for using GAS (see next).

7.5.2 Gene-Assisted Selection (GAS)

An alternative form of MAS is that using markers in strong LD with the QTN causing actual genetic variation, the ideal being for the markers to represent the actual QTN, whereby LD by definition equals zero. Such markers have been referred to as "functional" or "perfect" markers (e.g., Bagge et al. 2008 and references therein). The selection process that uses such markers has been termed "gene-assisted selection" (GAS), and has been previously described in the context of tree improvement (Wilcox et al. 2007). We note that some causative polymorphisms may not be physically close to transcribed coding sequences—such as the *Vgt-1* locus in maize involving a regulatory element (Salvi et al. 2007)—so the QTN themselves may not be within coding genes, but nonetheless influence coding sequences. This is further discussed below.

7.5.2.1 Marker Identification

There are various steps involved in identifying polymorphisms for GAS. Because it is currently impractical and too expensive to evaluate via association tests (see González-Martínez et al. 2011) all possible sequence variants in the species of interest, some pre-screening is necessary. However, it is as yet unknown which criteria increase most the probability of successful gene discovery, namely identifying causative sequences. Typically, a range of criteria are used. Wilcox et al. (2007) list eight of these:

1. Orthologous genes from model plant systems known to be involved in trait development (cf Mouradov et al. 1998; Walden et al. 1999);
2. Orthologous genes from other conifers and other woody perennials;
3. Endogenous genes with known or suspected roles in relevant biochemical pathways;
4. Transcript profiling, including combining microarrays and mapping ("eQTL mapping");
5. Proteomic profiling, similar to the above;
6. Expressed genes that co-localize with QTL in QTL mapping populations;
7. Genes shown to be associated with heritable variations in traits of interest in other species ("comparative mapping");
8. Genes shown to be involved in trait development in the species of interest via genetic transformation experimentation.

Of these criteria, 1–3 in particular depend on choice of candidate genes, and are likely to be very important for gene discovery in conifers because of the difficulties in conifers of obtaining both direct genomic information and good, timely phenotypic information.

As "omics" technologies improve, other approaches are being developed that will also be useful. These include metabolite profiling in segregating mapping and/or association populations ("mQTL" mapping), and advancements in DNA sequencing technologies. The latter is progressing very quickly, to a point where entire genomes of multiple genotypes could be cost-effectively resequenced. Examples of recent developments include ultra-high throughput single molecule DNA sequencing (Eid et al. 2009), which can sequence thousands of base pairs at a time.

Currently, relatively few markers linked to QTN have been identified —and even fewer independently identified (see Chapter 6)—owing to (1) relatively few genes (compared to other agronomically important crop species) being screened, (2) uncertainty as to optimal criteria for candidate-gene selection, and (3) limited statistical power of association tests (Ball 2005), particularly for low-heritability traits. Nonetheless, these results indicate that association genetics—when applied correctly—will identify sufficient polymorphisms for use in applied breeding. For a more comprehensive description of candidate-gene selection criteria, see Wilcox et al. (2007) and González-Martínez et al. (2011), and references therein.

7.5.2.2 Implications for Tree Breeding Practices

Implementation of technologies such as GAS in tree breeding will require modifications in practices. In addition to acquiring and integrating the various skills in molecular genetics, statistics and genomics, breeders will need to establish or access sufficiently large association tests. These will need to be sufficiently large to detect enough markers for cost-effective selection on an ongoing basis. For most quantitatively inherited characteristics, the majority of individual QTN effects are likely to be small, particularly for low-moderate heritability traits. Association tests will therefore need to be sufficiently powerful to detect such QTN, and therefore involve many hundreds to (preferably) thousands of replicated genotypes on multiple sites with good experimental design that accounts for within-site variation. Thus additional costs are incurred with association genetics, which need to be offset by additional gains, and/or other benefits from association tests. Alternatively, should GS/GWS be undertaken, "training" and validation populations will need to be developed alongside the standard breeding and testing activities.

So where can markers identified via association genetics be useful for selection? In principle, GAS can be applied at any level in the tree improvement hierarchy, virtually wherever phenotypic selection could be or is being practiced. Three generic areas of application of markers described by Stromberg et al. (1994) all apply: early selection, cheaper selection (particularly for expensive-to-measure traits), and increased selection intensity. These have been broadly described by Wilcox et al. (2007), and are currently being evaluated in more detail for various conifer species (P Wilcox and R Ball unpubl.; N Wheeler and T Byram in prep.). These studies show that there are many possibilities regarding where to use GAS, including some in combination with population-management applications. It appears likely however, that most applications will be in breeding and/or production/deployment populations. Increased selection intensity in mainline and/or elite populations is likely to be among the initial applications, on the basis that seedling-stage selection will occur for traits for which markers are available, followed by later-stage selection for breeding objectives that include traits where no markers are available. Marker-based breeding values could also be useful for the later-stage selection phases, or in single-stage selection where phenotypic records are not available for some traits. As breeders' confidence in MAS improves (despite the problems described below), we anticipate additional applications in production and/or deployment for species that are amenable to clonal propagation. Here, markers may be used either as an adjunct to phenotypic selection in clonal (or "varietal") screening, or as a selection step when establishing either stool-bed plants or cultures for clonally amplifying seed from top production-population parents (e.g., Wilcox et al. 2001). Other applications may also be implemented in time–provided sufficient markers can be cost-effectively attained and utilized.

7.5.2.3 General Issues and Problems

With GAS/association genetics, there are a number of challenges and issues that may restrict the extent to which markers can be detected, utilized, or perceived.

7.5.2.3.1 Causes of Functional Polymorphisms

In conifers, like virtually all other species, the actual variants causing trait variation are largely unknown. Trait variation can potentially be caused by many different categories of DNA sequence variants, not all of which occur in or adjacent to coding sequences. Moreover, they cannot necessarily be easily predicted based on knowledge of biochemical alterations. For example, synonymous polymorphisms coding for the same amino acids can have phenotypic significance (Chamary and Hurst 2009). Non gene-associated polymorphisms (e.g., Salvi 2007) will require full or near-full genomic sequence, which is gradually becoming a possibility, but with further delay in the identification of markers for GAS.

7.5.2.3.2 Statistical Issues

As mentioned above, experimental design and analytical methods are important. In terms of experimental design, large sample sizes are needed for detection of small-effect loci (e.g., Ball 2005). In addition, use of either clonal replication and/or breeding values derived from offspring records will effectively raise heritability and individual QTN effects, effectively increasing power. Once identified, marker effects need to be estimated to allow accurate predictions of genetic gain. Overestimates of marker effects at QTN will almost certainly occur with small sample sizes, an effect known as "selection bias" (see Beavis 1997; Ball 2001), thus effects will either need to be independently estimated or methods developed and applied that remove such bias. A further consideration is the likelihood of appreciable genetic sampling effects when working with small families and small elite populations. While working with such material is attractive, due to low cost, genetic sampling error can negate gains, so GAS/association genetics will need to be applied carefully—preferably informed by prior numerical simulation.

Statistical issues for genome wide association studies are specifically reviewed by Weir (2010), in relation to inbreeding, population structure, and relatedness among individuals within population units.

7.5.2.3.3 Non-additive Gene Effects

Average phenotypic effects of allelic substitutions constitute **additive effects** of alleles at individual loci. Through such effects phenotypes of offspring tend to be intermediate between their male and female parents' phenotypes. These effects form the basis of cumulative genetic gain resulting from generations of recurrent selection. However, additive effects of alleles can, in the presence of non-additive gene effects, be in part conditional upon allele frequencies.

Non-additive gene effects, which involve dominance and various classes of epistasis, are not readily captured permanently by traditional selection, although they can be readily captured in full if selection of clones for mass-propagation can be practiced satisfactorily, i.e., in clonal forestry.

With **dominance effects**, whereby phenotypes of heterozygotes tend to depart from means of the two homozygotes, the additive effects of individual alleles vary according to allele frequencies at the individual loci.

With **epistasis**, generating epistatic gene effects, the effect of an allelic substitution is conditional upon alleles at one or more other loci, which can be important. It can arise through redundancy among loci, with either repeats of the same gene and "saturation" effects, or through alternative biochemical pathways achieving the same or similar phenotypes. However, there are also possible synergisms between effects of alleles at different loci, which may best be captured under clonal forestry, and are potentially of great significance. Ideally, the breeder wants to know the basis of individual non-additive gene effects, but that will need accurate knowledge of gene expression. As indicated earlier in connection with QTL detection and MAS, use of association genetics to detect epistatic gene effects entails testing almost astronomic numbers of combinations of loci, with very great scope for false positive synergisms. Where such effects are small, and involve numerous loci, the problem can become intractable, as experience with human disorders has indicated (Weiss 2008).

7.5.2.3.4 Pleiotropy

This relates to where particular genes affect phenotypes for more than one trait. To some degree, this is probably almost universal. The tree breeder almost certainly encounters it in most of the observed between-trait genetic correlations, which must lead one to be prepared for individual genes to exercise pleiotropic effects. Pleiotropy may make it easier for the breeder to detect QTL or even QTN, because of simultaneous phenotypic associations among traits known to be genetically correlated.

Adverse genetic correlations, however, pose special problems for the breeder, limiting the genetic gain simultaneously achievable in the traits involved, and creating a great need for accurate economic-worth information on the respective traits (e.g., Burdon 2004). With polymorphisms of large phenotypic effect, which are likely to be used in MAS or GAS, adverse pleiotropic effects can pose a special danger. Adverse side-effects on field fitness of quality-related alleles are likely to be far more serious in conifers than in annual field crops, which will typically grow in more tightly controlled environments. Quite prolonged field testing may therefore be needed for using genes of major effects, which as such pose special risks of adverse side-effects. Even genes that are desirable for some wood processing (e.g., chemical pulping) or end-products may be undesirable for other processes or products, e.g., solid-wood products.

7.5.2.3.5 Marker-Assisted "Recovery" of Genotypes (MARG)

This approach involves using markers to (a) determine/confirm parentage of specific individuals of high genetic value, and (b) screen large numbers of full-sibs of such individuals to identify siblings most similar to the outstanding individual(s).

Occasionally in tree breeding programs, genotypes of exceptional genetic merit are identified. Examples include the well-known *Pinus taeda* L. clone 7–56, and *P. radiata* clone 850.055. Such individuals are often recognized as outstanding either as adolescent or—more often—mature trees. If species are prone to maturation-related effects—as with various commercially important conifers—then it is almost impossible to clonally propagate and mass-deploy clones of these genotypes. However, if parents of such genotypes were available, it should *in theory* be possible to generate full-sibs with genetic composition very similar to that of the outstanding genotype. Molecular markers could then be used to identify such individuals, on the basis of either (a) whole-genome similarity, or (b) similarity at functional loci that control the trait(s) for which the extreme genotype is known. Here, markers are useful because (i) they can be applied to screen large numbers of siblings more cheaply than phenotyping, and (ii) such screening can be undertaken very early, e.g., in emerging seedlings, long before field testing.

Simulations by PL Wilcox and RD Ball (2009 unpubl.) of the conifer genome indicated a very low probability of generating a sib with high levels of identity by function (IBF, referring to identity-by-descent at loci of *functional* polymorphisms) when whole-genome similarity is used as the basis for selection. However, when only functional loci are considered they found that high levels of IBF were possible *at the actual loci controlling trait variation*, particularly where relatively few loci controlled the variation. For example, for traits where five loci control within-family variation, the most similar 1% of siblings will have a minimum IBF of 90%. This decreased as the number of loci controlling trait variation increased: IBF dropped to 70% where 20 loci controlled within-family variation.

Genetic gains from MARG are likely to be slightly less than from clonal (or "varietal") forestry using proven clones, but greater than from full-sib family forestry using controlled pair-crosses. However, for MARG to be useful to tree breeders, several preconditions need to be met:

- Phenotyping methods and common-garden genetic tests available, to identify rare genotypes of high genetic merit;
- A catalog of functional polymorphisms known to affect important trait(s)—as would arise from association genetics research; and
- A suite of cost-effective technologies, including:
  - o high-throughput propagation system with the capability of generating tens of thousands to millions of copies of each genotype. Such a system should have little or no genotype specificity;
  - o means of producing large numbers of seed per full-sib family; and
  - o DNA extraction and genotyping.

Given these preconditions are not yet in place for most conifers, MARG is unlikely to be implemented now, but may be useful for the future in specific circumstances, particularly as association genetics experiments worldwide are developing catalogs of polymorphisms that affect trait variation.

7.5.3 Summary of Selection Applications

Widespread application of MAS is unlikely to come with conifers in the next few years. However, it may be introduced in specific breeding programs as (a) many more markers are identified via better designed experiments including better phenotypic data, and (b) implementation costs fall. Some of the best prospects exist in the area of capturing oligogenic disease resistance, in cases where resistance loci are reasonably straightforward to detect. Genomic selection may be used in the future, but if so is most likely to be restricted to small populations such as elites with the proviso that genetic load will not become problematic.

GAS is more likely to be implemented, albeit slowly, as markers are identified via ongoing association genetics experiments in various species. Candidate-gene selection methods that effectively identify likely functional polymorphisms are improving. Moreover, leveraging genomic information with other plants, and among conifer species, is also gathering momentum (e.g., Wilcox et al. 2007), and will become important. It is likely that GAS will be implemented in different ways in different breeding programs, with special applications such as MARG being considered on a case-by-case basis.

For the longer term, both GAS and GWS appear attractive, subject of course to affordability and availability of sufficient markers. These approaches are complementary in that while both allow for earlier selection they compensate for each other's disadvantages: GWS holds special promise for multi-trait selection, whereas GAS can potentially allow significant cost-effective increases in selection intensity.

7.6 Integrating Population-Management and Selection Applications

Tree breeders are faced with the challenge of integrating into individual breeding programs the multiple genotyping options (Table 7-1) for the various purposes described above. Currently, a number of breeding programs are beginning to extend marker applications beyond clonal and pedigree verification to more advanced applications such as polymix breeding (e.g., Lambeth et al. 2001; Kumar et al. 2006, 2007), greater use of open pollination (e.g., Dungey et al. 2009) and selection (N Wheeler and T Byram in prep.). Breeders therefore need to make choices about which genotyping technologies can best serve multiple purposes. Table 7-1 indicates likely complementarities between marker types, with SSRs and SNPs tending to be best suited to somewhat different purposes. However, current trends appear to be favoring SNPs, with their potential for further cost reductions, but indels, where important, may need to play a complementary role in future. Both have the potential to serve several purposes. For example, combining parentage verification, polymix breeding and GAS applications from the one marker kit is theoretically possible, and has the benefits of reducing genotyping costs.

Choices of which markers are used for what purpose, however, can also be influenced by the following considerations:

- Which markers have already been developed for the material in question. Development costs for some technologies are high (e.g., SSRs) although application-specific payoffs could be considerable.
- Purpose of greatest current or short-term importance.
- What genotyping has already been done; for example, where verifying genotypes and progeny of a finite number of parents is the prime issue, and the parents have been genotyped for SSRs, then SSRs would certainly be the marker type of current choice, even if quite different markers may be needed for the future.
- Numbers of genotypes to be assayed.
- The exact purpose(s) of marker use; for instance, quite different markers (albeit not necessarily of different types) are likely to be needed for characterizing populations versus genetic fingerprinting within populations.

7.7 Prospective Role of Epigenetic Markers

Epigenetics, in the form of developmentally orchestrated activation or inactivation of genes, can be a very important issue with conifer breeding. While the phenomenon of epigenetics pervades all differentiation of tissues and organs, it commands the attention of conifer breeders in the manifestations of maturation (or cyclophysis) and topophysis (White et al. 2007 and references therein). Maturation involves a progression of morphological changes from the fully juvenile seedling habit to an adult habit, with a threshold change in the onset of sexual reproduction. Associated with the morphological changes are various anatomical changes, often changes in phenology, and other physiological changes such as loss of ability to produce adventitious roots when cuttings are set. Even with successful propagation such changes can make graft scions or rooted cuttings behave like shoots in the upper crowns of mature trees.

It is pronounced in most genera of conifers. Topophysis derives from the part of a tree, and can be manifested in persistent plagiotropic (non-vertical) growth of scions or cuttings taken from branches. It is prevalent in *Abies* spp. and in *Pseudotsuga*, and extreme in *Araucaria*.

Both maturation and topophysis can be very troublesome for the breeder. Not only can these phenomena severely affect both the propagation success and subsequent performance of material, but they are also typically irreversible in conifers except in the course of seed production—which cannot preserve the parental genotypes. More radically than seeking simple reversibility of maturation, the breeder would like trees that would flower "on command and command only"; that would allow accelerated turnover of breeding generations, yet avert diversion of resources away from wood production, and provide a means of genetic containment for transgenes used in genetic engineering.

Epigenetics as such does not necessarily involve actual changes in DNA sequence. Still, the availability of genetic markers for the epigenetic state, be it maturation or topophysis, would be valuable. The breeder could know, without having to grow material in any specific environment, whether or to what extent either effect had been reversed or accelerated. That in turn would reveal whether the material was in a fit state for field deployment. And such information could be of great help in researching the conditions needed for achieving reversal in vegetative material. Assays of expressed sequence tags (ESTs) in cDNA (and the "downstream" proteins and other metabolites) provide the obvious measure of activity of the genes concerned, but application of ESTs will depend on a knowledge of what genes are involved in the epigenetic effects. Thus it will be an iterative process, using ESTs associated with the maturation state to help identify the genes concerned, and then to use ESTs for the identified genes as indicators of epigenetic state.

7.8 Comparison with Some Other Breeding Scenarios

Comparing the situation with other areas of plant and animal breeding is instructive. Cases we consider are:

- Introgression of genes from wild relatives in an annual crop.
- Ongoing breeding in an already intensively domesticated crop, namely maize.
- Dairy herd improvement.

In making comparisons, the following are relevant:

- Magnitude of QTL/QTN effects involved.
- Degree of linkage disequilibrium (which can be expressed quantitatively in numbers of base pairs and/or centiMorgan distances).

- Existing base of genomic information.
- Marginal costs (relative to routine management practices) of obtaining phenotypic data.
- Generation length versus relative economic value of the prospective genetic gains.

How these factors apply to the different cases is summarized in Table 7-2.

With introgression of genes from wild relatives of annual crops, there have been cases of rapid success (e.g., Dekkers and Hospital 2002), resulting from large QTL that could be easily detected and quantified together with very strong LD between existing cultivar material and the wild relative (e.g., Bernacchi et al. 1998; Fulton et al. 2000). As a bonus, strong epistasis can arise between cultivar- and wild-relative alleles in tomato. For instance, a wild relative that has green fruit, can contribute intense fruit color upon introgression into domesticated stocks (Grandillo et al. 1999); evidently there is gene conferring intense fruit color that is present but has its expression blocked by other DNA sequences in the wild relative. Introgression of major-gene resistance—in particular via marker-assisted backcrossing (MAB)—has been extensively applied in annual crops since the mid-1990s. By avoiding the need for phenotyping every successive generation, MAB substantially reduces the time required to successfully introgress resistance genes from donor germplasm into recipient commercial varieties.

In maize, a long history of domestication, often involving close inbreeding, has generated considerable LD, at least in commercial germplasm (Rafalski and Morgante 2004). In addition, good marker diversity exists, even though QTL effects tend to be modest. An abundance of existing genomic knowledge, and the strong institutional base for maize breeding are very important.

In dairy herd improvement, moves have been made to undertake at least provisional selection of bulls entirely on marker information (Hayes et al. 2009; Strauss 2010). This is in a context of detailed genomic information, significant LD resulting from breeding practices that have led to quite a small effective population size, and moderate QTL effects respite very few known QTN effects. Most importantly, detailed phenotypic data are collected in "real time" in operational production, with the producing herds representing the breeding population. Genomic data are also collected largely as an integral part of population management. This has important implications. The marginal costs of obtaining phenotypic data are essentially nil, and those of obtaining the genomic data for selection are limited. Moreover, provisional marker-based selections can be revised as soon as phenotypic performance data come to hand.

The situation for forest tree breeding is much less favorable. However, it can only improve as better knowledge is acquired, and there is much

Table 7-2 Qualitative comparisons among classes of breeding programs for major factors influencing potential for use of DNA markers in selection.

Scenario	Size of QTL/QTN	Level of linkage disequilibrium	Existing genomic information	Marginal costs of phenotyping	Generation time; economic value
Marker-assisted introgression in annual crops	Often very large	Very high	Considerable	Modest	Annual; Moderate-high
Maize breeding	Moderate to small?	Significant	Very detailed	Modest	Annual; Moderate-high
Dairy herd improvement	Moderate to small?	Significant	Considerable	Almost nil	3-5 years; Moderate-high
Tree breeding	Usually small	Usually minimal	Mainly very limited	High	>12 years; Low-moderate

scope for leveraging genomic information from other species, especially among forest trees themselves. Moreover, with the low level of genetic domestication of forest trees plenty of genetic diversity usually exists. Furthermore, the typical generation interval for forest trees, and the costs of producing and managing phenotypes strongly favor the quest for measures that accelerate genetic gain.

7.9 Institutional Challenges

Widespread uptake and application of markers as selection tools is yet to be achieved in conifer breeding programs, despite almost two decades of molecular marker research in a range of economically important species (see Plomion et al. 2007 and references therein). As well as the obstacles posed in conifers by the time frames, tree sizes, wind pollination and outbreeding, huge genomes, and dearth of major genes, institutional barriers currently exist that preclude effective uptake. These include:

- Funding streams and perceived competition for financial resources;
- Skill gaps;
- "Cultural" differences between molecular geneticists and operational breeders;
- Communication challenges associated with multidisciplinary research; and
- Ongoing costs associated with genotyping.

Without much increased total expenditure on genetic improvement the research on DNA technology and its applications may be at the expense of conventional breeding, especially if operational tree breeding is conducted within R & D budgets. This can entail opportunity costs in terms of short- to medium-term genetic gain, and disrupt longer-term field research and evaluation of genetic material. Moreover, developing association genetics can demand an even greater infrastructure of field plantings, in terms of family size and structuring of study populations, than does conventional breeding. Competition for resources between DNA-based research and conventional breeding can therefore create tensions: operational breeders having already demonstrated the efficacy of their approaches, have expressed frustrations with R & D investment in DNA marker-related research that has yet to demonstrate financial benefits, despite these technologies already delivering genetic gains in various agronomic species. Such divides greatly hinder the collaborative relationships that are needed for implementation.

Integrating skill sets and researchers from different backgrounds can create other challenges. Operating and employment realities can often differ between tree breeders and molecular geneticists. The latter tend to be publicly-funded scientists who generally seek to resource experiments (usually via competitive grants), conduct experiments, and publish results. However, operational breeders are primarily concerned about optimizing genetic gains with available resources, and often have significant financial investment from industry. Differing priorities and expectations associated with the differing funding sources can create difficulties, particularly in maintaining long-term, durable relationships.

Molecular technologies also require skill sets not hitherto needed in operational tree improvement. Wilcox et al. (2007) listed the prerequisites for implementation of GAS, many of which also apply to alternative approaches such as GWS/GS. These include access to, and understanding of:

- candidate gene detection technologies and associated genomics platforms (see Wilcox et al. 2007; Plomion et. al. 2007 and references therein).
- resequencing/SNP discovery technologies, and
- genotyping facilities and associated technologies for activities such as DNA extraction.

In regard to genotyping, many if not all marker technologies need to be modified and optimized for application in conifers, requiring access to the appropriate laboratory skills. In addition, advanced quantitative genetics skills are also needed for both experimental design and the analyses and interpretation of results. While such quantitative skills are essential in operational tree breeding, the content of molecular analyses differs. Incorporating such skills and technologies is difficult for the typically financially constrained tree breeding programs, so the skills are usually accessed via collaborations with molecular geneticists employed at publicly funded institutions—albeit with the challenges described above.

Associated with the above are the challenges typically encountered when integrating multiple scientific disciplines. Commonly used terminologies in tree breeding differ from those of molecular genetics and, in particular, genomics. Lack of effective translation can present barriers to understanding the various technologies, which in our experience has impeded effective consideration and uptake by breeders. Potential solutions include (a) up-skilling specific individuals (either breeders or molecular geneticists) who are then able to integrate multiple technologies, or (b) engaging individuals with sufficient knowledge and skills in both areas.

Allocation of resources and coordination of efforts between the DNA technology and conventional breeding is a stern R & D management challenge. There are both perceptions of competition for resources and a gap to bridge between different professional cultures. But the rewards for getting it right should be great indeed.

7.10 Overall Conclusions

Applications of molecular markers in conifer breeding fall into two main categories: population management and actual selection. A lesser, but still potentially significant application is in characterizing epigenetic states.

7.10.1 Population Management

For population management, molecular markers have already found a number of applications. Markers help characterize base-population material, and allow the breeder to infer the identity of base populations from which breeding programs were founded. Such information aids decision-making concerning the choice of base populations or possible infusion of new germplasm into breeding populations. Another application is characterizing breeding systems, whereby the breeder can better estimate genetic parameters from open-pollinated progenies, and know what levels of inbreeding can be tolerated by the species. An immediate and very tangible application, however, is verifying identity of clones and parentage. This serves to safeguard genetic gain, and has proved to be much needed. Such verification also helps assure the integrity of pedigree, with attendant benefits in avoiding unwanted inbreeding and erosion of the genetic base. More radical applications, which are now technically attainable, involve pedigree reconstruction, achieving all or most of the benefits of full pedigree information without total reliance on pair-crossing. This can be applied to polycross progenies, but the greatest potential saving arises with the breeder being able to rely on open pollination with little or no sacrifice of pedigree information. It should even become possible to recruit material from extremely large unpedigreed production populations derived from existing breeding populations, yet maintaining control of inbreeding and effective population size.

7.10.2 Selection Applications

Given the typical delays and cost in obtaining phenotypic expression, and long generation intervals, the early exploration of MAS for tree breeding was understandable. Yet payoffs have generally proved elusive. For selection applications, while time- and cost savings will eventually be most welcome, they have been slow in materializing. Indeed, several features of conifers make application of molecular markers for selection very challenging. There are the typical costs and delays of obtaining the high-quality phenotypic information whereby the effects of genes or chromosomal regions can be detected or verified. The typical outbreeding, and virtually wild state of the genetic systems, means a general lack of significant linkage disequilibrium outside specific pedigrees. That, and a general paucity of large QTL effects, severely limits the scope for effective marker-assisted (or marker-based) selection (MAS). Similarly, an apparent general paucity of large QTN effects, along with the problems posed by the enormous size of conifer genomes, creates major difficulties for developing association genetics as a basis for gene-assisted (or gene-based) selection (GAS). Moreover, institutional challenges arise in both allocation of resources and communication between the conventional breeders and those researching selection applications of genetic markers.

Despite these problems with selection applications, there is reason for optimism. Results described in Chapter 6 indicate markers can be identified, although the design of association tests will need to be addressed in the context of tree improvement imperatives. As with conventional breeding, the essentially wild state of conifer genomes means that populations typically have abundant genetic variability. The remarkable levels of orthology among distantly related plants, and the strong synteny within conifers, will surely help the breeder identify worthwhile candidate genes for gene discovery which will then provide a basis for GAS. Leveraging such knowledge from other species should also contribute to phenomics, whereby the pathways of initial gene action (transcriptomics—see Mackay and Dean 2011 and references therein) through proteomics and metabolomics to phenotypic expression can be elucidated. Breeding for disease resistance is a specific area where application of genetic markers holds special promise, given increasing evidence of existence resistance factors of large effect, and the fact that "pyramiding" genes for different resistance mechanisms can make resistance durable against genetic shifts in pathogens.

Finally, future applications of markers for selection in forest tree breeding may benefit from lessons learned in the interim from experience with smaller, shorter-lived plants.

7.10.3 Applications for Epigenetic Research

Epigenetics, involving developmentally orchestrated activation or inactivation of genes, falls outside coding-region genomics. However, it belongs within phenomics in general. Specifically, it is of interest for control of flowering and general maturation state, which are of interest to the breeder in their own right, and for making operational use of genetic engineering politically acceptable. For this, much more research is needed.

References

Ahuja MR (2001) Recent advances in molecular genetics of forest trees. Euphytica 121: 173–195.

Ahuja MR, Neale DB (2005) Evolution of genome size in conifers. Silvae Genet 54: 126–137.

Andrew RL, Peakall R, Wallis IR, Wood JT, Knight EJ, Foley WJ (2005) Marker-based quantitative genetics in the wild? The heritability and genetic correlation of chemical defenses in a tree. Genetics 171: 1989–1998.

Bagge M, Xia X, Lubberstedt T (2008) Functional markers in wheat: technical and economic aspects. Mol Breed 22(3): 319–328.

Bagnoli F, Fady B, Fineschi S, Ouddu-Moratorio S, Piotti A, Sebastiani F, Vendramin GG (2011) Neutral patterns of genetic variation and applications to in conifer species. In: C Plomion, J Bousquet, C Kole (eds) Genetics, Genomics and Breeding of Conifers. Science Publishers, Enfield, New Hampshire, USA, pp 141–195.

Ball RD (2001) Bayesian methods for quantitative trait loci mapping based on model selection: approximate analysis using the Bayesian information criterion. Genetics 159: 1351–1364.

Ball RD (2005) Experimental designs for reliable detection of linkage disequilibrium in unstructured random population studies. Genetics 170: 859–873.

Ball RD (2007) Statistical analysis and experimental design. In: ND Oraguzie, EH Rikkerink, SE Gardiner, N De Silva (eds) Association Mapping in Plants. Springer, New York, USA, pp 133–196.

Barbazuk WB, Emrich SJ, Chen HD, Li K, Schnable PS (2007) SNP discovery via 454 transcriptome sequencing. Plant J 51: 910–918.

Beavis WD (1997) QTL Analyses: power, precision and accuracy. In: AH Paterson (ed) Molecular Dissection of Complex Traits. CRC Press, Boca Raton, FL, USA, pp 145–159.

Bell JC, Powell M, Devey ME, Moran GF (2004) DNA profiling, pedigree lineage analysis and monitoring in the Australian breeding program of radiata pine. Silvae Genet 53: 130–143.

Bernacchi D, Beck-Bunn T, Emmatty D, Eshed Y, Inai S, Lopez J, Petiard V, Sayama H, Uhlig J, Zamir D Tanksley S (1998) Advanced backcross analysis of tomato. II. Evaluation of near-isogenic lines carrying single-donor introgressions for desirable QTL-alleles derived from *Lycopsersicon hirsutum* and *L. pimpinellifolium*. Theor Appl Genet 97: 170–180.

Bernardo R, Yu J (2007) Marker-assisted selection without QTL mapping: prospects for genome-wide selection for quantitative traits in maize. Maize Genet Coop Newsl 81: 26.

Borralho NMG (1994) Heterogeneous selfing rates and dominance effects in estimating heritabilities from open-pollinated progeny. Can J For Res 24: 1079–1082.

Brondani RV, Williams ER, Brondani C, Grattapaglia D (2006) A microsatellite based consensus map for species of *Eucalyptus* and a novel set of 230 microsatellite markers for the genus. BMC Plant Biol 6:20doi 10.1186/1471-2229-6-20 (Google Scholar 2009).

Brown GR, Bassoni DL, Gill GP, Fontana JR, Wheeler NC, Megraw RA, Davis MF, Sewell MM, Tuskan GA, Neale DB (2003) Identification of quantitative trait loci influencing wood property traits in loblolly pine (*Pinus taeda* L.). III. QTL verification and candidate gene mapping. Genetics 164: 1537–1546.

Brown GR, Gill GP, Kuntz RJ, Langley CH, Neale DB (2004) Nucleotide diversity and linkage disequilibrium in loblolly pine. Proc Natl Acad Sci USA 101: 15255–15260.

Burdon RD (1997) Genetic diversity for the future: Conservation or creation and capture? In: RD Burdon, JM Moore (eds) IUFRO '97 Genetics of Radiata Pine. Proc IUFRO/NZFRI Conf, 1–4 Dec and Workshop 5 Dec, Rotorua, New Zealand. NZ For Res Inst Bull No 203, pp 237–264.

Burdon RD (2001) Genetic diversity and disease resistance: some considerations for research, breeding and deployment. Can J For Res 31: 596–606.

Burdon RD (2004) Breeding goals: issues of goal-setting and applications. In: C Walter, MJ Carson (eds) Plantation Biotechnology for the 21st Century. Research Signpost, Trivandrum, Kerala, India, pp 101–118.

Burdon RD, Kumar S (2002) Stochastic modeling of the impacts of four generations of pollen contamination in unpedigreed gene resources. Silvae Genet 52: 1–7.

Burdon RD, Wilcox PL (2007) Population management: potential impacts of advances in genomics. New For 34: 187–206.

Burdon RD, Lstibůrek M (2010) Incorporating genetically modified traits into tree improvement programs. In: YA El-Kassaby (ed) Forests and Genetically Modified Trees, IUFRO Biotechnology Task Force State-of-Knowledge Report. IUFRO and FAO, Rome, Italy, pp 123–134.

Bus V, Gardiner S, Bassett H, Runaranga C, Rikkerink E (2000) Marker assisted selection for pest and disease resistance in the New Zealand apple breeding programme. Acta Hort 538: 541–547.

Bush RM, Smouse PE (1992) Evidence for the adaptive significance of allozymes in forest trees. New For 6: 179–196.

Butcher PS, Southerton S (2007) Marker-assisted selection in forestry species. In: E Guimaraes, J Ruane, BD Scherf, A Sonnino, JD Dargie (eds) Marker-Assisted Selection: Current Status and Future Perspectives in Crops, Livestock, Forestry and Fish. FAO, Rome, Italy, pp 283–305.

Caballero A (1994) Developments in the prediction of effective population size. Heredity 43: 557–579.

Cato SA, Richardson TE (1996) Inter- and intra-specific polymorphism at chloroplast SSR loci and the inheritance of plastids in *Pinus radiata* D. Don. Theor Appl Genet 93: 587–592.

Chamary JV, Hurst LD (2009) The price of silent mutations. Sci Am 300(6): 34–41.

Dauwe R, Robinson A, Mansfield SD (2011) Recent advances in proteomics and metabolomics in gymnosperms. In: C Plomion, J Bousquet, C Kole (eds) Genetics, Genomics and Breeding in Conifers. Science Publ, Enfield, NH, USA, pp 358–388.

Dean JFD (2011) Future prospects. In: C Plomion, J Bousquet, ,C Kole (eds) Genetics, Genomics and Breeding in Conifers. Science Publ, Enfield, NH, USA, pp 404–438.

Dekkers JCM, Hospital F (2002) The use of molecular genetics in the improvement of agricultural crops. Nat Rev Genet 3: 22–32.

Devey ME, Delfino-Mix A, Donaldson D, Kinloch BB, Neale DB (1995) Efficient mapping of a gene for resistance to white pine blister rust in sugar pine. Proc Natl Acad Sci USA 92: 2066–2070.

Dudley JW (1977) 76 generations of selection for oil and protein percentage in maize. In: O Kempthorne, TB Bailey (eds) Proc Int Conf on Quantitative Genetics. Iowa State Univ Press, Ames, Iowa, USA, pp 459–473.

Dungey HS, Brawner JT, Burger F, Carson M, Henson M, Jefferson P, Matheson AC (2009) A new direction for the *Pinus radiata* breeding strategy for the Radiata Pine Breeding Company. Silvae Genet 58: 28–39.

Edwards JD, Janda J, Sweeney MT, Gaikwad AB, Leung LB, Galbraith DW (2008) Development and evaluation of a high throughput, low-cost genotyping platform based on oligonucleotide microarrays in rice. Plant Meth 8: 13.

Eid J, Fehr A, Gray J, Luong K, Lyle J, Otto G, Peluso P, et al. (2009) Real-time DNA sequencing from single polymerase molecules. Science 323(5910): 133–138.

El-Kassaby YA, Lstibůrek M (2009) Breeding without breeding. Genet Res 91: 111–120.

Falconer DS, Mackay TFC (1996) Introduction to Quantitative Genetics, 4th edn. Longman, Harlow, UK.

Fowler DP, Lester DT (1970) The genetics of red pine. USDA Forest Service Research Paper WO-8.

Fulton TM, Grandillo SB, Beck-Bunn T, Fridman E, Frampton A, Lopez J, Petiard, Uhlig J, Zamir D, Tanksley SD (2000) Advanced backcross QTL analysis of a *Lycopersicon esculentum* x *Lycopersicon parviflorum* cross. Theor Appl Genet 100(7): 1025–1042.

Gea LD, McConnochie RM, Wynyard SS (2007) Parental reconstruction for breeding, deployment, and seed-orchard management of *Eucalyptus nitens*. NZ J For Sci 37: 23–36.

Gernandt DS, López GG, García SO, Liston A (2005) Phylogeny and classification of *Pinus*. Taxon 54: 29–42.

Gernandt DS, Willyard A, Syring JV, Liston A (2011) The conifers (Pinophyta). In: C Plomion, J Bousquet, C Kole (eds) Genetics, Genomics and Breeding in Conifers. Science Publ, Enfield, NH, USA, pp 1–39.

González-Martínez SC, Dillon S, Garnier-Géré P, Krutovsky K, Alía R, Burgarella C, Eckert AJ, García-Gil MR, Grivet D, Heuertz M, Jamarillo-Correa JP, Lascoux M, Neale DB, Savolainen O, Tsumura Y, Vendramin GG (2011) Patterns of nucleotide diversity in conifers. In: C Plomion, J Bousquet, C Kole (eds) Genetics, Genomics and Breeding in Conifers. Science Publ, Enfield, NH, USA, pp 239–275.

Grandillo S, Bernacchi D, Fulton TM, Zamir D, Tanksley SD (1999) Advanced backcross QTL analysis: a method for systematic use of exotic germplasm in the improvement of crop quality. In: GT Magnozza, E Porceddu, MA Pagnotta (eds) Genetics and Breeding for Crop Quality and Resistance. Kluwer Acad Publ, Dorcrecht, The Netherlands, pp 283–290.

Grattapaglia D (2007) Marker-assited selection in *Eucalyptus*. In: E Guimaraes, J Ruane, BD Scherf, A Sonnino, JD Dargie (eds). Marker-Assisted Selection: Current Status and Future Perspectives in Crops, Livestock, Forestry and Fish. FAO, Rome, Italy, pp 251–281.

Grattapaglia D, Resende MDV (2011) Genomic selection in forest tree Breeding. Tree Genet Genomes 7: 241–256.

Gunderson KL, Steemers FJ, Lee G, Mendoza LG, Chee MS (2005) A genome-wide scalable SNP genotyping assay using microarray technology. Nat Genet 37: 549–554.

Habier D, Fernando RL, Dekkers JCM (2009) Genomic selection using low-density marker panels. Genetics 182: 343–353.

Hayes B, Bowman PJ, Chamberlain AJ, Goddard ME (2009) *Invited review*: Genomic selection in dairy cattle: Progress and challenges. J Dairy Sci 92: 433–443.

Hill WG (2005) A century of corn selection. Science 307: 683–684.

Hodge GR, Volker PW, Potts BM, Owen JV (1996) A comparison of genetic information from open-pollinated and control-pollinated progeny information in two eucalypt species. Theor Appl Genet 92: 53–63.

Jaccoud D, Peng K, Feinstein D, Kilian A (2001) Diversity Arrays: a solid state technology for sequence information independent genotyping. Nucl Acids Res 29(4): e25.

Johnson GR, Wheeler NC, Strauss SH (2000) Financial feasibility of maker-aided selection. Can J For Res 30: 1942–1952.

Johnson R (2004) Marker-assisted selection. Plant Breed Rev 24(1): 293–309 (Cited by Wong and Bernardo 2008).

Jones N, Ougham H, Thomas H, Pašalinskienė I (2009) Markers and mapping revisited: finding your genes. New Phytol 183: 935–966.

Karhu A, Hurme P, Karjalainen M, Karvonen P, Karkkainen K, Neale D, Savolainen O (1996) Do molecular markers reflect patterns of differentiation in adaptive traits of conifers? Theor Appl Genet 93: 215–221.

Kinloch BB, Parks GK, Flower CW (1970) White pine blister rust: Simply inherited resistance in sugar pine. Science 167:193–195.

Krutovsky KV, Troggio M, Brown GR, Jermstad KD, Neale DB (2004) Comparative mapping in the Pinaceae. Genetics 168: 447–461.

Krutovsky KV, Elsik CG, Matvienko M, Kozik M, Neale DB (2006) Conserved ortholog sets in forest trees. Tree Genet Genomes 3: 61–70.

Kumar S (2004) Effect of selfing on various economic traits in *Pinus radiata* and some implications for breeding strategy. For Sci 50: 571–578.

Kumar S, Richardson TE (2005) Inferring relatedness and heritability using molecular markers in radiata pine. Mol Breed 15: 55–64.

Kumar S, Gerber S, Richardson TE (2006) In: CF Mercer (ed) Breeding for Success: Diversity in Action. Proc 13th Australasian Plant Breed Conf, Christchurch, New Zealand, 18–21 April pp 578–583.

Kumar S, Richardson TE, Gerber S, Gea LD (2007) Tesing for unequal parental contributions in using nuclear and chloroplast SSR markers in radiata pine. Tree Genet Genomes 3: 207–214.

Lambeth CC, Lee B-C, O'Malley D, Wheeler NC (2001) Polymix breeding with parental analysis of progeny: an alternative to full-sib breeding and testing. Theor Appl Genet 103: 930–943.

Lande R, Thompson R (1990) Efficiency of marker-assisted selection in the improvement of quantitative traits. Genetics 121: 185–199.

Li H, Ghosh S, Amerson H, Li B (2006) Major gene detection for fusiform rust resistance using Bayesian complex segregation analysis in loblolly pine. Theor Appl Genet 113: 921–929.

Lindgren D, Gea L, Jefferson P (1996) Loss of genetic diversity monitored by status number. Silvae Genet 45: 52–59.

Lindgren D, Gea LD, Jefferson PA (1997) Status number for measuring genetic diversity. For Genet 4: 69–76.

Mackay JJ, Dean DFD (2011) Transcriptomics. In: C Plomion, J Bousquet, C Kole (eds) Genetics, Genomics and Breeding in Conifers. Science Publ, Enfield, New Hampshire, USA, pp 323–357.

Mattick JS (2004) The hidden genetic program of complex organisms. Sci Am 291(4): 30–37.

Mergen F (1963) Evaluation of spontaneous, chemical and radiation-induced mutations in Pinaceae. In: Proc World Consultation on Forest Genetics and Tree Improvement, Stockholm, August . FAO, Rome, Italy, paper 1/1.

Meuwissen THE, Hayes BJ, Goddard ME (2001) Prediction of total genetic value using genome-wide dense marker maps. Genetics 157: 1819–1829.

Monroe D (2009) Genomic clues to DNA treasure sometimes lead nowhere. Science 325: 142–143.

Morgante M, Salamini F (2003) From Plant Genetics to Breeding Practice. Curr Opin Biotechnol 14: 214–219.

Morgante M, De Paoli E (2011) Toward the conifer genome sequence. In: C Plomion, J Bousquet, C Kole (eds) Genetics, Genomics and Breeding in Conifers. Science Publ, Enfield, NH, USA, pp 389–403.

Mouradov A, Glassick T, Hamdorf B, Murphy L, Fowler B, Marla S, Sederoff R (1999) *NEEDLY*, a *Pinus radiata* ortholog of *FLORICAULA/LEAFY* genes, expressed in both reproductive and vegetative meristems. Proc Natl Acad Sci USA 95: 6537–6542.

Mullin TJ, Andersson B, Bastien J-C, Beaulieu J, Burdon RD, Dvorak WS, King JN, Kondo T, Krakowski J, Lee SJ, McKeand SE, Pâques L, Raffin A, Russell JH, Skrøppa T, Stoehr M, Yanchuk A (2011) Economic importance, breeding objectives and achievements. In: C Plomion, J Bousquet, C Kole (eds) Genetics, Genomics and Breeding in Conifers. Science Publ, Enfield, NH, USA, pp 40–127.

Namkoong G (1966) Inbreeding effects on estimation of additive genetic variance. For Sci 12: 8–13.

Neale DB, Sederoff RR (1989) Paternal inheritance of chloroplast DNA and maternal inheritance of mitochondrial DNA in loblolly pine. Theor Appl Genet 77: 212–216.

Nelson CD, Kubisiak TL, Amerson HV (2010) Unravelling and managing fusiform rust resistance: a model approach for coevolved forest tree pathosystems. For Pathol 40: 64–72.

Paran I, Zamir D (2003) Quantitative traits in plants: beyond the QTL. Trends Genet 19(8): 303–306.

Pindo M, Vezzulli S, Coppola G, Cartwright DA, Zharkikh A, Velasco R, Troggio M (2008) SNP high-throughput screening in grapevine using the SNPlex genotyping system. BMC Plant Biol 8.

Plomion C, Chagné D, Pot D, Kumar S, Wilcox PL, Burdon RD, Prat D, Peterson DG, Paiva J, Vendramin GG, Sebastiani F, Nelson CD, Echt CS, Savolainen O, Kubisiak TL, Cervera MT, de Maria M, Islam-Faridi MN (2007) The pines. In: C Kole (ed) Genome Mapping and Molecular Breeding in Plants, vol 7: Forest Trees. Springer, Berlin, Heidelberg, Germany, pp 29–92.

Pollard KS, Salama SR, Lambert N, Lambot M-A, Coppens S, Pedersen JS, Katzman S, King B, Onodera C, Siepel A, Kern AD, Dehay C, Igel H, Ares M, Vanderhaeghen P, Haussler D (2006) An RNA gene expressed during cortical development evolved rapidly in humans. Nature 443(7108): 167–172.

Rafalski JA (2010) Association genetics in crop improvement. Curr Opin Plant Biol 13: 174–180.

Rafalski JA, Morgante M (2004) Corn and Humans: recombination and linkage disequilibrium in two genomes of similar size. Trends Genet 20(2): 103–111.

Ragoussis J (2006) Genotyping technologies for all. Drug Discov Today: Technol 3:115–122.

Remington DL, Thornsberry JM, Matsuoka Y, Wilson LM, Whitt SR, Doebly J, Kresovich S, Goodman MM, Buckler ES (2001) Structure of linkage disequilibrium and phenotypic associations in the maize genome. Proc Natl Acad Sci USA 98: 11479–11484.

Ritchie ME, Carvalho BS, Hetrick KN, Tavaré S, Irizarry RA (2009) R/Bioconductor software Illumina's Infinium whole-genotyping BeadChips. Bioinformatics 25: 2621–2623.

Ritland K, Krutovsky K, Tsumura Y, Pelgas B, Isabel N, Bousquet J (2011) Genetic mapping in conifers. In: C Plomion, J Bousquet, C Kole (eds) Genetics, Genomics and Breeding in Conifers. Science Publ, Enfield, NH, USA, pp 196–238.

Robinson C (1999) Making forest biotechnology a commercial reality. Nat Biotechnol 17: 27–30.

Rosvall O, Lindgren D, Mullin TJ (1998) Sustainability robustness and efficiency of a multi-generation breeding strategy based on within-family clonal selection. Silvae Genet 47: 307–321.

Russell JH, Ferguson DC (2008) Preliminary results from five generations of a western redcedar (*Thuja plicata*) selection study with self-mating. Tree Genet Genomes 4: 509–518.

Russell JH, Yanchuk AD, Burdon RD (2003) Inbreeding depression and variance structures for height and adaptation in self- and outcross *Thuja plicata* families in varying environments. For Genet 10: 171–184.

Salvi S, Sponza G, Morgante M, Tomes D, Niu X, et al. (2007) Conserved noncoding genomic sequences associated with a flowering-time quantitative trait locus in maize. Proc Natl Acad Sci USA 104 11376–11381

Stoeckli S, Modi K, Gessler C, Patocchi A, Jermini M, Dorn S (2008) QTL analysis for aphid resistance and growth traits in apple. Tree Genet Genomes 4: 833–847.

Strauss S (2010) Biotech breeding goes bovine. Nat Biotechnol 28(8): 540–543.

Strauss SH, Libby WJ (1987) Allozyme heterosis in radiata pine is poorly explained by overdominance. Am Nat 130: 879–890.

Stromberg LD, Dudley JW, Rufener GK (1994) Comparing conventional early generation selection with molecular marker assisted selection in maize. Crop Sci 34: 1221–1225.

Walden AR, Walter C, Gardner RC (1999) Genes expressed in *Pinus radiata* male cones include homologs to anther-specific and pathogenic response genes. Plant Physiol 121: 1103–1116.

Wang G.-L, Mackill DJ, Bonman JM, McCouch SR, Champoux MC, Nelson RJ (1994) RFLP mapping of genes conferring complete and partial resistance to blast in a durably resistant rice cultivar. Genetics 136: 1421–1434.

Wang, X-R, Torimaru T, Lindgren D, Fries A (2010) Marker-based parentage analysis facilitates low input 'breeding without breeding' strategies for forest trees. Tree Genet Genomes 6: 227–235.

Wei W-H, Knott S, Haley CS, de Koning D-J (2010) Controlling false positives in the mapping of epistatic QTL. Heredity 104: 401–409.

Weir BS (1996) Genetic Data Analysis II. Sinauer Assoc, Sunderland, MA, USA.

Weir BS (2010) Statistical genetic issues for genome-wide association studies. Genome 53: 859–875.
Weiss KM, (2008) Tilting at quixotic trait loci: an evolutionary perspective on genetic causation. Genetics 179: 1741–1756.
White TL, Adams WT, Neale DB (2007) Forest Genetics. CABI Publ, Wallingford, UK, and Cambridge, MA, USA.
Wilcox PL, Amerson HV, Kuhlman EG, Liu B-H, O'Malley DM, Sederoff RR (1996) Detection of a major gene for resistance to fusiform rust disease in loblolly pine by genomic mapping. Proc Natl Acad Sci. USA 93:3859–3864.
Wilcox PL, Richardson TE, Carson SD (1997) Nature of quantitative trait variation in *Pinus radiata*: insights from QTL detection experiments. In: RD Burdon, JM Moore (eds). IUFRO '97 Genetics of Radiata Pine. Proc IUFRO/NZFRI Conf, 1–4 Dec and Workshop 5 Dec Rotorua, New Zealand. NZ For Res Inst Bull No 203, pp 304–312.
Wilcox PL, Carson SD, Richardson TE, Ball RD, Horgan GP, Carter P (2001) Benefit-cost analysis of DNA marker-based selection in progenies of *Pinus radiata* and seed orchard parents. Can J For Res31: 2213–2224.
Wilcox PL, Echt CE, Burdon RD (2007) Gene-assisted selection: applications of association genetics for forest tree breeding. In: ND Oraguzie, EH Rikkerink, SE Gardiner, N De Silva (eds) Association Mapping in Plants. Springer, New York, USA, pp 211–247.
Williams CG (2009) Conifer Reproductive Biology, Springer, Dordrecht, Heidelberg, London, New York, USA.
Williams CG, Savolainen O (1996) Inbreeding depression in conifers: implications for breeding and testing. For Sci 42: 102–117.
Wong CK, Bernardo R (2008) Genomewide selection in oil palm: increasing selection gain per unit time and cost with small populations. Theor Appl Genet 116: 815–824.
Yanchuk AD (2001) A quantitative framework for breeding and conservation of forest tree genetic resources in British Columbia. Can J For Res 31(4): 566–576.
Yin TM, DiFazio SP, Gunter LE, Riemenschneider D, Tuskan GA (2004) Large-scale heterospecific segregation distortion in *Populus* revealed by a dense genetic map. Theor Appl Genet 109: 451–463.

8

Transcriptomics

John J. Mackay[1,*] and *Jeffrey F.D. Dean*[2]

ABSTRACT

RNA transcripts are the first discrete products on the path linking genomes to function and phenotype. Thus, characterization of the transcriptome, which is the sum total of all transcripts produced from the genome, establishes the cast of players working on the cellular stage to create a biological outcome. The enormity of conifer genomes has so far constrained researchers to focus their genomic aspirations on transcriptomes, and substantial bodies of sequence information have been accumulated to identify many if not most of the more abundant genes contributing to conifer growth and development. This chapter reviews the current status of our understanding of conifer transcriptomes and discusses how this information is being used to quantify the dynamics of the gene expression changes that define conifer responses to developmental cues, as well as environmental challenges. Instrumentation for nucleic acid sequencing and quantitation is changing rapidly, creating opportunities to study aspects of conifer RNA metabolism that were previously inaccessible. These new technologies, and some still on the horizon, are reviewed with respect to their current application to problems in conifer biology, and possibilities for future uses are discussed.

Keywords: Pinaceae, cDNA sequencing, expressed sequence tag, EST clustering, bioinformatics, transcriptional profiling, microarray, digital gene expression, wood formation, abiotic and biotic stress, gene families

[1]Center for Forest Research, Laval University, Québec City, Québec, Canada, G1V 0A6; e-mail: *John.mackay@sbf.ulaval.ca*

[2] Warnell School of Forestry and Natural Resources, University of Georgia, Athens, GA 30602, USA; e-mail: *jeffdean@uga.edu*

*Correspondind author

8.1 Gene Discovery and Catalog Development

Over a decade ago, sequencing of random cDNA clones to obtain expressed sequence tags (ESTs) took root as a new approach for gene discovery in forest trees. An EST is a single sequencing reaction obtained from a cDNA fragment or clone, usually from one of its termini (3′- or 5′-end). Loblolly pine was among the very first trees to be targeted (Allona et al. 1998). The approach had been successfully applied to several model organisms and as costs had begun to decrease it was applied to diverse organisms, including crop plants. Considering the exceedingly large size of conifer genomes (ranging from 20 to 30 Gigabases), only a small fraction of which likely contains protein coding information, it was acknowledged that sequencing numerous cDNAs clones was an efficient approach to characterizing portions of the genome that could most rapidly contribute to our understanding of tree biology, as well as provide new tools for tree improvement (Kirst et al. 2003). To the extent that cDNA clones represent faithful DNA copies of a majority of the RNA transcripts from a given tissue sample, deep sequencing of the cDNA pool has the potential to reveal a significant fraction of all of the coding sequences that contribute to the growth and development of the tissue, and to the extent that sampling can be extended to as many tissues and stages of development as possible, a catalog of all the actively expressed sequences in the organisms—the transcriptome—may be defined. Until such time as DNA sequencing technologies have improved enough to make characterization of very large genomes cost-efficient, the transcriptome will represent the window through which the conifer genome is most frequently accessed. Today, the transcriptome is at the heart of many basic and applied endeavors in conifer genomics.

8.1.1 Experimental Approaches

Transcriptome sequencing—also known as random cDNA sequencing or EST analysis—comes in many shapes and colors. One of the first large-scale initiatives was aimed at gene discovery related to wood formation in loblolly pine (Kirst et al. 2003). By randomly sequencing the cDNA copies of transcripts derived from secondary differentiating xylem at different stages of development, from different parts of the tree and from stressed trees, thousands of different coding sequences were discovered. This early success led to the sampling of more diverse tissues, in different physiological states or following treatments involving abiotic or biotic factors (Cairney and Pullman 2007; see Table 8-1). Early projects applied relatively standard methods for cDNA library synthesis, and randomly selected clones were sequenced most often from the 5′-end aiming to minimize costs, reduce degeneracy and maximize the recovery of coding sequences. It was

Table 8-1 Evolution of the number of Expressed Sequenced Tags in NCBI's dbEST for conifers and other selected gymnosperms.

Species Name	Common Name	Family (f) Order (o) (*non-conifers*)	dbEST entries (2010-08-01)[1]	dbEST entries (2009-08-20)	dbEST entries (2006-10-27)[2]
Chamaecyparis formosensis	Formosan or Taiwan cypress	Cupressaceae	805	805	503
Chamaecyparis obtusa	Hinoki Cypress or Hinoki	Cupressaceae	5897	5,897	5,830
Cryptomeria japonica	Japanese cedar, Sugi	Cupressaceae	56,645	56,645	16,230
Cycas rumphii	Cycads, Sumo	Cycadaceae (f) Cycadales (o)	21,997	21,997	8,058
Ginkgo biloba	Maidenhair tree, ginkgo	Ginkgoales (f) Ginkgoaceae (o)	21,590	21,590	6,248
Gnetum gnemon	Melinjo or Belinjo Bago Peesae	Gnetaceae (f) Gnetales (o)	10,724	10,724	4,234
Picea abies	Norway spruce	Pinaceae	14,224	10,217	7,935
Picea engelmannii x Picea glauca	BC Interior spruce (hybrid)	Pinaceae	28,174	28,170	28,170
Picea glauca	White spruce	Pinaceae	313,110	298,657	132,624
Picea sitchensis	Sitka spruce	Pinaceae	186,637	168,675	80,789
Total, all *Picea*			**542,149**	**505,719**	**249,518**
Pinus banksiana	Jack Pine	Pinaceae	36,379	—	—
Pinus contorta	Lodgepole Pine	Pinaceae	40,483	—	—
Pinus densiflora	Japanese Red Pine	Pinaceae	3,316	3,316	—
Pinus pinaster	Maritime Pine	Pinaceae	34,044	27,847	27,283
Pinus pinea	Stone pine	Pinaceae	289	290	289
Pinus radiata	Radiata pine	Pinaceae	7,538	6,757	—
Pinus rigida x Pinus x rigitaeda	Pitch x Loblolly Pine (hybrid)	Pinaceae	1,002	—	—
Pinus sylvestris	Scots pine	Pinaceae	666	660	—

Table 8-1 contd....

Table 8-1 contd....

Species Name	Common Name	Family (f) Order (o) (*non-conifers*)	dbEST entries (2010-08-01)[1]	dbEST entries (2009-08-20)	dbEST entries (2006-10-27)[2]
Pinus sylvestris /Heterobasidion annosum	Scots pine	Pinaceae	1,689	1,689	1,663
Pinus taeda	Loblolly Pine	Pinaceae	328,628	328,628	329,469
Total, all *Pinus*			**454,034**	**369,187**	**358,704**
Pseudotsuga menziesii var. menziesii	Douglas fir	Pinaceae	18,142	18,142	6,721
Taiwania cryptomerioides		Cupressaceae	2,220	2,220	—
Welwitschia mirabilis	Tree tumbo, welwitschia	Welwitschiaceae (f) Welwitschiales (o)	10,129	10,129	10,129
Grand Total			**1,144,332**	**1,023,055**	**666,175**

[1]All species with more that 200 entries in dbEST are listed.
[2]According to Cairney and Pullman (2007).

rapidly acknowledged that this relatively simple and straightforward approach could pave the way to numerous opportunities to investigate gene expression and gene function, to map genes and discover molecular markers linked to phenotypes of interest (Komulainen et al. 2003).

Over time, a variety of technical improvements have been applied to conifer cDNA sequencing to lower costs while improving rates of novel gene discovery. Transcript abundance varies widely for the genes expressed in any given RNA sample, and as a consequence cDNA library normalization methods (Bonaldo et al. 1996) have been applied to decrease the redundancy of abundant transcripts and enhance the likelihood of sampling rare transcripts (Ralph et al. 2006a, 2008). An early finding was that while many of the conifer gene sequences shared significant sequence similarity with gene sequences of known function from other plant species; a large fraction (30–40%) showed no such homology (e.g., Kirst et al. 2003; Pavy et al. 2005b). New techniques for cDNA synthesis have resulted in libraries that contain a higher proportion of full-length cDNA sequences, which has improved the likelihood of discovering the conserved coding regions that help to predict gene function (Ralph et al. 2008). Approaches that pair gene discovery with analyses of genotypic variability and diversity are also generating a lot of excitement (Le Dantec et al. 2004; Pavy et al. 2006). Depending on specific project objectives, cDNA libraries have been developed using clonally replicated genotypes exposed to various treatment conditions (Lorenz et al. 2006), and from parents used to develop pedigrees or from populations of unrelated trees (J MacKay et al. unpubl.; see *www.arborea.ca*).

8.1.2 EST Datasets Available for Conifer Species

Moderate to large-sized EST datasets are available for several conifer trees, mostly from the Pinaceae, but also from the Cupressaceae (Table 8-1). ESTs are available from a few other taxa, such as cycads and ginkgo, belonging to other orders within the gymnosperms. Of the approximately 1 million gymnosperm ESTs that may be downloaded from dbEST, over 90% are from conifers. A current list of present and past conifer gene discovery projects is available on the webpage of the Conifer Genome Network (CGN; *http://www.pinegenome.org/*). Most of the projects to date have focused on aspects of wood formation and properties because of the central ecological and economic importance of conifer wood production. Among the other unique or important biological processes targeted for large-scale gene discovery in conifers have been somatic embryogenesis (Cairney et al. 2006), responses to defoliating insects (e.g., Ralph et al. 2006a), and root responses to water stress (Lorenz et al. 2006).

The species with the largest number of ESTs available in NCBI's dbEST are *Pinus teada* (328,628), *Picea glauca* (298,657) and *Picea sitchensis* (168,675);

while the genus with the largest number of ESTs is *Picea* (505,719) (Table 8-1). These entries are for ESTs obtained using Sanger sequencing methods, and did not include any data obtained using next-generation sequencers. There are also several species, such as *Pinus pinaster*, *Cryptomeria japonicum*, and Douglas-fir, for which EST datasets ranging from a few hundred up to approximately 30,000 sequences are available. Among the research projects currently working to add to the publicly available EST resources for conifers is an effort at the US DOE Joint Genome Institute that has targeted the release of 2 million new sequences for *P. taeda*, in addition to 1 million for *Pseudotsuga menziesii*, and 0.5 million each for *Cedrus atlantica*, *Cephalotaxus harringtonia*, *Gnetum gnemon*, *Picea abies*, *Pinus lambertiana*, *Pinus palustris*, *Podocarpus macrophylla*, *Sciadopitys verticillata*, *Sequoia sempervirens*, *Taxus baccata*, *Wollemia nobilis* (J Dean, unpubl. data). While the JGI effort will generate relatively long sequences using the Roche GS-XLR sequencing platform (average read-length of 400–500 nt), the 1KP Project (*http://www.1kp-project.com/*) plans to use the Illumina GAII short-read platform (>75 nt paired-end reads) to generate complete transcriptome reconstructions for at least one species from every conifer family.

8.1.3 Clustering Methods

Regardless of the sequencing platform used, once ESTs have been obtained they are typically analyzed through various clustering procedures to determine how many unique transcript sequences have been sampled (Nagaraj et al. 2007). A molecular function may also be assigned to the putative proteins that they encode based on sequence similarity with known sequences found in public databases. The following is meant to provide an overview of methodologies and outcomes from large-scale EST clustering analyses. A recent review thoroughly describes current technologies for gathering sequence information, as well as sequence assembly algorithms and their limitations (Scheibye-Alsing et al. 2009).

Clustering aims to group ESTs (into contigs or clusters) according to quantitative measures of sequence similarity, usually for the purpose of identifying all transcripts arising from an individual gene. To accomplish this, cluster analyses employ a variety of bioinformatic subroutines that may be generally classified into pre-clustering, clustering, and post-clustering steps (Box 8.1). Complicating the process are both technical limitations related to cDNA preparation and sequencing, such as artefactual cDNA constructs (chimeras) and sequencing error, and biological variation, such as allelic variation and alternate splicing. Thus, it is not surprising that reports describing efforts to cluster conifer ESTs, most of which employed different methods and procedures, have yielded different outcomes (Kirst et al. 2003; Liang et al. 2007; Pavy et al. 2007; Ralph et al. 2008; Wegrzyn et al. 2009).

However, as sampling depth and sequence quality has increased, so too has the reliability of clusters for representing the true transcriptional products of expressed genes. Cluster analyses usually yield a set of "consensus" or "representative" sequences that are the result of concatenating the unique sequence information from all reads in a given cluster. Consensus sequences obtained through these in silico reconstructions are not acceptable for submission to public databases, like GenBank (NCBI) or Swissprot. Therefore, smaller, independent databases dedicated to conifer or forest tree genome analysis have been developed to provide researchers access to these consensus sequences as well as datamining tools with which to explore them. Examples of these databases include ConiferGDB (Liang et al. 2007), ForestTreeDB (Pavy et al. 2007), and TreeGenes (Wegrzyn et al. 2008).

Box 8.1 Major steps in clustering analysis of EST data.

1. *Pre-clustering steps.* The pre-clustering processes remove or identify and mask stretches of nucleotides that are external to the cDNA insert, including those derived from the cloning vector, primers and any linkers (adaptors) introduced during library manufacture. All or part of the polyA tail, as well as low complexity regions, may be masked or trimmed. Pre-clustering also includes basecalling in which quality scores are assigned to each nucleotide using basecaller software, such as Phred (Ewing and Green 1998, Ewing et al. 1998) or KB™ basecaller (Applied Biosystems). Stretches of DNA sequence having low quality scores are then either tagged or trimmed. For example, parameters for pre-clustering may set for a minimum Phred score threshold between 20 and 30, i.e., representing a maximum error rate of between 0.01 and 0.001, respectively. Additional filtering during pre-clustering may include removal of sequences that are clearly foreign to the genome under investigation, such as bacterial, fungal, human or even organellular sequences, as well as tagging of putative chimeric clones where obvious linker (adaptor) sequences appear in the middle of a putative cDNA insert. The pre-clustering steps thus identify and prepare high-quality cDNA sequences that pose minimal risk for artefactual clustering. These "cleaned" sequences are typically the information submitted to public EST databases.
2. *Clustering.* Clustering procedures use sequence alignment software, such as Phrap (*http://www.phrap.org*), CAP3 (Huang and Madan 1999), or others (e.g., Brown and Hudek 2004; Kim et al. 2008), to group sequences according to matrix scores that reflect nucleotide identity/similarity at a given position, as well as length of overlapping sequence. These groupings aim to provide a reasonable approximation of the genes and transcripts that have been sampled. Clustering steps are inherently challenging because each gene may be represented by different alleles and alternative transcripts, and in addition, closely related genes may share long stretches of highly similar sequence. These sequence polymorphisms may result in either under-clustering (separation of sequence variants from a single gene into several clusters) or over-clustering (grouping of sequences from related genes into a single

Box 8.1 contd....

Box 8.1 contd....

cluster). These difficulties can become particularly acute when assembling short sequences (e.g. under 400–500 bp). Although alignment parameters can be adjusted to minimize under- and over-clustering, there are no universal rules than can be applied across all situations. Thus, clustering results can vary in any iteration into which new sequences, new parameters or new algorithms have been introduced, which can make it difficult to perform detailed and direct comparisons between assemblies. Yet properly performed, clustering does order sequences into groups and sub-groups that give good representation of genes and gene variants (alleles and alternate transcripts) as a means to organize the information complexity.

3. *Post-clustering.* Post-clustering employs a variety of procedures, some still manual in nature, to identify poor quality or artefactual clusters and improve the overall quality of clustering analyses. Artefactual clusters may arise when chimeric cDNA clones are not detected and removed during pre-clustering, which results in non-overlapping sequences derived from different transcripts being clustered together to form contigs (Pavy et al. 2008a). Similarly, missed adapter or primer sequences that lead to artificial over-clustering can be detected during post-clustering routines. Sequence similarity analysis between clusters may be used to identify putative instances of under-clustering, particularly where alternative splicing events are involved (Cordonnier-Pratt et al. 2004; Murray et al. 2005).

8.1.4 Clustering Analyses

In recent years, species-specific EST cluster assemblies have become available for many plant species through NCBI's UniGene database (Sayers et al. 2009) and at PlantGDB (Dong et al. 2004). The NCBI's UniGene database has included three conifers, namely *P. taeda*, *P. glauca* and *P. sitchensis* (Table 8.2A), whereas the PlantGDB assemblies include *Cryptomeria japonica, Gnetum gnemon, Picea abies, Picea engelmannii x Picea glauca, Picea glauca, Picea sitchensis, Pinus pinaster, Pinus taeda, Pseudotsuga menziesii var. menziesii, and Welwitschia mirabilis* (Table 8.2B).

These UniGene builds and PlantGDB cluster assemblies are useful for comparing across species where ESTs were obtained at different times on different technology platforms and libraries were sampled to different depths; however, they are not updated continuously and so they may not incorporate all of the currently available ESTs. When comparing the UniGene builds, it is striking that the number of clusters is very similar among the three most deeply sequenced conifer species (ranging from 18.8 to 21.6 thousand UniGenes), even though the number of ESTs used to generate the clusters varied more than two-fold (from 121,000 to 293,000). Using a different approach, the PlantGDB has arrived at much larger and more variable numbers of clusters for the same three species (ranging from

29.1 to 72.8 thousand putative transcripts or PUTs) based on more recent but only slightly larger data sets than those used by the NCBI. Consistent with the significant differences between these two public databases, there is evidence to indicate that the bioinformatic approach used to create the UniGene builds (NCBI) has a tendency to collapse gene families (overclustering), not infrequently placing closely related family members in the same cluster (Frank and Ercal 2005).

Table 8-2 A. Number of clusters for conifers available in NCBI's Unigene Database[1].

Species	mRNAs	ESTs (3'reads/ 5'reads/ unknowns)	Total sequences in clusters	Number of Clusters	Clusters with mRNA & ESTs	Date of EST retrieval (dbEST)	Unigene Build
Picea glauca	79	124,913/ 140,355	265,347	22,472	65	31 Mar 2009	#13
Picea sitchensis	10,501	104,277/ 17,570	132,348	19,828	6,299	21 Oct 2009	#14
Pinus taeda	154	97,364/ 172,250/ 24,146	293,914	18,079	126	25 Sept 2009	#12

B. Number of clusters for conifers available Plant GDB[2].

Plant Species	GenBank EST + cDNA		PlantGDB assembly		
	Total available	Currently assembled	Total # of PUTs	Date	Version
Cryptomeria japonica	56,706	56,703	24,299	2008-09-13	167a
Gnetum gnemon	10,753	10,430	6,193	2008-01-25	163a
Picea abies	14,480	14,478	8,715	2010-01-14	175a
Picea engelmannii x Picea glauca	28,186	28,179	13,880	2007-01-26	157a
Picea glauca (white spruce)	332,879	321,190	48,619	2010-01-19	175a
Picea sitchensis (Sitka spruce)	199,673	186,161	31,054	2010-01-21	175a
Pinus banksiana	36,387	36,387	13,040	2010-05-25	177a
Pinus contorta	42,091	41,136	13,570	2010-01-19	175a
Pinus pinaster (cluster pine)	34,884	34,792	15,648	2010-05-25	177a
Pinus taeda (loblolly pine)	329,024	329,584	72,829	2007-01-26	157a
Pseudotsuga menziesii var. menziesii	14,354	14,353	9,857	2007-09-3	161a
Welwitschia mirabilis	10,137	10,134	6,606	2007-01-29	157a

[1]The Unigene database contains assemblies for species for with > 100,000 ESTs. Accessed August 2, 2010 (*http://www.ncbi.nlm.nih.gov/unigene*).
[2]The PlantGDB database contains assemblies for species for with > 10,000 ESTs. Accessed August 2, 2010 (*http://www.plantgdb.org/prj/ESTCluster/progress.php*).

Another public source for conifer sequence assemblies is ConiferGDB (Liang et al. 2007) (not related to PlantGDB) (*www.conifergdb.org/cgdb3p1/tiki-index.php*), which currently only contains information for *P. taeda*, but has plans to develop and release other species-specific clusters. In a recent update, ConiferGDB reclustered the same *P. taeda* ESTs used in the NCBI UniGene build (11) using the latest version of CAP3. Many of the UniGene clusters ended up subdivided to yield around 36,800 contigs and singletons that are best characterized as representing unique transcripts. Again, the different results of these EST clustering analyses highlight the impact of the analysis methods, dictated in part by the desired outcomes. Perhaps more importantly, they serve as a reminder that each gene may be represented by different transcript isoforms, such as variants arising from alternative splicing or alternative polyadenylation. Such variants may be clustered together or separated according to analysis goals, which may focus on defining genes or describing transcripts. For example, recent analyses of the *P. sitchensis* transcriptome identified around 46,700 putative transcripts (Ralph et al. 2008), while analyses of the *P. glauca* dataset identified 28-30,000 clusters purported to represent distinct genes (P Rigault pers. comm.). In a contrasting strategy, another public resource, the Plant Gene Indices (PGI) database (*http://compbio.dfci.harvard.edu/tgi/plant.html*), contains a catalog of sequence clusters generated using ESTs pooled for several species within the same genus (Quackenbush et al. 2001). The most recent PGI release reporting 80,494 unique sequences for spruce (42,051 multi-sequence clusters) and 61,864 unique sequences for pine (42,051 multi-sequence clusters). The larger numbers of clusters generated in these assemblies likely results in part from the fact that sequences from multiple species have been clustered together, but likely also reflects differences in clustering methods.

When different ESTs datasets are clustered using the same method, cluster size (number of sequences per cluster) is a useful indicator of relative sampling redundancy and transcriptome coverage (Fig. 8-1). For example, the number of clusters containing a single sequence (also termed singletons) is directly related to the depth of sequencing and overall redundancy in the data. The number of singletons in the three conifer species UniGene builds varied greatly, ranging from 65 in *P. taeda* to more than 4,000 in *P. glauca*, even though both EST collections contained a similar number of sequences (293,000 versus 252,000, respectively). These data reflect a lower redundancy in the *Picea glauca* EST dataset and indicate that transcriptome sampling in this species remains incomplete. It should be noted that the higher discovery rate in *P. glauca* is consistent with many of the *P. glauca* cDNA libraries being normalized, whereas the non-normalized libraries sampled for *P. taeda* would be expected to have higher degeneracy. It must also be kept in mind that randomly sampling of cDNA libraries prepared from bulk tissues only yields an "average" transcriptome for the most abundant cell types. The coupling of techniques for high-resolution sampling of specific cell types (e.g., laser-capture microdissection) with next-generation sequencers that can use minute amounts of input DNA has shown just how drastically different transcriptomes can be in proximal cell types (Emrich et al. 2007).

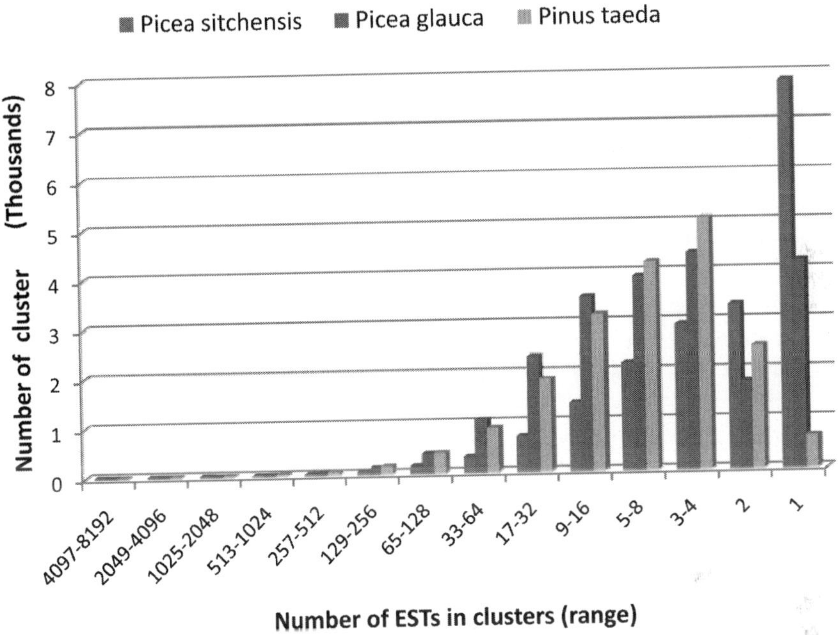

Figure 8-1 Histogram of cluster size's from NCBI Unigene database (builds from Table 2A).
Color image of this figure appears in the color plate section at the end of the book.

8.1.5 Full-length cDNA Sequencing

The full-length analysis of cDNA clones is a valuable approach to improve transcriptome characterization. FL-cDNA sequencing involves selecting individual clones after the clustering step (typically the clone with the longest insert in a cluster or contig) and then obtaining additional end and internal sequences through the use of custom primers (Ralph et al. 2008; see Fig. 8-1). When conducted on a large-scale, it can help to overcome some of the limitations or uncertainties of EST clustering. Because the entire sequence is derived from a single cDNA insert, whether it represents a partial or a complete coding sequence (CDS), it may be used to validate clustering outcomes (consensus sequences or TCs) and eliminate artefacts. Large-scale FL-cDNA analysis has seldom been undertaken in conifers because it is more tedious and costly than random-end sequencing. However, it has become more tractable as many high-throughput laboratories have acquired the capacity for routine handling of large collections of cDNA clones, and as the cost of oligonucleotide primer synthesis has decreased.

Ralph et al. (2008) reported the characterization of 6,464 FL-cDNAs from *P. sitchensis* selected from among a set of approximately 46,700 putative transcripts. This work has proven important for the accurate analysis of closely related members of genes families, which in turn was critical for assigning specific biochemical functions to closely related enzymes, such as diterpene synthase family members (Keeling et al. 2008). Despite the fact that this particular dataset appears populated with many short open reading frames (ORFs) or partial cDNAs (average ORF length is 616 nucleotides), it remains a valuable resource for future annotation of conifer genomes.

8.1.6 Datamining ESTs for Gene Family Analyses

EST data clusters have become the starting point for more efficient and comprehensive investigations of gene families in conifers. They have been used alongside or in combination with targeted PCR amplification method to collect more comprehensive gene family collections. A non-exhaustive list of the genes families that have been so far investigated include the arabinogalactan proteins (AGPs) (Zhang et al. 2000), cellulose synthases (*CesA*) (Nairn and Haselkorn 2005), dirigent proteins (Ralph et al. 2006b, 2007), the cytochrome P450 monooxygenases of the terpenoid oxygenase superfamily (Hamberger and Bohlmann 2006), expansins (Sampedro et al. 2006), auxin response factors (*Aux/IAAs*) (Goldfard et al. 2003), *KNOTTED-LIKE HOMEOBOX Class 1* (*KNOX1*) transcription factors (Guillet-Claude et al. 2004), and the R2R3-MYB transcription factors (Bedon et al. 2007).

Sequence and phylogenetic analyses in several of the above reports have proven useful for efforts to compare and contrast the evolution of gene families in gymnosperms and angiosperms (including *Arabidopsis thaliana*, poplar, grape vine, rice, and maize among others). As might be expected, these investigations indicate that conifer gene families appear to have followed evolutionary trajectories that are both partly overlapping and partly distinct from the angiosperms. While whole genome sequencing will be required to generalize conclusions from these studies, highlights from their findings speak to the potential for such studies to enhance our understanding of conifer biology. For example, all of the expressed sequences for the conifer *KNOX1* gene family fell into only one of the three *KNOX1* sub-family branches identified in angiosperms, but all of the conifer species had more numerous and more recent gene duplications within this branch (Guillet-Claude et al. 2004). The phylogenetic trees generated for AGPs and Aux/IAAs in *P. taeda*, as well as dirigent proteins in *Picea* spp., provided clear evidence that major gene duplication events have occurred since the angiosperm-gymnosperm split. The dirigent proteins (Ralph et al. 2006b, 2007), the expansin superfamily (Sampedro et al. 2006), and the R2R3-MYBs (Bedon 2007) all provide striking examples of entire subfamilies or subclades that are specific to conifers. And Pavy et al. (2005b) concluded that *P. glauca* expressed double the number of S-adenosylmethionine transferase (SAMT) genes than either *Arabidopsis* or rice. As several of these authors point out, these divergent evolutionary trajectories may well provide the basis for sub- or neo-functionalization of the new gene family members.

It is sometimes assumed that coniferous trees may have more in common with woody perennial angiosperms, such as poplar, than with herbaceous angiosperms, such as rice or *Arabidopsis*. The phylogenetic analyses of conifer gene family expansions have yet to yield evidence of closer or shared gene family evolution between gymnosperms and woody angiosperms. Large-scale comparison of spruce EST clusters (Pavy et al. 2005b) showed that the proportion of shared genes with *Arabidopsis* was nearly identical to the proportion shared with poplar, and that the overall level of sequence similarity in each of the comparisons was likewise equivalent. These observations that woody growth habit alone should not be taken as an indicator of similarities in the molecular regulatory networks between woody angiosperm species and coniferous trees are sensible in light of the fact that the character of secondary xylem (wood) formation has arisen and been lost from independent plant lineages on multiple occasions throughout the course of evolutionary history (Groover 2005; Rothwell et al. 2008).

8.1.7 Non-coding, Small RNAs

In recent years, several classes of small non-coding RNAs (ncRNAs) have been shown to accumulate in plants and animals where they contribute to transcriptional and post-transcriptional gene regulation (Bartel 2004; Zhang et al. 2006a, 2006b). The various classes—microRNAs (miRNA), short interfering RNAs (siRNA), heterochromatin siRNA, and trans-acting siRNAs (tasiRNA)—are distinguished by their function and mode of formation (Morin et al. 2008). The miRNAs, which range in length from 19 to 24 nt, are cleaved by precise mechanisms from highly conserved stem-loop structures (Meyers et al. 2008). They are a diverse class represented by several families involved in negative regulation of gene expression, often through the targeting of transcription factors, such as HD-zips and ARFs (Oh et al. 2008), as well as other types of protein-encoding transcripts. The first evidence for conifer miRNAs was reported by Axtell and Bartel (2005) who identified 11 conserved miRNAs in *Pinus resinosa* needles. Subsequently, 26 additional conserved miRNAs were described in rust-infected tissues of *P. taeda* rust (Lu et al. 2007). Conserved miRNAs have also been reported in *Pinus contorta* (Morin et al. 2008), *Taxus chinensis* (Qiu et al. 2009) and several other conifers (Dolgosheina et al. 2008).

A unique feature of conifer genome biology was highlighted by Dolgosheina et al. (2008) who showed that the populations of small RNAs present in tissue samples from a panel of gymnosperms (including several conifers) were structurally distinct from those that had been widely observed in angiosperm plants. These researchers showed that whereas angiosperm miRNAs accumulated in two predominant size-classes of 21 and 24 nt (Axtell et al. 2007), only the 21 nt size class of miRNAs was found to accumulate in gymnosperms. Large-scale sequencing of small RNAs in *P. contorta* (Morin et al. 2008) and *T. chinensis* (Qiu et al. 2009) confirmed that 21 nt miRNAs were the only size class to accumulate in these species. Since distinct mechanisms are employed to produce the different size classes of miRNAs (e.g., different members of the Argonaute protein family are involved), these results indicate that the miRNA synthesis mechanisms have diverged between these two plant phyla. The studies by Morin et al. (2008) and Qiu et al. (2009) identified additional novel sequences not encountered in angiosperms that could potentially represent gymnosperm-specific miRNAs. Among the apparent target sequences for these novel miRNAs were cellulose synthases and glutathione peroxidases among others. Limited EST datasets for these species and lack of a reference genome for conifers have restricted characterization of miRNAs that are putatively conifer-specific. Nonetheless, 51 novel small RNAs having the secondary structure typical of miRNAs were identified in *P. contorta* (Morin et al. 2008). Taken together, these findings depict a distinct evolutionary path for miRNA synthesis in conifers compared to angiosperms. The potential functional implications with respect to gene regulation remain largely to be described. In conifers, miRNAs have been implicated as potential regulators of gene expression during embryo growth and development (Oh et al. 2008), as well as during defense responses (Lu et al. 2007; Qiu et al. 2008). It stands to reason that these putative miRNAs could be linked to these and many other important biological processes in conifers.

8.2 Transcriptome Analyses

8.2.1 Wood Formation

Vascular tissue development and wood formation have been investigated by transcript profiling in a variety of experimental systems, including *Arabidopsis*, poplars and conifer trees. Results from these studies indicate that wood-forming tissues in gymnosperms and angiosperms share extensive conservation in terms of sequence similarity and expression profiles for their transcriptomes. However, large numbers of sequences are either uniquely expressed in the xylem of conifers or their sequences are so highly diverged that their relationship to angiosperm sequences cannot be established for comparisons.

8.2.1.1 Transcripts Preferentially Expressed in Xylem

A major approach for this line of inquiry has been to identify transcripts that accumulate preferentially in xylem compared to other tissues in order to define broad categories of genes that may play a role in wood formation. Yang et al. (2004) used a *P. taeda* cDNA microarray comprised of 1,500 unique sequences to identify 30 transcripts that accumulated preferentially in differentiating xylem compared to needles, embryos and megagametophytes. Pair-wise comparisons between xylem and each of these tissues identified many additional sequences that may also be associated with vascular tissue development. Pavy et al. (2008a) used microarrays that comprised 9,000–11,000 unique cDNAs to identify 360 *P. glauca* transcripts that were accumulated preferentially in differentiating secondary xylem compared to young needles and differentiating phloem. Computational methods were used to compare these spruce data to angiosperm xylem core gene sets of 52 sequences (Ko et al. 2006), and of 319 sequences (Zhao et al. 2005). An overlapping set of 31 transcripts were identified as conserved between *P. glauca* (gymnosperms) and these angiosperms, both in terms of sequence and preferential accumulation in xylem (Pavy et al. 2008a). The set included four of the six stem-specific markers from the AtGenExpress profiles for *A. thaliana* (Schmid et al. 2005), specifically a nodulin-like protein (AT1G21890), a protein kinase (AT1G24030), a thaumatin family protein (AT5G40020), and a glycogenin glucosyltransferase (AT3G18660). Suppression subtractive hybridization was used to identify a small collection of transcripts from the wood-forming tissues of Chinese fir, *Cunninghamia lanceolata*, and their preferential expression in xylem was confirmed using macroarray hybridization and quantitative RT-PCR (Wang et al. 2007).

Among the genes that show conserved preferential expression in xylem across both gymnosperms and angiosperms are enzymes and structural proteins related to cell wall biosynthesis, as well as signaling and regulatory proteins (e.g., Hertzberg et al. 2001; Demura et al. 2002; Oh et al. 2003; Yang et al. 2004; Nairn and Hazelkorn 2005; Friedmann et al. 2007; Pavy et al. 2008a). The cell wall structure-related transcripts include enzymes involved in cellulose, hemicellulose and lignin biosynthesis, as well as arabinogalactan proteins (AGPs), proline and hydroxyproline-rich proteins, expansins, xyloglucan endo-transglycosylases (XETs), glycosyl transferases (family 8, 43) and others. Yang et al. (2004) showed that many transcripts belonging to these gene families accumulated preferentially pine xylem compared to megagametophyte, needle or embryo tissues. In a serial analysis of gene expression (SAGE) profiling experiment using differentiated xylem samples from the crown and base of a 10-year old *P. taeda* tree, Lorenz and Dean (2002) found that alpha-tubulin, AGPs, ubiquitin and an aquaporin, along with several unidentified sequences, were among the most abundant transcript tags. Among the sequences preferentially expressed in spruce xylem was one similar to the *Arabidopsis cobra-like4* gene (Friedman et al. 2007; Pavy et al. 2008a), which is involved in cellulose and pectin-containing secondary cell wall biogenesis (Roudier et al. 2005). Zhao et al. (2005) showed that transcripts from genes linked to cellular communication/signal transduction, as well as transport facilitation, were over-represented in the secondary xylem of *Arabidopsis* hypocotyls. Similarly in spruce, Pavy et al. (2008a) found that most of the differentially expressed receptor protein kinases and carbohydrate transporters, as well as several of the major intrinsic proteins, peptide transporters and nodulin-related sequences (e.g., MtN21) accumulated preferentially in differentiating xylem. One of the few stem-specific expression markers recognized from AtGenExpress studies is a specific nodulin (Schmid et al. 2005); thus, it is of interest to note that the family of xylem-expressed nodulin genes appears to be more diverse in conifers.

Transcription factors (TFs) belonging to the ARF, AUX/IAA, MYB, HD-ZIP, WRKY and NAM families that have been linked to differentiation of secondary vascular tissues in herbaceous angiosperms (Demura and Fukuda 2007), as well as woody perennial angiosperms, including *Populus* (Hertzberg et al. 2001) and *Eucalyptus* (Paux et al. 2004), also have putative homologues expressed in xylem or stem tissues of conifers (Patzlaff et al. 2003a, 2003b; Bedon et al. 2007; Friedmann et al. 2007; Bomal et al. 2008; Pavy et al. 2008a). For example, 39 different transcription factor (TF) sequences were identified in white spruce xylem, including sequences that had not been previously linked to vascular tissue development, such as two sequences encoding TUBBY-like proteins (TLP), which belong to the F box proteins, and a *gras* family member similar to *gai* (Pavy et al. 2008a). Other TLPs are known to play roles in determining sensitivity to various environmental factors, such as salt, chilling, oxidative stress, or water-deficit (Lai et al. 2004). The gibberellin insensitive (*gai*) gene is known as a repressor of the gibberellin (GA) signaling pathway, but its potential role in xylem development remains to be determined.

Gymnosperms and angiosperms diverged over 300 million years ago (Magallón and Sanderson 2005). Although pine has been described as sharing a level of general genomic similarity with angiosperms (Kirst et al. 2003), xylem cell types differ quite profoundly between these taxonomic groups. It is not surprising then that many sequences expressed in the wood-forming tissues of conifers do not match the signatures of proteins of known or predicted function in other organisms (Kirst et al. 2003;

Pavy et al. 2005b). However, targeted searches using hidden Markov model (HMM) methods have uncovered conserved domains of unknown function (DUFs) among conifer transcripts that are preferentially expressed in xylem tissues. For example, sequences harboring DUF547 (AT5G60720) and DUF579 (AT5G67210) were linked to xylem tissues in both spruce (Pavy et al. 2008a) and *Arabidopsis* (Ko et al. 2006).

Ultimately, transcript profiling using developmental series has the potential to reveal the full spectrum of xylem-expressed genes and shed light on the timing of their expression relative to cellular differentiation events. A wood formation roadmap was developed for *Populus* using microarray analyses of finely dissected tangential microsections along an axis running from the cambial zone to an area of fully differentiated xylem (Hertzberg et al. 2001). The study established that genes encoding lignin and cellulose biosynthetic pathway enzymes, as well as several potential regulators of xylogenesis, were under strict stage-specific transcriptional control. A follow up analysis using a more comprehensive cDNA microarray comprised of over 13,000 unique sequences identified a broader set of stage-specific markers and potential regulators of cambial cell identity (Schrader et al. 2004). No similar studies have yet been reported for conifers, but Friedmann et al. (2007) compared gene expression in different portions of the terminal stem of Sitka spruce, contrasting regions of stem consisting mainly of primary xylem to those containing a well-defined layer of secondary xylem. Their study showed that the base segment containing secondary xylem accumulated higher levels of many of the transcripts associated with preferential xylem expression, and were similar to transcripts that accumulated in the later stages of *Populus* wood formation.

8.2.1.2 Variation in Wood Formation and Wood Properties

Wood properties are complex traits that are influenced by environmental conditions as well as developmental factors related to the cambial age, and as such they can vary extensively within a single tree (Zobel and Talbert 1984; Whiteman et al. 1996; Neale et al. 2002; Gonzalez-Martinez et al. 2007). The amplitude and rate of change leading to variations in wood properties may range from large and rapid to subtle and slow, and the processes may be linked or relatively independent of one another. It then stands to reason that understanding the molecular basis for variations in wood quality within trees could contribute significantly to the development of genetic markers with predictive capacity for wood properties. Xylem libraries representing different stages of differentiation, development or tissues under specific mechanical stress have been used for the production of ESTs (Kirst et al. 2003; Paiva et al. 2008a), which have been analyzed by digital gene expression approaches (Whetten et al. 2001; Pavy et al. 2005a) in addition to being used for the construction microarrays.

In conifers, tracheid cell wall thickness and wood density vary significantly across the seasonal developmental gradient from earlywood to latewood formed within a single growth ring, regardless of the cambial age (Plomion et al. 2001). Transcript profiling experiments have been performed on samples of differentiating xylem collected at different time points throughout the growth seasons for *P. taeda* (Egertsdotter et al. 2004) and *P. pinaster* (Le Provost et al. 2003). Both experiments followed only a few hundred sequences and identified relatively small sets of differentially expressed genes. Transcripts for a number of cell wall-related enzymes and structural proteins were found to accumulate during latewood formation, including *P. taeda* enzymes involved in lignin or cellulose biosynthesis, as well as a glycine-rich protein and an alpha-tubulin in *P. pinaster*. A transcript encoding a putative low molecular weight heat shock proteins that was preferentially expressed during *P. pinaster* latewood formation had a homolog identified in a *P. taeda* latewood library using digital profiling (Pavy et al. 2005a). This latter study also found that two putative dehydrin transcripts were more abundant in *P. taeda* latewood samples, but the water stress-inducible gene, *lp3*, accumulated to higher levels in earlywood. Overall, fewer transcripts were preferentially expressed in earlywood, and the transcripts from earlywood represented a greater diversity of functional categories, ranging from glyceraldehyde-3-phosphate dehydrogenase to proline-rich cell wall proteins, as well as numerous sequences without similarity to known proteins. Yang and Loopstra (2005) used microarrays and quantitative RT-PCR to show that timing of the earlywood to latewood transition, as well as expression patterns for many of the genes involved, were strongly influenced by seed source genetics. All of this suggests that transcript profiling with large gene sets may be needed to fully delineate the molecular events of the earlywood to latewood transition.

A developmental gradient related to cambial age was investigated in *P. pinaster* by collecting xylem samples at different points along the main stem of a 30 year-old tree (Paiva et al. 2008a). Six different xylem samples estimated to represent cambial ages ranging from 3 to 21 years were analyzed using a custom cDNA microarray comprised of 3,512 different gene sequences and in parallel using two-dimesional polyacrylamide gel electrophoresis (2-D PAGE). Differential transcript accumulation patterns identified four general profile classes, including two major classes comprised of transcripts preferentially expressed in base wood (93 genes) and crown wood (71). In addition to sequences already discussed, such as cell wall-related proteins and biosynthetic enzymes, a number of transcripts encoding products not usually associated with xylem formation were also recovered. These included many enzymes related to protein synthesis and energy production (up-regulated in crown wood), as well as various stress response genes (up-regulated in base wood) (Paiva et al. 2008a). However, an earlier study comparing mature (base) and juvenile (crown) wood in *P. taeda* had obtained some overlapping data, although the results were not reported for relative transcript levels (Whetten et al. 2001). For example, many stress related transcripts were more abundant in the mature wood (e.g., low molecular weight heat-shock proteins, MW-HSP). Paiva et al. (2008a) concluded that their transcript accumulation profiles were generally consistent with the protein profiling data and other phenotypic information, such as higher cellulose accumulation in base wood (*CesA* expression) and prolonged differentiation leading to thicker cell walls (possibly related to delayed apoptosis due to stress response gene expression). Another earlier study compared crown and base wood used serial analysis of gene expression (SAGE) to identify a large number of sequences tags that accumulated differentially in differentiating xylem below and within the live crown (Lorenz and Dean 2002). However, in that study, none of the differentially expressed tags could be assigned a function, largely due to the lack of annotated pine sequences available when the study was pursued.

Large variations in wood structure and composition are observed between normal wood and the compression wood that develops locally in portions of the stem under mechanical stress, typically on the underside of leaning trees and of branches (Du and Yamamoto 2007). Compression wood is characterized by accelerated growth rate (wider annual rings), thicker and more highly lignified cell calls, and altered cellulose microfibril angle, among other structural and chemical changes. Digital profiling of pine ESTs that occurred in xylem tissues forming compression versus normal wood identified many transcripts that were up-regulated in compression wood, including ones encoding putative AGPs, lignin biosynthetic enzymes, alpha-tubulins, and various other cell wall proteins (Whetten et al. 2001; Pavy et al. 2005a; Koutaniemi et al. 2007). Interestingly, AGPs represent one of the few classes of genes that are up-regulated in both the compression wood of conifers (Zhang et al. 2000) and the tension wood of angiosperm trees, such as poplar (Lafarguette et al. 2004). A transcript profiling study in *P. taeda* using a cDNA microarray comprised of 3,000 different ESTs arrived at similar conclusions, also identifying down-regulated transcripts that included an ABC transporter-like sequence, a pectinesterase, and a Phi1-like protein (Yang et al. 2004). An investigation of compression wood formation in *Chamaecyparis obtusa* showed that in trees grown at angles ranging from 10 to 50 degrees relative to vertical transcripts up-regulated in compression wood included laccase and various unknowns, while those that were down-regulated included a subtilisin-like protease and a beta-1, 3-glucanase protein (Yamashita et al. 2009). Similarly, different R2R3-MYB transcripts were either up-regulated along with lignin biosynthesis enzymes in differentiating compression wood or down-regulated in compression xylem from young *P. glauca* trees (Bedon et al. 2007).

Gion et al. (2005) analyzed the *P. pinaster* proteome in several xylem samples and found that there was an overall weak to moderate correlation between relative transcript abundance, based on EST frequency, as well as relative protein abundance determined from 2-D gel spot intensity. Pair-wise comparisons between different xylem samples (latewood vs. earlywood; juvenile vs. mature wood; compression vs. opposite wood) were consistent with findings from independent transcript analyses. For example, several of the proteins that appeared to accumulate in latewood were similar to drought-associated transcripts noted by Egertsdotter et al. (2004), while mature wood accumulated several stress-related proteins consistent with transcript sequences reported by Paiva et al. (2008a).

8.2.2 Biotic and Abiotic Stress

Transcriptional profiling has created a lot of interest in its potential for teasing out the molecular mechanisms determining susceptibility and resistance in plants, but useful results from such studies are only just starting to appear (Tan et al. 2009). In conifers, Myburg et al. (2006) used a rather small (1248 element) cDNA microarray to explore gene expression changes in susceptible and resistant classes of *P. taeda* seedlings challenged with a single aeciospore isolate of *Cronartium fusiforme*, the causative agent of fusiform rust. Although a few genes associated with cell wall formation were identified in the study as differentially expressed, the overall conclusions were not strong owing in large part to the small number genes included in the array. More recently, Adomas et al. (2008) used a somewhat larger (2109 element) *P. taeda* array to measure the responses in *P. sylvestris* seedlings challenged with saprotrophic (*Trichoderma aureoviride*), pathogenic (*Heterobasidion annosum*), or ectomycorrhizal (*Laccaria bicolor*) fungi. A variety of overlapping and divergent gene expression patterns were detected for the three classes of plant-fungus interaction. The ectomycorrhizal interaction was reported in greater detail in a second microarray study, and the results indicated that establishment of the symbiosis between *L. bicolor* and pine was quite comparable to the process that takes place between *L. bicolor* and various angiosperm hosts (Heller et al. 2008).

Insect pests represent the another major biotic stress faced by conifers, and a microarray comprised of approximately 5,500 unique spruce cDNA elements has been used to study the transcriptional profiles of *P. sitchensis* responding to mechanical wounding, or feeding by either spruce budworm (*Choristoneura occidentalis*) or white pine weevils (*Pissodes strobi*) (Ralph et al. 2006c). The study uncovered a large collection of genes that had previously been associated with wounding and pest responses, but also identified a variety of genes that had not previously been associated with such responses. More recently proteomic work has demonstrated significant concordance between transcriptional changes in these responses and the production of specific proteins (Lippert et al. 2007).

Drought stress is a significant concern for conifer forestry since it is frequently a primary contributor to losses during reforestation efforts. The amplified fragment length polymorphism (AFLP)-cDNA approach was used to identify a few hundred genes that showed differential expression in *P. pinaster* roots and hydroponically grown seedlings subjected to drought conditions (Dubos and Plomion 2003; Dubos et al. 2003). Lorenz et al. (2006) used digital tagging of ESTs from *P. taeda* cDNA libraries that varied by genotype and water status to identify genes that displayed genotype-specific changes in expression pattern under drought conditions. Interestingly, one genotype in the study, whose parents were selected from a region that seldom experiences drought conditions, showed lower overall tolerance to drought, and this was reflected in stronger up-regulation of drought stress-associated genes under drought conditions. This suggests that adaptive alleles for drought tolerance may not be well represented in this particular genotype or its parents. Watkinson et al. (2003) also looked at drought responses in *P. taeda* using a 2,173-element cDNA microarray and reported on physiological and metabolic responses observed during photosynthetic acclimation. Paiva et al. (2008b) used a microarray comprised of 3,512 unique cDNA elements from *P. pinaster* to examine the effect of drought conditions on gene expression over the earlywood to latewood transition. The authors noted a great deal of plasticity in the response of genes related to wood formation when confronted by the challenge posed by drought. Sathyan et al. (2005) used quantitative RT-PCR to investigate the expression of drought-responsive genes in Aleppo pine under varied levels of water stress.

Another area of conifer research where transcriptional profiling is in active use is the study of cold acclimation. Joosen et al. (2006) used a *P. taeda* cDNA microarray to assess differences in gene expression levels of apical bud or root tissues in *P. sylvestris* that had either been acclimated to cold conditions or subjected to an abrupt shift to near-lethal cold temperatures. Though limited in terms of the number of genes assessed, significant differences were seen in a number of genes and several appeared to be homologs of genes that had previously been associated with cold tolerance. Holliday et al. (2008) recently used a much larger (21,840 unique element) spruce cDNA array to study acclimation in three populations of *P. sitchensis*. More than 2000 genes were significantly up or down-regulated in response to cold acclimation and there was evidence that at least some of the responses possibly reflected local adaptations for the populations studied.

8.2.3 Embryogenesis and Other Developmental Processes

DNA arrays were in use to compare transcriptional profiles for zygotic and somatic embryos even before microarray technologies became widespread and readily accessible (Cairney et al. 1999, 2000). Once it was clearly demonstrated that microarrays constructed using cDNA sequences from *P. taeda* could be used effectively to monitor gene expression patterns in other members of the Pinaceae (van Zyl et al. 2002), pine microarrays were adopted for use in a variety of experiments directed toward the improvement of techniques for conifer somatic embryogenesis (Stasolla et al. 2004b). For example, the effects of various chemicals, such as reducing agents (Stasolla 2004a) and compounds affecting water potential (Stasolla et al. 2003a, b), on gene expression patterns and cultured embryo development were compared with the same phenomena in zygotic embryos in an effort to develop more effective media formulations for embryo development and maturation.

Other developmental processes that have been investigated using transcriptional profiling approaches include adventitious root development (Brinker et al. 2004) and seasonal bud burst (Yakovlev et al. 2006; Asante et al. 2009). These studies used microarrays fabricated with *P. taeda* cDNAs to assess gene expression in *P. contorta* and *P. abies*, respectively. Studies of MYB transcription factor effects on gene expression patterns in the developing tissues of transgenic spruce were performed using a microarray fabricated with *P. glauca* cDNA sequences (Bomal et al. 2008).

8.3 Technological Advances for Gene Discovery and Transcriptional Profiling

8.3.1 DNA Sequencing

8.3.1.1 New Platforms

In the short time since the first peer-reviewed report from 454 Life Sciences on use of their "next-generation" DNA sequencer to sequence and assemble the *Mycoplasma genitalium* genome from the data generated in a single run of their new instrument (Margulies et al. 2005) there has been a complete revolution in how we pursue genome studies. Massively parallel, short-read DNA sequencing platforms using sequencing-by-synthesis approaches (Illumina,GAII and ABI SOLiD) have further contributed to this revolution (Ansorge 2009), while even newer platforms promising single-molecule sequencing (e.g., Helicos and Pacific Biosciences) wait in the wings. Each of these platforms has its own inherent biases with respect to data output (Harismendy et al. 2009), but no more than was associated with Sanger sequencing when one considers library cloning bias (Osoegawa et al. 2007)

However, the various platform biases do not for the most part overlap, and hybrid approaches pairing the strengths of different platforms are already showing great promise for the rapid production of *de novo* genome sequences (DiGuistini et al. 2009)

These next-generation sequencer (NGS) or high-throughput sequencing (HTS) technologies are all characterized by the use of massively parallel approaches in which millions of sequencing reactions are performed simultaneously on specialized nano-scale fabricated supports. The scale of these reactions has reduced the price per base by orders of magnitude. For example, whereas current best costs for Sanger sequencing run about US$2.00/kB (US$0.50–1.00 per 500 nt read), the Roche GS-FLX Titanium platform (based on the original 454 Life Sciences technology and yielding approximately 1 million reads of 400 nt from a single run) delivers data at about US$0.03/kB. The GAII and SOLiD sequencers, both of which generate sequence in the range of up to 50 million reads of 70–120 nt during a single run, currently deliver data at less than US$0.0025/kB, which is almost a thousand-fold less than the cost delivered by Sanger sequencing. All three of these technologies continue to improve by extending read lengths and increasing the density of parallel reactions, which is a good thing since the next wave of technology platforms approaching the market have the purported potential to drop costs per base by another order of magnitude or more. These prospects are particularly exciting for the conifer community since they signal a rapidly approaching time when we will have completed reference genome sequences in hand for multiple members of the conifer family.

8.3.1.2 Data Assembly

Whether we limit the discussion to whole transcriptome sequencing (RNA-Seq) studies or consider strategies for sequencing a complete conifer genome using HTS approaches, data flow and assembly quickly become the critical bottleneck, and even data storage becomes an issue (Batley and Edwards 2009). The terabyte-scale datasets generated by the new short-read sequencing platforms are so large that it is cheaper to repeat a sequencer run than to store the data, not to mention the fact that few institutional data networks have the capacity or hardware to transmit such datasets between sequencers and bioinformatics clusters. The speed and accuracy with which DNA sequences can be assembled from fragments is dramatically impacted by read length, and assemblies based on short sequence reads from NGS platforms run into difficulties with repeated sequences, insertion/deletions, and other variations that occur at scales of a few tens of bases (Wall et al. 2009). However, as previously noted, new software and hybrid approaches based on multiple technology platforms are being adapted to circumvent these problems (DiGuistini et al. 2009).

Paired-end tag (PET) sequencing approaches in which both ends of each DNA fragment are sequenced have proven a useful means for addressing the problem of short-read assembly (Fullwood et al. 2009). PET sequencing relies of the circularization of DNA fragments, so paired-ends cannot be captured for sequences much over 5 kB in length. This is not a big issue for RNA-Seq studies where average cDNA length is more on the order of about 1 kB; however, it is an issue for researchers trying to use short-read sequencers for whole-genome shotgun (WGS) sequencing of organisms with large genomes. For RNA-Seq experiments, uniform fragmentation of the relatively short cDNAs can be difficult, but specialized instrumentation, such as the Covaris Adaptive Force Acoustics (AFA) system, can efficiently and reproducibly shear duplex DNA into random fragments as small as 50 bp.

8.3.1.3 Impact on Transcriptome Analyses

Carninci (2009) recently posed the provocative question, "Is sequencing enlightenment ending the dark age of the transcriptome?" Indeed, the new HTS platforms provide sufficient depth of coverage that one or two runs of a cDNA preparation are usually sufficient to generate a robust collection of full-length gene model assemblies for a given transcriptome, and identify at least portions of the many low abundance transcripts. Likewise, these new DNA sequencing platforms are enabling researchers to catalog and probe deeply into the new world of small ncRNAs (Forrest and Carninci 2009), as well as explore the full range transcript variations that arise from alternative splicing (Sultan et al. 2008). Indeed, these new technologies raise a multitude of possibilities for new approaches to transcriptional analyses in plants (Simon et al. 2009). No doubt we can look forward to a time in the near future when conifer researchers use NGS to identify RNAs brought down by immunoprecipitation of RNA-binding proteins, similar to a study recently reported for *Arabidopsis* (Terzi and Simpson 2009).

One of the more provocative suggestions elicited by development of the short-read sequencing platforms has been that hybridization-based approaches to quantifying gene expression, including DNA microarrays, might be nearing obsolescence (Coppee 2008). While early attempts to use sequence tag profiling were not very successful at identifying differentially expressed genes in juvenile and mature wood due to the lack of a sufficient cDNA sequence resource for loblolly pine that the tags could be associated with annotated gene sequences (Lorenz and Dean 2002), such efforts have a much greater potential for success provided the current depth of sequence available and the speed with which a robust transcriptome sequence can be assembled. Indeed, DNA sequence tag-based transcriptional profiling approaches appear to be enjoying a resurgence of interest (Ng et al. 2006;

Vega-Sanchez et al. 2007; Morrissey et al. 2009), and they should prove particularly useful for studies in species where the community is insufficient to support the development costs of microarrays.

8.3.2 *Other Approaches*

Of course, suggestions that the imminent demise of microarrays is at hand should be considered highly exaggerated. As we have previously recounted, even small, custom cDNA arrays have proven useful in studies of a wide variety of biological responses in diverse conifer species (e.g., Joosen et al. 2006; Adomas et al. 2008; Paiva et al. 2008a). The recent development of larger cDNA arrays (>20,000 spots) using *P. glauca* (Holliday et al. 2009), *P. taeda* (Lorenz et al. 2009), or *P. pinaster* (*http://www.picme.at/index.phtml?content=%2fcontent%2f00_products%2fpine Unigene.php*) sequences holds great promise for more robust findings from truly genome-scale transcriptional analyses. Microarrays fabricated from amplified cDNA sequences provide a means for quick entry into this technology arena and such arrays can be used productively across closely related species. However, it was pointed out earlier on that microarrays fabricated with synthetic oligonucleotides have the potential to provide higher resolution and more precise information when it comes to relating the expression of specific conifer genes to biological processes of interest (Cairney et al. 2000). Large collections of synthetic oligonucleotides are currently being used for genetic mapping of conifers using SNP markers (Pavy et al. 2008b; Eckert et al. 2009), and development efforts are underway to create oligonucleotide arrays appropriate for transcriptional profiling of conifers (J MacKay unpubl. data). A move to oligonucleotide arrays has implications for studies seeking to compare transcriptional responses across a range of species. So, for the future development of oligonucleotide arrays that could be used in comparative genomic studies, van de Mortel and Aarts (2006) have provided a useful commentary on sequence selection and design considerations. Another variation on the microarray platform that bears consideration for future conifer applications is the use of short oligonucleotides synthesized using modified nucleotide chemistries, such as locked nucleic acids (LNA) or peptide nucleic acids (PNA) for the profiling of small RNAs (Castoldi et al. 2008; Endoh et al. 2009).

Quantitative RT-PCR is not easily scalable to levels that might be considered genomic, but some groups are applying this approach to study the expression of a few tens of conifer genes in parallel (Bomal et al. 2008; Nairn et al. 2008). A community resource established to study transcription factors in *Medicago truncatula* provides an example of how this approach might be scaled up to quantify the expression of several hundred genes in parallel (Kakar et al. 2008).

With respect to the large-scale discovery of differentially expressed genes, approaches such as suppression subtractive hybridization (Wang et al. 2007) and AFLP-cDNA (Dubos and Plomion 2003; Aquea and Arce-Johnson 2008) still have a role to play in conifer genomic research.

8.4 Conclusions and Future Directions

Analyses of large-scale EST datasets obtained from random cDNAs provide the dominant genomic resource for conifer biology today. Thus, expressed gene sequences inferred from EST sequence assemblies are the cornerstone for ongoing investigations to elucidate the genetic control of traits of economic and ecological significance (Neale et al. 2002; Gonzalez-Martinez et al. 2006b). Populations of conifers and other forest trees typically have high levels of allelic diversity and heterozygosity (Gonzalez-Martinez et al. 2006a), and the numerous DNA sequence polymorphisms of these diverse alleles can be used to improve our understanding of the molecular basis for phenotypic variability in valuable traits, such as adaptation to climate, growth rates and response to environmental change (Krutovsky 2006). Establishing the links between genotype and phenotype, however, represents a significant challenge given our current incomplete knowledge of conifer transcriptomes and genomes. Most gene sequences are only partially known, and, as previously recounted, consensus sequences obtained through EST clustering are prone to uncertainty. This uncertainty, coupled with incomplete transcriptome sampling, severely impact the reliability and completeness of unigene assemblies, characterizations of genes families and, ultimately, affect our ability to classify sequence variants as distinct genes, alleles, or even alternate transcripts derived from the same gene. Greater certainty can be expected from full-length cDNA sequencing (e.g., Ralph et al. 2008), deeper transcriptome sequencing as enabled by the current next-generation sequencers, and most significantly from direct sequencing of one or more conifer genomes.

Despite the fact that genomics is rapidly changing our concept of what actually constitutes a gene (Portin 2009), the number of genes in the conifer genome remains an issue of interest and curiosity. Kinlaw and Neale (1997) reviewed the evidence indicating that many pine gene families are more complex than their angiosperms equivalents and concluded that, in general, more gene copies were to be expected in conifers. It remains unclear the extent to which the sequences they detected through genomic hybridization experiments (Southern blots) represented transcriptional units from which functional proteins were expressed. However, as previously discussed, recent studies tend to support the existence of more numerous unique transcripts arising from the gene families characterized to date. It is possible, though, that conifer gene family size could be inflated by the occurrence of transcriptionally active pseudogenes that do not produce functional proteins. It is also unknown the extent to which these observations, based on a small sampling of gene families, can be extrapolated to the entire genome. Indeed, some gene families appear to be simpler in conifers compared to angiosperms, e.g., *KNOX1* genes (Guillet-Claude et al. 2004). Finally, predicting the number of genes based on EST clustering entails a rather large margin of error considering the variable results yielded by clustering. Thus, recent estimates for the number of unique transcripts ranged from 18,000 to 36,000 in loblolly pine, and as high as 80,000 for composite assemblies comprised of sequences from several pine species. Even the limited attempts to address this question through direct analysis of genomic DNA have varied widely, with estimates ranging as high as 225,000 genes (Rabinowicz et al. 2005). From this it appears that for the time being, the number of genes in the conifer genome is likely to remain a point of philosophical discussion.

Data from different species in the Pinaceae provide evidence that conifer gene sequences are highly conserved and gene order is frequently collinear (Shepard and Williams 2008). This high degree of genome conservation across many different conifers suggests that comparative genomic methods and approaches will be highly useful for the rapid transfer of concepts and findings from one species to another. For example, highly conserved, single-copy genes have been used as the basis for syntenic genetic maps bridging spruce and pine, although reduced nucleotide heterozygosity in this class of genes reduces their attractiveness for mapping in pedigrees (Krutovsky et al. 2007; Liewlaksaneeyanawin et al. 2009). To realize the full power of this approach, we will need to use more complex gene families whose members harbor increased levels of heterozygosity. This, in turn, will require further studies to improve the resolution of paralogous and orthologous genes, as well as more uniform gene cataloging methods.

The issues addressed here point to major needs and future directions for conifer genomic research. Addressing issues such as clustering methods and gene cataloging, as well as gene annotation and nomenclature standards, will become both critical and urgent as new sequencing technologies expand the scale of EST sequencing and transcriptome characterization by several orders of magnitude. Signs indicate that the costs and rates of throughput for DNA sequencing will not be a bottleneck to conifer genomic studies much longer, but bioinformatics will remain a challenge for some time to come.

References

Adomas A, Heller G, Olson A, Osborne J, Karlsson M, Nahalkova J, van Zyl L, Sederoff R, Stenlid J, Finlay R, Asiegbu FO (2008) Comparative analysis of transcript abundance in *Pinus sylvestris* after challenge with a saprotrophic, pathogenic or mutualistic fungus. Tree Physiol 28: 885–897.

Allona I, Quinn M, Shoop E, Swope K, St. Cyr S, Carlis J, Riedl J, Retzel E, Campbell MM, Sederoff R, Whetten RW (1998) Analysis of xylem formation in pine by cDNA sequencing. Proc Natl Acad Sci USA 95: 9693–9698.

Ansorge WJ (2009) Next-generation DNA sequencing techniques. Nat Biotechnol 25: 195–203.

Aquea F, Arce-Johnson P (2008) Identification of genes expressed during early somatic embryogenesis in *Pinus radiata*. Plant Physiol Biochem 46: 559–568.

Asante DKA, Yakovlev IA, Fossdal CG, Timmerhaus G, Partanen J, Johnsen O (2009) Effect of bud burst forcing on transcript expression of selected genes in needles of Norway spruce during autumn. Plant Physiol Biochem 47: 681–689.

Axtell MJ, Bartel DP (2005) Antiquity of microRNAs and their targets in land plants. Plant Cell 17: 1658–1673.

Axtell MJ, Snyder JA, Bartel DP (2007) Common functions for diverse small RNAs of land plants. Plant Cell 19: 1750–1769.

Batley J, Edwards D (2009) Genome sequence data: management, storage, and visualization. BioTechniques 46: 333–336.

Bartel DP (2004) MicroRNAs: Genomics, biogenesis, mechanism, and function. Cell 116: 281–297.

Bedon F, Grima-Pettenati J, MacKay J (2007) Conifer R2R3-MYB transcription factors: sequence analyses and gene expression in wood-forming tissues of white spruce (*Picea glauca*). BMC Plant Biol 7: 17.

Bomal C, Bedon F, Caron S, Mansfield SD, Levasseur C, Cooke JEK, Blais S, Tremblay L, Morency MJ, Pavy N, Grima-Pettenati J, Séguin A, MacKay J (2008) Involvement of *Pinus taeda MYB1* and *MYB8* in phenylpropanoid metabolism and secondary cell wall biogenesis: a comparative *in planta* analysis. J Exp Bot 59: 3925–3939.

Bonaldo MF, Lennon G, Soares MB (1996) Normalization and subtraction: two approaches to facilitate gene discovery. Genome Res 6: 791–806.

Brinker M, van Zyl L, Liu W, Craig D, Sederoff RR, Clapham DH, von Arnold S (2004) Microarray analyses of gene expression during adventitious root development in *Pinus contorta*. Plant Physiol 135: 1526–1539.

Brown DG, Hudek AK (2004) New algorithms for multiple DNA sequence alignment. Lect. Notes Comput. Sci. 3240: 314–325.

Cairney J, Pullman GS (2007) The cellular and molecular biology of conifer embryogenesis. New Phytol 176: 511–536.

Cairney J, Xu NF, Pullman GS, Ciavatta VT, Johns B (1999) Natural and somatic embryo development in loblolly pine—Gene expression studies using differential display and DNA arrays. Appl Biochem Biotechnol 77: 5–17.

Cairney J, Xu NF, MacKay J, Pullman J (2000) *In vitro* plant recalcitrance transcript profiling: A tool to assess the development of conifer embryos. In Vitro Cell Dev Biol-Plant 36: 155–162.

Cairney J, Zheng L, Cowels A, Hsiao J, Zismann V, Liu J, Ouyang S, Thibaud-Nissen F, Hamilton J, Childs K, Pullman GS, Zhang YT, Oh T, Buell CR (2006) Expressed Sequence Tags from loblolly pine embryos reveal similarities with angiosperm embryogenesis. Plant Mol Biol 62: 485–501.

Carninci P (2009) Is sequencing enlightenment ending the dark age of the transcriptome? Nat Meth 6: 711–713.

Castoldi M, Schmidt S, Benes V, Hentze MW, Muckenthaler MU (2008) miChip: an array-based method for microRNA expression profiling using locked nucleic acid capture probes. Nat Protoc 3: 321–329.

Coppee JY (2008) Do DNA microarrays have their future behind them? Microbes Infect 10: 1067–1071.

Cordonnier-Pratt MM, Liang C, Wang HM, Kolychev DS, Sun F, Freeman R, Sullivan R, Pratt LH (2004) MAGIC Database and interfaces: an integrated package for gene discovery and expression. Comp Funct Genom 5: 268–275.

Demura T, Fukuda H (2007) Transcriptional regulation in wood formation. Trends Plant Sci 12: 64–70.

Demura T, Tashiro G, Horiguchi G, Kishimoto N, Kubo M, Matsuoka N, Minami A, Nagata-Hiwatashi M, Nakamura K, Okamura Y, Sassa N, Suzuki S, Yazaki J, Kikuchi S, Fukuda H (2002) Visualization by comprehensive microarray analysis of gene expression programs during transdifferentiation of mesophyll cells into xylem cells. Proc Natl Acad Sci USA 99: 15794–15799.

DiGuistini S, Liao NY, Platt D, Robertson G, Seidel M, Chan SK, Docking TR, Birol I, Holt RA, Hirst M, Mardis E, Marra MA, Hamelin RC, Bohlmann J, Breuil C, Jones SJM (2009) *De novo* genome sequence assembly of a filamentous fungus using Sanger, 454 and Illumina sequence data. Genome Biol 10: R94.

Dolgosheina EV, Morin RD, Aksay G, Sahinalp SC, Magrini V, Mardis ER, Mattsson J, Unrau PJ (2008) Conifers have a unique small RNA silencing signature. RNA 14: 1508–1515.

Dong Q, Schlueter SD, Brendel V (2004) PlantGDB, plant genome database and analysis tools. Nucl Acids Res 32: D354–D359.

Du S, Yamamoto F (2007) An overview of the biology of reaction wood formation. J Integr. Plant Biol 49: 131–143.

Dubos C, Plomion C (2003) Identification of water-deficit responsive genes in maritime pine (*Pinus pinaster* Ait.) roots. Plant Mol Biol 51: 249–262.

Dubos C, Le Provost G, Pot D, Salin F, Lalane C, Madur D, Frigerio JM, Plomion C (2003) Identification and characterization of water-stress-responsive genes in hydroponically grown maritime pine (*Pinus pinaster*) seedlings. Tree Physiol 23: 169–179.

Eckert AJ, Pande B, Ersoz ES, Wright NH, Rashbrook VK, Nicolet CM, Neale DB (2009) High-throughput genotyping and mapping of single nucleotide polymorphisms in loblolly pine (*Pinus taeda* L.). Tree Genet Genomes 5: 225–234.

Egertsdotter U, van Zyl LM, MacKay J, Peter G, Kirst M, Clark C, Whetten R, Sederoff R (2004) Gene expression during formation of earlywood and latewood in loblolly pine: expression profiles of 350 genes. Plant Biol 6: 654–663.

Emrich SJ, Barbazuk WB, Li L, Schnable PS (2007) Gene discovery and annotation using LCM-454 transcriptome sequencing. Genome Res 17: 69–73.

Endoh T, Kitamatsu M, Sisido M, Ohtsuki T (2009) PNA arrays for miRNA detection. Chem Lett 38: 438–439.

Ewing B, Green P (1998) Base-calling of automated sequencer traces using phred. II. Error probabilities. Genome Res 8: 186–194.

Ewing B, Hillier L, Wendl MC, Green P (1998) Base-calling of automated sequencer traces using phred. I. Accuracy assessment. Genome Res 8: 175–185.

Forrest ARR, Carninci P (2009) Whole genome transcriptome analysis. RNA Biol 6: 107–112.

Frank RL, Ercal F (2005) Evaluation of *Glycine max* mRNA clusters. BMC Bioinformat 6: S7.

Friedmann M, Ralph SG, Aeschliman D, Zhuang J, Ritland K, Ellis BE, Bohlmann J, Douglas CJ (2007) Microarray gene expression profiling of developmental transitions in Sitka spruce (*Picea sitchensis*) apical shoots. J Exp Bot 58: 593–614.

Fullwood MJ, Wei CL, Liu ET, Ruan YJ (2009) Next-generation DNA sequencing of paired-end tags (PET) for transcriptome and genome analyses. Genome Res 19: 521–532.

Gion JM, Lalanne C, Le Provost G, Ferry-Dumazet H, Paiva J, Frigerio JM, Chaumeil P, Barré A, de Daruvar A, Brach J, Barre A, de Daruvar A, Claverol S, Bonneu M, Sommerer

N, Negroni L, Plomion C (2005) The proteome of maritime pine wood forming tissue. Proteomics 5: 3731–3751.

Goldfarb B, Lanz-Garcia C, Lian ZG, Whetten R (2003) *Aux/IAA* gene family is conserved in the gymnosperm, loblolly pine (*Pinus taeda*). Tree Physiol 23: 1181–1192.

Gonzalez-Martinez SC, Ersoz E, Brown GR, Wheeler NC, Neale DB (2006a) DNA sequence variation and selection of tag single-nucleotide polymorphisms at candidate genes for drought-stress response in *Pinus taeda* L. Genetics 172: 1915–1926.

Gonzalez-Martinez SC, Krutovsky KV, NealeDB (2006b) Forest-tree population genomics and adaptive evolution. New Phytol 170: 227–238.

Gonzalez-Martinez SC, Wheeler NC, Ersoz E, Nelson CD, Neale DB (2007) Association genetics in *Pinus taeda* L. I. Wood property traits. Genetics 175: 399–409.

Groover AT (2005) What genes make a tree a tree? Trends Plant Sci 10: 210–214.

Guillet-Claude C, Isabel N, Pelgas B, Bousquet J (2004)The evolutionary implications of *knox-I* gene duplications in conifers: Correlated evidence from phylogeny, gene mapping, and analysis of functional divergence. Mol Biol Evol 21: 2232–2245.

Hamberger B, Bohlmann J (2006) Cytochrome P450 mono-oxygenases in conifer genomes: discovery of members of the terpenoid oxygenase superfamily in spruce and pine. Biochem Soc Trans 34: 1209–1214.

Harismendy O, Ng PC, Strausberg RL, Wang XY, Stockwell TB, Beeson KY, Schork NJ, Murray SS, Topol EJ, Levy S, Frazer KA (2009) Evaluation of next generation sequencing platforms for population targeted sequencing studies. Genome Biol 10: R32.

Heller G, Adomas A, Li GS, Osborne J, van Zyl L, Sederoff R, Finlay RD, Stenlid J, Asiegbu FO (2008) Transcriptional analysis of *Pinus sylvestris* roots challenged with the ectomycorrhizal fungus *Laccaria bicolor*. BMC Plant Biol 8: 19.

Hertzberg M, Aspeborg H, Schrader J, Andersson A, Erlandsson R, Blomqvist K, Bhalerao R, Uhlen M, Teeri TT, Lundeberg J, Sundberg B, Nilsson P, Sandberg G (2001) A transcriptional roadmap to wood formation. Proc Natl Acad Sci USA 98: 14732–14737.

Holliday JA, Ralph SG, White R, Bohlmann J, Aitken SN (2008) Global monitoring of autumn gene expression within and among phenotypically divergent populations of Sitka spruce (*Picea sitchensis*). New Phytol 178: 103–122.

Huang X, Madan A (1999) CAP3: A DNA sequence assembly program. Genome Res 9:868–877.

Joosen RVL, Lammers M, Balk PA, Bronnum P, Konings MCJM, Perks M, Stattin E, Van Wordragen MF, van der Geest AHM (2006) Correlating gene expression to physiological parameters and environmental conditions during cold acclimation of *Pinus sylvestris*, identification of molecular markers using cDNA microarrays. Tree Physiol 26: 1297–1313.

Kakar K, Wandrey M, Czechowski T, Gaertner T, Scheible WR, Stitt M, Torres-Jerez I, Xiao YL, Redman JC, Wu HC, Cheung F, Town CD, Udvardi MK (2008) A community resource for high-throughput quantitative RT-PCR analysis of transcription factor gene expression in *Medicago truncatula*. Plant Meth 4: 18.

Keeling CI, Weisshaar S, Lin RPC, Bohlmann J (2008) Functional plasticity of paralogous diterpene synthases involved in conifer defense. Proc Natl Acad Sci USA 105: 1085–1090.

Kim K, Kim M, Woo Y (2008) A DNA sequence alignment algorithm using quality information and a fuzzy inference method. Prog Nat Sci 18: 595–602.

Kinlaw CS, Neale DB (1997) Complex gene families in pine genomes. Trends Plant Sci 2: 356–359.

Kirst M, Johnson AF, Baucom C, Ulrich E, Hubbard K, Staggs R, Paule C, Retzel E, Whetten R, Sederoff R (2003) Apparent homology of expressed genes from wood-forming tissues of loblolly pine (*Pinus taeda* L.) with *Arabidopsis thaliana*. Proc Natl Acad Sci USA 100: 7383–7388.

Ko JH, Beers EP, Han KH (2006) Global comparative transcriptome analysis identifies gene network regulating secondary xylem development in *Arabidopsis thaliana*. Mol Genet Genom 276: 517–531.

Komulainen P, Brown GR, Mikkonen M, Karhu A, Garcia-Gil MR, O'Malley D, Lee B, Neale DB, Savolainen O (2003) Comparing EST-based genetic maps between *Pinus sylvestris* and *Pinus taeda*. Theor Appl Genet 107: 667–678.

Koutaniemi S, Warinowski T, Karkonen A, Alatalo E, Fossdal CG, Saranpaa P, Laakso T, Fagerstedt KV, Simola LK, Paulin L, Rudd S, Teeri TH (2007) Expression profiling of the lignin biosynthetic pathway in Norway spruce using EST sequencing and real-time RT-PCR. Plant Mol Biol 65: 311–328.

Krutovsky KV (2006) From population genetics to population genomics of forest trees: Integrated population genomics approach. Russ J Genet 42: 1088–1100.

Krutovsky KV, Elsik CG, Matvienko M, Kozik A, Neale DB (2007) Conserved ortholog sets in forest trees. Tree Genet Genomes 3: 61–70.

Lafarguette F, Leple JC, Dejardin A, Laurans F, Costa G, Lesage-Descauses MC, Pilate G (2004) Genes encoding fasciclin-like arabinogalactan proteins are specifically expressed during cotton fiber development. New Phytol 164: 107–121.

Lai CP, Lee CL, Chen PH, Wu SH, Yang CC, Shaw JF (2004) Molecular analyses of the Arabidopsis TUBBY—like protein gene family. Plant Physiol 134: 1586–1597.

Le Dantec LL, Chagné D, Pot D, Cantin O, Garnier-Géré P, Bedon F, Frigerio JM, Chaumeil P, Léger P, Garcia V, Laigret F, De Daruvar A, Plomion C (2004) Automated SNP detection in expressed sequence tags: statistical considerations and application to Maritime pine sequences. Plant Mol Biol 54: 461–470.

Le Provost G, Paiva J, Pot D, Brach J, Plomion C (2003) Seasonal variation in transcript accumulation in wood-forming tissues of maritime pine (*Pinus pinaster* Ait.) with emphasis on a cell wall glycine-rich protein. Planta 217: 820–830.

Liang C, Wang G, Liu L, Ji GL, Fang L, Liu YS, Carter K, Webb JS, Dean JFD (2007) ConiferEST: an integrated bioinformatics system for data reprocessing and mining of conifer expressed sequence tags (ESTs). BMC Genom 8: 134.

Liewlaksaneeyanawin C, Zhuang J, Tang M, Farzaneh N, Lueng GL, Cullis C, Findlay S, Ritland CE, Bohlmann J, Ritland K (2009) Identification of COS markers in the Pinaceae. Tree Genet Genomes 5: 247–255.

Lippert D, Chowrira S, Ralph SG, Zhuang J, Aeschliman D, Ritland C, Ritland K, Bohlmann J (2007) Conifer defense against insects: Proteome analysis of Sitka spruce (*Picea sitchensis*) bark induced by mechanical wounding or feeding by white pine weevils (*Pissodes strobi*). Proteomics 7: 248–270.

Lorenz WW, Dean JFD (2002) SAGE profiling and demonstration of differential gene expression along the axial developmental gradient of lignifying xylem in loblolly pine (*Pinus taeda*). Tree Physiol 22: 301–310.

Lorenz WW, Sun F, Liang C, Kolychev D, Wang HM, Zhao X, Cordonnier-Pratt MM, Pratt LH, Dean JFD (2006) Water stress-responsive genes in loblolly pine (*Pinus taeda*) roots identified by analyses of expressed sequence tag libraries. Tree Physiol 26: 1–16.

Lorenz WW, Yu YS, Simões M, Dean JFD (2009) Processing the loblolly pine PtGen2 cDNA microarray. JoVE. 25. *http://www.jove.com/index/details.stp?id=1182*, doi: 10.3791/1182.

Lu SF, Sun YH, Amerson H, Chiang VL (2007) MicroRNAs in loblolly pine (*Pinus taeda* L.) and their association with fusiform rust gall development. Plant J 51: 1077–1098.

Magallón SA, Sanderson MJ (2005) Angiosperm divergence times: the effect of genes, codon positions, and time constraints. Evolution 59: 1653–1670.

Margulies M, Egholm M, Altman WE, Attiya S, Bader JS, Bemben LA, Berka J, Braverman MS, Chen YJ, Chen Z, Dewell SB, Du L, Fierro JM, Gomes XV, Godwin BC, He W, Helgesen S, Ho CH, Irzyk GP, Jando SC, Alenquer MLI, Jarvie TP, Jirage KB, Kim JB, Knight JR, Lanza JR, Leamon JH, Lefkowitz SM, Lei M, Li J, Lohman KL, Lu H, Makhijani VB, McDade KE, McKenna MP, Myers EW, Nickerson E, Nobile JR, Plant R, Puc BP, Ronan MT, Roth GT, Sarkis GJ, Simons JF, Simpson JW, Srinivasan M, Tartaro KR, Tomasz A, Vogt KA,

Volkmer GA, Wang SH, Wang Y, Weiner MP, Yu P, Begley RF, Rothberg JM (2005) Genome sequencing in microfabricated high-density picolitre reactors. Nature 437: 376–380.

Meyers BC, Axtell MJ, Bartel B, Bartel DP, Baulcombe D, Bowman JL, Cao X, Carrington JC, Chen X, Green PJ, Griffiths-Jones S, Jacobsen SE, Mallory AC, Martienssen RA, Poethig RS, Qi Y, Vaucheret H, Voinnet O, Watanabe Y, Weigel D, Zhu JK. (2008) Criteria for annotation of plant microRNAs. Plant Cell 20: 3186–3190.

Morin RD, Aksay G, Dolgosheina E, Ebhardt HA, Magrini V, Mardis ER, Sahinalp SC, Unrau PJ. (2008) Comparative analysis of the small RNA transcriptomes of *Pinus contorta* and *Oryza sativa*. Genome Res 18: 571–584.

Morrissy AS, Morin RD, Delaney A, Zeng T, McDonald H, Jones S, Zhao Y, Hirst M, Marra MA (2009) Next-generation tag sequencing for cancer gene expression profiling. Genome Res 19: 1825–1835.

Murray CG, Larsson TP, Hill T, Bjorklind R, Fredriksson R, Schioth HB (2005) Evaluation of EST-data using the genome assembly. Biochem Biophys Res Comm 331: 1566–1576.

Myburg H, Morse AM, Amerson HV, Kubisiak TL, Huber D, Osborne JA, Garcia SA, Nelson CD, Davis JM, Covert SF, van Zyl LM (2006) Differential gene expression in loblolly pine (*Pinus taeda* L.) challenged with the fusiform rust fungus, *Cronartium quercuum* f. sp. *fusiforme*. Physiol Mol Plant Pathol 68: 79–91.

Nagaraj SH, Gasser RB, Ranganathan S (2007) A hitchhiker's guide to expressed sequence tag (EST) analysis. Brief Bioinformat 8: 6–21.

Nairn CJ, Haselkorn T (2005) Three loblolly pine CesA genes expressed in developing xylem are orthologous to secondary cell wall CesA genes of angiosperms. New Phytol 166: 907–915.

Nairn CJ, Lennon DM, Wood-Jones A, Nairn AV, Dean JFD (2008) Carbohydrate-related genes and cell wall biosynthesis in vascular tissues of loblolly pine (*Pinus taeda*). Tree Physiol 28: 1099–1110.

Neale DB, Sewell MM, Brown GR (2002) Molecular dissection of the quantitative inheritance of wood property traits in loblolly pine. Ann For Sci 59: 595–605.

Ng P, Tan JJS, Ooi HS, Lee YL, Chiu KP, Fullwood MJ, Srinivasan KG, Perbost C, Du L, Sung WK, Wei CL, Ruan Y (2006) Multiplex sequencing of paired-end ditags (MS-PET): a strategy for the ultra-high-throughput analysis of transcriptomes and genomes. Nucl Acids Res 34: e84.

Oh S, Park S, Han KH (2003) Transcriptional regulation of secondary growth in *Arabidopsis thaliana*. J Exp Bot 54: 2709–2722.

Oh TJ, Wartell RM, Cairney J, Pullman GS (2008) Evidence for stage-specific modulation of specific microRNAs (miRNAs) and miRNA processing components in zygotic embryo and female gametophyte of loblolly pine (*Pinus taeda*). New Phytol 179: 67–80.

Osoegawa K, Vessere GM, Shu CL, Hoskins RA, Abad JP, de Pablos B, Villasante A, de Jong PJ (2007) BAC clones generated from sheared DNA. Genomics 89: 291–299.

Paiva JA, Garcés M, Alves A, Garnier-Géré P, Rodrigues JC, Lalanne C, Porcon S, Le Provost G, Perez Dda S, Brach J, Frigerio JM, Claverol S, Barré A, Fevereiro P, Plomion C (2008a) Molecular and phenotypic profiling from the base to the crown in maritime pine wood forming tissue. New Phytol 178: 283–301.

Paiva JA, Garcés M, Alves A, Garnier-Géré P, Rodrigues JC, Lalanne C, Porcon S, Le Provost G, Perez Dda S, Brach J, Frigerio JM, Claverol S, Barré A, Fevereiro P, Plomion C (2008b) Plasticity of maritime pine (*Pinus pinaster*) wood-forming tissues during a growing season. New Phytol 179: 1080–1094.

Patzlaff A, McInnis S, Courtenay A, Surman C, Newman LJ, Smith C, Bevan MW, Mansfield S, Whetten RW, Sederoff RR, Campbell MM (2003a) Characterisation of a pine MYB that regulates lignification. Plant J 36: 743–754.

Patzlaff A, Newman LJ, Dubos C, Whetten RW, Smith C, McInnis S, Bevan MW, Sederoff RR, Campbell MM (2003b) Characterisation of *Pt*MYB1, an R2R3-MYB from pine xylem. Plant Mol Biol 53: 597–60.

Paux E, Tamasloukht M, Ladouce N, Sivadon P, Grima-Pettenati J (2004) Identification of genes preferentially expressed during wood formation in *Eucalyptus*. Plant Mol Biol 55: 263–280.

Pavy N, Laroche J, Bousquet J, MacKay J (2005a) Large-scale statistical analysis of secondary xylem ESTs in pine. Plant Mol Biol 57: 203–224.

Pavy N, Paule C, Parsons L, Crow JA, Morency MJ, Cooke J, Johnson JE, Noumen E, Guillet-Claude C, Butterfield Y, Barber S, Yang G, Liu J, Stott J, Kirkpatrick R, Siddiqui A, Holt R, Marra M, Seguin A, Retzel E, Bousquet J, MacKay J (2005b) Generation, annotation, analysis and database integration of 16,500 white spruce EST clusters. BMC Genom 6: 144.

Pavy N, Parsons LS, Paule C, MacKay J, Bousquet J (2006) Automated SNP detection from a large collection of white spruce expressed sequences: contributing factors and approaches for the categorization of SNPs. BMC Genom 7: 174.

Pavy N, Johnson JJ, Crow JA, Paule C, Kunau T, MacKay J, Retzel EF (2007) ForestTreeDB: a database dedicated to the mining of tree transcriptomes. Nucl Acids Res 35: D888–D894.

Pavy N, Boyle B, Nelson C, Paule C, Giguère I, Caron S, Parsons LS, Dallaire N, Bedon F, Bérubé H, Cooke J, Mackay J (2008a) Identification of conserved core xylem gene sets: conifer cDNA microarray development, transcript profiling and computational analyses. New Phytol 180: 766–786.

Pavy N, Pelgas B, Beauseigle S, Blais S, Gagnon F, Gosselin I, Lamothe M, Isabel N, Bousquet J (2008b) Enhancing genetic mapping of complex genomes through the design of highly-multiplexed SNP arrays: application to the large and unsequenced genomes of white spruce and black spruce. BMC Genom 9: 21.

Plomion C, Le Provost G, Stokes A (2001) Wood formation in trees. Plant Physiol 127: 1513–1523.

Portin P (2009) The elusive concept of the gene. Hereditas 146: 112–117.

Qiu DY, Pan XP, Wilson IW, Li FL, Liu M, Teng WJ, Zhang BH (2009) High throughput sequencing technology reveals that the taxoid elicitor methyl jasmonate regulates microRNA expression in Chinese yew (*Taxus chinensis*). Gene 436: 37–44.

Quackenbush J, Cho J, Lee D, Liang F, Holt I, Karamycheva S, Parvizi B, Pertea G, Sultana R, White J (2001) The TIGR Gene Indices: analysis of gene transcript sequences in highly sampled eukaryotic species. Nucl Acids Res 29: 159–164.

Rabinowicz PD, Citek R, Budiman MA, Nunberg A, Bedell JA, Lakey N, O'Shaughnessy AL, Nascimento LU, McCombie WR, Martienssen RA (2005) Differential methylation of genes and repeats in land plants. Genome Res 15: 1431–1440.

Ralph S, Oddy C, Cooper D, Yueh H, Jancsik S, Kolosova N, Philippe RN, Aeschliman D, White R, Huber D, Ritland CE, Benoit F, Rigby T, Nantel A, Butterfield YS, Kirkpatrick R, Chun E, Liu J, Palmquist D, Wynhoven B, Stott J, Yang G, Barber S, Holt RA, Siddiqui A, Jones SJ, Marra MA, Ellis BE, Douglas CJ, Ritland K, Bohlmann J (2006a) Genomics of hybrid poplar (*Populus trichocarpa x deltoides*) interacting with forest tent caterpillars (*Malacosoma disstria*): normalized and full-length cDNA libraries, expressed sequence tags, and a cDNA microarray for the study of insect-induced defences in poplar. Mol Ecol 15: 1275–1297.

Ralph S, Park JY, Bohlmann J, Mansfield SD (2006b) Dirigent proteins in conifer defense: gene discovery, phylogeny, and differential wound- and insect-induced expression of a family of DIR and DIR-like genes in spruce (*Picea* spp.). Plant Mol Biol 60: 21–40.

Ralph SG, Yueh H, Friedmann M, Aeschliman D, Zeznik JA, Nelson CC, Butterfield YS, Kirkpatrick R, Liu J, Jones SJ, Marra MA, Douglas CJ, Ritland K, Bohlmann J (2006c) Conifer defence against insects: microarray gene expression profiling of Sitka spruce (*Picea sitchensis*) induced by mechanical wounding or feeding by spruce budworms (*Choristoneura occidentalis*) or white pine weevils (*Pissodes strobi*) reveals large-scale changes of the host transcriptome. Plant Cell Environ 29: 1545–1570.

Ralph SG, Jancsik S, Bohlmann J (2007) Dirigent proteins in conifer defense II: Extended gene discovery, phylogeny, and constitutive and stress-induced gene expression in spruce (*Picea* spp.). Phytochemistry 68: 1975–1991.

Ralph SG, Chun HJ, Kolosova N, Cooper D, Oddy C, Ritland CE, Kirkpatrick R, Moore R, Barber S, Holt RA, Jones SJ, Marra MA, Douglas CJ, Ritland K, Bohlmann J (2008) A conifer genomics resource of 200,000 spruce (*Picea* spp.) ESTs and 6,464 high-quality, sequence-finished full-length cDNAs for Sitka spruce (*Picea sitchensis*). BMC Genom 9: 484.

Rothwell GW, Sanders H, Wyatt SE, Lev-Yadun S (2008) A fossil record for growth regulation: The role of auxin in wood evolution. Ann MO Bot Gard 95: 121–134.

Roudier F, Fernandez AG, Fujita M, Himmelspach R, Borner GH, Schindelman G, Song S, Baskin TI, Dupree P, Wasteneys GO, Benfey PN (2005) COBRA, an Arabidopsis extracellular glycosylphosphatidylinositol-anchored protein, specifically controls highly anisotropic expansion through its involvement in cellulose microfibril orientation. Plant Cell 17: 1749–1763.

Sampedro J, Carey RE, Cosgrove DJ (2006) Genome histories clarify evolution of the expansin superfamily: new insights from the poplar genome and pine ESTs. J Plant Res 119: 11–21.

Sathyan P, Newton RJ, Loopstra CA (2005) Genes induced by WDS are differentially expressed in two populations of aleppo pine (*Pinus halepensis*). Tree Genet Genomes. 1: 166–173.

Sayers EW, Barrett T, Benson DA, Bolton E, Bryant SH, Canese K, Chetvernin V, Church DM, Dicuccio M, Federhen S, Feolo M, Geer LY, Helmberg W, Kapustin Y, Landsman D, Lipman DJ, Lu Z, Madden TL, Madej T, Maglott DR, Marchler-Bauer A, Miller V, Mizrachi I, Ostell J, Panchenko A, Pruitt KD, Schuler GD, Sequeira E, Sherry ST, Shumway M, Sirotkin K, Slotta D, Souvorov A, Starchenko G, Tatusova TA, Wagner L, Wang Y, John Wilbur W, Yaschenko E, Ye J (2009) Database resources of the National Center for Biotechnology Information. Nucl Acids Res 37: D5–D15.

Scheibye-Alsing K, Hoffmann S, Frankel A, Jensen P, Stadler PF, Mang Y, Tommerup N, Gilchrist MJ, Nygård AB, Cirera S, Jørgensen CB, Fredholm M, Gorodkin J (2009) Sequence assembly. Comput Biol Chem 33: 121–136.

Schmid M, Davison TS, Henz SR, Pape UJ, Demar M, Vingron M, Schölkopf B, Weigel D, Lohmann JU (2005) A gene expression map of *Arabidopsis thaliana* development. Nat Genet 37: 501–506.

Schrader J, Nilsson J, Mellerowicz E, Berglund A, Nilsson P, Hertzberg M, Sandberg G (2004) A high-resolution transcript profile across the wood-forming meristem of poplar identifies potential regulators of cambial stem cell identity. Plant Cell 16: 2278–2292.

Shepherd M, Williams CG (2008) Comparative mapping among subsection Australes (genus *Pinus*, family Pinaceae). Genome 51: 320–331.

Simon SA, Zhai J, Nandety RS, McCormick KP, Zeng J, Mejia D, Meyers BC (2009) Short-read sequencing technologies for transcriptional analyses. Annu Rev Plant Biol 60: 305–333.

Stasolla C, van Zyl L, Egertsdotter U, Craig D, Liu WB, Sederoff RR (2003a) Transcript profiles of stress-related genes in developing white spruce (*Picea glauca*) somatic embryos cultured with polyethylene glycol. Plant Sci 165: 719–729.

Stasolla C, van Zyl L, Egertsdotter U, Craig D, Liu WB, Sederoff RR (2003b) The effects of polyethylene glycol on gene expression of developing white spruce somatic embryos. Plant Physiol 131: 49–60.

Stasolla C, Belmonte MF, van Zyl L, Craig DL, Liu W, Yeung EC, Sederoff RR (2004a) The effect of reduced glutathione on morphology and gene expression of white spruce (*Picea glauca*) somatic embryos. J Exp Bot 55: 695–709.

Stasolla C, Bozhkov PV, Chu TM, Van Zyl L, Egertsdotter U, Suarez MF, Craig D, Wolfinger RD, Von Arnold S, Sederoff RR (2004b) Variation in transcript abundance during somatic embryogenesis in gymnosperms. Tree Physiol 24: 1073–1085.

Sultan M, Schulz MH, Richard H, Magen A, Klingenhoff A, Scherf M, Seifert M, Borodina T, Soldatov A, Parkhomchuk D, Schmidt D, O'Keeffe S, Haas S, Vingron M, Lehrach H, Yaspo

ML (2008) A global view of gene activity and alternative splicing by deep sequencing of the human transcriptome. Science 321: 956–960.

Tan KC, Ipcho SVS, Trengove RD, Oliver RP, Solomon PS (2009) Assessing the impact of transcriptomics, proteomics and metabolomics on fungal phytopathology. Mol Plant Pathol 10: 703–715.

Terzi LC, Simpson GG (2009) *Arabidopsis* RNA immunoprecipitation. Plant J 59: 163–168.

van de Mortel JE, Aarts MGM (2006) Comparative transcriptomics—model species lead the way. New Phytol 170: 199–201.

van Zyl L, von Arnold S, Bozhkov P, Chen YZ, Egertsdotter U, MacKay J, Sederoff RR, Shen J, Zelena L, Clapham DH (2002) Heterologous array analysis in Pinaceae: hybridization of *Pinus taeda* cDNA arrays with cDNA from needles and embryogenic cultures of *P. taeda*, *P. sylvestris* or *Picea abies*. Comp Funct Genom 3: 306–318.

Vega-Sanchez ME, Gowda M, Wang GL (2007) Tag-based approaches for deep transcriptome analysis in plants. Plant Sci 173: 371–380.

Wall PK, Leebens-Mack J, Chanderbali AS, Barakat A, Wolcott E, Liang H, Landherr L, Tomsho LP, Hu Y, Carlson JE, Ma H, Schuster SC, Soltis DE, Soltis PS, Altman N, dePamphilis CW (2009) Comparison of next generation sequencing technologies for transcriptome characterization. BMC Genom 10: 347.

Wang GF, Gao Y, Yang LW, Shi JS (2007) Identification and analysis of differentially expressed genes in differentiating xylem of Chinese fir (*Cunninghamia lanceolata*) by suppression subtractive hybridization. Genome 50: 1141–1155.

Watkinson JI, Sioson AA, Vasquez-Robinet C, Shukla M, Kumar D, Ellis M, Heath LS, Ramakrishnan N, Chevone B, Watson LT, van Zyl L, Egertsdotter U, Sederoff RR, Grene R (2003) Photosynthetic acclimation is reflected in specific patterns of gene expression in drought-stressed loblolly pine. Plant Physiol 133: 1702–1716.

Wegrzyn JL, Lee JM, Tearse BR, Neale DB (2008) TreeGenes: A forest tree genome database. Int J Plant Genom 2008: 412875.

Wegrzyn JL, Lee JM, Liechty J, Neale DB (2009) PineSAP-sequence alignment and SNP identification pipeline. Bioinformatics 25: 2609–2610.

Whetten R, Sun YH, Zhang Y, Sederoff R (2001) Functional genomics and cell wall biosynthesis in loblolly pine. Plant Mol Biol 47: 275–291.

Whiteman PH, Cameron JN, Farrington A (1996) Breeding trees for improved pulp and paper production—A review. Appita J 49: 50–53.

Yakovlev IA, Fossdal CG, Johnsen O, Junttila O, Skroppa T (2006) Analysis of gene expression during bud burst initiation in Norway spruce via ESTs from subtracted cDNA libraries. Tree Genet Genomes. 2: 39–52.

Yamashita S, Yoshida M, Yamamoto H (2009) Relationship between development of compression wood and gene expression. Plant Sci 176: 729–735.

Yang SH, Loopstra CA (2005) Seasonal variation in gene expression for loblolly pines (*Pinus taeda*) from different geographical regions. Tree Physiol 25: 1063–1073.

Yang SH, van Zyl L, No EG, Loopstra CA (2004) Microarray analysis of genes preferentially expressed in differentiating xylem of loblolly pine (*Pinus taeda*). Plant Sci 166: 1185–1195.

Zhang BH, Pan XP, Cannon CH, Cobb GP, Anderson TA (2006a) Conservation and divergence of plant microRNA genes. Plant J 46: 243–259.

Zhang BH, Pan XP, Cobb GP, Anderson TA (2006b) Plant microRNA: a small regulatory molecule with big impact. Dev Biol 289: 3–16.

Zhang Y, Sederoff RR, Allona I (2000) Differential expression of genes encoding cell wall proteins in vascular tissues from vertical and bent loblolly pine trees. Tree Physiol 20: 457–466.

Zhao CS, Craig JC, Petzold HE, Dickerman AW, Beers EP (2005) The xylem and phloem transcriptomes from secondary tissues of the *Arabidopsis* root-hypocotyl. Plant Physiol 138: 803–818.

Zobel BJ, Talbert J (1984) Applied Forest Tree Improvement. John Wiley, New York, USA.

9

Recent Advances in Proteomics and Metabolomics in Gymnosperms

Rebecca Dauwe,[1,a,#] Andrew Robinson[1,b,#] and Shawn D. Mansfield[1,c,*]

ABSTRACT

Functional genomics in forestry is a research area in its infancy that has the capacity to significantly impact our understanding of the genetic and environmental control of tree development, and provide new methods to exploit the genetic variation in tree phenotypes. Tools made available through the analysis of tree genomes have the potential to have a profound effect on the capacity to improve forest productivity and monitor forest health, and revolutionize in the selection of trees lines for the future. This chapter discusses the emerging application of proteomics and metabolomics in gymnosperms.

Keywords: 2-DE; SDS-PAGE; genetic mapping; developmental biology; metabolite profiling; metabolic markers; GC-MS; LC-MS; high throughput; automation, somatic embryogenesis, wood formation

9.1 Introduction

In this era of plant systems biology, "high-throughput" and genome-wide research aims at establishing an integrated understanding of the biological processes occurring in all plant tissues. In that respect,

[1]Department of Wood Science, Faculty of Forestry, 4030-2424 Main Mall, Vancouver, BC, V6T 1Z4, Canada;
[a]e-mail: *rebecca.dauwe@u-picardie.fr*
[b]e-mail: *andrewrobinsonnz@gmail.com*
[c]e-mail: *shawn.mansfield@ubc.ca*
[#]these authors contributed equally
*Corresponding author

transcriptomics, which consists of the profiling of gene expression at the level of transcript accumulation, has become a widely adopted approach. After the successful development and application of transcriptomics, a new challenge constituted the post-transcriptional characterization of the biological processes. Among a range of "-omics" tools, proteomics and metabolomics have been developed to characterize plant tissues at the protein and the metabolite levels. Despite their biological, ecological and economical importance, intrinsic characteristics of gymnosperms, such as long generation times, large genome sizes, and recalcitrance to genetic transformation, have limited the establishment of gymnosperms as plant model systems, and as such have slowed down the development of most functional genomic research tools to evaluate gymnosperms. However, in recent years this has rapidly changed and in the following pages we describe and discuss the applications of proteomics and metabolomics in gymnosperm systems.

9.2 Proteomics

The term "proteome" defines the expressed protein complement of a genome and was, according to a review by Thiellement et al. (2002), first introduced in 1994 by Wilkins at a conference. The roots of this concept, however, date back to 1975 with the development of high resolution 2-dimensional polyacrylamide gel electrophoresis, abbreviated as 2-D PAGE or 2-DE. The first plant large-scale proteomic work was published on *Arabidopsis thaliana* (Kamo et al. 1995). During the following years, proteomics emerged as a complementary approach to the analysis of genome expression at the mRNA level.

9.2.1 *The Analytical Process*

Although plant proteomics studies have been conducted on no less than 35 species, most of these studies focus on the dicot model species, thale cress (*Arabidopsis thaliana*); or, on the monocot model species, rice (*Oryza sativa*) (Rossignol et al. 2006). The availability of the complete genome sequence of both these species facilitates protein identification. In contrast, few proteomic studies on conifers have been reported to date. This is related to their limited amenability as an experimental system (large physical size, large genome with highly repetitive character, long life cycle, recalcitrance for genetic transformation and regeneration *in vitro*, and difficulties in sample preparation for molecular and biochemical analyses), which has negatively influenced the development of molecular and biochemical studies in conifers in general. The low number of proteomic studies in conifers in particular, is largely due to the inherent difficulties involved in extracting proteins of sufficient quality from conifer tissues, and to the difficulties in protein identification in species for which the number of genomic DNA, expressed sequence tag (EST) or protein sequences available in public databases, is limited. However, gymnosperms are perennials with a long life cycle and have features and processes, like cold hardiness and wood formation, which cannot be approached via a herbal model plant. Therefore, efforts are underway to push the limits of the traditional proteomics techniques and consequently enhance the possibilities of non-model plant-, including gymnosperm-, proteomics (Carpentier et al. 2008).

9.2.1.1 Platforms

Two-dimensional polyacrylamide gel electrophoresis (2-DE) coupled with MS has been the most extensively used platform in plant proteome analyses. To analyze simple proteomes, such as the proteome of juniper pollination drops, even 1-dimensional sodium dodecyl sulfate-polyacrylamide gel electrophoresis (SDS-PAGE), proved to generate good quality data (Wagner et al. 2007). 2-DE, which separates denatured proteins based on their charge (pI), by isoelectric focusing (IEF) in the first dimension, and based on their relative molecular weight (MW), by SDS-PAGE in the second dimension, results in a high resolution, quantitative image of intact proteins that provides a good overview of different isoforms and post-translational modifications. However, the technique cannot meet the "full-genome" objectives of proteomics, analogous to, for example, what microarray analysis presents in transcriptomics. 2-DE covers only a subset of the most abundant proteins, with a poor representation of membrane, high molecular weight, and basic proteins. Furthermore, the fraction of the proteome that is studied is determined by the precipitation protocol and the 2-DE separation technique (for example the pH range of the first dimension). A major disadvantage of the use of 2-DE as a high-throughput method, is that this technique is difficult to automate, and the technical variation is therefore difficult to control. In an attempt to realize automated high-throughput proteomics, gel-free protein separation methods (Berg et al. 2006; Pirondini et al. 2006; Roe and Griffin 2006; Whitelegge et al. 2006; Zolla 2006) and "second generation" proteomic techniques are currently being developed and refined. These include MudPIT (multidimensional protein identification technology) and quantitative proteomics techniques such as DIGE (differential in gel electrophoresis), ICAT (isotope-coded affinity tags), iTRAQ (isobaric tag for relative and absolute quantitation), and SILAC (stable isotope labeling by amino acids in cell culture) (Amme et al. 2006; Basu et al. 2006; Bayer et al. 2006; Jones et al. 2006; Komatsu et al. 2006; Lilley and Dupree 2006), all of which have been developed and applied recently in plant biology research but remain unexploited today in gymnosperm studies. Despite the rapid development of these different methods for protein extraction and separation (in plants and other organisms), it is generally agreed that, to permit the analysis of entire proteomes, a combination of different methodologies is still needed (Jorrin et al. 2007).

9.2.1.2 Protein Extraction and Solubilization

Protein extraction from plant samples is usually challenging due to the low protein content relative to the large quantities of secondary compounds and proteolytic enzymes, which accumulate in the central vacuole and are released upon tissue disruption. The isolation of proteins from gymnosperm tissues is particularly troublesome because of the abundance of polysaccharides, pigments and phenolics, which interferes with the efficiency of isolation/purification. Phenolic compounds reversibly bind to proteins by hydrogen bonding and irreversibly by oxidation followed by covalent condensations (Loomis and Battaile 1966), leading to charge heterogeneity and streaking in the gels. Therefore, the 2-DE resolution is strongly influenced by the procedure of sample preparation. Much effort has been invested in the establishment of 2-DE sample preparation methods for plants. An elaborate and comprehensive overview of the problems associated with plant protein sample preparation for 2-DE and the different optimizations has been reported recently in the literature (Carpentier et al. 2008). The number of protocol optimizations that were tested on gymnosperm samples remains very low (Valcu and Schlink 2006; Wang et al. 2006b). Wang et al. (2006b) reported a universal protocol for protein extraction from recalcitrant plant tissues, which obtained electrophoretic separation of proteins for a wide range of tissues. Specifically, in the case of tissues containing high levels of phenolic compounds, such as pine aged leaves, pulverizing plant tissues with polyvinylpolypyrrolidone (PVPP, 0.05 g/g tissue) was suggested to help remove phenolic compounds (Wang et al. 2006b).

Apart from the optimization of the extraction protocol, protein solubilization is also a critical factor. In order to separate proteins under denaturing conditions in the first dimension of a 2-DE analysis, proteins are solubilized in the presence of high concentrations of chaotropes, a reductant and a neutral detergent. The use of a detergent in conjunction with chaotropes is decisive for the subset of proteins that can be analyzed. Efficient solubilization of proteins and 2-DE pattern resolution strongly depends on the chaotrope and detergent in the extraction buffer. Valcu and Schlink (2006) tested various chaotrope and detergent combinations and showed that, for optimal separation of proteins from woody plant samples, the extraction buffers have to be optimized for each type of sample by adding an adapted combination of chaotrope and detergent. For example, buffers optimized for beech roots, containing two chaotropes (urea and thiourea) and a sugar-based detergent, could be used for the extraction and separation of proteins from spruce roots only after optimizations (not specifically reported). Valcu and Schlink (2006) also reported an extraction buffer containing two chaotropes and two detergents (7 M urea, 2 M thiourea, 2% CHAPS, 2% SB3–10) that effectively separated spruce needle proteins.

9.2.1.3 Protein Identification

The identification of proteins separated by 2-DE (or 1-DE) is obtained from mass spectrometry [peptide mass fingerprinting (PMF) or de novo sequencing], or occasionally by Edman degradation. The difficulty and the success rate of the identification are dependent on the availability of genomic or transcript sequences. As a consequence, the progress in proteomics is an extent of the progress of genomics studies. The unusually large genomes of gymnosperms, ranging in size from 2.1 to 37.0 Gb (1C values), in contrast to angiosperms, which have much smaller genome sizes of ~0.6 Gb (Leitch et al. 2005), and their highly repetitive character, create a natural barrier hampering complete genomic sequencing efforts. At present there are no available conifer whole genome sequences. However, sequencing of loblolly pine (*Pinus taeda*) and white pine (*Pinus strobus*) are in progress under the Conifer Genome Network (*http://pinegenome.org*), and EST collections are available. Large-scale EST sequencing projects have been initiated for loblolly pine (Allona et al. 1998) and maritime pine (*Pinus pinaster*) (Le Dantec et al. 2004) and gymnosperm EST databases are accessible (*http://pinetree.ccgb.umn.edu; http://cbi.labri.fr/outils/SPAM/index.php*). In the common databases (dbEST, NCBI), a total of 1,079,016 conifer ESTs are available, with loblolly pine (328,756 ESTs), white spruce (*Picea glauca*; 313,110 ESTs), and Sitka spruce (*Picea sitchensis*; 186,637 ESTs) as the most abundant species (as of 27 August 2010). Interestingly, 12–25% of the isolated ESTs from woody xylem showed no similarity to sequences in the databases and may have unique functions in wood formation (Canovas et al. 2004).

As an alternative to the genome sequence, Lippert and coworkers demonstrated the value and feasibility of employing elaborate species-specific EST data, to aid in protein identification in large-scale proteomics studies in spruce (Lippert et al. 2005; Lippert et al. 2007). In a proteomic study of somatic embryogenesis in white spruce, a six-frame translated EST database was presented, containing over 45,000 ESTs and representing a 21,631 unigene sequence set from three species (white, sitka and interior spruce) (*http://www.treenomix.com*), available as GenBank accession numbers CO203068–CO258618), which, in combination with GenBank, improved the rate of protein identification in white spruce embryo tissue from 38 to

62% when compared with using GenBank data alone (Lippert et al. 2005). For the interpretation of protein sequence data collected in a following proteomic study of sitka spruce defense against white pine weevils (*Pissodes strobi* Peck), Lippert et al (2007) relied on a translated protein database that contained 164,621 predicted peptide sequences, derived from a total of 249,149 nucleotide sequences from sitka spruce, white spruce and interior spruce, available from previous research programs (Pavy et al. 2005; Ralph et al. 2006), and an additional 7,317 conifer proteins from various species obtained from NCBI. This database allowed the identification of 68.5% of the queried proteins, as compared to 31.7% when using the NCBI database, and, in contrast to the previous study, there appeared no additional benefit in employing the larger NCBI database. The improved protein identification results of the second, weevil response study, which focused on bark tissue, in comparison with the somatic embryogenesis study, was attributed not only to the size of the database used (5-times more ESTs), but also to the fact that ESTs from bark tissue were enriched in the database, whereas embryonic tissue was not represented in the EST database at the time of the somatic embryogenesis study. In a proteomic study of wood forming tissue in maritime pine, the use of an EST database of modest size but highly enriched in pine xylem ESTs (18,254 *Pinus pinaster* ESTs and 59,447 *Pinus taeda* xylem ESTs) resulted in a 67.9% identification of proteins by LC ESI-MS/MS (Gion et al. 2005). These examples illustrate the importance of deep and biologically relevant EST and FL-cDNA sequencing as an essential resource for proteome analysis when no whole genome sequence is available. The importance of having sequence data from the same species as studied by proteomics, results from the fact that a single amino acid substitution will, in most cases, change the mass of a peptide enough to prevent its being identified by MS data interpretation software. On the other hand, a drawback of the use of EST databases for identification is that, because of their restricted length, often only 400–600 bp (of which up to a third can represent untranslated regions), ESTs cannot be used to identify proteins by their peptide mass fingerprints (Lahm and Langen 2000), and peptide sequencing by MS/MS must be used. This has been confirmed in a proteome analysis of maritime pine xylem (Gion et al. 2005). In that study, the success rate of identification by MS/MS was high (68%), whereas identification by MALDI-TOF MS, using the same EST databases as used for the identification by MS/MS, had a success rate of only 16%.

In most gymnosperm spp., the situation is much less straightforward: with poorly characterized genomes and no considerable EST sequence collection available, cross-species identification becomes the only option for protein identification. If the corresponding gene of the investigated organism is unknown, actual identification of the protein senso stricto is not possible, and the goal is to find the most similar gene in a closely related organism. In such cases, sequence databases from closely related organisms can be interrogated by BLAST similarity searching, provided that a reasonable amount of amino acid substitution or deletion can be tolerated. Cross-species identification is further complicated by the evaluation of the multiple BLAST results obtained for the multiple candidate sequences derived from multiple MS/MS spectra derived from the same protein in an LC-MS/MS experiment. Therefore, under such circumstances, much of the identification success depends on the BLAST strategy and validation algorithms employed (Valledor et al. 2008). Recent improvements in the algorithms for database searching, offer the ability to exploit the information of the connectivity between multiple peptide MS/MS spectra, and to account for genetic variability. The usefulness of the novel Paragon algorithm developed by Applied Biosystems (Shilov et al. 2007), was presented by Valledor et al. (2008), in a proteomic analysis of *Pinus radiata* needles. The algorithm led to the identification of 77% of 115 spots, which is a remarkably high percentage given that for *Pinus radiata* only a limited sequence database of 164 ESTs and 158 non-redundant protein sequences was available. This study clearly demonstrates the power of using the information of the connectivity between peptides derived from the same protein for protein identification in non-model species. However, the newer gel-free proteomics strategies represent bottom-up shotgun approaches where protein extracts are first digested and the complex peptide mixtures generated are analyzed by LC-MS/MS. In such peptide-based separation techniques, such connectivity information is lost. Haynes and Roberts reviewed the possibilities of using a shotgun approach in plants and acknowledged that a completely sequenced genome is essential for a peptide-based separation, and that shotgun proteomics is currently only applicable in model plants (Haynes and Roberts 2007). 2-DE, as a protein based separation technique where the connectivity between the protein derived peptides is preserved, permits a comparison of multiple peptides per protein as a diagnostic assembly, and therefore remains the most powerful proteomics approach for most gymnosperm species, although gel-free approaches may be promising for the few gymnosperm species with elaborate EST databases available.

9.2.2 Biological Objectives

Proteomics is becoming a powerful technology, successfully used in plant research to investigate different biological processes, from growth and development to responses to biotic or abiotic stimuli, as well as to understand gene function, to characterize particular genotypes, and to analyze food traceability and substantial equivalence in transgenic crops (Thiellement et al. 1999; Rossignol 2001; Thiellement et al. 2002; Jorge et al. 2005; Jorrin et al. 2006, 2007).

9.2.2.1 2-DE as a Source of Genetic Markers

In conifers, the ability to reveal hundreds of proteins by 2-DE, was first exploited in genetic mapping and population genetic studies. The earliest studies were conducted in haploid megagametophytes of maritime pine (Bahrman and Damerval 1989; Gerber et al. 1993; Plomion et al. 1995). The megagametophyte is a haploid tissue surrounding the embryo in conifer seeds, and, because it provides distinguishing characteristics between allelic forms of protein loci, offers a unique opportunity to analyze the genetic inheritance and linkage of genes affecting protein phenotypes. In these genetic studies, three types of protein polymorphisms that can be detected on 2-DE gels were analyzed: position shifts, presence/absence variations, and intensity variations. The position shifts were identified as iso-electric differences caused by amino acid substitutions. Spots that differed in their IE point but had similar molecular weights and quantities, were thus considered allelic products of a structural gene and consequently were treated as a single spot. Spots with qualitative variation (position shifts and presence/absence variation) in 2-DE patterns of Maritime pine have been shown to be under monogenic control and to correspond to allelic variations (Bahrman and Damerval 1989; Gerber et al. 1993; Plomion et al. 1995; Plomion et al. 1997) (reviewed in deVienne et al. 1996). The quantitative polymorphisms are of particular interest because they have been shown to correlate with quantitative trait variation (Leonardi et al. 1991).

Genetic linkage maps are important tools for marker-assisted selection (Soller and Beckmann 1983). For proteins that show allelic variation, the encoding genes can be mapped on the chromosomes. In Maritime pine, genetic maps have been constructed using a combination of 2-DE protein markers and DNA-based markers, amplified fragment length polymorphisms (AFLPs) and random amplified polymorphic DNA (RAPD). Although DNA-based markers can be used to quickly saturate a genome, protein markers can offer several advantages. In conifer DNA, the coding region may not represent more than a small percent of the total genome, and protein markers allow the localization of exclusively coding DNA, whereas DNA markers essentially fall within repetitive, mostly non-coding DNA (Plomion et al. 1997). Moreover, protein markers may provide candidate proteins to understand the biological function of quantitative trait loci (QTLs). Proteins are directly involved in biochemical processes and therefore constitute more informative markers when compared to DNA-based markers. Associations between protein loci and phenotypic traits, which have for example been reported for seed weight and growth related traits in Maritime pine (Gerber et al. 1997), suggest that some of these proteins are responsible for the variation in the phenotypic traits.

In megagametophytes originating from a single Maritime pine tree and in a subsequent study, from 18 trees, linkage relationships for respectively 119 and 65 loci affecting protein spot polymorphisms were analyzed (Bahrman and Damerval 1989; Gerber et al. 1993). Plomion et al. (1995, 1997) and Costa et al. (2000) mapped 66 protein markers in total from Maritime pine megagametophytes and needles on the Maritime pine genome. The first reported genetic map was based on RAPD markers and protein markers assayed on megagametophytes (1*n*) (Plomion et al. 1995). To enable the localization of protein markers scored in diploid needle tissue on this haploid map by direct co-segregation analysis, Plomion et al. (1997) genotyped both haploid and diploid tissues with the same DNA markers. A second map was constructed based on RAPD and AFLP markers scored on megagametophytes, and proteins from diploid needle tissue were mapped on this haploid map using a QTL detection strategy based on the quantification of protein accumulation (Costa et al 2000). Both maps were finally aligned and the result can be viewed at *http://www.pierroton.inra.fr/genetics/*. The protein loci were well distributed throughout the map and interspersed with DNA-based markers (RAPDs and AFLPs).

Based on the same types of protein polymorphisms revealed by 2-DE, several population genetic studies were performed with Maritime pine. Bahrman et al. (1994) used 2-DE of megagametophyte proteins to evaluate the genetic variability existing within and between Maritime pine populations collected from seven geographical origins. Approximately, 845 to 870 spots per gel were scored for presence/absence, from which 154 were invariable over all 42 megagametophytes analyzed, and more than 84% were variable. Based on this information, three main groups (Atlantic, Mediterranean and North African) could be distinguished. In a single Maritime pine population (Les Landes, France), Bahrman and Petit (1995) examined the genetic variation of proteins in three organs (needle, bud and pollen) of 18 unrelated trees. Of the 902 spots detected on 2-DE, 245 were polymorphic showing (1) presence/absence, (2) position, or (3) intensity variation. This study clearly showed that organ-specific proteins are more variable between genotypes than organ-nonspecific proteins (56.0% *vs.* 18.4%), and that the level of genetic variability depends on the organ or tissue (lowest in needles, highest in buds). Petit et al. (1995) showed that proteins revealed by 2-DE displayed a similar level of genetic differentiation among populations than isozyme and terpenic loci, indicating the absence (or similar level) of selection acting on these three classes of loci.

9.2.2.2 *Gymnosperm Development*

In the early proteomic-based studies, comparisons between 2-DE gel patterns were performed visually by superimposing the dried gels. The development of image-analysis software that enabled 2-DE gel spot quantification and spot matching across gels, has opened the possibilities for comparison of a large number of spots in a large number of samples, and thus contributed significantly in making 2-DE a higher-throughput technology. The first use of a computer-assisted system for 2-DE gel analysis in gymnosperms was reported in 1998, in a proteomic study of drought stress in Maritime pine, where a total of 78 2D gels (corresponding to four different treatments) and about 1,000 reproducible spots were analyzed (Costa et al. 1998). The implementation of sensitive and rapid methods for identification of proteins has transformed the technique of 2-DE, so far largely used in a descriptive way as a source of genetic markers, into a powerful tool for functional analysis. In the first large scale gymnosperm protein characterization project, 28 xylem and 35 needle proteins from Maritime pine were characterized by internal microsequencing (Edman degradation) of proteins excised from 2-D gels (Costa et al. 1999). The application of mass spectrometry (MALDI-TOF for PMF or LC-MS/MS for *de novo* sequencing), together with the availability of thousands of sequenced cDNA ESTs, has enabled researchers to perform routine identification of proteins excised from 2-DE gels (see *Protein identification*). The characterization of proteomes gained a functional dimension by focusing on specialized tissues and by comparing these specific proteomes. Proteomic analysis of pollination drops in gymnosperms led to the identification of thaumatin-like proteins that may be involved in pathogen defense in ovules (O'Leary et al. 2007; Wagner et al. 2007). Comparison of the needle and the xylem proteomes in Maritime pine showed that most 2-DE spots were common to xylem and needle and the authors interpreted the presence of the same proteins in well-differentiated organs as "house-keeping proteins" (e.g., actin, HSP70, glutamine synthetase) (Costa et al. 1999). In contrast, proteins accumulating only in particular tissues are probably present in specialized cells and have specific functions. For example, the needle proteins were enriched in carbohydrate metabolism and photosynthetic enzymes (Costa et al. 1999; Valledor et al. 2008). An illustrative example of a functional analysis in gymnosperms based on comparison of the proteomes of specific tissues, involved a study of the proteomes in differentiating xylem in different types of wood: early, late, juvenile, mature, compression, and opposite wood (Gion et al. 2005). The identity of the proteins that were specifically over-expressed in mature wood suggested the up-regulation of mechanisms that contribute to the delay of programmed cell death, and therefore prolonged cell wall deposition, resulting in higher wood density, characteristic of mature wood. Proteins that were specifically highly expressed in late wood, suggested a physiological status similar to drought stress in that type of wood tissue. Proteins that were over-expressed in compression wood suggested that the typical color of compression wood results in part from the biosynthesis of anthocyanins, and some proteins that were under-expressed in compression wood suggested that molecular mechanisms determining cell shape and cell size are disturbed in gravity-stimulated tissue. Interestingly, this study also showed that seasonal effect dominated the tissue type effect on the control of protein accumulation. A more detailed physiological study was performed using the analysis of the proteome changes, in parallel with lignin and cellulose content changes, which accompanied the formation of compression wood in a range of Maritime pine xylem samples, subject to gradually increasing growth strain (Plomion et al. 2000). Functional relationships between proteins were traced via correlation analysis clustering of similar protein expression patterns along the gradient of gravity-stimulated stressed xylem tissue. A small cluster of traits, showing the strongest positive correlation with the growth strain, comprised the phenotypic trait, lignin content, an ethylene forming enzyme. A larger cluster, positively correlated with growth strain, included mainly lignification proteins. These results support the suggestion that ethylene plays a role in compression wood differentiation by inducing the expression of enzymes involved in lignification. Furthermore, a transcription factor identified in the protein cluster positively correlated with growth strain, was established as a candidate protein for controlling compression wood differentiation. Such proteomic expression analyses can thus lead to detailed functional hypotheses, which are, taking into account the direct involvement of the proteins in the biological processes, more relevant as compared to transcriptomics based hypotheses.

Only a few proteomic studies have focused on the changes associated with gymnosperm growth and development (Fernando et al. 2005; Lippert et al. 2005). A study of the molecular programs involved in pollen tube development was performed in eastern white pine (*Pinus strobus*) via a comparative analysis of the proteomes in 2-day-old pollen tubes and in ungerminated grains (Fernando et al. 2005). The differential proteome in the developing pollen tubes revealed the involvement of both cell wall formation and stress/defense responses. A study of the embryo developmental process during somatic embryogenesis was performed in white spruce, by recording the proteomic changes occurring across four stages of somatic embryo maturation, and revealed the involvement of a broad variety of biological processes (Lippert et al. 2005).

9.2.2.3 *Response to Biotic Stress*

Plants have developed sophisticated mechanisms to cope with various types of stress. In the last decade, proteomics has been widely adopted, next to the more conventional transcriptomics approach, to study the molecular responses of plants toward both biotic and abiotic stress stimuli. For example, in gymnosperms, a comparison of the proteomic changes in Sitka spruce bark tissue, subject to feeding by white pine weevils or mechanical wounding, has been reported (Lippert et al. 2007). This study revealed that changes in protein levels occur as early as 2 hours following the onset of insect feeding, and that the differential proteome induced by weevil feeding is similar to the differential proteome induced by simple mechanical wounding.

9.2.3 *Added Value of Proteomics to Transcriptomics*

Although proteomics is still not genome-wide, it certainly can offer added value to the analysis of genome expression at the mRNA level. The biological functions in a plant cell are executed by proteins rather than by mRNA, and several post-transcriptional and post-translational mechanisms, such as translation rate, half-lives of mRNAs and proteins, protein modifications and intercellular protein trafficking, dictate that far more information influencing plant phenotype is contained in the proteome as compared to the transcriptome.

A few studies in gymnosperm have analyzed the correlation between protein levels and the corresponding transcript levels (Gion et al. 2005; Lippert et al. 2007). A weakly positive correlation between mRNA and protein abundance has been reported in differentiating xylem samples in Maritime pine (Gion et al. 2005). The Pearson correlation coefficient between the protein levels, as quantified on 2-DE gels, and the number of corresponding ESTs in a cDNA library generated from the same samples, was 0.46, and dropped to 0.31 if the most highly expressed transcripts and proteins were not considered (Gion et al. 2005). Lippert et al. (2007) conducted a comparison of protein expression and cDNA microarray profiles on Sitka spruce bark induced by mechanical wounding or feeding by white pine weevils. The correlation between the proteomic and corresponding microarray datasets was surprisingly low. Of the 71 differentially accumulating proteins that were identified (which were represented on the microarray), only 10 were reported to show agreement at the level of transcript abundance, while 17 were in opposition and 44 showed no change in transcript abundance.

In the same study, Lippert et al. (2007) showed the importance of post-translational modifications (PTM) in the response to weevil damage or wounding in Sitka spruce, further highlighting the complementary nature of transcriptome and proteome analyses. For nine proteins in Sitka spruce bark, two to four different isoforms were reported, showing distinct patterns of differential expression induced by wounding or weevil damage (Lippert et al. 2007). For example, in response to weevil feeding, a single isoform (pH 6.8, 20kDa) was down-regulated and three new acidic isoforms (pH 5.2 + 0.2, 23–27 kDa) were induced. Post-translational phosphorylation induced by weevil feeding was proposed to be the cause of the pI shift. Further challenges of crucial functional importance will thus be the development of high throughput techniques for the identification of the protein molecular species owing to protein maturation.

9.2.4 Databases

Along with the development of reproducible high-throughput techniques, an enormous amount of data is expected to be produced via functional proteomics programs. The exploitation of these data will largely depend on the development and organization of databases.

The Maritime pine database constitutes the first public proteome database dealing with forest trees (*http://www.pierroton.inra.fr/genetics/2D*) (Costa et al. 1999). This database consists of scanned gels of Maritime pine needle and xylem tissue with hyperlinked spots, which allow one to retrieve sequence data and, for certain proteins, the location on a linkage map and the behavior in drought environment, from the position of protein markers in 2-DE gels, and vice versa.

In 2005, a web-based plant 2-DE database "PROTICdb" (*http://moulon.inra.fr/~bioinfo/PROTICdb*) was established to store, track, query and compare plant proteome data, and is freely available upon request (Ferry-Dumazet et al. 2005). Maritime pine proteomes of differentiating xylem, corresponding to different developmental stages and treatments, have been stored in "PROTICdb" and are also publicly available on the website *http://cbib1.cbib.u-bordeaux2.fr/Protic/Protic/home/index.php*, with the information concerning plant material and experimental conditions, protocols for extraction, electrophoresis, staining and digitalization, mass spectrometry techniques, and details concerning the identification of the protein spots and the query databases (Gion et al. 2005). On the other hand, all protein sequences derived from the translation of the coding sequences that have been submitted to the public nucleic acid database (EMBL/GenBank/DDBJ) are integrated into the UniProt Knowledgebase (UniProtKB) (*http://www.uniprot.org*). A query for gymnosperm (*Coniferopsida*) proteins, performed in UniProtKB as of 28 August 2008, resulted in only 108 gymnosperm proteins for which the existence has been proven on the protein level. Proteins in the database were identified in *Pinaceae* (78), *Taxus* (17), and *Cupressaceae* (13). This corresponds to less than 1% of the total of 14,104 predicted gymnosperm proteins in the database. The gymnosperm proteins represent only a very small fraction (less than 2%) of the 6,052 plant proteins (*Spermatophyta*), with proven existence on the protein level, in the database. For illustrative comparison, 2,497 (41%) of the proven plant proteins in the database are derived from *Arabidopsis thaliana*.

9.2.5 *Statistical Analysis*

2-DE has been criticized for its low reproducibility when a direct comparison of different gels is performed (Saravanan and Rose 2004; Hunt et al. 2005; Ruebelt et al. 2006). Analytical variation in gel patterns and spot quantification results from both experimental procedures (protein extraction, IEF separation, SDS-PAGE, gel staining-destaining) (Saravanan and Rose 2004; Hunt et al. 2005; Ruebelt et al. 2006), and hardware/software accuracy during post-electrophoretic analysis of the 2-DE protein profiles (image acquisition and analysis) (Wheelock and Buckpitt 2005; Wheelock and Goto 2006). Therefore, deep statistical analysis is required when analyzing differential protein abundances. A number of recent papers dealing with comparative proteomic studies (differential expression) in gymnosperms choose rather arbitrary criteria for considering a difference as being biologically relevant [1.5-fold change at $p < 0.05$ (Lippert et al. 2007); 2-fold change at $p < 0.01$ (Valcu et al. 2008)]. In order to have an appropriate idea of the biological relevance of change, however, a reference is needed. In order to set up a metric for further comparative analyses as a reference, a detailed analysis of both biological and analytical variation for the system of study is required. Therefore, a 2-DE reference map (pI 5–8, Mr 10–100 kDa) for physiologically mature *Pinus radiata* needles was constructed based on 10 independent protein extracts from homogeneous needles from the same branch and 12 needle protein extracts from distinct trees (Valledor et al. 2008). Based on this study, the average analytical and biological variability for 2-DE analysis of mature pine needles was determined to be 31 and 42%, respectively, and an optimal number of seven biological repeats for future comparative proteomic studies were proposed (Valledor et al. 2008).

9.3 Metabolomics

In systems biology, the term "metabolome" refers to the complete set of small molecules (i.e., metabolites) that participate in or are products of metabolic reactions within an organism or tissue. Hence, "metabolomics" is concerned with the identification and quantification of those molecules in order to advance biological understanding and the development of novel biomarkers. As an eventual product of gene expression under the influence of environment, cellular metabolism is the immediate progenitor of phenotype and, as such, the relationships between phenotypic and metabolomic traits are potentially more coherent than for the genomic, transcriptomic and proteomic counterparts. However, in comparison to the other "omics", for which rapid technological advances have been seen, the emergence of analytical and software tools for the comprehensive analysis of the metabolome has been slow. Whereas the genome, transcriptome and proteome are each comprised of a single class of polymeric molecules, the metabolome exhibits an enormous degree of physico-chemical molecular variety such that no current instrument platform is capable of analyzing all metabolites. The consequent need to employ a series of preparative and analytical techniques to (imperfectly) span the metabolome, and the technical difficulty of merging disparate and/or overlapping data generated by these diverse means have restricted the full potential of metabolomics to date. Despite this, significant technological advances have occurred over the last decade, and broad-scale metabolomics studies involving gymnosperm species have begun to appear in the literature.

Metabolomics analyses have been broadly classified as either "targeted" or "non-targeted". Targeted analysis, otherwise known as "metabolite profiling", typically focuses on quantifying a defined group of metabolites that are related by either a metabolic pathway or molecule class. These studies require a higher degree of a priori knowledge as far as compound identity and interrelationship are concerned, and in their most refined form become "target analysis"—the measurement of one or very few metabolites to serve as, for example, phenotypic biomarkers. Conversely, non-targeted analysis aims to measure as broad a range of metabolites as possible, with the intention of creating a global metabolic fingerprint. In the first instance, global fingerprinting is not so concerned with the metabolites' identity and absolute abundance as it is with their relative abundance and interrelationships, and aims primarily to classify samples based on metabolic "features". Ultimately, the reductive approaches commonly employed in these analyses usually lead to the identification of subsets of discriminating metabolites whose abundances correlate with phenotypic traits of interest, followed by attempts to identify those compounds so that their biological significance may be rationalized. Whereas broad-scale metabolomics is a recent development of the last 10 or so years, targeted analysis of metabolism has a much longer history. Although, due to their narrow focus, it is arguable that targeted analyses are not metabolomics in the strict sense, they do comprise the origin from which non-targeted, global metabolomics approaches have been derived with the assistance of advancing technology. As such, there is obvious interdependency between targeted metabolite profiling and non-targeted metabolic fingerprinting, which has arisen from their shared ultimate objective—that being an improved biological understanding and diagnostic capabilities. Because this conceptual bridge exists, recent metabolomics research in plants has frequently fallen into a middle ground in terms of the degree of prior knowledge of the identity and role of the metabolites being analyzed, the breadth of metabolites being analyzed and the basis for their inclusion. Clearly, the scale and rationality of analyses does not allow a practical distinction between modern metabolomics and historical metabolic analyses to be made. In reality, it is a new mentality that defines metabolomics—one under which powerful new analytical tools, abundant computing power and powerful data-handling software have made it conceivable to tackle metabolic questions at the whole organism or tissue level, with an emphasis on deconvoluting biological complexity. As such, this discussion will focus on recent research that shares this opinion, but reflects on classic-style analyses where relevant.

9.3.1 *The Analytical Process*

Practical metabolomics is concerned with measuring and analyzing metabolite pools in an attempt to understand metabolic networks and develop biological markers. Although the rate of flux through metabolic pathways would certainly be a more informative and a robust measure of metabolic activity in some cases, there is currently no practical way in which to measure this for individual metabolites on a broad scale. As such, the field of metabolomics must remain content with using the more easily measured phenomenon of metabolite pooling as a slightly ambiguous indicator of metabolic activity, although even with this limitation there remains a great deal to discuss regarding metabolomics infrastructure.

9.3.1.1 Analytical Tools

A variety of analytical tools are available for the generation of metabolite profiles or fingerprints, with specific tools being more appropriate for the determination of metabolites having particular physico/chemical properties. In this regard, the analysis of the gymnosperm metabolome requires no special consideration over that of other plants, with chromatography, mass spectrometry and nuclear magnetic resonance (NMR) spectroscopy being the analytical mainstays across the field.

Gas chromatography (GC) is the chromatographic technique of choice for the analysis of smaller (MW < ~1,000) molecules, owing to its broad specificity and high resolution. There are several approaches to sample introduction available for GC, including the evaporation of liquid phase extracts in the injector, direct pyrolysis of samples to generate volatiles of both extractable and structural metabolites, and the injection of naturally occurring volatiles obtained from contained sample headspace. The pyrolysis and headspace approaches both have the benefit of not requiring lengthy sample preparation prior to analysis. Alternatively, high pressure liquid chromatography (HPLC) is useful for the separation of molecules too large for GC, although the range of metabolites that may be collectively analyzed with a single column/eluent setup is typically limited to a specific molecular class. Both techniques have undergone remarkable improvements in recent times, with the emergence of ultra-high pressure liquid chromatography (U-HPLC) and ultra-fast gas chromatography, offering significant increases in resolution and sample processing efficiency that promise to assist the progressive development of large scale, non-targeted metabolomics.

Chromatographic separation systems require an attached quantitative detector, and there are many types in widespread use. While single channel detection such as FID for GC and ECD or single wavelength UV/Vis for HPLC will provide information regarding metabolite retention time and abundance, these detectors provide little, if any, assistance in identifying particular unknown metabolites beyond their general molecular class. It is for this reason that mass spectrometers, with the extensive molecular structural information that they provide, have gained popularity as detection systems for metabolomics. Quadrupoles, ion traps and time-of-flight (TOF) analyzers are all readily available as detectors for chromatographic analyses, although different approaches are required for the introduction of analytes into the respective mass spectrometer depending on whether gas or liquid phase chromatography, or direct sample insertion is used. The mass spectral data generated can be used to deconvolute signals from co-eluting metabolites, and combined with retention indices to provide relatively straightforward identification of analyte molecules via comparison with standard compounds or large compound libraries. Furthermore, when this approach is unsuccessful, MS^n analysis in appropriate spectrometers can often help to identify compounds in the absence of verified standards.

Without metabolite identification, all that can be provided by chromatographic analysis is a metabolic fingerprint, and while potentially useful for distinguishing between distinct metabolic systems, a fingerprint alone is not at all informative of underlying biological relationships. As alluded to above, the identification of metabolites based on mass spectral and retention index matches with standard compounds is commonplace. Although it is possible to assemble standard compound mass spectral libraries for LC/MS, the high degree of instrument-dependant variation in fragmentation patterns has meant that universally compatible libraries are not available. For GC/MS, however, the standard use of a 70eV potential for electron ionization and molecule fragmentation has meant that, not only can fragmentation pattern and retention index libraries be constructed, but they can also be shared between instruments and research groups. This has led to the publishing of extensive commercial and freely distributed libraries of mass spectra, and while commercial libraries represent an extremely broad range of molecules (e.g., the 2008 NIST library contains more than 190,000 compounds including various states of derivatization), smaller, free libraries such as that provided by the Golm Metabolome Database (GMD) (Kopka et al. 2005) are tailored specifically to the needs of plant metabolomics, are less redundant and frequently have more utility. Still, many compounds resolved from metabolite profiles are not present in these libraries, and continue to elude identification. The ongoing expansion of mass spectral library resources is of paramount importance, as at present, the process of compound identification constitutes a major limiting factor in the plant metabolomics field.

Nuclear magnetic resonance spectroscopy is a popular alternative to chromatography/mass spectrometry for resolving compounds from complex mixtures, with the sub-class of metabolomics employing this technique being known as "metabonomics". Biological NMR spectroscopy usually exploits the magnetic properties of ^{1}H or ^{13}C nuclei. The different proton or carbon nuclei in a molecule resonate at slightly different frequencies due to differences in local chemical environment, so particular compounds have characteristic nuclear resonance patterns for specific nuclei. Thus, a 1D NMR spectrum can provide information on the number and type of ^{1}H or ^{13}C nuclei in a mixture of metabolites, and from this the identity and relative contributions of individual metabolites may be resolved. One of the major benefits of NMR spectroscopy is that it is non-destructive, and as such, samples may be analyzed repeatedly over the course of a study, or studied in other ways once NMR analysis is complete.

9.3.1.2 Sample Preparation

While some of the analytical tools employed in metabolomics permit the determination of metabolite composition with minimal sample preparation, others require the extraction of metabolites from the specific tissue, prior to analysis. Most often, this involves tissue disruption followed by some form of liquid solvent extraction, and depending on the circumstances, crude extracts may need to undergo further preparatory work.

Because gymnosperms are not often the preferred plant model for novel method development, the preparation of metabolite extracts from gymnosperm tissue has almost completely involved protocols derived from those already established for other plant species. For example, the preparation of polar and/or non-polar metabolites for GC-based metabolomics in plant species has usually employed variations of the extraction and derivatization protocols for *Arabidopsis* published by Fiehn et al (2000a, b). The extraction is based on a dual-phase water/methanol/chloroform extraction that yields polar metabolites in the water/methanol phase and non-polar metabolites in the methanol/chloroform phase; subsequent derivatization of metabolites, to increase volatility and raise the high-mass cut-off of GC, involves the protection of carbonyl moieties by reaction with an alkoxyamine hydrochloride, followed by the elimination of acidic protons by reaction with a trimethylsilylating agent [e.g., *N*-methyl-*N*-trimethylsilyltrifluoroacetamide (MSTFA)]. For metabolites in the non-polar fraction a methanol/chloroform-based trans-methylation of hydrocarbon chains is also carried out prior to other derivatization reactions. In gymnosperms, adjustment of the procedure has been limited to varying the temperature and duration of the extractions and derivatizations, and the documented optimization of these parameters for developing xylem of loblolly pine (Morris et al. 2004) clearly illustrated the importance of optimizing extraction conditions for specific tissues, ensuring that the process has enough stringency to achieve good metabolite extraction, but is not so harsh as to cause degradation of labile compounds. The susceptibility of the metabolite profile to variation in sample handling and analytical conditions is a known problem of metabolomics that demands consistent processing in order for comparable datasets to be generated from individual samples or sample batches.

LC/MS-based analyses of soluble metabolites have not yet been reported in gymnosperm metabolomics, even though liquid chromatography offers the ability to profile the metabolites of greater molecular weight than may be analyzed by gas chromatography, and without the need for derivatization. In these analyses, simple solvent-based extractions akin to those described for GC-based analyses would be appropriate, but should involve further metabolite partitioning (of, for example, phenolics) and enrichment so that an adequate signal-to-noise ratio is achieved in LC/MS. The specific extraction of membrane phospholipids employed for LC/MS-based "lipidomics" analyses in gymnosperms also uses a protocol developed for *Arabidopsis* (Welti et al. 2002; Yang et al. 2007). In this, isopropanol with butylated hydroxytoluene (BHT) is used as the primary solvent, with various mixtures of chloroform, water and methanol with BHT used for subsequent, exhaustive tissue extraction. Combined extracts are washed with KCl solution, then with purified water, and then evaporated prior to resuspension in chloroform or a chloroform/methanol mixture.

9.3.1.3 Data Processing and Analysis

The collation of metabolite profile data from a number of samples is required prior to statistical analysis, but can pose a serious technical challenge. In chromatograms of complex biological samples the partial or complete co-elution of metabolites is a frequent occurrence that can limit biological resolution and introduce error into downstream data analyses, if left unaddressed. Fortunately, however, the process of deconvoluting signals from co-eluting metabolites is possible with multi-channel data, as is generated in mass spectrometry. In addition, inter-sample variation in the metabolite separation domain (which in gas and liquid chromatography is time based) is caused by fluctuations in temperature ramps, eluent gradients, column pressure or flow rates, and means that the retention time of any given metabolite is never one exact value across all runs. When an experiment involves a large set of samples and metabolites, manual collation becomes impractical, and the task must be handed to automated software that decides whether peaks in multiple samples represent the same or different compounds, based on retention time windows and mass spectral matching. The need to resolve these issues has led to the development of a number of commercial and free software tools that are capable of these tasks, with the more accomplished of these capable of both deconvolution and collation. Notable non-commercial examples include NIST AMDIS (for deconvolution only), MetAlign (Tikunov et al. 2005) and the complex yet highly capable XCMS (Smith et al. 2006). While there are reports of automated peak collation (AR Robinson et al. 2008 unpubl.), profile deconvolution has not yet been reported in the gymnosperm metabolomics literature, although it is certainly only a matter of time before it is.

Data analysis in metabolomics has advanced at a considerable rate, with the ongoing introduction of statistical analyses and other calculative tools to the field. Most statistical tools have been applied with a reductive perspective. Classic, univariate tests between means, such as Student's t-test, the F-test and more robust incarnations like Tukey's "honestly significant difference" (HSD) test have been used to individually identify metabolites exhibiting genotype- or treatment-related differences in abundance. Although useful, these tests deal with each metabolite as an isolated entity, and are unable to take the interdependence of the components of metabolite profiles into account. Multivariate analyses are better suited to this task. The default statistical tools of metabolomics are principal components analysis (PCA) and hierarchical cluster analysis (HCA), and many analyses are limited to the use of these two techniques. Both are useful for comparing complete profiles from multiple samples, and generate diagrammatic outputs that are visually appealing and easily interpreted. Although PCA does provide some information regarding the particular metabolites responsible for any distinction between sample classes, neither PCA nor HCA are very diagnostic, because they do not provide calculated measures of the relationships between metabolite profiles and, for example, phenotypic traits. Canonical correlation analysis (CCA) is one method that can assist in defining the relationships between two sets of variables, such as metabolites and quantitative phenotypic traits. Essentially, CCA identifies groups of variables in one set that are correlated to groups of variables in the other, and indicates the relative contributions of individual variables to the relationship. However, in cases where diagnostics are an objective, techniques that generate models for the prediction of specific traits on the basis of metabolite profiles are required. To this end, multiple discriminate analysis (MDA) is useful for distinguishing samples by class (e.g., genotype, species), while partial least squares regression (PLSR) and the less conventional Bayesian stepwise modeling procedure are powerful techniques for modeling quantitative phenotypic traits (e.g., the total lignin content of wood).

There are numerous other data analysis approaches that have been applied with success in metabolomics, but which have not been reported for studies of gymnosperms. Of these, neural networking and metabolic pathway scaffolding are of particular note. Neural networking, using software such as Pajek (Batagelj and Mrvar 2002) uses graphical marker size and proximity to generate a visual image of the relationship between the abundances of different metabolites. In pathway scaffolding and annotation, using software such as MapMan (Thimm et al. 2004), metabolites of interest are arranged on classic-style biochemical pathway diagrams and their individual contributions to a given relationship with, for example, a phenotypic trait, are represented by a heatmap output. It would be exciting to see the use of such tools in gymnosperm-related research, as their output can make considerable contributions to the understanding of biological relationships.

9.3.2 Biological Objectives

Metabolomics techniques are readily applicable to many aspects of plant biology, and in gymnosperms, analyses have targeted a variety of subjects across a broad range of genera including *Pinus* (pine), *Picea* (spruce), *Larix* (larch), *Pseudosuga* (Douglas-fir), *Thuja* (cedar) and *Taxus* (yew).

9.3.2.1 Somatic Embryogenesis

Somatic embryogenesis (SE) is now well-established as an effective approach to clonally propagating germplasm in gymnosperms, although the frequent loss of regenerative capacity in embryogenic cultures continues to have considerable impact on the efficiency and stability of embryo production. In a recent study by Robinson et al. (2009), the interaction between metabolic composition, physiological state, genotype, and embryogenic capacity in five genotypes of loblolly pine SE cultures was explored, using GC/MS to analyze solvent-extracted metabolites from embryonyl tissue in the filamentous phase of development. By adopting a stepwise variable selection procedure, based on minimizing the Bayesian information criterion (BIC), it was possible to model, and then accurately predict the eventual embryogenic productivity of these cultures, based on the abundance of a select subset of predictor metabolites.

From a biological perspective, the metabolic structure of the model implied that variation in culture regenerative capacity was closely linked to the physiological transition of cultures from the proliferation phase to the maturation phase of development (i.e., from organogenic metabolic sink to storage sink), with the degree to which this transition had emerged prior to transfer of cultures from proliferation to maturation medium relating strongly to the efficiency of mature embryo production. The propensity of cultures to advance into this transition appeared to relate to nutrient uptake and allocation *in vivo*, as indicated by positive correlations between culture productivity and the major carbon (sucrose) and nitrogen (glutamine) pools. It also appeared that developmental progression was related to the tolerance and response of cultures to stress during the proliferation phase, as indicated by negative correlations between culture productivity and the pools of osmoprotectants and osmoprotectant precursors such as proline, arabitol, and serine. This research clearly demonstrates the power of metabolomics in generating diagnostic phenotypic markers and providing insight into biological processes, and was also the first reported application of stepwise modeling in metabolomics.

9.3.2.2 Seed Biology

NMR-based metabolomics has been applied with success to the study of seed composition, metabolism and viability in industrially important conifer species. To date, it is the only metabolomics approach to have been applied in this field. Terskikh et al (2005) used standard and magic angle spinning (MAS) ^{13}C NMR spectroscopy to assess the mobile (i.e., liquid phase) component of seeds from several species of the Pinacea and Cupressaceae families. The mobile compounds identified in mature, dry seeds were predominantly associated with aliphatic methylene carbon and olefin carbon (as well as some glycerol and carbonyl carbon shifts), which generally corresponded to fatty acid oils and terpene molecules. Differences in fatty acid composition could be observed between species, with notable separation between those of different genera, and especially between the two families. The resolution was even high enough that four species of *Pinus* could be differentiated by chemotaxonomic analyses of NMR-determined oleic, pinolenic and linoleic acid contents. In addition, the break of seed dormancy and subsequent metabolic changes in western white pine were studied. The transition from the mature dry, to a fully imbibed state was accompanied by the emergence of chemical shifts associated with hydrated sucrose, which is a putative osmoprotectant during seed development and subsequent to the break of dormancy. Then, upon germination and with post-germination development, the progressive depletion of seed oil reserves was clearly evident in the NMR spectrum—especially in the olefin carbon region corresponding to triglycerides. An emergence of free amino acids—predominantly arginine and asparagine at the time of analysis—was also observed, presumably resulting from the cleavage of storage protein reserves.

The quantity and quality of the reserve held by a seed at the time of germination is critical to seedling emergence and vigor, and unfortunately the seeds of many conifers exhibit severe storage-related degradation. As a continuation of their earlier work on the metabolic composition of coniferous seeds, Terskikh et al. (2008) worked to demonstrate a correlation between changes in ^{13}C NMR spectra and the rate of deterioration in stored seed batches of western red cedar. They observed that a decline in germination capacity due to storage was accompanied by a correlative decrease and broadening of the resonances associated with triglyceride reserves, and furthermore that the proportion of polyunsaturated fatty acids in the triglyceride oil mixture decreased sharply. It was contended that lipid peroxidation, oxidative polymerization, and consequent solidification of storage oils was responsible for these changes, and at least partly responsible for the observed decrease in seed viability. The general relationship between NMR intensity and germination rate was determined to be hyperbolic, and modeled accordingly.

The foundation-laying research described not only demonstrates that metabonomics can help to improve our understanding of conifer seed biology, but also that it is a very promising diagnostic tool for effective breeding and propagation of valuable gymnosperm species. Within this context, the non-destructive nature of NMR makes it an ideal tool for analyzing precious plant material and monitoring stored germplasm. Research in this field has an exciting future, with a clear need and an enormous potential for the industrial application of NMR-based seed analysis in the coming years.

9.3.2.3 Wood Formation

Another active area of metabolomics research in gymnosperms is related to xylem development, wood biosynthesis and wood attributes. The primary application of metabolomics in the study of xylem/wood formation has been in differentiating between metabolic systems giving rise to wood with different physical or chemical properties, within individuals or species. As a simple example, Morris et al. (2004) conducted a straightforward GC/FID- and GC-MS-based metabolomic analysis of the developing xylem of loblolly pine trees, representing two families that produce wood with ~45% and ~50% alpha cellulose content. A small set of the largest metabolites found in the GC/FID chromatograms were analyzed by PCA, which loosely clustered and partially separated the samples of the two families. Both primary and secondary metabolites associated with wood formation were implicated in this loose distinction, including citric acid, shikimic acid, glucose and fructose. Although limited in terms of sample count and metabolic scope, this research did set the scene for the more comprehensive experiments that followed. To support mounting chemical and structural evidence, and their hypothesis that juvenile and compression woods of conifers were not as similar as had previously been suggested, Yeh et al. (2006) used GC/FID, GC/MS, PCA and HCA of polar metabolites extracted from developing xylem to distinguish between the metabolism involved in biosynthesis of variant wood forms in juvenile loblolly pine. Tight clustering and clean separation of sample treatment groups in PCA and HCA analyses of a set of 25 highly and moderately abundant metabolites showed that normal, wind-exposed, compression and opposite wood formation were each accompanied by different, characteristic metabolite profiles. The separation between normal and compression wood in PCA was largely due to the influence of lignin precursors such as shikimic acid, *p*-glucocoumaryl alcohol and coniferin, and the shifts in the metabolite profile for compression wood, relative to the control, involved significant increases in these lignin precursors as well as several carbohydrates. This finding concurs with the increase in lignin content and altered lignin composition typically observed in the compression wood of gymnosperms. Such examples demonstrate the effective use of metabolomics to rapidly identify the distinguishing components of closely related metabolic systems, which can then be related back to distinctive phenotypic traits.

Large-scale GC/MS-based metabolomics, studying the influence of genetic and environmental factors on wood quality traits in gymnosperms, was performed by Robinson et al. (2007). The scope of this research was broad. It involved the analysis of 139 polar metabolites from developing xylem across 181 individual Douglas-fir trees, including siblings from ten high-performance breeding families replicated on two sites with markedly different precipitation regimes ("very wet" and "very dry"). As well as the metabolites, whole-tree measurements (height, diameter and volume) and a set of quantitative phenotypic wood traits (density, microfibril angle, chemical composition, and fiber morphology) were included in the analysis. The series of reductive statistical approaches employed demonstrated an overall coherence between the genetic, metabolic, environmental, and phenotypic elements associated with wood formation. Factor analysis (FA), multivariate discriminate analysis (MDA) and canonical discriminate analysis (CDA) distinguished between samples based on both family and site, indicating that both genetic and environmental factors affected wood-forming metabolism in this scenario. However, it was clear from these analyses and the generally low heritabilities calculated for the metabolic and phenotypic traits that the variance observed in metabolite profiles was primarily due to the influence of environment (i.e., site). Canonical correlation analysis (CCA) was able to dissect the relationships between the metabolite profiles and quantitative phenotypic traits, and identified strong correlations between metabolite pools related to major components of cell wall biosynthesis [including cellulose (glucose and fructose), hemicellulose (xylose, arabinose and maltose), and lignin (quinic acid, shikimic acid and coniferin)] and phenotypic indicators of growth (diameter, height and volume), cell morphology (microfibril angle, fiber length and fiber coarseness), and cell wall chemistry. Although logistically and technically challenging, this type of unified metabolomics, in which genetic, metabolic, phenotypic and environmental elements are analyzed in concert, is set to become a powerful aid to our detailed understanding of tree growth and wood biosynthesis.

9.3.2.4 Plant Cell Culture

The metabolomics research reported for gymnosperm plant cell cultures varies from the classic analysis of intermediary metabolites in biochemical pathways. In two studies of *Taxus* cell lines, Yang et al. (2007, 2008) studied the composition of membrane glycerophospholipids in liquid suspension-cultured tissue through the progression of cellular development and apoptosis/taxol biosynthesis—both with and without the addition of chemical elicitors of taxol production. A specific protocol for phospholipid extraction (as described above) was employed, and LC/ESI/MS analysis and direct injection MS^n scanning resolved ~100 molecular species which could be assigned to different phospholipid classes. In the first study (Yang et al. 2007), it was shown that the phospholipid composition in apoptotic cells was markedly different than in living cells, with significant increases observed in species of phosphatidic acid (PA) and lysophosphatidylcholine (lysoPC), and decreases in phosphatidylcholine (PC) species. These changes were reflected in a PCA, which cleanly separated apoptotic tissues from living tissue, and attributed the separation mainly to the phospholipid classes indicated. These changes were concurrent with increases in the activity of phospholipase D (PLD), which is responsible for the conversion of PC to PA, and the authors suggested that the metabolism of these membrane phospholipids may regulate the processes of apoptosis and taxol production in at least some Taxus species. In a second study (Yang et al. 2008), it was found that non-apoptotic and apoptotic elicitors of taxol biosynthesis (methyl jasmonate (MeJa) and Ce^{4+}, respectively) had different effects on total phospholipid content. While no significant change in the total phospholipid content was seen with exposure of cultured tissue to MeJa, Ce^{4+} caused a considerable decrease over the course of 72 hours. In a PCA, samples from the control and elicitor treatments clustered and completely separated from one another under the first component, in which separation was attributed to similar differences in PC, PA and lysoPC, as seen in the first study. The notably differential effects of MeJa and Ce4+ on membrane phospholipid content and composition were consistent with their different effects with regard to programmed cell death, and with consideration of enzymatic data available, it was suggested that the two compounds elicit taxol production via alternate mechanisms for the induction of the jasmonic acid pathway by which taxol is produced.

9.3.2.5 Fragrance

The extractives content of wood has implications for the survival of a tree, but also for human utilization. One interesting study investigated the fragrance compounds of six, highly prized odorous and durable coniferous woods grown in Taiwan. Wang et al. (2006a) used solid-phase microextraction (SPME) and GC/MS to generate non-biased fragrance composition profiles for these woods at room temperature (30°C). With the aid of mass spectra databases, GC retention indices based on alkane standards, and authentic standard compounds, a total of 46 compounds were identified in the fragrances, across these species. It was found that fragrance compositions varied considerably from solvent-extracted essential oil compositions previously reported in the literature—to the extent that many extractive compounds previously believed to be important components of fragrance in specific species (e.g., δ-cardinene, δ-cardinol, β-eudesmol and copaene in *Cryptomeria japonica* (Chang et al. 2003)) were not detected as volatiles at all. Species-related variation in odor was associated with variation in fragrance chemical composition, and principal components analysis and the nearest neighbor cluster analysis were performed to determine the similarity/disparity between species. These multivariate approaches resolved three groups of two species, with each group based on shared chemical skeleton classes, and in doing so demonstrated the potential for further chemotaxonomic classification of gymnosperm species on the basis of volatile emissions.

9.4 Concluding Remarks

The usefulness of both proteomics and metabolomics in different applications of gymnosperm research has repeatedly been shown. Proteomics, for example has permitted the discovery of genetic markers, which can be used in breeding programs, especially if a protein marker is associated with an economically relevant phenotypic trait. Secondly, both proteomics and metabolomics can contribute to the functional characterization of cellular processes associated with developmental and/or environmental changes. However, the potential of proteomics and metabolomics is far from being fully exploited in gymnosperms (and in plants in general, particularly as compared to yeast and humans). Practically all proteomics reports evaluating gymnosperms are based on 2-DE/MS, a work-flow which focuses on highly abundant proteins and excludes certain classes of proteins based on hydrophobicity and pH. The abundance range and physico-chemical diversity of proteins make a whole genome approach a much greater challenge when compared to comparable analyses based on transcript abundance. However, the future availability of more elaborate gymnosperm sequence data (ESTs and ultimately full genome sequences), together with gel-free and second-generation proteomics techniques, is expected to collectively result in an explosion in proteomic-based research in gymnosperms. While proteomics has already had considerable attention in many plant species, metabolomics is clearly a branch of biological science still in its infancy, with a great deal yet to be realized. However, with the sample analysis and data processing technology currently available, the application of metabolomics in gymnosperms could achieve a great deal more than has been done to date. Further application in the areas outlined in this chapter, as well as expansion into other facets of biology, such as plant/pathogen interactions or environmental response, is indeed possible.

As compared to transcriptomics reports, the reports of differentially expressed proteins are more easily comprehensible due to the considerable added value of the direct involvement in the ongoing biochemical processes. Differential metabolomes can be interpreted in direct relation with the phenotype, but the biological interpretation often remains difficult, due to the complex causes of differential metabolite accumulation and the effects of metabolite channeling. The integrated analysis of metabolomics, genomics, transcriptomics and proteomics data in future research is an exciting prospect for gymnosperm biology that is already being realized in alternative plant systems. The research conducted to date confirms proteomics and metabolomics to be powerful scientific approaches and highly applicable in the context of gymnosperm biology, and systems biology in general.

References

Allona I, Quinn M, Shoop E, Swope K, St Cyr S, Carlis J, Riedl J, Retzel E, Campbell MM, Sederoff R, Whetten RW (1998) Analysis of xylem formation in pine by cDNA sequencing. Proc Natl Acad Sci USA 95: 9693–9698.

Amme, S, Matros A, Schlesier B, Mock HP (2006) Proteome analysis of cold stress response in *Arabidopsis thaliana* using DIGE-technology. J Exp Bot 57: 1537–1546.

Bahrman N, Damerval C (1989) Linkage relationships of loci controlling protein amounts in Maritime pine (*Pinus pinaster* Ait). Heredity 63: 267–274.

Bahrman N, Petit RJ (1995) Genetic-polymorphism in Maritime pine (*Pinus pinaster* Ait) assessed by 2-dimensional gel-electrophoresis of needle, bud, and pollen proteins. J Mol Evol 41: 231–237.

Bahrman N, Zivy M, Baradat P, Damerval C (1994) Organization of the variability of abundant proteins in 7 geographical origins of Maritime pine (*Pinus pinaster* Ait). Theor Appl Genet 88: 407–411.

Basu U, Francis JL, Whittal RM, Stephens JL, Wang Y, Zaiane OR, Goebel R, Muench DG, Good AG, Taylor GJ (2006) Extracellular proteomes of *Arabidopsis thaliana* and *Brassica napus* roots: analysis and comparison by MudPIT and LC-MS. Plant Soil 286: 357–376.

Batagelj V, Mrvar A (2002) Pajek—Analysis and visualization of large networks. In: Lecture Notes in Computer Science 2265: 477–478.

Bayer EM, Bottrill AR, Walshaw J, Vigouroux M, Naldrett MJ, Thomas CL, Maule AJ (2006) *Arabidopsis* cell wall proteome defined using multidimensional protein identification technology. Proteomics 6: 301–311.

Berg M, Parbel A, Pettersen H, Fenyo D, Bjorkesten L (2006) Reproducibility of LC-MS-based protein identification. J Exp Bot 57: 1509–1514.

Canovas FM, Dumas-Gaudot E, Recorbet G, Jorrin J, Mock HP, Rossignol M (2004) Plant proteome analysis. Proteomics 4: 285–298.

Carpentier SC, Panis B, Vertommen A, Swennen R, Sergeant K, Renaut J, Laukens K, Witters E, Samyn B, Devreese B (2008) Proteome analysis of non-model plants: A challenging but powerful approach. Mass Spectrom Rev 27: 354–377.

Chang ST, Wang SY, Kuo YH (2003) Resources and bioactive substances from taiwania (*Taiwania cryptomerioides*). J Wood Sci 49: 1–4.

Costa P, Bahrman N, Frigerio JM, Kremer A, Plomion C (1998) Water-deficit-responsive proteins in Maritime pine. Plant Mol Biol 38: 587–596.

Costa P, Pionneau C, Bauw G, Dubos C, Bahrmann N, Kremer A, Frigerio J-M, Plomion C (1999) Separation and characterization of needle and xylem Maritime pine proteins. Electrophoresis 20: 1098–1108.

Costa P, Pot D, Dubos C, Frigerio J M, Pionneau C, Bodenes C, Bertocchi E, Cervera M-T, Remington DL, Plomion C (2000) A genetic map of Maritime pine based on AFLP, RAPD and protein markers. Theor Appl Genet 100: 39–48.

deVienne D, Burstin J, Gerber S, Leonardi A, LeGuilloux M, Murigneux A, Beckert M, Bahrman N, Damerval C, Zivy M (1996) Two-dimensional electrophoresis of proteins as a source of monogenic and co-dominant markers for population genetics and mapping the expressed genome. Heredity 76: 166–177.

Fernando DD, Lazzaro MD, Owens JN (2005) Growth and development of conifer pollen tubes. Sex Plant Reprod 18: 149–162.

Ferry-Dumazet H, Houel G, Montalent P, Moreau L, Langella O, Negroni L, Vincent D, Lalanne C, de Daruvar A, Plomion C, Zivy M, Joets J (2005) PROTICdb: A web-based application to store, track, query, and compare plant proteome data. Proteomics 5: 2069–2081.

Fiehn O, Kopka J, Trethewey RN, Willmitzer L (2000a) Identification of uncommon plant metabolites based on calculation of elemental compositions using gas chromatography and quadrupole mass spectrometry. Anal Chem 72: 3573–3580.

Fiehn O, Kopka J, Doermann P, Altmann T, Trethewey RN, Willmitzer L (2000b) Metabolite profiling for plant functional genomics. Nat Biotechnol 18: 1157–1161.

Gerber S, Rodolphe F, Bahrman N, Baradat P (1993) Seed-protein variation in Maritime pine (*Pinus pinaster* Ait) revealed by 2-dimensional electrophoresis—genetic determinism and construction of a linkage map. Theor Appl Genet 85: 521–528.

Gerber S, Lascoux M, Kremer A (1997) Relation between protein markers and quantitative traits in Maritime pine (*Pinus pinaster* Ait). Silvae Genet 46: 286–291.

Gion JM, Lalanne C, Le Provost G, Ferry-Dumazet H, Paiva J, Chaumeil P, Frigerio JM, Brach J, Barre A, de Daruvar A, Claverol S, Bonneu M, Sommerer N, Negroni L, Plomion C (2005) The proteome of Maritime pine wood forming tissue. Proteomics 5: 3731–3751.

Haynes PA, Roberts TH (2007) Subcellular shotgun proteomics in plants: Looking beyond the usual suspects. Proteomics 7: 2963–2975.

Hunt SMN, Thomas MR, Sebastian LT, Pedersen SK, Harcourt RL, Sloane AJ, Wilkins MR (2005) Optimal replication and the importance of experimental design for gel-based quantitative proteomics. J Proteome Res 4: 809–819.

Jones AME, Bennett MH, Mansfield JW, Grant M (2006) Analysis of the defence phosphoproteome of *Arabidopsis thaliana* using differential mass tagging. Proteomics 6: 4155–4165.

Jorge I, Navarro RM, Lenz C, Ariza D, Porras C, Jorrin J (2005) The holm oak leaf proteome: Analytical and biological variability in the protein expression level assessed by 2-DE and protein identification tandem mass spectrometry de novo sequencing and sequence similarity searching. Proteomics 5: 222–234.

Jorrin JV, Maldonado AM, Castillejo MA (2007) Plant proteome analysis: A 2006 update. Proteomics 7: 2947–2962.

Jorrin JV, Rubiales D, Dumas-Gaudot E, Recorbet G, Maldonado A, Castillejo MA, Curto M (2006) Proteomics: a promising approach to study biotic interaction in legumes. A review. Euphytica 147: 37–47.

Kamo M, Kawakami T, Miyatake N, Tsugita A (1995) Separation and characterization of *Arabidopsis thaliana* proteins by 2-dimensional gel-electrophoresis. Electrophoresis 16: 423–430.

Komatsu S, Zang X, Tanaka N (2006) Comparison of two proteomics techniques used to identify proteins regulated by gibberellin in rice. J Proteome Res 5: 270–276.

Kopka J, Schauer N, Krueger S, Birkemeyer C, Usadel B, Bergmuller E, Dormann P, Weckwerth W, Gibon Y, Stitt M, Willmitzer L, Fernie AR, Steinhauser D (2005) GMD@CSB.DB: The Golm Metabolome Database. Bioinformatics 21: 1635–1638.

Lahm HW, Langen H (2000) Mass spectrometry: a tool for the identification of proteins separated by gels. Electrophoresis 21: 2105–2114.

Le Dantec L, Chagne D, Pot D, Cantin O, Garnier-Gere P, Bedon F, Frigerio JM, Chaumeil P, Leger P, Garcia V, Laigret F, de Daruvar A, Plomion C (2004) Automated SNP detection in expressed sequence tags: statistical considerations and application to Maritime pine sequences. Plant Mol Biol 54: 461–470.

Leitch IJ, Soltis DE, Soltis PS, Bennett DM (2005) Evolution of DNA amounts across land plants (Embryophyta). Ann Bot 95: 207–217.

Leonardi A, Damerval C, Hebert Y, Gallais A, Devienne D (1991) Association of protein amount polymorphism (Pap) among maize lines with performances of their hybrids. Theor Appl Genet 82: 552–560.

Lilley KS, Dupree P (2006) Methods of quantitative proteomics and their application to plant organelle characterization. J Exp Bot 57: 1493–1499.

Lippert D, Zhuang J, Ralph S, Ellis DE, Gilbert M, Olafson R, Ritland K, Ellis B, Douglas CJ, Bohlmann J. (2005) Proteome analysis of early somatic embryogenesis in *Picea glauca*. Proteomics 5: 461–473.

Lippert D, Chowrira S, Ralph SG, Zhuang J, Aeschliman D, Ritland C, Ritland K, Bohlmann J (2007) Conifer defense against insects: Proteome analysis of Sitka spruce (*Picea sitchensis*) bark induced by mechanical wounding or feeding by white pine weevils (*Pissodes strobi*). Proteomics 7: 248–270.

Loomis WD, Battaile J (1966) Plant phenolic compounds and the isolation of plant enzymes. Phytochemistry 5: 423–438.

Morris CR, Scott JT, Chang HM, Sederoff RR, O'Malley D, Kadla JF (2004) Metabolic profiling: A new tool in the study of wood formation. J Agri Food Chem 52: 1427–1434.

O'Leary SJB, Poulis BAD, von Aderkas P (2007) Identification of two thaumatin-like proteins (TLPs) in the pollination drop of hybrid yew that may play a role in pathogen defence during pollen collection. Tree Physiol 27: 1649–1659.

Pavy N, Paule C, Parsons L, Crow JA, Morency MJ, Cooke J, Johnson JE, Noumen E, Guillet-Claude C, Butterfield Y, Barber S, Yang G, Liu J, Stott J, Kirkpatrick R, Siddiqui A, Holt R, Marra M, Seguin A, Retzel E, Bousquet J, MacKay J (2005) Generation, annotation, analysis and database integration of 16,500 white spruce EST clusters. BMC Genom 6: 144–162.

Petit RJ, Bahrman N, Baradat P (1995) Comparison of genetic differentiation in Maritime pine (*Pinus pinaster* Ait) estimated using isozyme, total protein and terpenic loci. Heredity 75: 382–389.

Pirondini A, Visioli G, Malcevschi A, Marmiroli N (2006) A 2-D liquid-phase chromatography for proteomic analysis in plant tissues. J Chromatogr B 833: 91–100.

Plomion C, Bahrman N, Durel CE, O'Malley DM (1995) Genomic mapping in *Pinus pinaster* (Maritime pine) using RAPD and protein markers. Heredity 74: 661–668.

Plomion C, Costa P, Bahrman N, Frigerio JM (1997) Genetic analysis of needle proteins in Maritime pine 1 Mapping dominant and co-dominant protein markers assayed on diploid tissue, in a haploid-based genetic map. Silvae Genet 46: 161–165.

Plomion C, Pionneau C, Brach J, Costa P, Baillères H (2000) Compression wood-responsive proteins in developing xylem of Maritime pine (*Pinus pinaster* Ait.). Plant Physiol 123: 959–969.

Ralph SG, Yueh H, Friedmann M, Aeschliman D, Zeznik JA, Nelson CC, Butterfield YSN, Kirkpatrick R, Liu J, Jones SJM, Marra MA, Douglas CJ, Ritland K, Bohlmann J (2006) Conifer defence against insects: microarray gene expression profiling of Sitka spruce (*Picea sitchensis*) induced by mechanical wounding or feeding by spruce budworms (*Choristoneura occidentalis*) or white pine weevils (*Pissodes strobi*) reveals large-scale changes of the host transcriptome. Plant Cell Environ 29: 1545–1570.

Robinson AR, Ukrainetz NK, Kang KY, Mansfield SD (2007) Metabolite profiling of Douglas-fir (*Pseudotsuga menziesii*) field trials reveals strong environmental and weak genetic variation. New Phytol 174: 762–773.

Robinson AR, Dauwe R, Cullis I, Mansfield SD (2009) Predicting the regenerative capacity of conifer somatic embryogenic culture by metabolite profiling. Plant Biotechnol J 7:952–963.

Roe MR, Griffin TJ (2006) Gel-free mass spectrometry-based high throughput proteomics: Tools for studying biological response of proteins and proteomes. Proteomics 6: 4678–4687.

Rossignol M (2001) Analysis of the plant proteome. Curr Opin Biotechnol 12: 131–134.

Rossignol M, Peltier J-B, Mock HP, Matros A, Maldonado AM, Jorrín JV (2006) Plant proteome analysis: A 2004–2006 update. Proteomics 6: 5529–5548.

Ruebelt MC, Leimgruber NK, Lipp M, Reynolds TL, Nemeth MA, Astwood JD, Engel KH, Jany KD (2006) Application of two-dimensional gel electrophoresis to interrogate alterations in the proteome of genetically modified crops. 1. Assessing analytical validation. J Agri Food Chem 54: 2154–2161.

Saravanan RS, Rose JKC (2004) A critical evaluation of sample extraction techniques for enhanced proteomic analysis of recalcitrant plant tissues. Proteomics 4: 2522–2532.

Shilov IV, Seymour SL, Patel AA, Loboda A, Tang WH, Keating SP, Hunter CL, Nuwaysir LM, Schaeffer DA (2007) The paragon algorithm, a next generation search engine that uses sequence temperature values and feature probabilities to identify peptides from tandem mass spectra. Mol Cell Proteom 6: 1638–1655.

Smith CA, Want EJ, O'Maille G, Abagyan R, Siuzdak G (2006) XCMS: Processing mass spectrometry data for metabolite profiling using nonlinear peak alignment, matching, and identification. Anal Chem 78: 779–787.

Soller M, Beckmann JS (1983) Genetic-polymorphism in varietal identification and genetic-improvement. Theor Appl Genet 67: 25–33.

Terskikh VV, Feurtado JA, Borchardt S, Giblin M, Abrams SR, Kermode AR (2005) *In vivo* C-13 NMR metabolite profiling: potential for understanding and assessing conifer seed quality. J Exp Bot 56: 2253–2265.

Terskikh VV, Zeng Y, Feurtado JA, Giblin M, Abrams SR, Kermode AR (2008) Deterioration of western red cedar (*Thuja plicata* Donn ex D. Don) seeds: protein oxidation and *in vivo* NMR monitoring of storage oils. J Exp Bot 59: 765–777.

Thiellement H, Bahrman N, Damerval C, Plomion C, Rossignol M, Santoni V, de Vienne D, Zivy M (1999) Proteomics for genetic and physiological studies in plants. Electrophoresis 20: 2013–2026.

Thiellement H, Zivy M, Plomion C (2002) Combining proteomic and genetic studies in plants. J Chromatogr B 782: 137–149.

Thimm O, Blasing O, Gibon Y, Nagel A, Meyer S, Kruger P, Selbig J, Muller LA, Rhee SY, Stitt M (2004) MAPMAN: a user-driven tool to display genomics data sets onto diagrams of metabolic pathways and other biological processes. Plant J 37: 914–939.

Tikunov Y, Lommen A, de Vos CHR, Verhoeven HA, Bino RJ, Hall RD, Bovy AG (2005) A novel approach for nontargeted data analysis for metabolomics. Large-scale profiling of tomato fruit volatiles. Plant Physiol 139: 1125–1137.

Valcu CM, Schlink K (2006) Efficient extraction of proteins from woody plant samples for two-dimensional electrophoresis. Proteomics 6: 4166–4175.

Valcu CM, Lalanne C, Muller-Starck G, Plomion C, Schlink K (2008) Protein polymorphism between 2 *Picea abies* populations revealed by 2-dimensional gel electrophoresis and tandem mass spectrometry. J Hered 99: 364–375.

Valledor L, Castillejo MA, Lenz C, Rodriguez R, Canal MJ, Jorrin J. (2008) Proteomic analysis of *Pinus radiata* needles: 2-DE map and protein identification by LC/MS. J. Proteome Res 7: 2616–2631.

Wagner RE, Mugnaini S, Sniezko R, Hardie D, Poulis B, Nepi M, Pacini E, von Aderkas P (2007) Proteomic evaluation of gymnosperm pollination drop proteins indicates highly conserved and complex biological functions. Sex Plant Reprod 20: 181–189.

Wang SY, Wang YS, Tseng YH, Lin CT, Liu CP (2006a) Analysis of fragrance compositions of precious coniferous woods grown in Taiwan. Holzforschung 60: 528–532.

Wang W, Vignani R, Scali M, Cresti M (2006b) A universal and rapid protocol for protein extraction from recalcitrant plant tissues for proteomic analysis. Electrophoresis 27: 2782–2786.

Welti R, Li WQ, Li MY, Sang YM, Biesiada H, Zhou HE, Rajashekar CB, Williams TD, Wang XM (2002) Profiling membrane lipids in plant stress responses—Role of phospholipase D alpha in freezing-induced lipid changes in *Arabidopsis*. J Biol Chem 277: 31994–32002.

Wheelock AM, Buckpitt AR (2005) Software-induced variance in two-dimensional gel electrophoresis image analysis. Electrophoresis 26: 4508–4520.

Wheelock AM, Goto S (2006) Effects of post-electrophoretic analysis on variance in gel-based proteomics. Exp Rev Proteom 3: 129–142.

Whitelegge JP, Laganowsky A, Nishio J, Souda P, Zhang HM, Cramer WA (2006) Sequencing covalent modifications of membrane proteins. J Exp Bot 57: 1515–1522.

Yang S, Qiao B, Lu SH, Yuan YJ (2007) Comparative lipidomics analysis of cellular development and apoptosis in two *Taxus* cell lines. Biochim. Biophys. Acta Mol. Cell Biol Lipids 1771: 600–612.

Yang S, Lu SH, Yuan YJ (2008) Lipidomic analysis reveals differential defense responses of *Taxus cuspidata* cells to two elicitors, methyl jasmonate and cerium (Ce4+). Biochim. Biophys. Acta Mol Cell Biol Lipids 1781: 123–134.

Yeh TF, Morris CR, Goldfarb B, Chang HM, Kadla JF (2006) Utilization of polar metabolite profiling in the comparison of juvenile wood and compression wood in Loblolly pine (*Pinus taeda*). Tree Physiol 26: 1497–1503.

Zolla L (2006) Liquid extraction-ultracentrifugation-liquid chromatography-mass spectrometry: A potent tool for separation and identification of thylakoid membrane proteins. Curr Anal Chem 2: 139–155.

10

Toward the Conifer Genome Sequence

Michele Morgante[1,2,*] and *Emanuele De Paoli*[1,3]

ABSTRACT

Large genome size and highly repetitive DNA content have thus far posed considerable difficulties for both structural and functional genomic studies in coniferous species. As a result, the understanding of conifer genome structure and evolution is still deficient compared to the enormous progress in angiosperms genomics since the sequencing of the *Arabidopsis* genome. However, the development of high-throughput DNA sequencing technologies has enabled to complete large-scale sequencing efforts more cost-effectively and in a shorter time than previously possible, making the sequencing of a conifer genome an attractive opportunity to fill the gap. While two different conifer genome projects have been recently embarked, emerging data from preliminary studies are providing interesting insights into the characteristics of conifer genomes, especially with respect to the composition and evolution of transposable elements that populate them. This chapter will review the state of the art of DNA sequence analysis in conifers and other gymnosperms with emphasis on the interesting deviations from the current model of higher-plant genome evolution that these species are revealing.

Keywords: genome sequencing, genome evolution, transposable elements, conifer, gymnosperms, *Picea*, *Pinus*, *Ginkgo*

[1]Dipartimento di Scienze Agrarie ed Ambientali, Università di Udine, Via delle Scienze 208, 33100 Udine, Italy; e-mail: *michele.morgante@uniud.it*.
[2]Istituto di Genomica Applicata, Parco Scientifico e Tecnologico di Udine, Via Linussio 51, 33100 Udine, Italy.
[3]Current address: Istituto Agrario di San Michele all'Adige, Vie E. Mach 1, 38010 San Michele all'Adige, Italy; e-mail: *emanuele.depaoli@iasma.it*.
*Corresponding author

10.1 Introduction

Our understanding of plant biology and evolution has been greatly aided by the recent advances in DNA sequencing technology and the development of comparative methods for genome analysis. Since year 2000, when the first genomic sequence from a plant species, *Arabidopsis*, was released (Lin et al. 1999; Mayer 1999; Salanoubat et al. 2000; Tabata et al. 2000; Theologis et al. 2000), the nucleotide sequence of other 14 plant organisms has been made publicly available, albeit with different degrees of completion. Moreover, at the time this manuscript is being written, a total of 80 genome sequencing projects aimed at determining the nuclear genome sequences of as many land plant species or varieties have been publicly announced. While only two of them are deemed thoroughly completed (*Arabidopsis* and rice), 15 are at the assembly stage and the remaining 65 are in progress (NCBI data updated on March 2nd 2010, *http://www.ncbi.nlm.nih.gov/genomes/static/gpstat.html*). Notably, despite the immense ecological role and economical value of conifers, only one of these ongoing projects is committed to the deep characterization of a conifer genome, namely the Pine Genome Initiative (PGI) (*http://pinegenomeinitiative.org/*), which has become a reality almost a decade after the beginning of the genomic era in plant science. In parallel, however, a European consortium led by Sweden has recently been granted to sequence the genome of Norway spruce (*http://www.upsc.se/Networks/Networks/sprucegenome.html*) and our group at the University of Udine has developed extensive genomic resources for spruce, by the deep sequencing of genomic libraries from the genomes of four *Picea* species, the annotation of bacterial artificial chromosome (BAC) clones from the genome of Norway spruce and the characterization of its repetitive components.

Unquestionably, a major reason for such a delayed effort and at the same time one of the principal challenges facing gymnosperm genomics is the large size (Fig. 10-1) and repetitiveness of their genomes, which poses considerable difficulties for both structural and functional genomic studies. In consideration of the large costs of de novo sequencing, which until a few years ago could not take advantage of high-throughput parallel sequencing methods, the decision to sequence a large plant genome has been always a serious question that needed to carefully balance scientific interest, social and economic benefits with the impact on public funds and human resources. Before the development of next-generation DNA sequencing technologies, accompanied by technical simplification and drop in the sequencing costs, gymnosperms did not meet these common sense criteria. As a result, while reduced-representation approaches (mainly expressed sequence tag (EST) and cDNA sequencing efforts described in Chapter 8) have been used to alleviate the redundant nature of their genomes and focus on gene discovery and gene expression, many essential questions about the origin of gymnosperm genomes, their organization and evolution are yet to be answered.

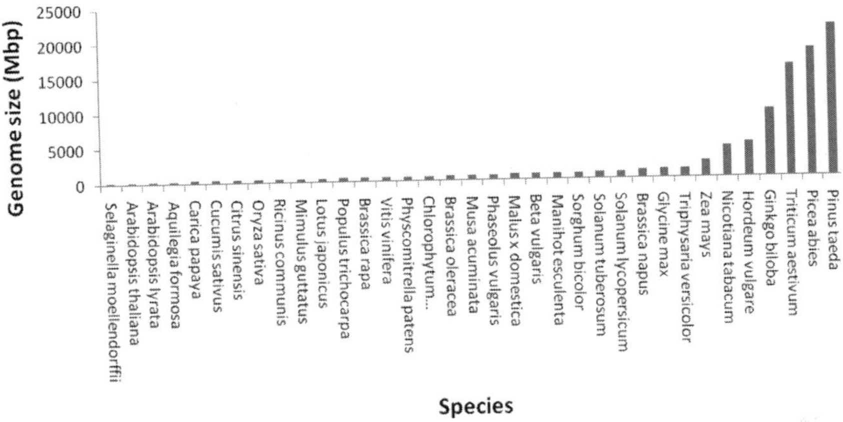

Figure 10-1 Selection of plant genome sizes. Genome size is provided for a subset of species that are object of genomic sequencing. Size information is from NCBI (*http://www.ncbi.nlm.nih.gov/genomes/static/gpstat.html*) with the exception of gymnosperm data from the Kew Plant DNA C-values Database (*http://data.kew.org/cvalues/*).

Color image of this figure appears in the color plate section at the end of the book.

The opportunity to revive such a fascinating topic of plant science with modern methodology has become fairly accessible and the preliminary results already available these days anticipate very interesting insights into plant evolution. This chapter will review the state of the art of DNA sequence analysis in gymnosperms with emphasis on their repetitive genomic components. The high expectations from the ongoing sequencing efforts will also be substantiated by pointing out the important deviations from the current model of higher-plant genome evolution that gymnosperms may reveal.

10.2 Characteristics of Gymnosperm Genomes

Gymnosperms diverged from the angiosperm lineage (i.e., flowering plants) in the paleozoic era around 250–300 million years ago. Since then, flowering plants thrived and differentiated enormously, thereby originating the extraordinary variability we can observe between the several living taxa. On the contrary, gymnosperms split into four extant orders (Coniferales, Cycadales, Ginkgoales, and Gnetales) right after the divergence from angiosperms and later radiated in a relatively small number of species. The conservation of ancestral traits did not prevent these plants from dominating vast areas of land, including most part of boreal forests in the Northern Hemisphere, until present. However, the limited morphological differentiation developed over a remarkable time span won gymnosperms the definition of "botanical fossil". Interestingly, it is believed that the slow rate of change observed at phenotypic level may reflect a general retention of ancient genomic features as well as a narrow range of genetic variation between species. Indeed, what we have been learning about gymnosperm genomes thus far has strongly supported this view, as described below.

Most of gymnosperms have very large genomes whose physical size is on average 18.11 Gbp per haploid genome (standard deviation 7.48 Gbp). The overall gamut of known genome sizes ranges between 2.21 and 35.28 Gbp. However, 88% of the 207 species where DNA content was estimated have 1C greater than 10 Gbp and 39% greater than 20 Gbp, which is almost six-fold larger than the human genome or more (*http://data.kew.org/cvalues/*). The smallest genome sizes are observed in the genus *Gnetum*, which includes 30-35 species of tropical evergreen trees, shrubs and lianas, while the largest genomes belong to the members of the Pinaceae family (pines), the most common and economically important taxonomic group among the four gymnosperm orders.

Among the land plants that are being deeply characterized, wheat (*Triticum aestivum*) exhibits a genome size comparable to the average for gymnosperms (~17 Gbp). However, while in wheat as well as in several other middle- or large-genome crops polyploidization has significantly contributed to the current DNA content, polyploidy is quite infrequent in conifers, occurred only in a few Cupressaceae (Ahuja and Neale 2005), and was never demonstrated within the Pinaceae family, even if evidences of gene duplication were reported (Kinlaw and Neale 1997). In addition, gymnosperms have a nearly constant diploid chromosome number of $2n$ = 18–24, an indication of limited chromosome rearrangements. Conservation of chromosome number and genome size is particularly striking within the Pinaceae family, where the diploid chromosome number is constant at 24 and the haploid genome size is on average 22.90 Gbp (standard deviation 6.02 Gbp). The high levels of synteny and macrocolinearity, together with the lack of evidence for large duplicated linkage groups from mapping data (Krutovsky et al. 2004), rule out extensive genomic rearrangements as major factors causing genome expansion in these species.

Indeed, the origin of large genome sizes in gymnosperms has to be mainly ascribed to the dramatic amplification of noncoding DNA. This was initially suggested by data from reassociation kinetics that showed 75% of a conifer genome to be repetitive (Rake et al. 1980). Later, the primary factor causing inflation of gymnosperm genomes was definitively recognized in the large amount of repetitive elements that populated them, in particular long terminal repeat (LTR) retrotransposons (Friesen et al. 2001; Morse et al. 2009). Retrotransposons (or retroelements or class I transposable elements) are mobile genetic elements that move to new chromosomal locations in a copy-and-paste fashion via an RNA intermediate; as a result, they are capable of increasing the genome size of their host in very short time intervals. One of the five major orders, namely the LTR retrotransposons, are the most abundant and widespread class of eukaryotic transposable elements and represent the major constituents of plant genomes as well (Flavell et al. 1992; Kumar and Bennetzen 2000; Wicker and Keller 2007). Using a reduced-representation shotgun sequencing approach, De Paoli and collaborators have estimated that 77% of a spruce genome (*Picea abies*) sequence is redundant and 42% of it consists of LTR-retrotransposons (E De Paoli et al. unpubl. data), although this may be an underestimate of the actual retrotransposon content, due to the difficulty to annotate transposable elements de novo from limited amount of contiguous sequence. Indeed, a large fraction of the genomic library analyzed, consisting of both repetitive and supposedly single-copy elements, is yet to be characterized (Fig. 10-2).

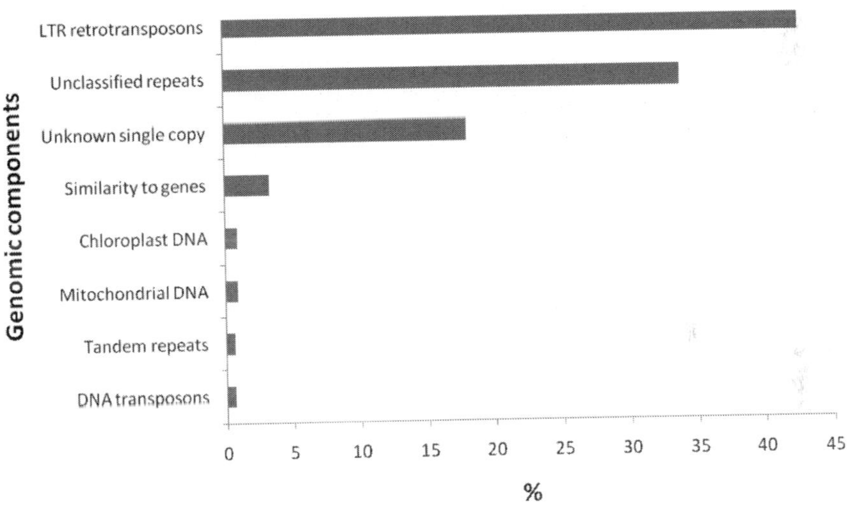

Figure 10-2 Estimate of Norway spruce genome composition. Principal genomic components of Norway spruce genome estimated based on the annotation of a 9 Mbp small insert genomic library (E. De Paoli et al. unpubl. data).

Color image of this figure appears in the color plate section at the end of the book.

10.3 The Paradox of Genome Complexity

Reassociation kinetics and most sequencing information available for conifers clearly points to a very large repetitive component in their genome. Nevertheless, in apparent contrast with these observations are estimates of genome complexity suggesting that the nature of conifer genomes may be not appropriately described by a neat contraposition between low-copy genic component and repetitive fraction. Sequence complexity is defined as the minimal amount of DNA sequences that represents a genome or, in other words, the combined length of all of the single-copy DNA sequences plus one copy of each repetitive sequence (Peterson et al. 2002). Cot analyses, which enable the separation of low copy genomic features (mainly protein-coding genes) from high-copy sequences (e.g., transposable elements), were used to assess sequence complexity before the introduction of high-throughput parallel sequencing approaches. These early studies revealed that genome complexity values varied between 13 and 77% among the angiosperm species analyzed and similarly between 24 and 71% among gymnosperms (Morse et al. 2009). This result is remarkable if one considers that the large sizes of gymnosperm genomes imply absolute values of complexity (or single-copy component) equal to several thousand megabases. Reasoning that this fraction of the genome should be relatively consistent in size among species, the values found in gymnosperm seem to be much larger than the minimum amount of genic components and regulatory elements required for biological functions. Interestingly, *Gymny* retrotransposons were found both in the high-copy and low-copy fraction of the loblolly pine genome and derivatives or rearranged copies of this retrotransposon family were deemed more frequent than full-length elements (Morse et al. 2009). This observation suggests that extensive rearrangements of repetitive sequences might have contributed to the apparent excess of low-repetitive kinetic component of the genome, an hypothesis already proposed by Elsik and Williams (2000).

10.4 Characterization of Retrotransposons as Major Genomic Components

While substantial progress has been made in unveiling the structure and organization of angiosperm genomes, the features of conifer genomes have been poorly investigated thus far. However, particular emphasis has been put in the characterization of the LTR-retrotransposons and other transposable elements that make up most part of these genomes (Kamm et al. 1996; Kossack and Kinlaw 1999; L'Homme et al. 2000; Friesen et al. 2001; Stuart-Rogers and Flavell 2001; Rocheta et al. 2007; Morse et al. 2009).

The LTR retrotransposon order of retroelements includes two main superfamilies called after two types of transposable elements initially identified in the yeast *Saccharomyces cerevisiae*, Ty1-*copia* (Pseudoviridae) and Ty3-*gypsy* (Metaviridae). These groups exhibit similar structure but differ from each other in both their degree of sequence similarity and the order of encoded gene products (Xiong and Eickbush 1990). Their direct long terminal repeats (LTRs) can range from a few hundred bp to over 5 kb in size, do not encode any known proteins, but contain the promoters and terminators associated with the transcription of retrotransposons during the transposition process. Within the two LTRs there is a region that encodes two proteins, specified by two genes called *gag* and *pol*. The *gag* gene encodes capsid-like proteins involved in maturation and packaging of retrotransposon RNA into a form suitable for integration into the genome. *Pol* encodes a multidomain protein and the single domains are arranged in different order in the Ty1-*copia* and the Ty3-*gypsy* elements respectively. These domains include in particular a reverse transcriptase and a RNase H activity that are required for replication/transposition of the retrotransposon as well as an integrase that allows the DNA form of the retrotransposon to insert into a new chromosomal location.

Both Ty1-*copia* and the Ty3-*gypsy* superfamilies have been detected in conifers and in other gymnosperm orders, but their relative proportions are generally not known. In our knowledge, the genome of Norway spruce (*Picea abies*) is the only one whose composition has been estimated by low-coverage shotgun sequencing (E De Paoli et al. unpubl. data). In this case, Ty3-*gypsy* elements were found to be twice as abundant as the Ty1-*copia* elements (25% vs. 13%). Thus far, no single prominent retrotransposon family has been recognized that may account for much of the extraordinary expansion of this genome in a fashion similar to what happened in some angiosperm species, i.e., in the cotton genus (Hawkins et al. 2006). On the other hand, the contribution of a single family to the sequence space can be impressive in absolute terms even if relatively modest relative to the overall genome size. For instance, by screening a collection of ~18,000 BAC clones from loblolly pine, Morse et al. (2009) estimated that the *Gymny* Ty3-*gypsy* retrotransposon family accounts for a genomic space at least as large as the *Arabidopsis* genome, namely ~135 Mb but corresponding to only ~0.6% of the host genome.

A few other retrotransposon families have been described in detail thus far, including *TPE1* in *Pinus elliottii* var. *elliottii* (Kamm et al. 1996), *IFG7* in *Pinus radiata* (Kossack and Kinlaw 1999), *PpRT1* in *Pinus pinaster* (Rocheta et al. 2007), *Gymny* in *Pinus taeda* (Morse et al. 2009) and *Alisei* in *Picea abies* (E De Paoli et al. unpubl. data). Notably, *TPE1* is the only representative of *copia*-like elements. Besides, non-LTR retrotransposons like long interspersed nuclear elements (LINEs) and DNA (class II) transposons have also been found in the spruce and pine genomes (Friesen et al. 2001; E De Paoli et al. unpubl. data), albeit at low frequency as in other large-genome plants.

10.5 Genomic Distribution and Arrangement of Retrotransposoans

Very few studies have investigated the genomic organization of retrotransposons in conifers by fluorescent *in situ* hybridization (Kamm et al. 1996; Frengen et al. 1999; Morse et al. 2009). All of them showed that individual element families were widespread across the chromosomes, consistent with dispersion and amplification via an RNA intermediate. Preferential localization in centromeric, pericentromeric or telomeric regions was observed but was not consistent for all of the families analyzed, in contrast to the tendency for small genomes like *Arabidopsis* (Peterson-Burch et al. 2004).

To have a more detailed look of how the genomic landscape of a conifer may be organized, De Paoli et al. (unpubl. data) sequenced and annotated four random BAC clones from a library of Norway spruce nuclear genome, obtaining the nucleotide sequence of four large genomic regions ranging between 82 and 124 kb in size. Annotation resulted in the identification of 12 full-length LTR-retrotransposons with intact ends and 16 incomplete retrotransposons that exhibited ill-defined or truncated boundaries but still presented extensive LTR sequences in the ends. Several retrotransposon fragments and minor DNA stretches with similarity to LTR-retrotransposons were observed as well. The arrangement of retrotransposons denoted extensive transposition activity in the regions and was similar to that observed in maize (SanMiguel et al. 1996) with several elements inserted into others with up to three levels of nesting. All of these elements lay among repetitive sequences of unknown function that were identified based on similarity to a shotgun genomic library. The detection of truncated retrotransposon remnants and of sequences showing low but detectable similarities to transposable elements between the nested structures suggested that extensive stratification of the genomic landscape had occurred over time, thereby determining severe rearrangements and impairments of previously inserted elements (Fig. 10-3).

The divergence within each pair of LTRs was used to estimate absolute and relative ages between retrotransposons in the four genomic regions. At the time an element inserts into the genome, the LTRs are 100% identical, unless early rearrangements take place. As time passes, mutations occur within the LTRs at a rate approximating the host's mutation rate and accumulate to an extent that is proportional to the age of the element (SanMiguel et al. 1998). The nucleotide distance between sister LTRs showed a large variation between the outermost data points in the range, representing a time span of at least 70 million years (based on a substitution rate of 1.31×10^{-9} mutations per year per nucleotide, which corresponds to the upper bound of the proposed range of synonymous mutation rates in pine (Willyard et al. 2007)). The most recent insertion dates back to 8 million years ago, which is prior to most of the retrotransposon insertions detectable in angiosperm genomes. The other insertion ages varied uniformly between 8 and 81 million years from present and were evenly distributed across the overall time span, with the only exception that Ty1-*copia* elements appeared to be significantly older than Ty3-*gypsy*.

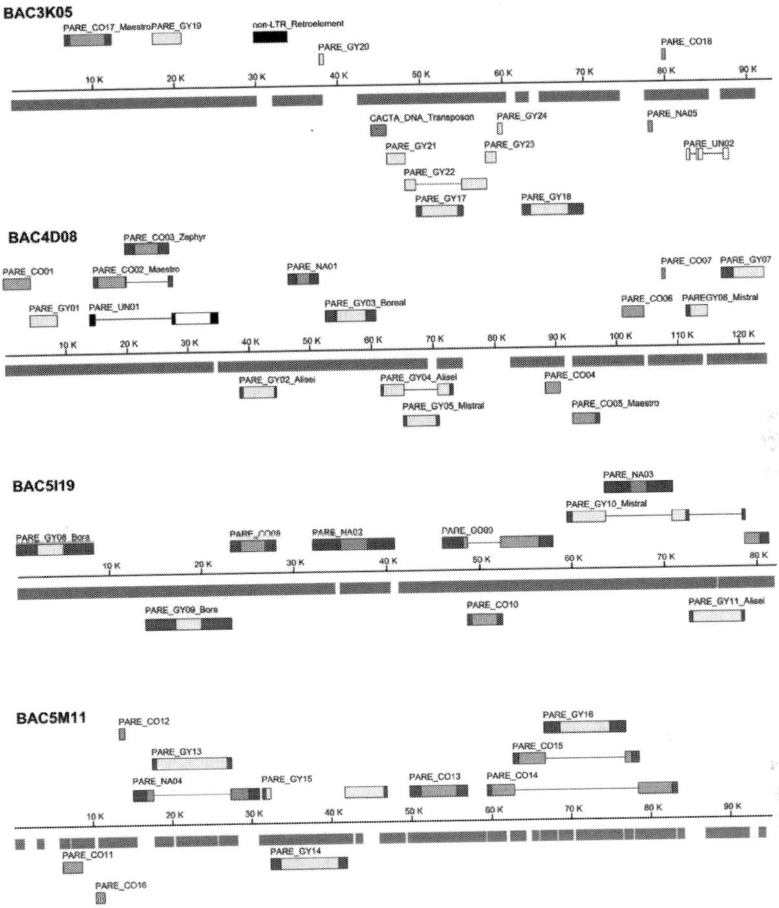

Figure 10-3 Annotation of four genomic regions (BAC) randomly selected from the genome of Norway spruce. Green boxes, LTR-retrotransposons of the *copia* superfamily; yellow boxes, LTR-retrotransposons of the *gypsy* superfamily; light-blue boxes, LTRs of *copia* elements; dark-blue boxes, LTRs of *gypsy* elements; white boxes, uncharacterized LTR-retrotransposons (with black LTRs when annotated), grey boxes, non-autonomous LTR-retrotransposons (with dark-grey LTRs); black box, non-LTR retrotransposon. The red bar indicates repetitive sequences; PARE, ***Picea abies*** **repetitive element**; GY, *gypsy*; CO, *copia*; UN, unidentified; NA, non-autonomous. From E. De Paoli et al. (unpubl. data).

Color image of this figure appears in the color plate section at the end of the book.

Overall, these observations pointed to a steady but slow accumulation of retrotransposons, not counteracted by substantial removal of older elements; therefore, they have posed an important exception to the model of retrotransposon life-cycle in higher-plants, which is strongly based on the better investigated angiosperms. In the latter species, DNA removal mechanisms like unequal homologous recombination and illegitimate recombination are responsible for a rapid turnover of transposable elements, clearly reflected by the lack of full-length retrotransposons as old as those observed in conifers (Marillonnet and Wessler 1998; Wicker et al. 2001; Devos et al. 2002; Ma et al. 2004). Should these results represent a common feature of conifer genomes, the definition of "botanic fossils" would find definitive support from molecular observations and perhaps the new definition of "genomic fossils" would appropriately render this peculiarity.

10.6 Conservation of Retrotransposons in Conifers and Beyond

Two comprehensive surveys dating back to 2001 pioneered the investigation of the phylogenetic relations among gymnosperm retrotransposons (Friesen et al. 2001; Stuart-Rogers and Flavell 2001). Both studies used PCR amplification and Southern blot hybridization analyses to characterize diversity and evolution of Ty3-*gypsy* and/or Ty1-*copia* LTR-retrotransposons. Stuart-Rogers and Flavell (2001) analyzed a collection of 35 different reverse transcriptase gene fragments from Ty1-*copia* elements of the genome of *Picea abies* (Norway spruce) and found strong cross-hybridization levels between elements not only within the *Picea* genus but also across larger evolutionary gaps up to and including Ginkgo, which is separate from modern conifers by a minimum of 250 million years. Some homologous elements (e.g., *Tpa28* and *Tpa13*) showed strong nucleotide sequence similarity between *Picea* and Ginkgo (> 80%), an extent that had never been observed for angiosperm retrotransposons, even across relatively small taxonomic boundaries. Similarly, Friesen and coworkers (Friesen et al. 2001) sequenced 165 reverse transcriptase gene domains of both Ty3-*gypsy* and Ty1-*copia* LTR-retrotransposons from the same genome and detected strong hybridization of clones from *Picea abies* to DNA of Ginkgo and *vice versa* from Ginkgo to *Picea* and *Pinus*. In addition to these studies, the sequence conservation of specific retrotransposon families has been investigated in greater detail in a few cases, resulting in similar conclusions. For instance, the analysis of *PpRT1*, the first complete gypsy-LTR retrotransposons detected and characterized in maritime pine (*Pinus pinaster*) (Rocheta et al. 2007), revealed extensive sequence conservation in the LTRs among *Picea abies*, *Pinus cembra*, *Cedrus atlantica* and *Gingko biloba*.

We also identified and described a LTR-retrotransposon family in Norway spruce, called *Alisei*. Elements of this family were conserved enough in sequence to be detected by interspecific hybridization of genomic DNA in a panel of six gymnosperm species including Gingko and five conifers of the Pinaceae family. Then, members of *Alisei* were cloned and the average degree of sequence divergence was estimated both within species and between species in pairwise comparisons focusing on the region encoding the INT (integrase) gene. The whole picture showed a surprising high level of sequence conservation, not only within conifers, but also between conifers and Ginkgo. Moreover, in most cases the nucleotide differences between species were similar or even lower than the within-species divergences (E De Paoli et al. unpubl. data).

In consideration of the very distant evolutionary relationship between the species investigated in these studies, it is surprising to find a high degree of sequence conservation in putatively non-functional repetitive DNA. It is a matter of fact that there is still no conclusive evidence about the origin of this similarity. Germ line vertical transmission of retroelement lineages that arose early in plant evolution, strong purifying selection or the less supported horizontal transmission are non-mutually excluding mechanisms that have been proposed (Friesen et al. 2001; Stuart-Rogers and Flavell 2001). Vertical transmission is the regular manner in which eukaryotes inherit their retroelements; then transposon extinction normally occurs through sequence degeneration by point mutations and/or removal mechanisms that may result in interspecific differences with respect to copy number and genomic location. Indeed, the monophyletic grouping of all types of retrotransposons analyzed by Friesen et al. (2001) and the timing of insertions revealed by De Paoli et al. (unpubl. data) suggest that the current diversity of retrotransposons in conifers and other gymnosperms may be the consequence of common ancestry and early rapid radiation before the split that generated the extant orders. According to this model of early generation, the differential amplification and loss of retroelements, possibly not accompanied by substantial lineage-specific differentiation, would be the major factor shaping diversity among the repetitive components of gymnosperm genomes. Until now, interspecific DNA hybridization analyses have proved consistent with this scenario, showing that it is the relative abundance of single retrotransposon families rather than the general composition of these genomes to change. A notable exception to this is the restriction of the *Gymny* retrotransposon family to the subgenus *Pinus* of the loblolly pine (Morse et al. 2009), which might anticipate more similar cases to be found by genome sequencing.

Though vertical transmission is a parsimonious explanation and is well established in all eukaryote branches, it is still not yet thoroughly understood why there should be such a limited sequence variation among noncoding repetitive elements compared to the scenario of much higher diversification among angiosperms (Flavell et al. 1992a; Flavell et al. 1992b; Gribbon et al. 1999). An important clue to address this question comes from the slow mutation rate found in conifers. Fossil calibration of molecular divergence over 11 nuclear loci yielded an absolute silent mutation rate estimate in the range of $0.70–1.31 \times 10^{-9}$ for the *Pinus* nuclear genome, which is approximately 4- to 20-fold slower than in angiosperms (Willyard et al. 2007). Although this rate is apparently not slow enough to justify the limited divergence between Pinaceae and Ginkgo, it should be clarified that at least in the case of the *Alisei* retrotransposons, PCR amplification using non-degenerated primers might have led to a bias in favor of the least divergent elements, revealing only a fraction of the overall diversity distribution that can be theoretically expected from the accumulation of mutations since the separation of the species.

An alternative hypothesis to explain the observed sequence conservation would evoke purifying selection acting on specific element(s) of a retrotransposon family and affecting the copies generated from the selected template(s). Indeed, the analysis of *Tpa28* reverse transcriptase fragments revealed a degree of sequence conservation as high as 85.6% between *Picea* and Ginkgo, corresponding to an estimated overall nucleotide substitution rate of $0.61–1.31 \times 10^{-9}$, which is roughly similar to that seen for nuclear protein-coding genes (Stuart-Rogers and Flavell 2001). Besides, the ratios of non-synonymous substitution to synonymous substitutions (K_a/K_s) between the *Tpa28* elements in *Picea* and Ginkgo seem to support the possibility of strong purifying selection for this family. Although this explanation cannot be completely ruled out, many elements sequenced thus far are non-functional due to the presence of stop codons and frameshift mutations. Therefore, it is unlikely that selection and convergence of function could account for similarities between several sequences belonging to different families of elements. Moreover, in a preliminary analysis of deep sequencing data from three genomes within the *Picea* genus (*P. glauca*, *P. mariana* and *P. rubens*), we have estimated that 20–25% of each dataset was homologous to a repertoire of repetitive sequences from the genome of Norway spruce with a degree of nucleotide similarity equal or higher than 90% (E De Paoli et al. unpubl. data). Thus, there is emerging evidence that the extent of sequence similarity among conifers does go beyond the conservation of just a few families of repetitive elements.

For analogous reasons, the horizontal-transmission of sequences between species through viruses and insects in recent times also seems unlikely, albeit intriguing. This phenomenon has been reported and supported by good evidence both in animals (Robertson and Lampe 1995; Jordan et al. 1999; Terzian et al. 2000) and in plants (Fortune et al. 2008). Nevertheless, while between-species transfer is considered a rare event that only affects small DNA fragments at a time, multiple transmissions of abundant genetic material between different species (not only between *Picea* and Ginkgo but also from or to several other Pinaceae) is deemed quite improbable, although it cannot be completely ruled out for limited genomic components.

10.7 Conclusions

Despite the limited amount of genomic information from gymnosperms made available thus far, a few pioneering studies reviewed in this chapter have clearly delineated a series of genomic features that make these species, in particular conifers, very attractive for a deep sequencing analysis:

1) The overall architecture of these genomes appears to be extraordinarily conserved with a fairly constant chromosome number across distantly-related taxonomic groups. Polyploidy has played a minor role in their evolution and major chromosome rearrangements are unlikely to have occurred, despite the long evolutionary history of these species.
2) The redundant component of the genome, mainly consisting of LTR-retrotransposons, is clearly the major factor underlying the large genome size, but no prominent repetitive families that alone could account for the great expansion of the genome have been identified yet.
3) Cytological analyses and the detailed annotation of large genomic regions has revealed a dispersed distribution of repetitive elements over the chromosomes and a complex genomic landscape dominated by LTR-retrotransposons arranged in multilayer nested structures.
4) The low-repetitive and single-copy component largely exceeds the amount of genomic information that is sufficient for full biological functionality in known small-genome angiosperms. Old and diverged copies of retrotransposon elements seem to significantly contribute to this fraction.
5) Compared to the scenario observed in angiosperms, LTR-retrotransposons present an unusual degree of conservation across large taxonomic divides. Whether the limited differentiation was caused only by the slow mutation rate, by selective restrains or a combination of the two, is still unclear. However, the extent of sequence conservation, which involves a large fraction of the conifer genomes, rules out horizontal transfer as a likely explanation and support the simple mechanism of vertical transmission accompanied by a slow rate of molecular evolution.
6) Timing of insertions demonstrates the retention of very ancient retrotransposon elements, which is a rare event in angiosperms. This suggests a much slower turnover of repetitive features in gymnosperm species, possibly caused by insufficient removal of redundant elements.

Several questions are left to be addressed, among which: (i) how diversified is the overall population of retrotransposons in a conifer genome? (ii) What are the dynamics, timing and extent of retrotransposon proliferation since the early divergence of the extant gymnosperm orders? (iii) Which are the factors underlying the differentiation between species in the light of limited sequence diversification? (iv) Is the pine genome in an expanding phase?

Many of these points cannot be resolved without a comprehensive comparative analysis of different species. Therefore, the prospected sequencing of two conifer genomes in parallel, those of *Pinus taeda* (loblolly pine) and *Picea abies* (Norway spruce) respectively, is an excellent opportunity to investigate the above findings in greater detail and integrate with new insights the current model of higher-plant genome evolution.

References

Ahuja MR, Neale DB (2005) Evolution of genome size in conifers. Silvae Genet (54): 126–137.

Devos KM, Brown JK, Bennetzen JL et al. (2002) Genome size reduction through illegitimate recombination counteracts genome expansion in *Arabidopsis*. Genome Res 12(7): 1075–1079.

Elsik CG, Williams CG (2000) Retroelements contribute to the excess low-copy-number DNA in pine. Mol Gen Genet 264(1-2): 47–55.

Flavell AJ, Dunbar E, Anderson R, Pearce SR, Hartley R, Kumar A (1992a) Ty1-copia group retrotransposons are ubiquitous and heterogeneous in higher plants. Nucl Acids Res 20(14): 3639–3644.

Flavell AJ, Smith DB, Kumar A (1992b) Extreme heterogeneity of Ty1-copia group retrotransposons in plants. Mol Gen Genet 231(2): 233–242.

Fortune PM, Roulin A, Panaud O (2008) Horizontal transfer of transposable elements in plants. Commun Integr Biol 1(1): 74–77.

Frengen E, Weichenhan D, Zhao B, Osoegawa K, van Geel M, de Jong PJ (1999) A modular, positive selection bacterial artificial chromosome vector with multiple cloning sites. Genomics 58(3): 250–253.

Friesen N, Brandes A, Heslop-Harrison JS (2001) Diversity, origin, and distribution of retrotransposons (gypsy and copia) in conifers. Mol Biol Evol 18(7): 1176–1188.

Gribbon BM, Pearce SR, Kalendar R, Schulman AH, Paulin L, Jack P et al. (1999) Phylogeny and transpositional activity of Ty1-copia group retrotransposons in cereal genomes. Mol Gen Genet 261(6): 883–891.

Hawkins JS, Kim H, Nason JD, Wing RA, Wendel JF et al. (2006) Differential lineage-specific amplification of transposable elements is responsible for genome size variation in Gossypium. Genome Res 16(10): 1252–1261.

Jordan IK, Matyunina LV, McDonald JF (1999) Evidence for the recent horizontal transfer of long terminal repeat retrotransposon. Proc Natl Acad Sci USA 96(22): 12621–12625.

Kamm A, Doudrick RL, Heslop-Harrison JS, Schmidt T (1996) The genomic and physical organization of Ty1-copia-like sequences as a component of large genomes in *Pinus elliottii* var. elliottii and other gymnosperms. Proc Natl Acad Sci USA 93(7): 2708–2713.

Kinlaw CS, Neale DB (1997) Complex gene families in pine genomes. Trends in Plant Science (2): 356–359.

Kossack DS, Kinlaw CS (1999) IFG, a gypsy-like retrotransposon in *Pinus* (Pinaceae), has an extensive history in pines. Plant Mol Biol 39(3): 417–426.

Krutovsky KV, Troggio M, Brown GR, Jermstad KD, Neale DB (2004) Comparative mapping in the Pinaceae. Genetics 168(1): 447–461.

Kumar A, Bennetzen JL (2000) Retrotransposons: central players in the structure, evolution and function of plant genomes. Trends Plant Sci 5(12): 509–510.

L'Homme Y, Séguin A, Tremblay FM (2000) Different classes of retrotransposons in coniferous spruce species. Genome 43(6): 1084–1089.

Lin X, Kaul S, Rounsley S, Shea TP, Benito MI, Town CD et al. (1999) Sequence and analysis of chromosome 2 of the plant *Arabidopsis thaliana*. Nature 402(6763): 761–768.

Ma J, Devos KM, Bennetzen JL (2004) Analyses of LTR-retrotransposon structures reveal recent and rapid genomic DNA loss in rice. Genome Res 14(5): 860–869.

Marillonnet S, Wessler SR (1998) Extreme structural heterogeneity among the members of a maize retrotransposon family. Genetics 150(3): 1245–1256.

Mayer K, Schüller C, Wambutt R, Murphy G, Volckaert G, Pohl T et al. (1999) Sequence and analysis of chromosome 4 of the plant *Arabidopsis thaliana*. Nature 402(6763): 769–777.

Morse AM, Peterson DG, Islam-Faridi MN, Smith KE, Magbanua Z, Garcia SA et al. (2009) Evolution of genome size and complexity in *Pinus*. PLoS One 4(2): e4332.

Peterson-Burch BD, Nettleton D, Voytas DF (2004) Genomic neighborhoods for *Arabidopsis* retrotransposons: a role for targeted integration in the distribution of the Metaviridae. Genome Biol 5(10): R78.

Peterson DG, Wessler SR, Paterson AH (2002) Efficient capture of unique sequences from eukaryotic genomes. Trends Genet 18(11): 547–550.

Rake AV,Miksche JP, Hall RB, Hansen KM (1980) DNA reassociation kinetics for four conifers. Can J Genet Cytol 22: 69–79.

Robertson HM, Lampe DJ (1995) Recent horizontal transfer of a mariner transposable element among and between Diptera and Neuroptera. Mol Biol Evol 12(5): 850–862.

Rocheta M, Cordeiro J, Oliveira M, Miguel C (2007) PpRT1: the first complete gypsy-like retrotransposon isolated in *Pinus pinaster*. Planta 225(3): 551–562.

Salanoubat M, Lemcke K, Rieger M, Ansorge W, Unseld M, Fartmann B et al. (2000) Sequence and analysis of chromosome 3 of the plant *Arabidopsis thaliana*. Nature 408(6814): 820–822.

SanMiguel P, Tikhonov A, Jin YK, Motchoulskaia N, Zakharov D, Melake-Berhan A et al. (1996) Nested retrotransposons in the intergenic regions of the maize genome. Science 274(5288): 765–768.

SanMiguel P, Gaut BS, Tikhonov A, Nakajima Y, Bennetzen JL et al. (1998) The paleontology of intergene retrotransposons of maize. Nat Genet 20(1): 43–45.

Stuart-Rogers C, Flavell AJ (2001) The evolution of Ty1-copia group retrotransposons in gymnosperms. Mol Biol Evol 18(2): 155–163.

Tabata S, Kaneko T, Nakamura Y, Kotani H, Kato T, Asamizu E et al. (2000) Sequence and analysis of chromosome 5 of the plant *Arabidopsis thaliana*. Nature 408(6814): 823–826.

Terzian C, Ferraz C, Demaille J, Bucheton A (2000) Evolution of the Gypsy endogenous retrovirus in the Drosophila melanogaster subgroup. Mol Biol Evol 17(6): 908–914.

Theologis A, Ecker JR, Palm CJ, Federspiel NA, Kaul S, White O et al. (2000) Sequence and analysis of chromosome 1 of the plant *Arabidopsis thaliana*. Nature 408(6814): 816–820.

Wicker T, Keller B (2007) Genome-wide comparative analysis of copia retrotransposons in Triticeae, rice, and *Arabidopsis* reveals conserved ancient evolutionary lineages and distinct dynamics of individual copia families. Genome Res 17(7): 1072–1081.

Wicker T, Stein N, Albar L, Feuillet C, Schlagenhauf E, Keller B (2001) Analysis of a contiguous 211 kb sequence in diploid wheat (*Triticum monococcum* L.) reveals multiple mechanisms of genome evolution. Plant J 26(3): 307–316.

Willyard A, Syring J, Gernandt DS, Liston A, Cronn R (2007) Fossil calibration of molecular divergence infers a moderate mutation rate and recent radiations for *Pinus*. Mol Biol Evol 24(1): 90–101.

Xiong Y, Eickbush TH (1990) Origin and evolution of retroelements based upon their reverse transcriptase sequences. EMBO J 9(10): 3353–3362.

11

Future Prospects

Jeffrey F.D. Dean

ABSTRACT

Rapid advancement of the "omic" sciences has resulted from the recent development of new technologies driven in large part by investment from the human biomedical arena. Thus, in considering what might lie ahead for conifer genomics, current and near-future genomic research in human biomedicine provides a good yardstick. This chapter relates current expectations for near-future development and improvements in technology platforms for nucleic acid sequencing, high-throughput genotyping, microarrays, and metabolomics. Subsequent discussion covers how these new technologies might be used to improve our understanding of fundamental conifer biology. Prospects for using the findings from conifer genomics studies to address practical problems for the forest products industry as well as problems related to the health and management of conifer forests are also discussed. Finally, some potential barriers to progress are noted along with suggestions for overcoming them.

Keywords: advanced breeding; biofuels; bioinformatics; carbon sequestration; comparative genomics; ecological genomics; forest health, gene conservation; phylogenetics; secondary products; technology platforms; third-generation sequencing

11.1 Introduction

Niels Bohr once said, "Prediction is very difficult, especially of the future", but in thinking about what possibilities the future might hold for improving our understanding of conifer biology, particularly as viewed through the twin lenses of genomics and bioinformatics, it might be better to think in terms of a quotation widely attributed to the science fiction writer, William Gibson: "The future is here. It's just not evenly distributed yet."

Warnell School of Forestry and Natural Resources, University of Georgia, Athens, GA 30602, USA; e-mail: *jeffdean@uga.edu*

The modern field of genomics may be said to have originated with the 1988 National Research Council study entitled, "Mapping and Sequencing the Human Genome" (NRC 1988), which laid groundwork for US contributions to the international efforts that culminated in the release of two draft human genome sequences in 2001 (IHGSC 2001; Venter et al. 2001). From that start, the development and first application of novel genomic technologies has been driven, not surprisingly, by the plans and needs of human biomedical research. Over time, these futuristic technologies become more evenly distributed, eventually finding their way to applications in plant and agricultural research, including forestry and conifer biology. As a result of this natural progression and redistribution, predictions and plans set forth in the recent past to guide the application of genomics to future biomedical research and now coming to fruition provide a realistic glimpse of what we should expect ahead in the field of conifer genomics.

Collins et al. (2003) distilled the input from numerous meetings and panels of experts in preparing a roadmap for future genomics research in human biomedicine for the US National Human Genome Research Institute (NHGRI). This chapter draws heavily on themes set forth in that report in considering future directions and prospects for research in conifer genomics. With this in mind, it may be useful to note some of the hopes for "quantum leaps" mentioned in that report as technological touchstones to "provoke creative dreaming" (Collins et al. 2003):

- the ability to determine a genotype at very low cost, allowing an association study in which 2,000 individuals could be screened with about 400,000 genetic markers for US$10,000 or less;
- the ability to sequence DNA at costs that are lower by four to five orders of magnitude than the current cost, allowing a human genome to be sequenced for US$1,000 or less;
- the ability to synthesize long DNA molecules at high accuracy for US$0.01 per base, allowing the synthesis of gene-sized pieces of DNA of any sequence for as little as US$10–20;
- the ability to determine the methylation status of all the DNA in a single cell; and
- the ability to monitor the state of all proteins in a single cell in a single experiment.

Some of these quantum leaps are well on their way to becoming fully realized only six years after the Collins report was released, which serves to emphasize just how fast the discipline is evolving and just how difficult it is to make truly bold and insightful predictions.

11.2 Trends in Technological Advances

11.2.1 DNA/RNA Sequencing Platforms

The year after the Collins report was published, the NHGRI announced its first round of support to develop a next-generation of DNA sequencing technologies. Funding was released in two packets, one for near-term technologies that were targeting whole human genome sequences for US$100,000, and one for riskier technologies shooting to realize whole genomes for US$1,000 (*http://www.genome.gov/10000368*). Amongst the awards made in 2004 were those for projects at 454 Life Sciences Corp. and Agencourt Bioscience Corp., which led to some of the current next-generation DNA sequencing platforms (GS-FLX and SOLiD) that have been described elsewhere in this book (see Chapter 7). Subsequent rounds of funding from NHGRI have gone to companies such as Illumina Inc., Pacific Biosciences (nee Nanofluidics), and Helicos Bioscience Corp. among others, and these companies, too, either have next-generation sequencers already on the market or have plans to begin shipping instruments within the next year. The Helicos (shipping by the end of 2009) and Pacific Biosciences (projected release in 2010) platforms are of particular interest because they are both "single-molecule" sequencers (no amplification of input nucleic acid required). The Pacific Biosciences instrument is also of interest for its purported ability to routinely yield long sequence reads (>1–3 Kb). Recent review articles describe the new sequencing technologies in some detail and also discuss some of the opportunities they present for genomics researchers (Gupta 2008; Mardis 2008; Ansorge 2009).

The current generation of high-throughput sequencing (HTS) platforms has already reduced the cost of DNA sequencing by two to three orders of magnitude, and upcoming platforms are likely to reduce the cost by at least another order of magnitude. The impact that this will have on conifer genomics can be illustrated by considering the estimated costs for sequencing a reference genome for conifers then and now. Prior to a 2003 workshop organized to discuss possibilities for sequencing the loblolly pine (*Pinus taeda*) genome (*http://dendrome.ucdavis.edu/lpgp/*), rough estimates for the cost of sequencing alone (based on dideoxy-dye terminator chemistry and 96-capillary array systems) were in excess of US$500 million . Now, only six years later, the author was recently presented with an estimate of US$4–5 million to generate a *de novo* assembly of the pine genome based on 75 nt (nucleotide) paired-end reads from a GA-II platform (Illumina) read to a depth of about 50× coverage of the haploid genome. Allowing for the fact that no similar *de novo* assembly of a large eukaryotic genome has yet been described in the literature and noting that the assembly promised would not have been complete, but would have been instead more likely in the form of 80,000–100,000 contigs of 200–300 kb in length, the estimate still provides a useful benchmark for just how far and how quickly we have come. One takeaway message is that one or more full-length reference genome(s) for conifers seem(s) a very real possibility within the next five years.

11.2.2 Genotyping Platforms

The genetic systems of humans and conifer trees share a great deal of similarity from both being relatively undomesticated outcrossing species with long generation times. Approaches requiring controlled crossing experiments to link traits with genes are, as a consequence, difficult in both species, and this has led to substantial interest and investment in whole-genome association (WGA) studies (Seng and Seng 2008). Several technology platforms have been developed for WGA studies, some of which provide for the simultaneous determination of as many as a million single-nucleotide polymorphisms (SNPs), and most likely further development of the existing platforms will include efforts to increase the density of oligonucleotides and decrease the reaction volumes (Ragoussis 2009). However, there is still a need to reduce costs for the large-scale platforms, possibly through implementation of strategies for multiplexing samples, as well as for the smaller, custom systems. If HTS costs continue to drop dramatically, it is conceivable that genotyping by whole-genome sequencing might become feasible for some conifer studies. That said, genotyping by whole-genome sequencing is not likely in the near future for conifers.

The current platforms for WGA studies have some technical limitations, as do the statistical designs most commonly used to analyze WGA datasets (Sebastiani et al. 2009). The general absence of rare alleles from current WGA platforms is a concern, but techniques like COLD-PCR (co-amplification at lower denaturation temperature-PCR) have been specifically developed to identify minority alleles from mixed populations in a high-throughput fashion (Li and Makrigiorgos 2009). Zheng et al. (2009) described a multiplexed array-based resequencing pipeline that also permits high-throughput discovery of rare alleles for association genetics studies.

11.2.3 Microarrays and Transcriptional Profiling

DNA microarrays are likely to remain a widely used technology, particularly for studies involving species that have previously been targeted in large genome sequencing programs and consequently have deep sequence resources available (Nygaard and Hovig 2009). However, despite the power that whole-genome tiling arrays have for illuminating transcriptional activity across the breadth of the genome (Yazaki et al. 2007), and notwithstanding breakthroughs that enable placement of increased numbers of oligonucleotides in smaller spaces (Wheelan et al. 2008), it is highly unlikely whole genome arrays will ever be developed for a conifer.

Platforms that allow for the simple, rapid and inexpensive development of custom oligonucleotide arrays (e.g., Cheng and Chen 2009) will provide opportunities to adapt microarray approaches to targeted studies of discrete biological functions, such as apoptosis (Swidzinski et al. 2002). Progress in the development of DNA sensors could improve the cost, sensitivity and reproducibility of microarray experiments by eliminating the need to label or even reverse-transcribe RNA prior to hybridization (Cagnin et al. 2009).

However, as suggested in Chapter 7, multiplexed DNA tag sequencing on HTS platforms will likely become the technique of choice for transcriptional profiling in species that do not already have established microarray resources available.

11.2.4 Proteomics

Despite the fact that high-throughput proteomics analysis is significantly more difficult than HTS, substantial improvements in the technology have been made, as discussed elsewhere in this volume (see Chapter 8). However, proteomics remains a technology platform that has not yet been fully exploited with respect to conifer genomics. Efforts to integrate proteomic analyses with genomic and phenomic data look promising for building gene network models that can explain the emergence of specific phenotypes (Gstaiger and Aebersold 2009). Such approaches should improve the predictability of genotype-phenotype relationships, which, if applied to conifers, would enable tree improvement programs to make more efficient crosses and selections. There can be little doubt that the future will bring greater activity in the area of proteomics with regard to conifer biology.

While advanced mass-spectrometry platforms are garnering the most attention with respect to proteomics approaches in plant biology, protein microarrays are just beginning to see application in these fields. A review by Joos and Bachmann (2009) provides an overview of recent developments in the use of protein microarrays. The use of a protein microarray to probe mitogen-activated protein kinase (MPK) target networks in *Arabidopsis* provides a fine example of how this technology could be used to study signaling pathways in conifers (Popescu et al. 2009). Similarly, Espina et al. (2009) coupled laser-microdissection of specific cell types with reverse-phase protein microarrays to measure activated signal pathway molecules in miniscule samples. It would be very interesting to see this approach applied to the differentiation events in conifer xylem cells undergoing transition from earlywood to latewood or normal to compression wood. Protein microarrays have also started to gain some attention as a platform for affinity purification of proteins, peptides and small molecules (Kwon et al. 2009).

11.2.5 Metabolomics

As noted in reviews by Harrigan et al. (2007) and Fernie and Schauer (2009), metabolomics is fast becoming a critical tool for the rapid assessment of chemical characteristics that can be used as the basis for selective breeding. Thus, metabolomic, transcriptomic, and phenotype profiles may all be interrogated with respect to genetic markers in a segregating population to identify valuable quantitative trait loci (QTLs) (Kliebenstein 2009b). And even though these approaches have only been largely applied to a handful of model species, their use for the study of naturally occurring variants is providing new insights into plant adaptation (Alonso-Blanco et al. 2009; Bundy et al. 2009). It will be truly illuminating to see the findings of such studies when the organisms under study are perennial and genetically diverse, keystone species, such as the conifers that dominate ranges as large as the boreal forests of the Northern Hemisphere. The recent study of metabolite profiles in Douglas-fir by Robinson et al. (2007), which demonstrated a stronger link to environment than to genetics, suggests how this type of research may have important implications for our understanding of conifer biology.

Metabolomics and transcriptomics have also been fruitfully paired as the basis for generating putative biosynthetic pathways and gene networks through a correlative reasoning process sometimes called "guilt-by-association" (Saito et al. 2008). This approach has started to see broad application to the study of secondary products production in a variety of species for which little or no genomic sequence information is available (Yonekura-Sakakibara and Saito 2009). Studies in which this approach is being used to probe the multitude of terpenoid and oleoresin defense compounds produced in conifers are underway, and the results from such studies will undoubtedly prove useful for the selection of trees that are more resistant to insects and disease.

11.2.6 Genotypes and Populations

A confounding factor that has until recently plagued many researchers interested in using transcriptional profiling, proteomics and/or metabolomics to understand the mechanisms of growth and development in conifers has been the difficulty in obtaining uniform genetic material (e.g., inbred lines or clonal propagules). Now that commercial-scale propagation of varietal conifers, i.e., clonal genotypes propagated via somatic embryogenesis, has become a reality (Sutton 2002; Pait 2005;

Sorensson 2006), researchers may purchase a ready supply of genetically identical trees for laboratory experimentation. Bettinger et al. (2009) have discussed a wide array of operational considerations for varietal loblolly pine forestry in the southeastern US, and many of these considerations, such as matching varietals to specific environments, determining their resistance to pests and disease, and the monitoring of wild conifer populations for any changes in overall genetic diversity due to widespread planting of varietal trees, have elements that could and should be addressed using the genomic approaches currently available or under development.

For researchers attempting to integrate genomic datasets obtained from clonal trees or species representatives in widely dispersed common garden studies with higher-scale information, such as population or ecosystem data, geographic information systems (GIS) have proven highly useful. For example, a GIS system was developed to manage breeding populations and provenance trials for 11 major commercial species of forest trees in British Columbia (Hamann et al. 2004). Substantial investments have been made in establishing breeding populations and provenance trials of many conifer species over several decades, but the full value of these genetic resources has yet to be realized, in part because of the difficulty in monitoring the individual trees over large scales in time and space. Further investment to organize and make available in GIS databases the information about these genetic resources would yield enormous payoffs by bringing these rich sources of unique biological materials to the attention of genomic researchers. Extending this line of inquiry beyond collections of breeding material, Kozak et al. (2008) have discussed how integration of GIS-based environmental data, along with new spatial tools, can transform evolutionary studies and provide new insights into the ecological causes of evolutionary patterns. Along similar lines, GIS underpins a new subdiscipline called geographical genetics, which is being applied to the conservation of forest trees (Pautasso 2009), and open-source software, such as GenGIS, has been developed to assist researchers interested in analyzing their genomic data in the context of geospatial information (Parks et al. 2009).

Vaughan et al. (2007) noted that a critical feature in the recent progress enjoyed in crop domestication has been the availability of well-characterized germplasm resources housed in a global network of genetic resource centers. The germplasm in these gene banks has provided critical research materials for understanding domestication in addition to serving as the source of new traits for crop improvement programs. Impressive progress in this field in recent years is transforming plant breeding into "crop engineering" to meet our desires for increased crop yield with the minimum environmental impact, an approach that Vaughan et al. (2007) termed "super-domestication". If a future is to be realized in which purpose-grown domesticated conifers fulfill their utmost potential for providing biomaterials that satisfy human needs in a sustainable manner, then the establishment and maintenance of a properly inventoried system of international gene banks for conifer germplasm should become a focus for the conifer breeding community. Establishment of such centers will require the cooperation of all entities in the private, academic and governmental sectors having a stake in the future of conifer forestry.

11.3 Prospects for Understanding Conifer Genomes

Much of today's research in conifer genomics remains focused on linear sequencing and the cataloging of genomic subunits, but a time is soon coming when completed reference genomes for conifers will be available. The challenge will then shift to understanding the multi-dimensional interaction networks that translate linear genomic sequence into biological action. A thorough understanding of the fundamental mechanisms linking genomes to biological outcomes will be necessary for the efficient application of genomic information to the problem, such as preservation of conifer genetic diversity in the face of climate change or domestication of conifers for a more efficient forest products industry.

11.3.1 Characterizing the Genome

11.3.1.1 Transcriptomes and Gene Space

As noted in Chapter 7, efforts to fully inventory the conifer transcriptome are moving ahead, particularly with respect to mRNA transcripts encoding functional proteins. One of the most important challenges ahead will be to increase the resolution of such studies to a level where transcriptome dynamics can be sampled at the level of individual cells. Nelson et al. (2008) have reviewed approaches for capturing specific cells for "omic" analyses, and the power of these techniques to define transcriptional networks that define individual cell types can be seen from the study of Jiao et al. (2009) who used laser-microdissection and microarray profiling to probe transcriptional networks defining 40 distinct cell types in rice. Laser-assisted microdissection has also been discussed with respect to micrometabolomic studies (Moco et al. 2009). With respect to understanding the complete transcriptional and biochemical profiles associated with wood formation, it is interesting to contemplate the potential of coupling single-cell RNA-Seq analysis (Tang et al. 2009) with laser-microdissection and capture of lignifying xylem cells (Ruel et al. 2009). If this were done by cutting and pooling cells at the different steps along the radial files of secondary xylem in conifer stem sections, similar to the approach followed by Goue et al. (2008) to study *Populus* wood formation by microarray analysis, it would provide an unprecedented level of detail for understanding the process of wood formation in conifers.

Studies of the small non-coding RNAs (ncRNAs) in conifers are lagging somewhat behind the work on mRNAs, but this is to be expected since a more complete knowledge of the complete mRNA inventory, not to mention a complete conifer genome, are needed for the full interpretation of ncRNA targets and functions. Adding a further layer of complication to understanding transcriptomes has been our recent appreciation that transcription is very much a stochastic process, so that almost any particular stretch of genomic DNA may be targeted by RNA polymerase II at some point in time (Berretta and Morillon 2009). This will no doubt be true for conifers as it is for other species, and understanding how this contributes to genomic biology and regulation of gene expression will be a major challenge for the future.

There are a number of other biological processes, as well as some technical limitations, that also constrain our current ability to interpret information from the conifer transcriptome. Alternative splicing of mRNA transcripts occurs in virtually all eukaryotic cells (Soller 2006). Lorenz and Dean (2002) noted the possibility that alternative splicing could explain some of the differences seen in sequence tag profiles between juvenile and mature wood in the lignifying xylem of loblolly pine, and alternative splicing has also been discussed for its tendency to complicate clustering of pine expressed sequence tag (EST) sequences (Lorenz et al. 2006). More recently, Tai et al. (2007) demonstrated that alternative splicing likely plays a role in fatty acid biosynthesis related to cold tolerance in spruce, and Fischerova et al. (2008) noted that alternative splicing might be affecting a transcription factor involved in embryogenesis of *Picea abies*. Alternative splicing is undoubtedly an important process that must be understood if we are to gain a full appreciation of conifer gene expression, but typically genomic (gene space) sequence information must be available to compare with transcribed sequences in order to positively identify alternative splicing events. While comprehensive identification of alternatively spliced transcripts will not be possible until a reference genome sequence is available for conifers, it should be possible to use the new genomic sequence capture techniques (e.g., Gnierke et al. 2009) to selectively retrieve gene space DNA for comparison with cDNA sequence information.

RNA editing, a post-transcriptional event that usually occurs as a cytosine to uracil conversion in mRNA or an adenosine to inosine conversion in tRNAs, occurs in the organellular genomes of most, if not all, terrestrial plants, including conifers (Cattaneo 1991; Glaubitz and Carlson 1992; Bass 2002). Recent evidence also suggests that RNA editing may play a role in modifying the regulatory activity of microRNAs (Ohman 2007), and some researchers believe that RNA editing may even serve as a driver for adaptive evolution (Jobson and Qiu 2008; Gommans et al. 2009). While progress has been made in developing algorithms to predict RNA editing target sites in silico (e.g., Du et al. 2009), one exciting prospect for the near future is the potential to use the RNA sequencing capability of the Helicos sequencing platform to directly detect modified nucleotides, such as inosine (Ozsolak et al. 2009). Because the Helicos system can sequence RNA transcripts directly and does not require prior cDNA production, it will almost certainly identify previously unappreciated transcriptional products that can not be efficiently reverse-transcribed. There is similar interest in the potential for direct sequencing of methylated genomic DNA on the Helicos platform as a means to study its role in epigenetic regulation of the genome (Gupta 2008; Ansorge 2009).

11.3.1.2 Whole Genome Sequencing

Despite the earlier discussion of cost estimates for generating a draft conifer genome sequence based on the greatly improved throughput capabilities of current next-generation DNA sequencers, it may be premature to suggest that such a project should be undertaken at this moment. A number of experienced groups are piloting studies that seek to establish efficient routes for incorporating these new technologies into pipelines for *de novo* assembly of large genomes. One recent study described how detailed physical maps combined with a system for shotgun-sequencing of pooled bacterial artificial chromosomes (BACs) using next-generation pyrosequencing could be used to assemble high-quality *de novo* genome sequences for rice (Rounsley et al. 2009). There was a tradeoff between depth of the multiplexed BAC pools and quality of the resultant assembly, and creation of the requisite detailed physical maps still presents an enormous challenge for genomes on the scale of those characterizing conifers. Further leveraging of novel approaches, such as the use of short mate-paired reads from the ends of genomic DNA fragmented into different size classes (Chaisson et al. 2009), as well as assisted parallel assembly using genomic sequence information from closely related species (Gnerre et al. 2009), will be required if the massively parallel short-read sequencing systems are to play a significant role in generating the first conifer reference genome sequence. However, even with the inclusion of such approaches, efficient genome assembly algorithms are likely to remain a bottleneck for using short-read sequences for de novo assemblies of large eukaryotic genomes (Pop 2009). Consequently, it is likely that full-scale sequencing of one or more conifer reference genomes will be delayed until single-molecule long-read systems, the so-called third-generation DNA sequencers such as the Pacific Biosciences platform, become commercially available.

Until then, most genome sequence information for conifers will be generated using one or more of the available genome partitioning approaches (Turner et al. 2009), including the sequence capture technique mentioned in the previous section (Gmierke et al. 2009). Chromatin immunoprecipitation coupled with next-generation sequencing (ChIP-Seq) is another approach that has recently realized great success in the study of mammalian gene promoters and genome structure (Park 2009). To the extent that the necessary reagents, such as antibodies having specific affinity for conifer DNA-binding proteins (e.g., transcription factors or RNA polymerase II), are available, ChIP-Seq experiments could be used to begin examination of the higher-order structure of conifer genomes.

11.3.1.3 Genome Variation and Structure

Of the 650 or so known species of conifer (see Chapters 1 and 3), only three or four are recognized as naturally reproducing polyploids—tetraploid *Juniperus chinensis* ($2n=4x=44$) and *Fitzroya cupressoides* ($2n=4x=44$), as well as hexaploid *Sequoia sempervirens* ($2n=3x=66$) (Libby et al. 1969; Ahuja 2005). There are a handful of other recognized polyploid conifers that are maintained as horticultural selections, but all appear to have been chemically induced or generated as interspecific or intergeneric hybrids, and none of these appear to exist in sustained natural populations. The latter two examples are members of the monophyletic redwood group of the Cupressaceae, and it may not be a coincidence that the genomes of the other two members of this group, *Sequoiadendron giganteum* and *Metasequoia glyptostroboides*, are among the smallest known conifer genomes at around 10–12 Gb (Ahuja 2009). While the almost uniformly diploid nature of conifer genomes bodes well for prospects to rapidly generalize findings from one conifer genome to other conifer species, when we consider the strong tendency of angiosperm genomes to exist as polyploids, the rarity of conifer polyploids raises the question of whether fundamental mechanisms for genome maintenance differ significantly between conifers and angiosperms. Although there are good arguments for why members of the Pinaceae should be targeted for the production of reference genome sequences, *S. giganteum* and *M. glyptostroboides* represent interesting targets, not only for their small genomes, but also for the possibility that comparative studies with the genomes of their polyploidy relatives might lead to a better understanding of the processes involved in generating and maintaining polyploidy in plants.

Small-interfering RNAs (siRNAs) are involved in gene silencing via DNA methylation (Verdel et al. 2009). This type of heterochromatin modification is brought about in most organisms through a highly conserved set of mechanisms; however, conifers display what is so far a unique siRNA profile (Dolgosheina et al. 2008), indicating that further studies of the epigenetic mechanisms present in conifers are warranted. Monteuuis et al. (2008) noted that there were differences in levels of genomic DNA methylation associated with shoot juvenility in *S. giganteum*. Given the importance of juvenility for such applications as rooted cuttings and the production of precocious strobili, it seems likely that future studies of siRNA profiles and DNA methylation patterns in conifers might be expected to yield further insight into control of the juvenile state.

Conifer genomes are frequently described as remarkable for their stability in size and chromosome number (Williams 2009). However, the older literature is replete with reports of intraspecific karyotype variation in conifers, including some studies that suggested linkage of such variability with environmental factors as well as specific tissue types (e.g., Davies et al. 1997; Murray 1998). In addition, some *Picea* species are well known to harbor highly heterochromatic supernumerary chromosomes (Williams 2009). Karyotype variation is well known amongst angiosperm species where it is most frequently associated with differences in chromosome number (Kato et al. 2005); however, it can also be the result of chromosome rearrangements (e.g., Mohanty et al. 2004). While it seems likely that many older reports of karyotype variation in conifers could eventually be dismissed as artifacts resulting from flawed protocols (Hesemann 1980), the observations of Pelgas et al. (2006) noting apparent instabilities in the *Pseudotsuga menziesii* genome in comparison with several other members of the Pinaceae suggest that additional efforts are warranted in this area.

11.3.2 *Comparative Genomics*

11.3.2.1 Inbreeding, Hybrids and Introgression

Current breeding schemes used for conifer tree improvement must constantly balance an imperative for aggressive selection of productivity characters, such as diameter growth and disease resistance, with the potential for inbreeding depression, which occurs as the size of the breeding population is increasingly constrained (White et al. 2007). On the other hand, purposeful inbreeding of conifers has also been advocated in the past by various researchers (e.g., as noted in Sniezko and Zobel 1988), and Williams and Savolainen (1996) reviewed a number of justifications for establishing inbred lines of conifers. With the advent of genomic analyses, such as whole-genome scans, the potential for inbred conifer lines to provide useful insights into gene family structure and gene function is greatly magnified. Using such approaches, even the slow-growing and misshapen progeny normally discarded from conifer inbreeding experiments become potentially informative with respect to gene and allele functions.

Although the accumulated deleterious alleles in most conifer species of commercial interest make inbreeding difficult, this is not true for all species. Through a series of genetic bottleneck events during the glaciation periods of the Pleistocene, red pine (*Pinus resinosa*) appears to have purged deleterious alleles from its genome to the extent that the species is self-fertile (Fowler 1964). Not surprisingly, *P. resinosa* also displays minimal genetic variation across its entire population (Boys et al. 2005). Although this lack of genetic diversity carries potentially negative ramifications for the ability of this species to deal with a broad scope of environmental challenges (Igic et al. 2008), the unique features of the *P. resinosa* genome and its reproductive biology would seem to carry significant opportunities for conifer genome researchers. Certainly, the discrimination of homologs from paralogs amongst the transcriptional products of gene families should be easier in *P. resinosa*. The species would also seem to carry interesting potential for facilitating forward-genetic as well as transgenic approaches to gene function determination (Kumar and Fladung 2003). Evidently, western red cedar (*Thuja plicata*) also has higher than expected levels of self-fertility, likely resulting from having passed through a genetic bottleneck (El-Kassaby et al. 1994). However, *T. plicata* has also retained some post-pollination mechanisms to promote outcrossing (O'Connell and Ritland 2005). It remains to be seen whether any conifers outside the Pinaceae display significant levels of self-fertility that could be exploited in model systems for studies of conifer gene function.

Interspecific conifer hybrids are used in commercial forestry operations in many parts of the world (Dungey 2001). Molecular markers and genetic mapping approaches using interspecific and intergeneric crosses have proven useful for identifying valuable traits in other large plant genomes, particularly those characterizing the grasses (Jones et al. 2009). Strauss et al. (1992) suggested some time ago that marker-aided selection, while fairly difficult to implement in traditional tree improvement programs, might be more easily implemented in breeding programs that use interspecific hybrids. Recognizing that the growth and yield traits of greatest value for production forestry are the integrated result of tens or hundreds of genes working together, Grattapaglia et al. (2009) have more recently advocated genome-wide or genomic selection (GS) as a more efficient means of breeding for such traits. With this in mind, Novaes et al. (2009) recently demonstrated how microarray analysis of a pseudo-backcross population of hybrid *Eucalyptus* could be used to rapidly identify multiple QTLs for several valuable phenotypic traits in parallel. The same reductions in the price of DNA sequencing that made such studies possible for *Eucalyptus*, will no doubt lead others to apply the same techniques to conifer hybrids.

At the level of genome structure and dynamics, it will also be interesting to apply the new genomic technologies to studies of the genome rearrangements that occur during meiotic segregation in conifer hybrids. For example, meiotic products from an F_1 hybrid of *Pinus elliottii* var. *elliottii* and *Pinus caribaea* var. *hondurensis* displayed substantial large-scale synteny with *P. taeda* and the two parents, but numerous small-scale disruptions, particularly inversions, were also present (Shepherd and Williams 2008).

Genomic approaches are also being used to investigate introgression events in conifers, particularly with respect to organellar genomes. For example, Liston et al. (2007) noted introgression of chloroplast genome sequence from whitebark pine to sugar pine, while Peng and Wang (2008) noted complex introgression patterns for chloroplast genomes in *Thuja*, recommending that future studies use low-copy nuclear gene sequences to track evolutionary history within the genus. Different patterns of inheritance were recently reported for the chloroplast, mitochondria, and nuclear genomes in pine (Tsutsui et al. 2009), which may have profound implications for phylogenetic studies that include conifers. Multiplex approaches to the shotgun sequencing of organellar genomes on next-generation sequencing platforms are being pioneered using conifer chloroplasts (Cronn et al. 2008), and the results from such studies will no doubt help to clarify the correct phylogenetic positioning of the conifers with respect to other major plant taxa.

11.3.2.2 Phylogenetics

Conifers hold an important place in the evolution of seed plants, and for many years were considered one of four distinct paraphyletic groups (the others being Gnetales, Cycads, and Gingko) comprising the gymnosperms (Doyle 1998; Palmer et al. 2004). However, placement of the Gnetales within this group has always been controversial (as reviewed in Burleigh and Matthews 2004). In recent years, molecular sequence data has led researchers to propose some highly controversial rearrangements of older phylogenies that were based solely on morphometric and developmental process characters (Chaw et al. 1997, 2000; Bowe et al. 2000; Burleigh and Mathews 2004). One of the most surprising of these new phylogenies places the Gnetales as a sister group to the Pinaceae, fully within the Coniferales (Gugerli et al. 2001). Although this particular phylogenetic topology has gained acceptance in some quarters of the systematics community (Doyle 2006), several groups have urged caution because the proposed topologies are highly sensitive to errors and bias in both the molecular and the quantitative characters on which they are based (Rydin and Källersjö 2002; Palmer et al. 2004). As an example of one potential source of error for DNA-based phylogenies that may be confounding this case, work in loblolly pine suggests the Pinaceae have somewhat lower rates of nucleotide substitution than is typical of most other plants (Brown et al. 2004 and Chapter 5 in this book), while the Gnetales have been characterized (at least in comparison to other gymnosperms) as having relatively high rates of substitution (Chaw et al. 1997; Burleigh and Mathews 2004). Differential rates of nucleotide substitutions between species have been identified as a potential major source of error for phylogenies based on DNA sequence information (Rydin and Källersjö 2002). In another example, Won and Renner (2003) have documented horizontal transfer of mitochondrial gene sequences between members of the angiosperms and the Gnetales in the past five million years, which calls into question characters based on mitochondrial DNA sequence from these species, such as were used in the study by Gugerli et al. (2001). Because sequence data is at present so sparse for plant species outside the angiosperms, many taxonomists recommend reliance on morphological and developmental characters of both extant and extinct species to establish broad phylogenies, while limiting use of DNA sequence information to tests of fine structure within well-recognized groups (Stace 2005; Hilton and Bateman 2006). Given this evidence, increased taxon sampling and the development of additional characters, both molecular and morphometric, will be needed to clarify whether or not the Gnetales should be placed within the Coniferales as a sister group to the Pinaceae or returned to their more familiar phylogenetic position(s) somewhere outside the Coniferales. Efforts to expand the DNA sequence information available for a broader spectrum of the gymnosperms should help identify the rarer genomic changes that can clarify broad-scale phylogenetic relationships (Rokas and Holland 2000; Burleigh and Matthews 2004).

11.3.3 Ecological Genomics

As the dominant species in many forest ecosystems, conifers have profound effects on other species through both physical and chemical processes. Likewise, both the biotic and abiotic environments affect conifer metabolism, as well as growth and development. Metlen et al. (2009) have reviewed the dynamics of secondary product pathway responses in plants to biotic and abiotic challenges. Through their end-products, these pathways are obviously critical for plant survival, and are thus an attractive target for studies coupling transcriptomic and metabolomic analyses. For example, Harding et al. (2005) used functional genomics to understand how foliar phenolic metabolites in *Populus* hybrids changed with growth and development, and noted the relationship with nutrient dynamics in soil. A synthesis of this work with results from many other studies explained some of these processes at the ecosystem level and pointed to areas in need of further research (Bailey et al. 2009). The studies of Ralph et al. (2006) are an excellent example of how such studies should be approached in conifers, and similar studies in the future will no doubt provide us with greater insight into the secondary product pathways and their function in a wide variety of conifers.

Provocative experimental results have recently emerged concerning the abilities of plants to recognize kin and how this impacts competition for resources (e.g., Milla et al. 2009). None of the experiments reported to date appear to have used genomic or even molecular approaches to understand the genetic components of this phenomenon. Given the obvious implications for conifer plantation forestry, particularly with the prospect of using varietals (clonal) versus outcrossed seedlings, this would appear to be a line of investigation that could produce interesting and practical results if applied to conifers.

Soil ecology plays a fundamental role in the sustainability of forest ecosystems, but the research on rhizosphere dynamics in forests has so far been insufficient to provide a predictive level of understanding of how rhizosphere communities operate and interact with host trees (Johnston and Crossley 2002). Transcriptomic, proteomic and metagenomic techniques are currently being applied to rhizosphere characterization from the level of single cells to entire microbial communities (Sorensen et al. 2009). The recent completion of a genome sequence for the ectomycorrhizal fungus, *Lacaria bicolor* (Martin et al. 2008), opens the door to a multitude of genomic approaches for studies of the ectomycorrhizal interactions between conifers and these beneficial fungi (Martin and Nehls 2009). Similarly, the completion of genome sequences for a variety of phytopathogens sets the stage for increased studies of the molecular mechanisms that fungi employ for causing diseases of conifers (Bhadauria et al. 2009). Tan et al. (2009) have recently reviewed the impact that transcriptomic, proteomic and metabolomic technologies are having on fungal phytopathology research (Tan et al. 2009).

Conifers, like other organisms, benefit from an internal ecosystem. We are only just beginning to appreciate the importance of these internal ecosystems for organismal biology, and the National Institutes of Health (NIH) recently launched the Human Microbiome Project to identify and characterize the roles that endosymbiotic organisms play in human health (Phillips 2008; Proal et al. 2009). Endosymbiotic bacteria and fungi also comprise complex communities in plants (Rosenblueth and Martinez-Romero 2006; Andreote et al. 2009), and have been recognized as the source of a variety of bioactive compounds whose formation was once attributed solely to plant metabolism (as noted in Guo et al. 2008). There have been studies describing endosymbiont communities in conifers (Chanway et al. 2000; Izumi et al. 2008); and in one recent piece of work, endosymbiont inoculation was demonstrated to help protect western white pine (*P. monticola*) against the pathogen that causes white pine blister rust (*Cronartium ribicola*) (Ganley et al. 2008). Just as the latest genomic technologies for microarrays and DNA sequencing are opening new areas of research into human-microflora interactions (Huyghe et al. 2008; Guazzaroni et al. 2009; Petrosino et al. 2009), so too are these technologies likely to modify our perception and appreciation of the microflora that inhabit the interior and exterior of conifers.

11.4 Prospects for Application of Conifer Genomics

11.4.1 Advanced Breeding

The comprehensive treatise on forest genetics produced by White et al. (2007) contains a thorough examination of advanced breeding strategies for forest trees, including conifers. In an excellent complement to that work, Varshney et al. (2009) recently reviewed the latest generation of high-throughput DNA sequencing platforms with respect to their potential impact on future approaches to the accelerated breeding of crop plants. These genomic approaches to plant improvement have been taken up more quickly for some crops, thus tomato currently provides some of the most exciting examples of "breeding by design" (Barone et al. 2009). These same approaches will undoubtedly be adapted for use in future conifer breeding programs.

11.4.1.1 Genomic Selection for Tree Improvement

High level of genetic diversity reside in most conifer populations, which suggests that enormous potential exists for selection of desirable traits if only we can identify and follow the specific alleles efficiently and economically. Association genetics, as reviewed by Neale and Savolainen (2004) and discussed in detail in Chapter 5, moves in the right direction by eschewing the inherent biases of candidate gene approaches and using high-throughput analyses of high resolution molecular markers. González-Martinez et al. (2007) demonstrated the association genetics approach in identifying loci associated with wood property traits in loblolly pine, and other researchers are applying this approach to studies of disease resistance and susceptibility. Kliebenstein (2009a) recently reviewed the use of eQTL analysis in association genetics studies and noted how the combination could contribute to identification of the mechanistic basis for quantitative traits. The results to date suggest that association genetics has potential for becoming a useful tool in conifer tree improvement efforts. However, because the magnitude of the effect on particular traits from any individual association has in most cases been small, it remains to be seen whether economy and efficiency can be maintained by this approach when breeding for complex, quantitative traits.

The advent of more efficient technologies for molecular marker genotyping led Meuwissen et al. (2001) to propose genomic selection (GS) or genome-wide s election (GWS) as an approach to rapidly capture most of the segregating variation underlying complex phenotypes. Recognizing that growth and yield, as well as many of the other traits of greatest value for production forestry, are the integrated result of tens or hundreds of genes working together, Grattapaglia et al. (2009) have more recently advocated genome-wide or genomic selection (GS) as a more efficient means of breeding for such traits. In contrast to marker-assisted selection or association genetics, GS is based on the development of a predictive model for performance that does not require the time and cost of advance work to establish genotype-phenotype associations (Grattapaglia and Resende, 2010). What the GS approach does require is a large number of markers distributed across the entire genome in such a way that all genes are in linkage disequilibrium with at least some of the markers. A training population is then phenotyped for the trait(s) of interest and genotyped to yield a dataset for the model, which is subsequently used to generate genomic breeding expected values for the selection candidates. The accuracy of the approach depends on the extent of linkage disequilibrium between the markers and the QTLs, the number of individuals in the training population, the heritability of the trait in question, and the distribution of QTL effects (number of loci and size effects). The first two conditions are under the control of the breeder, and despite the short range of decay reported for linkage disequilibrium in forest trees (Neale and Savolainen 2004; Ingvarsson 2008), simulations indicate that marker densities as low as 2 markers/cM could be used in GS breeding regimes if the effective breeding populations contain fewer than 30 individual genotypes (Grattapaglia and Resende 2010). Thus, the genomic selection approach to forest trees improvement is most likely to be picked up first by breeding programs that have already focused their efforts on a reduced number of elite genotypes selected from previous rounds of breeding.

11.4.1.2 Transgenic Conifers

Transgenic approaches to the development of conifers having traits of value to industry, such as resistance to herbicides, pests and diseases, as well as modified wood chemistry and fiber structure, have been pursued with great success from the late 1980s (Henderson and Walter 2006). Where particular traits, such as resistance to newly introduced insects or diseases, do not exist or show little variability in existent populations, there are currently no alternatives to transgenic approaches. But despite the demonstrated feasibility and utility of such genetic manipulations in a number of important commercial species, as noted in several chapters of the book edited by Williams (2006), a variety of ecological and societal concerns have so far limited the widespread release of trees modified in this manner.

Walter and coworkers have written several thoughtful reviews covering a variety of topics related to the production and use of transgenic conifers in commercial forestry (Walter et al. 2002; Walter 2004; Henderson and Walter 2006). Nehra et al. (2005) covered the topic with respect to valuable commercial targets in both hardwood and softwood tree species; however, Strauss et al. (2009) describe how various international agreements are being used to block not only the release, but even the environmental testing of such trees. To circumvent such issues, Flachowsky et al. (2009) relate how transgenic approaches directed at reproduction in trees might be used to speed tree improvement without necessarily resulting in the environmental release of transgenic plants. Another interesting report describes how a systems biology approach linking changes in transcriptional and metabolomic profiles might be used in assessing transgene risk assessment (Kiambi et al. 2008).

Despite the uncertainties surrounding the release of transgenic trees for commercial forestry, their use in research, particularly with respect to discovery of gene function, will remain critical. Thus, we should expect to see many more studies in the future similar to that reported by Bomal et al. (2008) who performed transcriptional profiling experiments on transgenic spruce expressing a transcription factor thought to function in vascular tissue development. The primary concern to genomic researchers wishing to use transgenesis as a tool, however, is that conifer transformation remains expensive and difficult, and if lack of public acceptance for transgenic trees leads companies to abandon further transgenic efforts, then such work will remain a difficult province left to a few large, well-funded research groups. Perhaps as the companies currently engaged in development of transgenic conifers move their production efforts to genotypes having superior traits for the production of wood products, they might be persuaded to release as a research platform readily transformable cell lines that do not necessarily produce trees of commercial value.

11.4.1.3 Secondary Products

Pitch, sap and resin harvested from pines and other conifers, known collectively as oleoresins or gum naval stores, were at one time more valuable as a commodity than the wood produced by these trees (Perry 1968). Although many of these products were displaced by cheaper petroleum distillates, there remains a large market for the sulfated tall oil derivatives that are captured as by-products from the Kraft pulping process (Coppen and Hone 1995). Bioenergy considerations, as well as forest health issues, are leading researchers to reconsider whether the terpenoid secondary products produced in large quantity by conifers might provide a value-added by-product that could make the economics more favorable for their recovery (Kelkar et al. 2006; Bohlmann and Keeling 2008). Different conifer species produce a wide variety of mixtures of these compounds, and relatively little is known about their biosynthetic pathways (Otto and Wilde 2001). Properly applied, the previously discussed genomics platforms could rapidly change our understanding of these pathways, and provide information that could be used to select trees for optimized production of these value-added chemicals.

In addition to an abundance of resinous terpenoid compounds, conifers synthesize a vast array of other secondary metabolites, the biosynthetic pathways for most of which are virtually unknown. These secondary metabolites, so-called phytochemicals, serve to protect plants against pests and pathogens, and so frequently have activities against such organisms (Wink 2003; Powell 2009). These compounds have long been the basis for pharmacology, and natural products chemistry remains a cornerstone for advances in human health (Ilic et al. 2002; Saklani and Kutty 2008). The opportunities for recovery of such valuable chemicals for forest tree species, including conifers, have previously been the topic of authoritative reviews (e.g., Pearl 1965; Anderson 1967; Goldstein 1975). Secondary compounds isolated from conifers are currently being used or tested for anticancer (e.g., Saarinen et al. 2000; Cragg and Newman 2005; Chien et al. 2008) and antimicrobial activities (e.g., Valimaa et al. 2007; Lee et al. 2009b), as well as activities against specific human cell surface receptors (e.g., Nakane et al. 2000; Cui et al. 2008). Other researchers are looking at the production in conifers of compounds that could be used to control problems in lipid metabolism (e.g., Lee et al. 2004; Li et al. 2007), as well as polyphenol antioxidants that can serve as nutraceuticals (e.g., Santos-Buelga et al. 2000; Rasmussen et al. 2005; Pietarinen et al. 2006). In fact, work in this latter area led to the recent awarding of the 25th Marcus Wallenberg Prize to Dr. Bjarne Holmbom for his work in developing novel methods to recover useful chemicals from forest tree biomass (*http://www.mwp.org/index.cfm?PageAction=ReadMore&id=30*). New strategies for exploiting these chemicals as valuable co-products for forest products operations have been advanced (Turley et al. 2006), and high-throughput "omic" technologies should make it possible to accelerate the identification and optimization of superior genotypes for their production (Yonekura-Sakakibara et al. 2009).

11.4.1.4 Biofuels

After a recent series of oil price hikes, the likes of which had not been seen since the early 1970s, there has again been renewed interest in the potential for developing energy resources from biomass, particularly in the form of liquid biofuels, for reducing current reliance on petroleum and other fossil fuels. Peter (2008) has reviewed the potential for using pine species for a bioenergy market in the southern US. There has been a good deal of interest in fermentative approaches to converting pine biomass to ethanol (e.g., Araque et al. 2008; Frederick et al. 2008), although lignin content and structure, as well as various secondary metabolites in conifer biomass, tend to be inhibitory to current fermentation reactions (Pienkos and Zhang 2009). Syngas production facilities utilizing the Fischer-Tropsch process to synthesize liquids fuels (see Shulz 1999 for a brief review) are also attracting attention for softwood biomass conversion, particularly for the prospect of coupling them with pulp and paper production facilities to create so-called biorefineries (Consonni et al. 2009; Digman et al. 2009; Jegannathan et al. 2009). Considering the thermal content of typical conifer biomass compared to other lignocellulosic materials (Demirbas and Demirbas 2009), however, there are reasonable arguments to be made that the most efficient use for conifer biomass in energy generation is either through burning or co-firing. In fact, wood pellets for home heating appears to be the fastest growing component of the bioenergy market for conifer biomass (Heinimo and Junginger 2009; Samuelsson et al. 2009).

With respect to the potential for genomic technologies to impact the future uses of conifers in the biofuels and bioenergy arena, the most likely prospects are through the previously discussed research related to improved growth and yield characteristics of improved trees. However, should fermentative approaches to liquid fuels production from conifer biomass take hold, then research directed at better understanding the biosynthesis of lignin and inhibitory secondary compounds will likely receive increased attention.

11.4.1.5 Carbon Sequestration

The prospects for global climate change from increasing levels of atmospheric carbon dioxide are driving efforts to find new ways to capture and sequester carbon. Forests comprise the greatest terrestrial repository for carbon, and Groover (2007) discussed the new imperatives for forest biotechnology with respect to both conservation and management of natural forests, as well as development of trees for applications to biofuels production and carbon sequestration. Genetic control of carbon allocation to conifer roots, as well as the lignification process that makes root tissues recalcitrant to degradation, would be good targets for genomic investigations to increase the utility of conifers as carbon sinks. Yang et al. (2009) recently reported on a comparative genomics study that identifies gene products expressed specifically in the roots of *Populus* trees, which the authors felt might have potential for improving carbon sequestration.

11.4.2 Conservation and Biodiversity

Some time ago, Ledig (1988) set forth arguments for preserving the genetic diversity of forest tree species, and described various programs whose purpose was to make sure this diversity was not lost. Global climate change was (and is) a major concern, and some of the more extreme scenarios provided the basis for arguments that efforts for in situ preservation of genetic diversity might require human-facilitated movement of species to areas outside their native range (Ledig and Kitzmiller 1992). Obviously, should such extreme measures become necessary, then genomic assessment will be indispensable for identifying genotypes whose preservation would maximize population diversity and adaptive potential, and subsequently verifying their persistence in collections over time. The approaches and concerns discussed by Pautasso (2009) with respect to geographical genetics would have great relevance for any such conservation effort. Although, as noted by Pautasso (2009), tropical and austral forest species are in dire need of much greater study, and thus present a dilemma for any conservation efforts launched in response to climate change.

Until recently, a major concern for gene conservationists contemplating assisted relocation of imperiled populations has been the necessity of using quantitative characters, such as neutral genetic markers, to identify adaptive variations (Crandall et al. 2000; Reed and Frankham 2001; McKay and Latta 2002). The common garden studies necessitated by this approach are expensive to establish, to the extent that they cannot even be considered for the minor species that are often of greatest concern. However, advances in genomic technologies are making it possible to identify that part of the genetic heritage (genes and their polymorphisms) that contributes directly to adaptation and other important quantitative traits, and the ability to perform so-called "genome typing" will greatly improve conservation efforts (Luikart et al. 2003; Storz 2005). Thus, Namroud et al. (2008) recently demonstrated how genome scans could be used for the identification of gene families, candidate genes and their specific single-nucleotide polymorphisms related to ecological differentiation. It should also be possible to use such marker systems to monitor the maintenance of adaptive variation in advanced generation breeding populations.

Recently, quantitative evidence has accumulated to demonstrate how species richness and genetic diversity contribute to the resilience of forests for resisting environmental threat (DeClerck et al. 2006). Fischer et al. (2009) have discussed how the concept of resilience may be integrated with optimization for conservation to bring about enduring conservation outcomes. With these concepts in mind, it would be of interest to apply genomic approaches to understanding the organismal responses on which such ecosystem resilience is based. Procaccini et al. (2007) have outlined how genomic approaches could be applied to develop an understanding of the interaction mechanisms that contribute to the resilience of seagrass communities. Ungerer et al. (2008) have described similar approaches in a concept they call ecological genomics. These ideas could and should be directed at conifer forest ecosystems using the available genomic tools. The resulting information would greatly facilitate our ability to conserve and protect our forest resources in the face of environmental change and the threat of invasive pests and pathogens.

11.5 Barriers to Progress

Research in all of the biological sciences is becoming increasingly complex as information accumulates at rates too great for individuals to manage, and as measurement technologies generate amounts of data that can only be retained and manipulated in silico. As a consequence, biological researchers increasingly face an imperative to become good collaborators or risk being left on the sidelines. The current global conifer genomics research community still numbers only a few hundred people, most of whom are active collaborators with peers within the community. However, as the genomics platforms become more mature and more specialized, it is imperative that the community extend its collaborative embrace not just to genome specialists working on other organisms, but also to forest ecologists and forest tree breeders, specialists in environmental restoration and pulp and paper chemists. All of these specialties are potential consumers of information from genomic studies of conifers. To ensure that future research on conifer genomics remains relevant, the forest tree genomics community must stay engaged with these other communities.

11.5.1 Bioinformatics Integration and Compatibility

Bioinformatics is central to our ability to infer and understand relationships between DNA sequence information, gene expression patterns, protein structure, and metabolite abundance data collected from the growing diversity of high-throughput analysis platforms. Smedley et al. (2008) have discussed the bioinformatic challenges that can hinder efforts to integrate data from different technology platforms stored in widely distributed, customized databases. Ontologies and semantic web services are just some of the tools that bioinformatics specialists are using to overcome these issues (Antezana et al. 2009). However, it also critical that there be frequent interactions and dialog directly between the bioinformaticians who collect, organize, and query the data and the biologists who frame the questions and attempt to interpret the output from datamining exercises (Lee et al. 2009a). For bioinformaticians, Baxter et al. (2006) have provided concise and cogent advice on preparations for new software projects, and Bolchini et al. (2009) describe how usability metrics can be used to evaluate and improve software interfaces. At the same time, biologists are reminded that the language and thought processes employed by bioinformatics professionals are in many ways distinct from those of biological researchers, and it is imperative that they take this into consideration as they share information and ideas during software or query development (Penders et al. 2008).

11.5.2 Data Sharing

Fundamental to the rapid advancement of our understanding of the biological sciences through genomic technologies has been the mandate(s) from governmental funding agencies that raw data be placed in public repositories at the earliest possible date, often before publication of any peer-reviewed research papers. The rapid, prepublication release of data is a phenomenon that began with the Human Genome Project, and while it has represented a disconcerting departure from past practices for many researchers, it has had a profound effect on the speed of advancement in the genomic sciences. The rationale and expectations for prepublication release of genomic data were recently the focus of a meeting of international granting agencies, and a paper produced from that summit outlines new and stronger expectations from these funding agencies for rapid release of data and makes clear that this is to become the norm for handling data from government-sponsored research projects (Toronto International Data Release Workshop Authors 2009).

The logic for this paradigm shift is obvious. As data resources grow, all communities benefit. As just one example of how such information can be used in new and interesting ways, Brady and Provart (2009) describe the use of large datasets in online repositories to stimulate hypothesis-driven plant biology research. Yet even as they draw upon these public resources for inspiration, many research groups choose not to release their data to these repositories for fear of being scooped for publication or for the possibility that some small piece of potentially valuable intellectual property could be lost. This despite arguments that large genomics datasets contain many more potential publications than any individual researcher is likely to perceive in a career. The intellectual property argument is even weaker, given the fact that only a small percentage of all the gene patents ever awarded have been profitable, and given the knowledge that any practical utilization of patented tree genes for forest tree improvement will need to deal with such issues as getting the gene into an elite line of a production species from a provenance that will grow in a specific locale. Even if these parameters can be satisfied, it will be 10 to 20 years before the effect can be verified, not to mention any profit realized. The assumption that great profits will result from the wholesale patenting of tree genes is bankrupt from the outset. Thus, to the extent that researchers can overcome reflexive self-interest and release datasets to the public repositories, we will all gain from speedier advancement of genomic science.

11.5.3 Annotation

Attwood (2000) noted some time ago that our ability to make sense of functional genomics data was being compromised by poor annotation and conflicting nomenclature. This is not a problem that will disappear rapidly, given the exponential growth of new sequence information, the retention of legacy datasets and databases, and the tight state of research funding. Some intermediary databases and software tools have been developed to address the problem in part (Benoit 2005), while some have proposed new conventions for naming proteins and genes that would eliminate many of the uncertainties that now exist (Schluter et al. 2009). Yet problems would still exist because pipelines developed to handle gene annotation in an automated fashion so often draw their information from the public repositories in which poorly annotated legacy genes reside (Liu et al. 2008; Liang et al. 2009). At some point in the near future, the conifer genomics community will need to establish a supervised database of high-quality gene models that are manually curated with respect to their assigned names and functions.

11.5.4 Economics and Funding

Libby et al. (1969) once noted that an estimated 50 to 75% of the research information potentially available from forest genetics research has been lost due to personnel changes, administrative inconsistencies, and damage due to the occurrence of low-probability disasters. Despite the advice they presented to combat this information loss through development of institutes of forest genetics where major experiments could be carried as line projects, and record-keeping could be standardized and institutionalized, it seems clear that forest genetics research remains a highly inefficient enterprise. Perhaps even more worrisome, however, has been the almost continuous erosion of the funding that has supported both long-term investments in forest tree genetic infrastructure and short-term investments in conifer research and development. In the current economic cycle, fewer and fewer companies enjoy sufficient financial health and stability that they continue to support the private-public partnerships of the tree improvement cooperatives. Unfortunately, there is no evidence to suggest that increases in public funding will be forthcoming to help cover maintenance of the irreplaceable genetic stock and progeny trials that were established under the purview of these cooperatives and that are essential to genomic research. Genomic technologies and approaches hold great promise for opening a window on conifer biology and enabling us to make quantum leaps in selecting trees for superior performance, but sustained funding at reasonable levels will be imperative if we are to realize the benefits of the many years of hard work and patient waiting.

References

Ahuja MR (2005) Polyploidy in gymnosperms: Revisited. Silvae Genet 54: 59–69.

Ahuja MR (2009) Genetic constitution and diversity in four narrow endemic redwoods from the family Cupressaceae. Euphytica 165: 5–19.

Alonso-Blanco C, Aarts MGM, Bentsink L, Keurentjes JJB, Reymond M, Vreugdenhil D, Koornneef M (2009) What has natural variation taught us about plant development, physiology, and adaptation? Plant Cell 21: 1877–1896.

Anderson AB (1967) Silvichemicals from the forest. Econ Bot 21: 15–30.

Andreote FD, Azevedo JL, Araujo WL (2009) Assessing the diversity of bacterial communities associated with plants. Braz J Microbiol 40: 417–432.

Ansorge WJ (2009) Next-generation DNA sequencing techniques. New Biotechnol 25: 195–203.

Antezana E, Kuiper M, Mironov V (2009) Biological knowledge management: the emerging role of the Semantic Web technologies. Brief Bioinformat 10: 392–407.

Araque E, Parra C, Freer J, Contreras D, Rodriguez J, Mendonca R, Baeza J (2008) Evaluation of organosolv pretreatment for the conversion of *Pinus radiata* D. Don to ethanol. Enz Microb Technol 43: 214–219.

Attwood TK (2000) Genomics—The babel of bioinformatics. Science 290: 471–473.

Bailey JK, Schweitzer JA, Ubeda F, Koricheva J, LeRoy CJ, Madritch MD, Rehill BJ, Bangert RK, Fischer DG, Allan GJ, Whitham TG (2009) From genes to ecosystems: a synthesis of the effects of plant genetic factors across levels of organization. Phil Trans Roy Soc B 364: 1607–1616.

Barone A, Di Matteo A, Carputo D, Frusciante L (2009) High-throughput genomics enhances tomato breeding efficiency. Curr Genom 10: 1–9.

Bass BL (2002) RNA editing by adenosine deaminases that act on RNA. Annu Rev Biochem 71: 817–846.

Baxter SM, Day SW, Fetrow JS, Reisinger SJ (2006) Scientific software development is not an oxymoron. PLOS Comput Biol 2: 975–978.

Benoit G (2005) Bioinformatics. Annu Rev Info Sci Tech 39: 179–218.

Berretta J, Morillon A (2009) Pervasive transcription constitutes a new level of eukaryotic genome regulation. EMBO Rep 10: 973–982.

Bettinger P, Clutter M, Siry J, Kane M, Pait J (2009) Broad implications of southern United States pine clonal forestry on planning and management of forests. Int For Rev 11: 331–345.

Bhadauria V, Banniza S, Wei Y, Peng Y-L (2009) Reverse genetics for functional genomics of phytopathogenic fungi and *Oomycetes*. Comp Funct Genom 2009: Art 380719.

Bohlmann J, Keeling CI (2008) Terpenoid biomaterials. Plant J 54: 656–669.

Bolchini D, Finkelstein A, Perrone V, Nagl S (2009) Better bioinformatics through usability analysis. Bioinformatics 25: 406–412.

Bomal C, Bedon F, Caron S, Mansfield SD, Levasseur C, Cooke JEK, Blais S, Tremblay L, Morency MJ, Pavy N, Grima-Pettenati J, Seguin A, MacKay J (2008) Involvement of

Pinus taeda MYB1 and MYB8 in phenylpropanoid metabolism and secondary cell wall biogenesis: a comparative *in planta* analysis. J Exp Bot 59: 3925–3939.
Bowe LM, Coat G, dePamphilis CW (2000) Phylogeny of seed plants based on all three genomic compartments: Extant gymnosperms are monophyletic and Gnetales' closest relatives are conifers. Proc Natl Acad Sci USA: 97: 4092–4097.
Boys J, Cherry M, Dayanandan S (2005) Microsatellite analysis reveals genetically distinct populations of red pine (*Pinus resinosa* Pinaceae). Am J Bot 92: 833–841.
Brady SM, Provart NJ (2009) Web-queryable large-scale data sets for hypothesis generation in plant biology. Plant Cell 21: 1034–1051.
Brown GR, Gill GP, Kuntz RJ, Langley CH, Neale DB (2004) Nucleotide diversity and linkage disequilibrium in loblolly pine. Proc Natl Acad Sci USA 101: 15255–1526.
Bundy JG, Davey MP, Viant MR (2009) Environmental metabolomics: a critical review and future perspectives. Metabolomics 5: 3–21.
Burleigh JG, Mathews S (2004) Phylogenetic signal in nucleotide data from seed plants: Implications for resolving the seed plant tree of life. Am J Bot 91: 1599–1613.
Cagnin S, Caraballo M, Guiducci C, Martini P, Ross M, SantaAna M, Danley D, West T, Lanfranchi G (2009) Overview of electrochemical DNA biosensors: new approaches to detect the expression of life. Sensors 9: 3122–3148.
Cattaneo R (1991) Different types of messenger RNA editing. Annu Rev Genet 25: 71–88.
Chaisson MJ, Brinza D, Pevzner PA (2009) *De novo* fragment assembly with short mate-paired reads: Does the read length matter? Genome Res 19: 336–346.
Chaw SM, Zharkikh A, Sung HM, Lau TC, Li WH (1997) Molecular phylogeny of extant gymnosperms and seed plant evolution: Analysis of nuclear 18S rRNA sequences. Mol Biol Evol 14: 56–68.
Chanway CP, Shishido M, Nairn J, Jungwirth S, Markham J, Xiao G, Holl FB (2000) Endophytic colonization and field responses of hybrid spruce seedlings after inoculation with plant growth-promoting rhizobacteria. For Ecol Manag 133: 81–88.
Cheng JY, Chen HY (2009) Microfluidic ARray Synthesizer (MArS) for rapid preparation and hybridization of custom DNA microarray. Biotechnol Bioeng 104: 400–407.
Chien SC, Chen CC, Chiu HL, Chang CI, Tseng MH, Kuo YH (2008) 18-nor-Podocarpanes and podocarpanes from the bark of *Taiwania cryptomerioides*. Phytochemistry 69: 2336–2340.
Collins FS, Green ED, Guttmacher AE, Guyer MS (2003) A vision for the future of genomics research. Nature 422: 835–847.
Consonni S, Katofsky RE, Larson ED (2009) A gasification-based biorefinery for the pulp and paper industry. Chem. Eng. Res. Design 87: 1293–1317.
Coppen JJW, Hone GA (1995) Gum naval stores: turpentine and rosin from pine resin, FAO, Rome, Italy, *http://www.fao.org/docrep/V6460E/ v6460e00.HTM*.
Cragg GM, Newman DJ (2005) Plants as a source of anti-cancer agents. J Ethnopharm 100: 72–79.
Crandall KA, Bininda-Emonds ORP, Mace GM, Wayne RK (2000) Considering evolutionary processes in conservation biology. Trends Ecol Evol 15: 290–295.
Cronn R, Liston A, Parks M, Gernandt DS, Shen R, Mockler T (2008) Multiplex sequencing of plant chloroplast genomes using Solexa sequencing-by-synthesis technology. Nucl Acids Res 36:e122, doi:10.1093/nar/gkn502.
Cui YM, Yasutomi E, Otani Y, Yoshinaga T, Ido K, Sawada K, Ohwada T (2008) Design, synthesis and characterization of podocarpate derivatives as openers of BK channels. Bioorg Med Chem Lett 18: 5197–5200.
Davies BJ, O'Brien IEW, Murray BG (1997) Karyotypes, chromosome bands and genome size variation in New Zealand endemic gymnosperms. Plant Syst Evol 208: 169–185.
DeClerck FAJ, Barbour MG, Sawyer JO (2006) Species richness and stand stability in conifer forests of the Sierra Nevada. Ecology 87: 2787–2799.
Demirbas T, Demirbas C (2009) Fuel properties of wood species. Energ Sourc Rec Util Environ Effect 31: 1464–1472.

Digman B, Joo HS, Kim DS (2009) Recent progress in gasification/pyrolysis technologies for biomass conversion to energy. Environ Prog Sust Energ 28: 47–51.
Dolgosheina EV, Morin RD, Aksay G, Sahinalp SC, Magrini V, Mardis ER, Mattsson J, Unrau PJ (2008) Conifers have a unique small RNA silencing signature. RNA 14: 1508–1515.
Doyle JA (1998) Phylogeny of vascular plants. Annu Rev Ecol Syst 29: 567–599.
Doyle JA (2006) Seed ferns and the origin of angiosperms. J Torr Bot Soc 133: 169–209.
Du PF, Jia LY, Li YD (2009) CURE-Chloroplast: A chloroplast C-to-U RNA editing predictor for seed plants. BMC Bioinformat 10: Art 135.
Dungey HS (2001) Pine hybrids—a review of their use performance and genetics. For Ecol Manag 148: 243–258.
El-Kassaby YA, Russell J, Ritland K (1994) Mixed mating in an experimental population of western red cedar, *Thuja plicata*. J Hered 85: 227–231.
Espina V, Wulfkuhle J, Liotta LA (2009) Application of laser microdissection and reverse-phase protein microarrays to the molecular profiling of cancer signal pathway networks in the tissue microenvironment. Clin Lab Med 29: 1–13.
Fernie AR, Schauer N (2009) Metabolomics-assisted breeding: a viable option for crop improvement? Trends Genet 25: 39–48.
Fischer J, Peterson GD, Gardner TA, Gordon LJ, Fazey I, Elmqvist T, Felton A, Folke C, Dovers S (2009) Integrating resilience thinking and optimisation for conservation. Trends Ecol Evol 24: 549–554.
Fischerova L, Fischer L, Vondrakova Z, Vagner M (2008) Expression of the gene encoding transcription factor PaVP1 differs in *Picea abies* embryogenic lines depending on their ability to develop somatic embryos. Plant Cell Rep 27: 435–441.
Flachowsky H, Hanke MV, Peil A, Strauss SH, Fladung M (2009) A review on transgenic approaches to accelerate breeding of woody plants. Plant Breed 128: 217–226.
Fowler DP (1964) Effects of inbreeding in red pine, *Pinus resinosa* Ait. Silvae Genet 14: 37–46.
Frederick WJ, Lien SJ, Courchene CE, DeMartini NA, Ragauskas AJ, Iisa K (2008) Co-production of ethanol and cellulose fiber from Southern Pine: A technical and economic assessment. Biomass Bioenergy 32: 1293–1302.
Ganley RJ, Sniezko RA, Newcombe G (2008) Endophyte-mediated resistance against white pine blister rust in *Pinus monticola*. For. Ecol. Manag. 255: 2751–2760.
Glaubitz JC, Carlson JE (1992) RNA editing in the mitochondria of a conifer. Curr Genet 22: 163–165.
Gnerre S, Lander ES, Lindblad-Toh K, Jaffe DB (2009) Assisted assembly: how to improve a *de novo* genome assembly by using related species. Genome Biol 10: Art R88.
Gnirke A, Melnikov A, Maguire J, Rogov P, LeProust EM, Brockman W, Fennell T, Giannoukos G, Fisher S, Russ C, Gabriel S, Jaffe DB, Lander ES, Nusbaum C (2009) Solution hybrid selection with ultra-long oligonucleotides for massively parallel targeted sequencing. Nat Biotechnol 27: 182–189.
Goldstein IS (1975) Potential for converting wood into plastics. Science 189: 847–852.
Gommans WM, Mullen SP, Maas S (2009) RNA editing: a driving force for adaptive evolution? Bioessays 31: 1137–1145.
Gonzalez-Martinez SC, Wheeler NC, Ersoz E, Nelson CD, Neale DB (2007) Association genetics in *Pinus taeda* L. I. Wood property traits. Genetics 175: 399–409.
Goue N, Lesage-Descauses MC, Mellerowicz EJ, Magel E, Label P, Sundberg B (2008) Microgenomic analysis reveals cell type-specific gene expression patterns between ray and fusiform initials within the cambial meristem of *Populus*. New Phytol 180: 45–56.
Grattapaglia D, Resende MDV (2010) Genomic selection in forest tree breeding. Tree Genet Genomes (in press).
Grattapaglia D, Plomion C, Kirst M, Sederoff RR (2009) Genomics of growth traits in forest trees. Curr Opin Plant Biol 12: 148–156.
Groover AT (2007) Will genomics guide a greener forest biotech? Trends Plant Sci 12: 234–238.

Gstaiger M, Aebersold R (2009) Applying mass spectrometry-based proteomics to genetics, genomics and network biology. Nat Rev Genet 10: 617–627.

Guazzaroni ME, Beloqui A, Golyshin PN, Ferrer M (2009) Metagenomics as a new technological tool to gain scientific knowledge. Wor J Microb Biotechnol 25: 945–954.

Gugerli F, Sperisen C, Buchler U, Brunner L, Brodbeck S, Palmer JD, Qiu YL (2001) The evolutionary split of Pinaceae from other conifers: Evidence from an intron loss and a multigene phylogeny. Mol Phylogenet Evol 21: 167–175.

Guo B, Wang Y, Sun X, Tang K (2008) Bioactive natural products from endophytes: A review. Appl Biochem Microb 44: 136–142.

Gupta PK (2008) Single-molecule DNA sequencing technologies for future genomics research. Trends Biotechnol 26: 602–611.

Hamann A, Aitken SN, Yanchuk AD (2004) Cataloguing *in situ* protection of genetic resources for major commercial forest trees in British Columbia. For Ecol Manag 197: 295–305.

Harding SA, Jiang HY, Jeong ML, Casado FL, Lin HW, Tsai C-J (2005) Functional genomics analysis of foliar condensed tannin and phenolic glycoside regulation in natural cottonwood hybrids. Tree Physiol 25: 1475–1486.

Harrigan GG, Martino-Catt S, Glenn KC (2007) Metabolomics, metabolic diversity and genetic variation in crops. Metabolomics 3: 259–27.

Heinimo J, Junginger M (2009) Production and trading of biomass for energy—An overview of the global status. Biomass Bioenergy 33: 1310–1320.

Henderson AR, Walter C (2006) Genetic engineering in conifer plantation forestry. Silvae Genet 55: 253–262.

Hesemann CU (1980) Cytophotometrical measurement of nuclear-DNA content in some coniferous and deciduous trees. Theor Appl Genet 57: 187–191.

Hilton J, Bateman RM (2006) Pteridosperms are the backbone of seed-plant phylogeny. J Torr Bot Soc 133: 119–168.

Huyghe A, Francois P, Charbonnier Y, Tangomo-Bento M, Bonetti EJ, Paster BJ, Bolivar I, Baratti-Mayer D, Pittet D, Schrenzel J (2008) Novel microarray design strategy to study complex bacterial communities. Appl Environ Microb 74: 1876–1885.

Igic B, Lande R, Kohn JR (2008) Loss of self-incompatibility and its evolutionary consequences. Intl J Plant Sci 169: 93–104.

Ilic N, Poulev A, Borisjuk N, Brinker A, Moreno DA, Ripoll C, Yakoby N, O'Neal JM, Cornwell T, Pastor I, Fridlender B (2002) Plants and human health in the twenty-first century. Trends Biotechnol 20: 522–531.

Ingvarsson PK (2008) Multilocus patterns of nucleotide polymorphism and the demographic history of *Populus tremula*. Genetics 180: 329–340.

International Human Genome Sequencing Consortium (2001) Initial sequencing of and analysis of the human genome. Nature 409: 860–921.

Izumi H, Anderson IC, Killham K, Moore ERB (2008) Diversity of predominant endophytic bacteria in European deciduous and coniferous trees. Can J Microbiol 54: 173–179.

Jegannathan KR, Chan ES, Ravindra P (2009) Harnessing biofuels: A global Renaissance in energy production? Renewab Sustainab Energy Rev 13: 2163–2168.

Jiao YL, Tausta SL, Gandotra N, Sun N, Liu T, Clay NK, Ceserani T, Chen MQ, Ma LG, Holford M, Zhang HY, Zhao HY, Deng XW, Nelson T (2009) A transcriptome atlas of rice cell types uncovers cellular, functional and developmental hierarchies. Nat Genet 41: 258–263.

Jobson RW, Qiu YL (2008) Did RNA editing in plant organellar genomes originate under natural selection or through genetic drift? Biol Dir 3: Art 43.

Johnston JM, Crossley DA (2002) Forest ecosystem recovery in the southeast US: soil ecology as an essential component of ecosystem management. For Ecol Manag 155: 187–203.

Jones N, Ougham H, Thomas H, Pasakinskiene I (2009) Markers and mapping revisited: finding your gene. New Phytol 183: 935–966.

Joos T, Bachmann J (2009) Protein microarrays: potentials and limitations. Front Biosci 14: 4376–4385.

Kato A, Vega JM, Han FP, Lamb JC, Birchler JA (2005) Advances in plant chromosome identification and cytogenetic techniques. Curr Opin Plant Biol 8: 148–154.

Kelkar VM, Geils BW, Becker DR, Overby ST, Neary DG (2006) How to recover more value from small pine trees: Essential oils and resins. Biomass Bioenergy 30: 316–320.

Kiambi DK, Fortin M, Stromvick M (2008) Linking transcript profiles to metabolites and metabolic pathways: A systems biology approach to transgene risk assessment. Plant Omics 1: 26–36.

Kliebenstein D (2009a) Quantitative genomics: Analyzing intraspecific variation using global gene expression polymorphisms or eQTLs. Annu Rev Plant Biol 60: 93–114.

Kliebenstein DJ (2009b) Advancing genetic theory and application by metabolic quantitative trait loci analysis. Plant Cell 21: 1637–1646.

Kozak KH, Graham CH, Wiens JJ (2008) Integrating GIS-based environmental data into evolutionary biology. Trends Ecol Evol 23: 141–148.

Kumar S, Fladung M (2003) Forest tree transgenesis and functional genomics: From fast forward to reverse genetics. Silvae Genet 52: 229–232.

Kwon K, Grose C, Pieper R, Pandya GA, Fleischmann RD, Peterson SN (2009) High quality protein microarray using *in situ* protein purification. BMC Biotechnol 9: Art 72.

Ledig FT (1988) The conservation of diversity in forest trees—Why and how should genes be conserved. Bioscience 38: 471–479.

Ledig FT, Kitzmiller JH (1992) Genetic strategies for reforestation in the face of global climate change. For Ecol Manag 50: 153–169.

Lee ES, McDonald DW, Anderson N, Tarczy-Hornoch P (2009a) Incorporating collaboratory concepts into informatics in support of translational interdisciplinary biomedical research. Intl J Med Inform 78: 10–21.

Lee JH, Lee BK, Kim JH, Lee SH, Hong SK (2009b) Comparison of chemical compositions and antimicrobial activities of essential oils from three conifer trees; *Pinus densiflora, Cryptomeria japonica*, and *Chamaecyparis obtusa*. J Microb Biotechnol 19: 391–396.

Lee JW, Lee KW, Lee SW, Kim IH, Rhee C (2004) Selective increase in pinolenic acid (all-cis-5,9,12–18: 3) in Korean pine nut oil by crystallization and its effect on LDL-receptor activity. Lipids 39: 383–387.

Li J, Makrigiorgos GM (2009) COLD-PCR: a new platform for highly improved mutation detection in cancer and genetic testing. Biochem Soc Trans 37: 427–432.

Li W, Dai RJ, Yu YH, Li L, Wu CM, Luan WW, Meng WW, Zhang XS, Deng YL (2007) Antihyperglycemic effect of *Cephalotaxus sinensis* leaves and GLUT-4 translocation facilitating activity of its flavonoid constituents. Biol Pharm Bull 30: 1123–1129.

Liang CZ, Mao L, Ware D, Stein L (2009) Evidence-based gene predictions in plant genomes. Genome Res 19: 1912–1923.

Libby WJ, Stettler RF, Seitz FW (1969) Forest genetics and forest-tree breeding. Annu Rev Genet 3: 469–494.

Liston A, Parker-Defeniks M, Syring JV, Willyard A, Cronn R (2007) Interspecific phylogenetic analysis enhances intraspecific phylogeographical inference: a case study in *Pinus lambertiana*. Mol Ecol 16: 3926–3937.

Liu Q, Crammer K, Pereira FCN, Roos DS (2008) Reranking candidate gene models with cross-species comparison for improved gene prediction. BMC Bioinformat 9: Art 433.

Lorenz WW, Dean JFD (2002) SAGE Profiling and demonstration of differential gene expression along the axial developmental gradient of lignifying xylem in loblolly pine (*Pinus taeda*). Tree Physiol 22: 301–310.

Lorenz WW, Sun F, Liang C, Kolychev D, Wang HM, Zhao X, Cordonnier-Pratt M-M, Pratt LH, Dean JFD (2006) Water stress-responsive genes in loblolly pine (*Pinus taeda*) roots identified by analyses of expressed sequence tag libraries. Tree Physiol 26: 1–16.

Luikart G, England PR, Tallmon D, Jordan S, Taberlet P (2003) The power and promise of population genomics: from genotyping to genome typing. Nat Rev Genet 4:981–994.

Mardis ER (2008) Next-generation DNA sequencing methods. Annu Rev Genom Hum Genet 9: 387–402.

Martin F, Nehls U (2009) Harnessing ectomycorrhizal genomics for ecological insights. Curr Opin Plant Biol 12: 508–515.

Martin F, Aerts A, Ahren D, Danchin EGJ, Duchaussoy F, Gibon J, Kohler A, Lindquist E, Pereda V, Salamov A, Shapiro HJ, Wuyts J, Blaudez D, Buee M, Brokstein P, Canback B, Cohen D, Courty PE, Coutinho PM, Delaruelle C, Detter JC, Deveau A, DiFazio S, Duplessis S, Fraissinet-Tachet L, Lucic E, Frey-Klett P, Fourrey C, Feussner L, Gay G, Grimwood J, Hoegger PJ, Jain P, Kilaru S, Labbe J, Lin YC, Legue V, Le Tacon F, Marmeisse R, Melayah D, Montanini B, Muratet M, Nehls U, Niculita-Hirzel H, Oudot-Le Secq MP, Peter M, Quesneville H, Rajashekar B, Reich M, Rouhier N, Schmutz J, Yin T, Chalot M, Henrissat B, Kues U, Lucas S, Van de Peer Y, Podila GK, Polle A, Pukkila PJ, Richardson PM, Rouze P, Sanders IR, Stajich JE, Tunlid A, Tuskan G, Grigoriev IV (2008) The genome of *Laccaria bicolor* provides insights into mycorrhizal symbiosis. Nature 452: 88-U7.

McKay JK, Latta RG (2002) Adaptive population divergence: markers, QTL, and traits. Trends Ecol Evol 17: 285–291.

Metlen KL,Aschehoug ET, Callaway RM (2009) Plant behavioural ecology: dynamic plasticity in secondary metabolites. Plant Cell Environ 32: 641–653.

Meuwissen TH, Hayes BJ, Goddard ME (2001) Prediction of total genetic value using genome-wide dense marker maps. Genetics 157: 1819–1829.

Milla R, Forero DM, Escudero A, Iriondo JM (2009) Growing with siblings: a common ground for cooperation or for fiercer competition among plants? Proc Roy Soc B 276: 2531–2540.

Moco S, Schneider B, Vervoort J (2009) Plant Micrometabolomics: The analysis of endogenous metabolites present in a plant cell or tissue. J Proteome Res 8: 1694–1703.

Mohanty IC, Mahapatra D, Mohanty S, Das AB (2004) Karyotype analyses and studies on the nuclear DNA content in 30 genotypes of potato (*Solanum tuberosum*) L. Cell Biol Int 28: 625–633.

Monteuuis O, Doulbeau S, Verdeil JL (2008) DNA methylation in different origin clonal offspring from a mature *Sequoiadendron giganteum* genotype. Trees Struc Funct 22: 779–784.

Murray BG (1998) Nuclear DNA amounts in gymnosperms. Ann Bot 82: 3–15.

Nakane S, Tanaka T, Satouchi K, Kobayashi Y, Waku K, Sugiura T (2000) Occurrence of a novel cannabimimetic molecule 2-sciadonoylglycerol (2-eicosa-5 ′,11 ′,14 ′-trienoylglycerol) in the umbrella pine *Sciadopitys verticillata* seeds. Biol Pharm Bull 23: 758–761.

Namroud MC, Beaulieu J, Juge N, Laroche J, Bousquet J (2008) Scanning the genome for gene SNPs involved in adaptive population differentiation in white spruce. Mol Ecol 17: 3599–3613.

National Research Council (1988) Mapping and Sequencing the Human Genome. National Academies Press, Washington, DC, USA.

Neale DB, Savolainen O (2004) Association genetics of complex traits in conifers. Trends Plant Sci 9: 325–330.

Nehra NS, Becwar MR, Rottmann WH, Pearson L, Chowdhury K, Chang SJ, Wilde HD, Kodrzycki RJ, Zhang CS, Gause KC, Parks DW, Hinchee MA (2005) Forest biotechnology: Innovative methods, emerging opportunities. In Vitro Cell Dev Biol Plant 41: 701–717.

Nelson T, Gandotra N, Tausta SL (2008) Plant cell types: reporting and sampling with new technologies. Curr Opin Plant Biol 11: 567–573.

Novaes E, Osorio L, Drost DR, Miles BL, Boaventura-Novaes CRD, Benedict C, Dervinis C, Yu Q, Sykes R, Davis M, Martin TA, Peter GF, Kirst M (2009) Quantitative genetic analysis of biomass and wood chemistry of *Populus* under different nitrogen levels. New Phytol 182: 878–890.

Nygaard V, Hovig E (2009) Methods for quantitation of gene expression. Front Biosci 14: 552–569.

O'Connell LM, Ritland K (2005) Post-pollination mechanisms promoting outcrossing in a self-fertile conifer, *Thuja plicata* (Cupressaceae). Can J Bot 83: 335–342.

Ohman M (2007) A-to-I editing challenger or ally to the microRNA process. Biochimie 89: 1171–1176.

Oshlack A, Wakefield MJ (2009) Transcript length bias in RNA-seq data confounds systems biology. Biol Direct 4: Art 14.
Otto A, Wilde V (2001) Sesqui-, Di-, and triterpenoids as chemosystematic markers in extant conifers—A review. Bot Rev 67: 141–238.
Ozsolak F, Platt AR, Jones DR, Reifenberger JG, Sass LE, McInerney P, Thompson JF, Bowers J, Jarosz M, Milos PM (2009) Direct RNA sequencing. Nature 461: 81–U73.
Pait JA (2005) Production and deployment of conifer varietal germplasm. In: R Kellison, S McCord, KMA Gartland KMA (eds) Forest Biotechnology in Latin America. Proc Workshop Biotecnologia Forestal, 2–5 March 2004, Concepcion, Chile. Institute of Forest Biotechnology, Raleigh, NC, USA, pp 41–48.
Palmer JD, Soltis DE, Chase MW (2004) The plant tree of life: An overview and some points of view. Am J Bot 91: 1437–1445.
Parks DH, Porter M, Churcher S, Wang SW, Blouin C, Whalley J, Brooks S, Beiko RG (2009) GenGIS: A geospatial information system for genomic data. Genome Res 19: 1896–1904.
Park PJ (2009) ChIP-seq: advantages and challenges of a maturing technology. Nature Rev Genet 10: 669–680.
Pautasso M (2009) Geographical genetics and the conservation of forest trees. Persp Plant Ecol Evol Syst 11: 157–189.
Pearl IA (1965) Silvichemicals products of the forest. J For 63: 163–167.
Pelgas B, Beauseigle S, Achere V, Jeandroz S, Bousquet J, Isabel N (2006) Comparative genome mapping among *Picea glauca*, *P. mariana* x *P. rubens* and *P. abies*, and correspondence with other Pinaceae. Theor Appl Genet 113: 1371–1393.
Penders B, Horstman K, Vos R (2008) Walking the line between lab and computation: The "Moist" zone. Bioscience 58: 747–755.
Peng D, Wang XQ (2008) Reticulate evolution in *Thuja* inferred from multiple gene sequences: Implications for the study of biogeographical disjunction between eastern Asia and North America. Mol. Phylogenet Evol 47: 1190–1202.
Perry P (1968) Naval-stores industry in the Old South, 1790–1860. J South Hist 34: 509–526.
Peter GF (2008) Southern pines: a resource for bioenergy. In: W Vermerris (ed) Genetic Improvement of Bioenergy Crops. Springer, Berlin, Germany, pp 397–419.
Petrosino JF, Highlander S, Luna RA, Gibbs RA, Versalovic J (2009) Metagenomic pyrosequencing and microbial identification. Clin Chem 55: 856–866.
Phillips K (2008) Human Microbiome Project launched by NIH. Lancet Inf Dis 8: 93–93.
Pienkos PT, Zhang M (2009) Role of pretreatment and conditioning processes on toxicity of lignocellulosic biomass hydrolysates. Cellulose 16: 743–762.
Pietarinen SP, Willfor SM, Ahotupa MO, Hemming JE, Holmbom BR (2006) Knotwood and bark extracts: strong antioxidants from waste materials. J Wood Sci 52: 436–444.
Pop M (2009) Genome assembly reborn: recent computational challenges. Brief Bioinformat 10: 354–366.
Popescu SC, Popescu GV, Bachan S, Zhang ZM, Gerstein M, Snyder M, Dinesh-Kumar SP (2009) MAPK target networks in *Arabidopsis thaliana* revealed using functional protein microarrays. Genes Dev. 23: 80–92.
Powell RG (2009) Plant seeds as sources of potential industrial chemicals, pharmaceuticals, and pest control agents. J Nat Prod 72: 516–523.
Proal AD, Albert PJ, Marshall T (2009) Autoimmune disease in the era of the metagenome. Autoimmun Rev 8: 677–681.
Procaccini G, Olsen JL, Reusch TBH (2007) Contribution of genetics and genomics to seagrass biology and conservation. J Exp Mar Biol Ecol 350: 234–259.
Ragoussis J (2009) Genotyping technologies for genetic research. Annu Rev Genom Human Genet. 10: 117–133.
Ralph SG, Yueh H, Friedmann M, Aeschliman D, Zeznik JA, Nelson CC, Butterfield YSN, Kirkpatrick R, Liu J, Jones SJM, Marra MA, Douglas CJ, Ritland K, Bohlmann J (2006) Conifer defence against insects: microarray gene expression profiling of Sitka spruce

(*Picea sitchensis*) induced by mechanical wounding or feeding by spruce budworms (*Choristoneura occidentalis*) or white pine weevils (*Pissodes strobi*) reveals large-scale changes of the host transcriptome. Plant Cell Environ 29: 1545–1570.

Rasmussen SE, Frederiksen H, Krogholm KS, Poulsen L (2005) Dietary proanthocyanidins: Occurrence, dietary intake, bioavailability, and protection against cardiovascular disease. Mol Nutr Food Res 49: 159–174.

Reed DH, Frankham R (2001) How closely correlated are molecular and quantitative measures of genetic variation? A meta-analysis. Evolution 55: 1095–1103.

Robinson AR, Ukrainetz NK, Kang KY, Mansfield SD (2007) Metabolite profiling of Douglas-fir (*Pseudotsuga menziesii*) field trials reveals strong environmental and weak genetic variation. New Phytol 174: 762–773.

Rokas A, Holland PWH (2000) Rare genomic changes as a tool for phylogenetics. Trends Ecol Evol 15: 454–459.

Rosenblueth M, Martinez-Romero E (2006) Bacterial endophytes and their interactions with hosts. Mol Plant-Microbe Interact 19: 827–837.

Rounsley S, Marri PR, Yu Y, He R, Sisneros N, Goicoechea JL, Lee SJ, Angelova A, Kudrna D, Luo M, Affourtit J, Desany B, Knight J, Niazi F, Egholm M, Wing RA (2009) *De novo* next generation sequencing of plant genomes. Rice 2: 35–43.

Ruel K, Berrio-Sierra J, Derikvand MM, Pollet B, Thevenin J, Lapierre C, Jouanin L, Joseleau JP (2009) Impact of CCR1 silencing on the assembly of lignified secondary walls in *Arabidopsis thaliana*. New Phytol 184: 99–113.

Rydin C, Kallersjo M (2002) Taxon sampling and seed plant phylogeny. Cladistics 18: 485–513.

Saarinen NM, Warri A, Makela SI, Eckerman C, Reunanen M, Ahotupa M, Salmi SM, Franke AA, Kangas L, Santti R (2000) Hydroxymatairesinol, a novel enterolactone precursor with antitumor properties from coniferous tree (*Picea abies*). Nutr. Cancer Intl J. 36: 207–216.

Saito K, Hirai MY, Yonekura-Sakakibara K (2008) Decoding genes with coexpression networks and metabolomics—'majority report by precogs'. Trends Plant Sci 13: 36–43.

Saklani A, Kutty SK (2008) Plant-derived compounds in clinical trials. Drug Disc Today 13: 161–171.

Samuelsson R, Thyrel M, Sjostrom M, Lestander TA (2009) Effect of biomaterial characteristics on pelletizing properties and biofuel pellet quality. Fuel Proc Technol 90: 1129–1134.

Santos-Buelga C, Scalbert A (2000) Proanthocyanidins and tannin-like compounds—nature, occurrence, dietary intake and effects on nutrition and health. J Sci Food Agri. 80: 1094–1117.

Schluter H, Apweiler R, Holzhutter HG, Jungblut PR (2009) Finding one's way in proteomics: a protein species nomenclature. Chem Cent J 3: Art 11.

Schulz B (1999) Short history and present trends of Fischer–Tropsch synthesis. Appl Catal A 186: 3–12

Sebastiani P, Timofeev N, Dworkis DA, Perls TT, Steinberg MH (2009) Genome-wide association studies and the genetic dissection of complex traits. Am J Hematol 84: 504–515.

Seng KC, Seng CK (2008) The success of the genome-wide association approach: a brief story of a long struggle. Eur J Hum Genet 16: 554–564.

Shepherd M, Williams CG (2008) Comparative mapping among subsection Australes (genus *Pinus*, family Pinaceae). Genome 51: 320–331.

Smedley D, Swertz MA, Wolstencroft K, Proctor G, Zouberakis M, Bard J, Hancock JM, Schofield P (2008) Solutions for data integration in functional genomics: a critical assessment and case study. Brief Bioinformat 9: 532–544.

Sniezko RA, Zobel BJ (1988) Seedling height and diameter variation of various degrees of inbred and outcross progenies of loblolly pine. Silvae Genet 37: 50–60.

Soller M (2006) Pre-messenger RNA processing and its regulation: a genomic perspective. Cell Mol Life Sci 63: 796–819.

Sorensen J, Nicolaisen MH, Ron E, Simonet P (2009) Molecular tools in rhizosphere microbiology-from single-cell to whole-community analysis. Plant Soil 321: 483–512.

Soresson C (2006) Varietal pines boom in the US South. N Z J For 51: 34–40.
Stace CA (2005) Plant taxonomy and biosystematics—does DNA provide all the answers? Taxon 54: 999–1007.
Storz JF (2005) Using genome scans of DNA polymorphism to infer adaptive population divergence. Mol Ecol 14: 671–688.
Strauss SH, Lande R, Namkoong G (1992) Limitations of molecular marker-aided selection in forest tree breeding. Can J For Res 22: 1050–1061.
Strauss SH, Tan HM, Boerjan W, Sedjo R (2009) Strangled at birth? Forest biotech and the Convention on Biological Diversity. Nat Biotechnol 27: 519–527.
Sutton B (2002) Commercial delivery of genetic improvement to conifer plantations using somatic embryogenesis. Ann For Sci 59: 657–661.
Swidzinski JA, Sweetlove LJ, Leaver CJ (2002) A custom microarray analysis of gene expression during programmed cell death in *Arabidopsis thaliana*. Plant J 30: 431–446.
Tai HH, Williams M, Iyengar A, Yeates J, Beardmore T (2007) Regulation of the beta-hydroxyacyl ACP dehydratase gene of *Picea mariana* by alternative splicing. Plant Cell Rep 26: 105–113.
Tan KC, Ipcho SVS, Trengove RD, Oliver RP, Solomon PS (2009) Assessing the impact of transcriptomics, proteomics and metabolomics on fungal phytopathology. Mol Plant Pathol.10: 703–715.
Tang FC, Barbacioru C, Wang YZ, Nordman E, Lee C, Xu NL, Wang XH, Bodeau J, Tuch BB, Siddiqui A, Lao KQ, Surani MA (2009) mRNA-Seq whole-transcriptome analysis of a single cell. Nat Meth 6: 377–U86.
Toronto International Data Release Workshop Authors (2009) Prepublication data sharing. Nature 461: 168–170.
Tsutsui K, Suwa A, Sawada K, Kato T, Ohsawa TA, Watano Y (2009) Incongruence among mitochondrial, chloroplast and nuclear gene trees in *Pinus* subgenus *Strobus* (Pinaceae). J Plant Res 122: 509–521.
Turley DB, Chaudhry Q, Watkins RW, Clark JH, Deswarte FEI (2006) Chemical products from temperate forest tree species—Developing strategies for exploitation. Indust Crop Prod 24: 238–243.
Turner EH, Ng SB, Nickerson DA, Shendure J (2009) Methods for genomic partitioning. Annu Rev Genom Hum Genet 10: 263–284.
Ungerer MC, Johnson LC, Herman MA (2008) Ecological genomics: understanding gene and genome function in the natural environment. Heredity 100: 178–183.
Valimaa AL, Honkalampi-Hamalainen U, Pietarinen S, Willfor S, Holmbom B, von Wright A (2007) Antimicrobial and cytotoxic knotwood extracts and related pure compounds and their effects on food-associated microorganisms. Int J Food Microbiol 115: 235–243.
Varshney RK, Nayak SN, May GD, Jackson SA (2009) Next-generation sequencing technologies and their implications for crop genetics and breeding. Trends Biotechnol 27: 522–530.
Vaughan DA, Balazs E, Heslop-Harrison JS (2007) From crop domestication to super-domestication. Ann Bot 100: 893–901.
Venter JC, Adams MD, Myers EW, Li PW, Mural RJ, Sutton GG, Smith HO, Yandell M, Evans CA, Holt RA, Gocayne JD, Amanatides P, Ballew RM, Huson DH, Wortman JR, Zhang Q, Kodira CD, Zheng XH, Chen L, Skupski M, Subramanian G, Thomas PD, Zhang J, Gabor Miklos GL, Nelson C, Broder S, Clark AG, Nadeau J, McKusick VA, Zinder N, Levine AJ, Roberts RJ, Simon M, Slayman C, Hunkapiller M, Bolanos R, Delcher A, Dew I, Fasulo D, Flanigan M, Florea L, Halpern A, Hannenhalli S, Kravitz S, Levy S, Mobarry C, Reinert K, Remington K, Abu-Threideh J, Beasley E, Biddick K, Bonazzi V, Brandon R, Cargill M, Chandramouliswaran I, Charlab R, Chaturvedi K, Deng Z, Di Francesco V, Dunn P, Eilbeck K, Evangelista C, Gabrielian AE, Gan W, Ge W, Gong F, Gu Z, Guan P, Heiman TJ, Higgins ME, Ji RR, Ke Z, Ketchum KA, Lai Z, Lei Y, Li Z, Li J, Liang Y, Lin X, Lu F, Merkulov GV, Milshina N, Moore HM, Naik AK, Narayan VA, Neelam B, Nusskern D, Rusch DB, Salzberg S, Shao W, Shue B, Sun J, Wang Z, Wang A, Wang X, Wang J, Wei M, Wides R, Xiao C, Yan C, Yao A, Ye J, Zhan M, Zhang W, Zhang H, Zhao Q,

Zheng L, Zhong F, Zhong W, Zhu S, Zhao S, Gilbert D, Baumhueter S, Spier G, Carter C, Cravchik A, Woodage T, Ali F, An H, Awe A, Baldwin D, Baden H, Barnstead M, Barrow I, Beeson K, Busam D, Carver A, Center A, Cheng ML, Curry L, Danaher S, Davenport L, Desilets R, Dietz S, Dodson K, Doup L, Ferriera S, Garg N, Gluecksmann A, Hart B, Haynes J, Haynes C, Heiner C, Hladun S, Hostin D, Houck J, Howland T, Ibegwam C, Johnson J, Kalush F, Kline L, Koduru S, Love A, Mann F, May D, McCawley S, McIntosh T, McMullen I, Moy M, Moy L, Murphy B, Nelson K, Pfannkoch C, Pratts E, Puri V, Qureshi H, Reardon M, Rodriguez R, Rogers YH, Romblad D, Ruhfel B, Scott R, Sitter C, Smallwood M, Stewart E, Strong R, Suh E, Thomas R, Tint NN, Tse S, Vech C, Wang G, Wetter J, Williams S, Williams M, Windsor S, Winn-Deen E, Wolfe K, Zaveri J, Zaveri K, Abril JF, Guigó R, Campbell MJ, Sjolander KV, Karlak B, Kejariwal A, Mi H, Lazareva B, Hatton T, Narechania A, Diemer K, Muruganujan A, Guo N, Sato S, Bafna V, Istrail S, Lippert R, Schwartz R, Walenz B, Yooseph S, Allen D, Basu A, Baxendale J, Blick L, Caminha M, Carnes-Stine J, Caulk P, Chiang YH, Coyne M, Dahlke C, Mays A, Dombroski M, Donnelly M, Ely D, Esparham S, Fosler C, Gire H, Glanowski S, Glasser K, Glodek A, Gorokhov M, Graham K, Gropman B, Harris M, Heil J, Henderson S, Hoover J, Jennings D, Jordan C, Jordan J, Kasha J, Kagan L, Kraft C, Levitsky A, Lewis M, Liu X, Lopez J, Ma D, Majoros W, McDaniel J, Murphy S, Newman M, Nguyen T, Nguyen N, Nodell M, Pan S, Peck J, Peterson M, Rowe W, Sanders R, Scott J, Simpson M, Smith T, Sprague A, Stockwell T, Turner R, Venter E, Wang M, Wen M, Wu D, Wu M, Xia A, Zandieh A, Zhu X (2001) The sequence of the human genome. Science 291: 1304–1351.

Verdel A, Vavasseur A, Le Gorrec M, Touat-Todeschini L (2009) Common themes in siRNA-mediated epigenetic silencing pathways. Int J Dev Biol 53: 245–257.

Walter C (2004) Genetic engineering in conifer forestry: Technical and social considerations. In Vitro Cell Dev Biol Plant 40: 434–441.

Walter C, Charity J, Grace L, Hofig K, Moller R, Wagner A (2002) Gene technologies in *Pinus radiata* and *Picea abies*: tools for conifer biotechnology in the 21st century. Plant Cell Tiss Org Cult 70: 3–12.

Wheelan SJ, Murillo FM, Boeke JD (2008) The incredible shrinking world of DNA microarrays. Mol Biosys 4: 726–732.

White TL, Adams WT, Neale DB (2007) Forest Genetics. CABI Publ, Cambridge, MA, USA.

Williams CG (2006) Landscapes, genomics and transgenic conifers. Springer, Berlin Heidelberg, Germany; New York, USA.

Williams CG (2009) Conifer Reproductive Biology, Springer,Berlin, Germany.

Williams CG, Savolainen O (1996) Inbreeding depression in conifers: Implications for breeding strategy. For Sci 42: 102–117.

Wink M (2003) Evolution of secondary metabolites from an ecological and molecular phylogenetic perspective. Phytochemistry 64: 3–19.

Won H, Renner SS (2003) Horizontal gene transfer from flowering plants to *Gnetum*. Proc Natl Acad Sci USA 100: 10824–10829.

Yang XH, Jawdy S, Tschaplinski TJ, Tuskan GA (2009) Genome-wide identification of lineage-specific genes in *Arabidopsis*, *Oryza* and *Populus*. Genomics 93: 473–480.

Yazaki J, Gregory BD, Ecker JR (2007) Mapping the genome landscape using tiling array technology. Curr Opin Plant Biol 10: 534–542.

Yonekura-Sakakibara K, Saito K (2009) Functional genomics for plant natural product biosynthesis. Nat Prod Rep 26: 1466–1487.

Zheng JB, Moorhead M, Weng L, Siddiqui F, Carlton VEH, Ireland JS, Lee L, Peterson J, Wilkins J, Lin S, Kan ZY, Seshagiri S, Davis RW, Faham M (2009) High-throughput, high-accuracy array-based resequencing. Proc Natl Acad Sci USA 106: 6712–6717.

Index

C

D

H

I

J

K

L

Q

R

S

T

Color Plate Section

Chapter 1

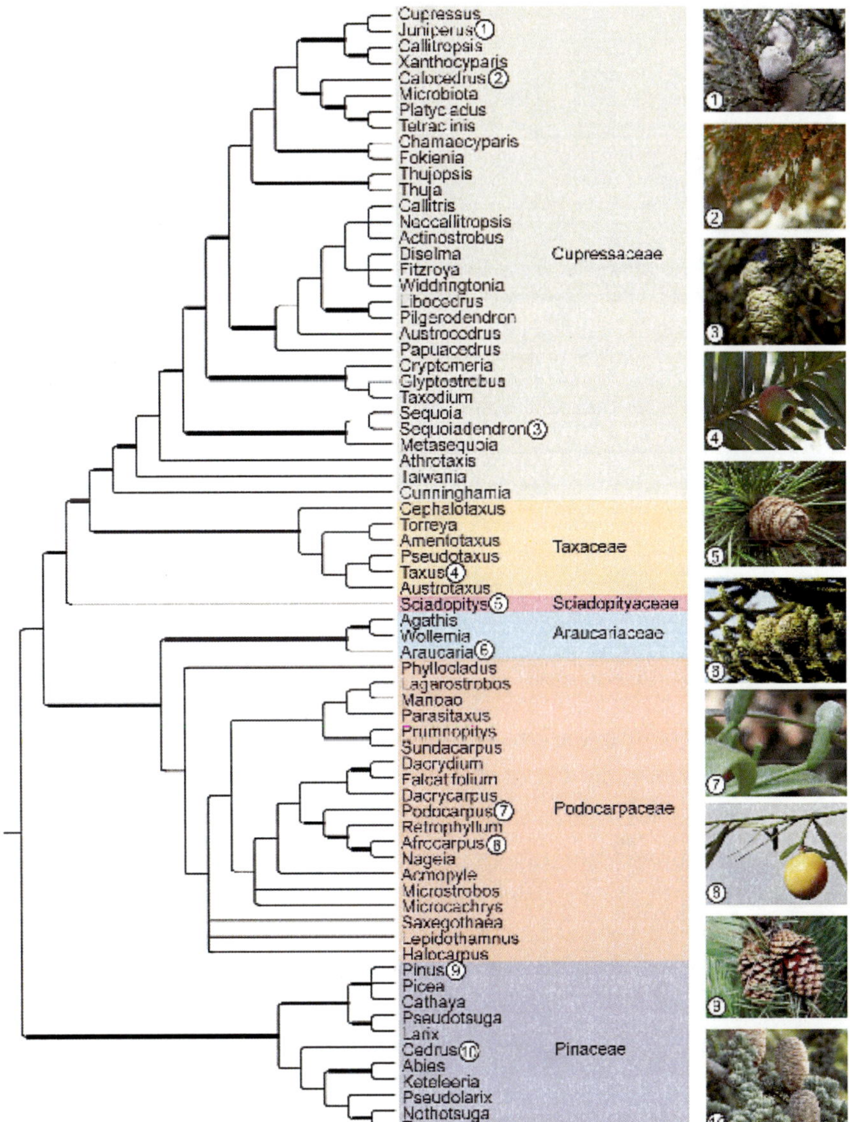

Figure 1-1 Conifer phylogenetic tree. A representation of our current understanding of intergeneric relationships.

Chapter 3

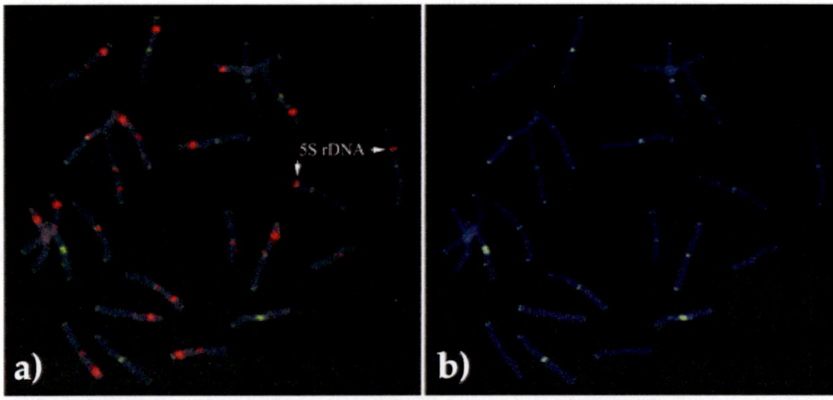

Figure 3-1 Fluorescent *in situ* hybridization images of ribosomal DNAs (18S-28S rDNA, 5S rDNA) and telomere (ATRS) probes on *Pinus echinata* somatic metaphase chromosomes: a) superimposed images of DAPI, Cy3 (red signals, 18S and 5S rDNA sites) and FITC (green signals, ATRS sites) filters; b) super imposed images of DAPI and FITC filters.

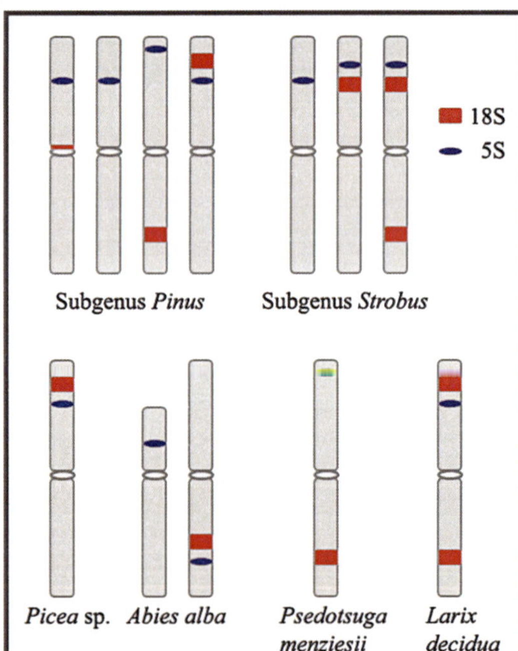

Figure 3-2 Diagrammatic representation of 18S and 5S rDNA loci in different Pinaceae genera. All 18S and 5S rDNA patterns reported in the less extensively studied genera (i.e., *Picea*, *Abies*, *Pseudotsuga* and *Larix)* are present in the more extensively studied genus *Pinus* (subgenera, *Pinus* and *Strobus*).

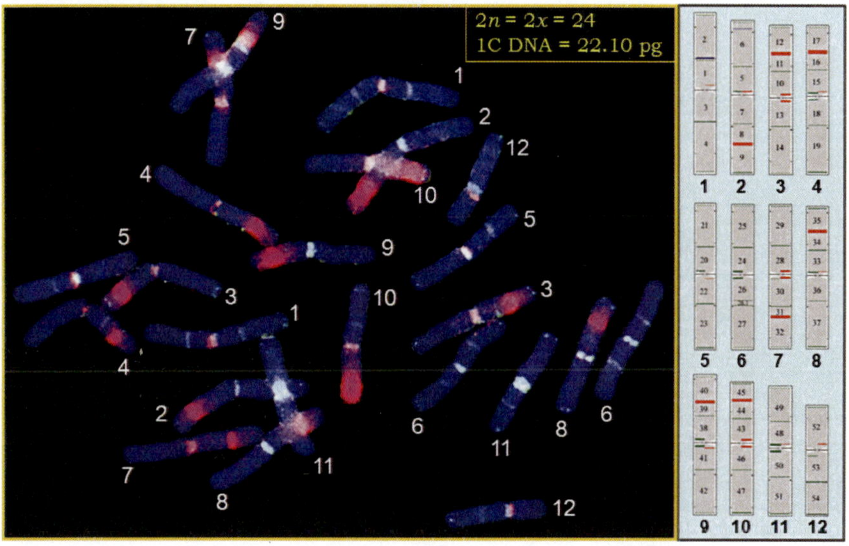

Figure 3-3 Fluorescent *in situ* hybridization image of *Pinus taeda* somatic metaphase chromosomes probed with 18S-28S rDNA (red signals) and *Arabidopsis*-type telomere repeat sequence (green signals). Numbers from 1 to 12 enumerate homologous chromosome pairs. The ideogram of *Pinus taeda* in the right hand side box is based on 108 readings of each measurement (see Islam-Faridi et al. 2007 for details).

Chapter 5

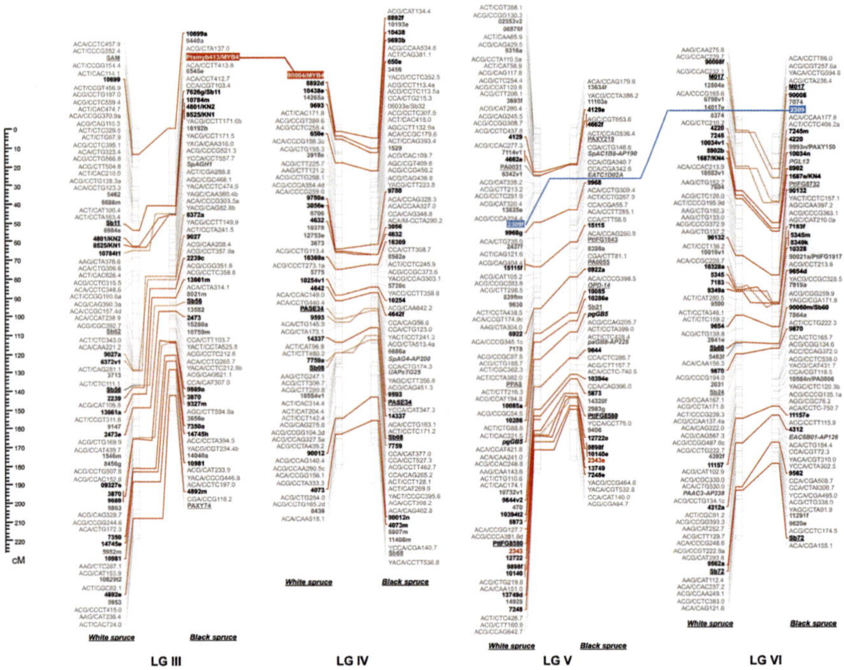

Figure 5-3 Comparison of homologous linkage groups between white spruce (*Picea glauca*) and black spruce (species complex *Picea mariana* × *P. rubens*).

Chapter 6

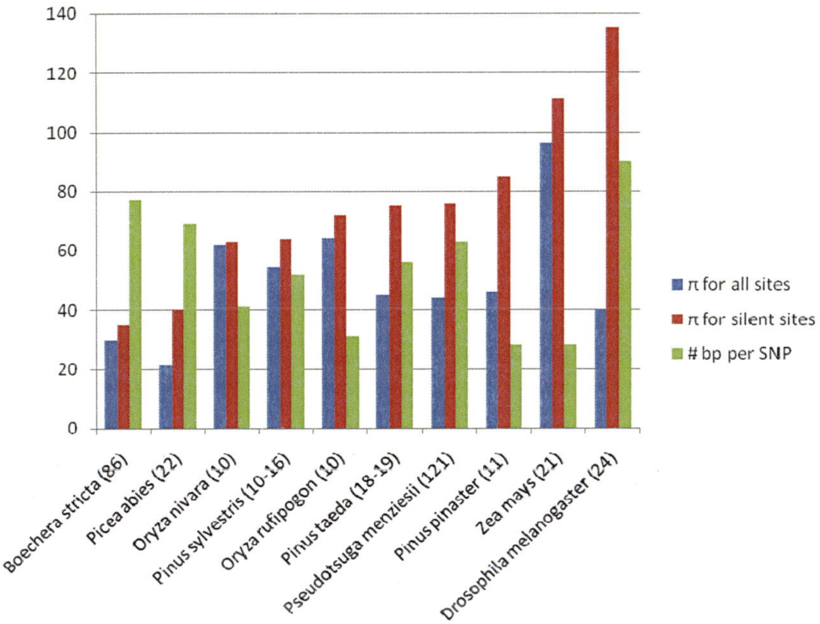

Figure 6-1 Nucleotide diversity estimates for all and silent (synonymous and noncoding) sites, and number of base pairs per SNP for studies where 10 or more loci were studied. The number of loci is in parentheses. Estimates for *Pinus taeda* and *Pinus sylvestris* were averaged for two or more studies. See Table 6.1 for references.

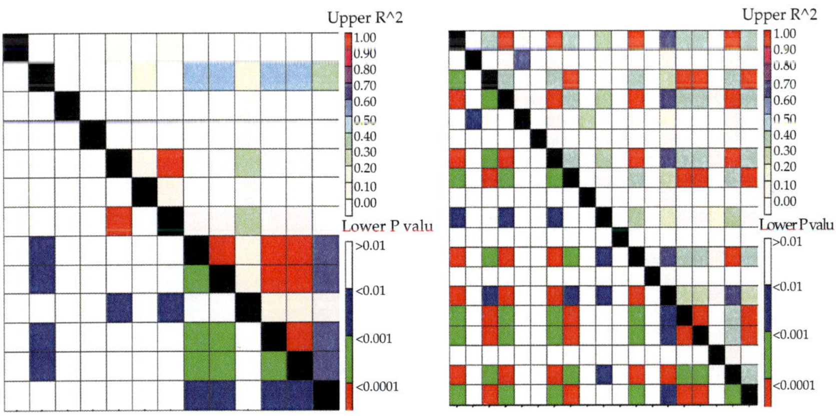

Figure 6-2 LD plots for *dhn1* (left) and *sod-chl* (right) candidate genes for drought in loblolly pine. A LD block is apparent in the lower right part of *dhn1* while LD is distributed more regularly in *sod-chl*.

Chapter 8

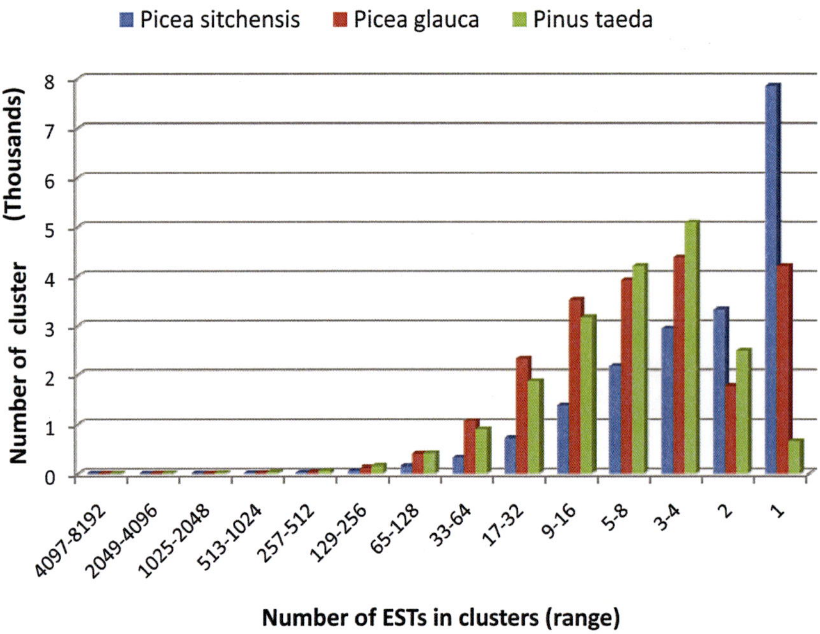

Figure 8-1 Histogram of cluster size's from NCBI Unigene database (builds from Table 2A).

Chapter 10

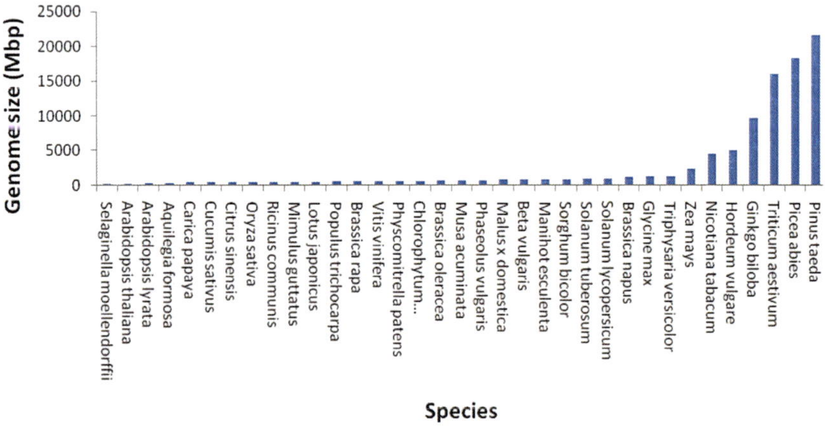

Figure 10-1 Selection of plant genome sizes. Genome size is provided for a subset of species that are object of genomic sequencing. Size information is from NCBI (*http://www.ncbi.nlm.nih.gov/genomes/static/gpstat.html*) with the exception of gymnosperm data from the Kew Plant DNA C-values Database (*http://data.kew.org/cvalues/*).

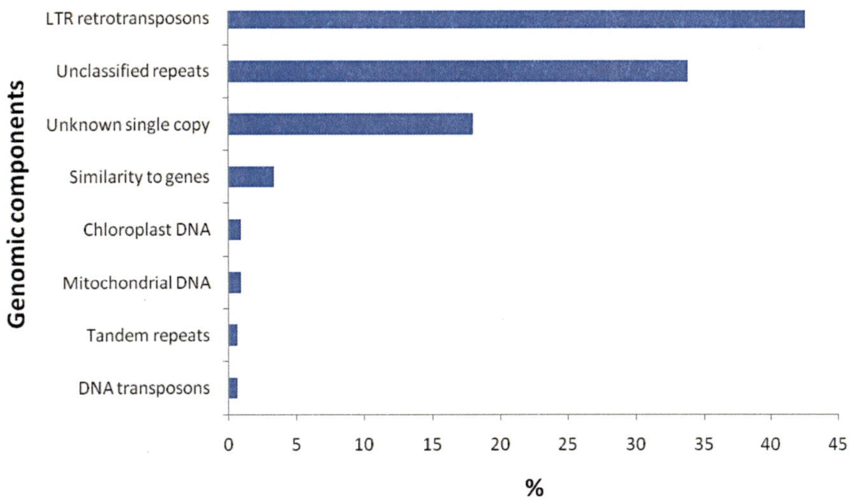

Figure 10-2 Estimate of Norway spruce genome composition. Principal genomic components of Norway spruce genome estimated based on the annotation of a 9 Mbp small insert genomic library (E. De Paoli et al. unpubl. data).

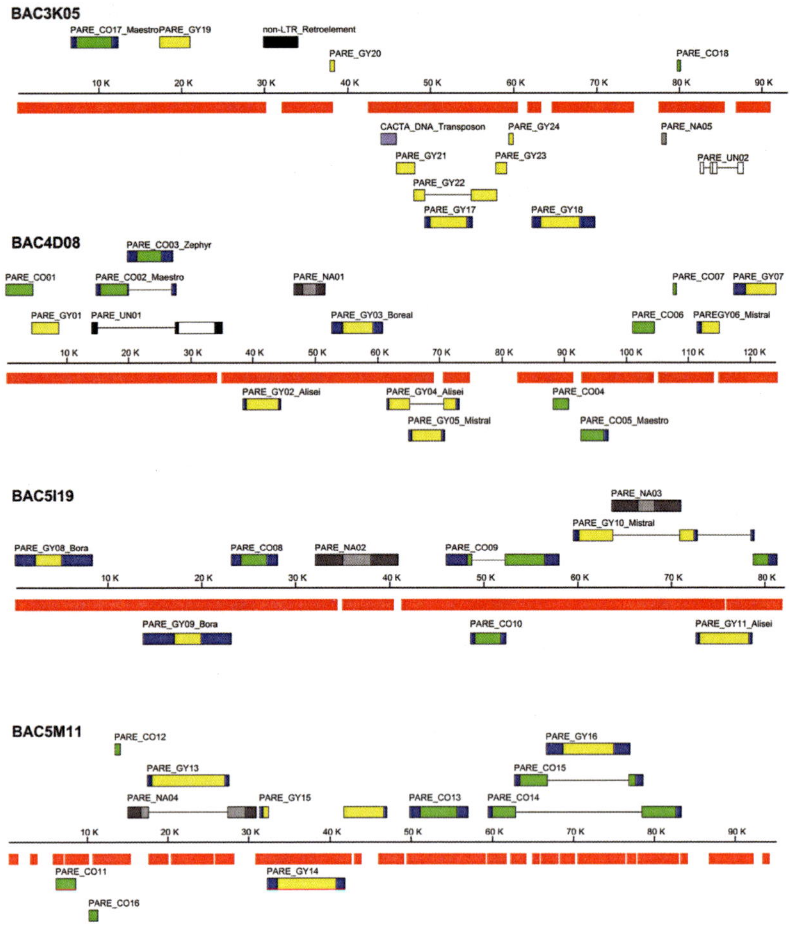

Figure 10.3 Annotation of four genomic regions (BAC) randomly selected from the genome of Norway spruce. Green boxes, LTR-retrotransposons of the *copia* superfamily; yellow boxes, LTR-retrotransposons of the *gypsy* superfamily; light-blue boxes, LTRs of *copia* elements; dark-blue boxes, LTRs of *gypsy* elements; white boxes, uncharacterized LTR-retrotransposons (with black LTRs when annotated), grey boxes, non-autonomous LTR-retrotransposons (with dark-grey LTRs); black box, non-LTR retrotransposon. The red bar indicates repetitive sequences; PARE, *Picea abies* repetitive element; GY, *gypsy*; CO, *copia*; UN, unidentified; NA, non-autonomous. From E. De Paoli et al. (unpubl. data).

DUDEN

WISSEN >> ÜBEN >> TESTEN

Deutsch

Lösungsheft

5.
Klasse

1 Wortarten

ÜBUNG 1 **S. 7**
Peter, Daniel, Eva, die Großmutter, die Klingel, der Nachmittag, die Freude, die Oma, die Liebe, die Kinder, der Kuchen, das Haus, das Kleid, die Enkel, die Geschenke, die Tür

ÜBUNG 2 **S. 7**
die Jungen, die Blumen, das Mädchen, das Paket, die Dame, das Geschenk, das Buch, die Tiere, der Tag, das Lexikon

ÜBUNG 3 **S. 7**
Lebewesen: Peter, Daniel, Eva, die Großmutter, die Oma, die Kinder, die Enkel, die Jungen, die Blumen, das Mädchen, die Dame, die Tiere
Dinge: die Klingel, der Kuchen, das Haus, das Kleid, die Geschenke, die Tür, das Paket, das Geschenk, das Buch, das Lexikon
Begriffe: der Nachmittag, die Freude, die Liebe, der Tag

ÜBUNG 4 **S. 8**
siehe unten

ÜBUNG 5 **S. 8**
Lösungsvorschlag:
auf -e: Fische, Beine, Scherze, Dinge, Ringe, Stifte, Zweige, Schweine, Hunde usw.
auf -(e)n: Zangen, Ohren, Kerzen, Herzen, Katzen, Gabeln, Fabeln, Blumen usw.
auf -er: Kinder, Rinder, Bilder, Felder, Bretter
auf -s: Omas, Autos, Handys, DVDs usw.
mit Umlaut: Büsche, Bücher, Gänse, Räder, Blätter, Tücher, Bäume, Dächer, Körbe, Vögel usw.
wie im Singular: Messer, Löffel, Eimer, Mädchen, Gürtel

ÜBUNG 6 **S. 9**
die Pizzen / Pizzas – die Figuren – die Firmen – die Späße – die Zwiebacke / Zwiebäcke – die Alben – die Babys – die Drosseln – die Messer – die Ängste – die Staaten – die Lexika / Lexiken

ÜBUNG 7 **S. 9**
die Kinder – Neutrum, Plural
die Großmutter – Femininum, Singular
das Geschenk – Neutrum, Singular
den Nachmittag – Maskulinum, Singular
ihren Enkeln – Maskulinum, Plural
die Kinder – Neutrum, Plural
das Tierlexikon – Neutrum, Singular
eine Idee – Femininum, Singular
einen Ausflug – Maskulinum, Singular
den Zoo – Maskulinum, Singular

ÜBUNG 8 **S. 9**
2. die Großmutter – Nominativ
3. über das (schöne) Geschenk – Akkusativ
über den (wunderbaren) Nachmittag – Akkusativ
4. Mit **wem** verbrachte sie den Nachmittag? – mit ihren Enkeln – Dativ
5. **Wen** will sie überraschen? – die Kinder – Akkusativ

ÜBUNG 4	Geschlecht: Genus	Femininum	Maskulinum	Neutrum
Einzahl: Singular	1. Nominativ	die Mutter	der Mann	das Kind
	2. Genitiv	der Mutter	des Mannes	des Kindes
	3. Dativ	der Mutter	dem Mann	dem Kind
	4. Akkusativ	die Mutter	den Mann	das Kind
Mehrzahl: Plural	1. Nominativ	die Mütter	die Männer	die Kinder
	2. Genitiv	der Mütter	der Männer	der Kinder
	3. Dativ	den Müttern	den Männern	den Kindern
	4. Akkusativ	die Mütter	die Männer	die Kinder

6. **Wer oder was** brachte sie auf eine Idee? – das Tierlexikon – Nominativ
7. Auf **wen oder was** brachte sie das Tierlexikon? – auf eine Idee – Akkusativ
8. **Wen oder was** würde sie machen? – einen Ausflug (in den Zoo) – Akkusativ

ÜBUNG 9 S. 10

Er (Pers.) verlässt **sein (Poss.)** Zuhause, weil **seine (Poss.)** Eltern ... hat **er (Pers.)** Probleme: **Seine (Poss.)** Noten ... **seinen (Poss.)** Mitschülern versteht **er (Pers.)** sich ... denkt **er (Pers.)**: **Ich (Pers.)** hau ab! **Wir (Pers.)** lernen ... **Ihr (Pers.)** könnt ... bleibt **er (Pers.)** ... **Er (Pers.)** findet ... erleben **sie (Pers.)** ...

ÜBUNG 10 S. 10

witziger Wörter (Genitiv) – seinen Ohren (Dativ) – den Spitznamen (Akkusativ) – den Hausmeister (Akkusativ) – jede Kleinigkeit (Akkusativ) – seinen Späßen (Dativ) – den Spitznamen (Akkusativ) – Sein Vater (Nominativ) – seiner dummen Scherze (Genitiv) – schlechte Noten (Akkusativ)

ÜBUNG 11 S. 12

	Einzahl: Singular	Mehrzahl: Plural
1. Fall: Nominativ	das kleine Kind	die kleinen Kinder
2. Fall: Genitiv	des kleinen Kindes	der kleinen Kinder
3. Fall: Dativ	dem kleinen Kind	den kleinen Kindern
4. Fall: Akkusativ	das kleine Kind	die kleinen Kinder

ÜBUNG 12 S. 12

Ergänzung in dieser Reihenfolge:
besten – aufregenden – letzten – großen – schlechtes – fürchterlichen – freundliche – erfahrene – spannendes – kleines – köstlicher – erlebnisreichen – wunderschöne

ÜBUNG 13 S. 12

Lösungsvorschlag:
Meine Deutschlehrerin ist eine große Frau mit blonden, gelockten Haaren. Ihre Kleidung ist modisch und sie trägt eine rote Brille auf ihrer markanten Nase. Ihre Augen sind dunkelgrün und ihre Augenbrauen schmal. Sie lächelt meist freundlich und ist zu allen Schülerinnen und Schülern gerecht. Sie gibt nur wenig Hausaufgaben auf, allerdings kontrolliert sie immer sehr gründlich, ob wir sie auch eigenständig und gewissenhaft erledigt haben.

ÜBUNG 14 S. 13

Adjektiv	Bezugswort	Grundform
alter	Mann	alt
engen	Wohnung	eng
schönes	Wetter	schön
treuen	Hund	treu
kleine	Stadt	klein
nötigsten	Dinge	nötig
meisten	Spaß	viel
alten	Hund	alt
schönen	Tages	schön
enge	Gasse	eng
kleiner	Junge	klein
prächtigen	Knochen	prächtig

ÜBUNG 15 S. 13

Positiv	Komparativ	Superlativ
trüb	trüber	am trübsten
früh	früher	am frühesten
fein	feiner	am feinsten

ÜBUNG 16 S. 13

Adjektive: super, prima, gut, einzigartig
steigerbar: gut – Paul kann seine Leistung noch steigern.

ÜBUNG 17 **S. 14**
Ergänzung in dieser Reihenfolge:
bestes (*von:* gut) Kleid – unverheiratete Tante – fernen Afrika – leckeren Kuchen – verwilderten Vorgarten – vielen Jahren – benachteiligten Kindern – langes Gesicht – große Schwester – mehr (*von:* viel) Zeit – fantastische Pläne – neue Schule – wenigen Monaten – anstrengenden, aber sicher erfolgreichen Schulzeit

ÜBUNG 18 **S. 17**
Lösungsvorschlag:
nachgeben – hintergehen – überschreiten – mitkommen – niederschlagen – beistehen – entgegenlaufen – vorlesen

ÜBUNG 19 **S. 17**
Singular: ich lauf**e**, du läuf**st**, er / sie / es läuf**t**
Plural: wir lauf**en**, ihr lauf**t**, sie lauf**en**

ÜBUNG 20 **S. 17**
1. = 2. Person Singular
2. = 2. Person Plural
3. = 1. Person Singular
4. = 3. Person Plural
5. = 3. Person Singular
6. = 1. Person Plural

ÜBUNG 21 **S. 18**
ich liebe – wir reden – ihr tanzt – du fragst – sie schreiben

ÜBUNG 22 **S. 18**
Perfekt mit *sein*: ich bin gelaufen, ich bin gegangen, ich bin gereist, ich bin geflogen
Perfekt mit *haben*: ich habe gehabt, ich habe gedacht, ich habe gegessen, ich habe gehalten

ÜBUNG 23 **S. 18**
Blau zu unterstreichen sind die Sätze:
wie es gewöhnlich ist – was er normalerweise nur an Schultagen tut – Üblicherweise stellt meine Mutter die Milch in die Kühlschranktür – Bis jetzt: Meine Mutter betritt freudestrahlend das Zimmer, öffnet das Fenster und säuselt mir ins Ohr: „Willst du heute gar nicht mehr aufstehen? Das Frühstück ist fertig!" Verwundert hebe ich den Kopf: War alles nur ein schlechter Traum?
Alle anderen Sätze müssen rot unterstrichen sein.

ÜBUNG 24 **S. 19**
1. Z **2.** Z **3.** G **4.** Z **5.** G

ÜBUNG 25 **S. 19**
ging – flehte – helfen – jagt – antwortete – kommt – tun – tobte – fängt – vergräbt

ÜBUNG 26 **S. 19**
fragen: frag(e)!, fragt!
essen: iss!, esst!
lesen: lies!, lest!
geben: gib!, gebt!
lachen: lach(e)!, lacht!
reden: rede!, redet!

ÜBUNG 27 **S. 20**
Lösungsvorschlag:
Frag bitte deine Mutter! – Frag deine Mutter mal! – Frag zuerst deine Mutter! usw.

ÜBUNG 28 **S. 20**
Ich **kam** gerade aus der Schule und **schaute** mich noch mal nach meinem Freund Fabian um. Deshalb **bemerkte** ich das herankommende Auto auch erst gar nicht. Natürlich **blieb** ich an der roten Ampel stehen. Da **merkte** ich plötzlich, dass hinter mir die anderen Schulkinder **anfingen** zu drängeln. Ich **ging** natürlich gleich auf die Seite, aber irgendwie **schaffte** es der Basti nicht mehr, sich zu halten, und **wurde** voll auf die Straße **geschubst**. Tja, und da **kam** das Auto ...

ÜBUNG 29 **S. 20**
In dieser Reihenfolge einzukreisen sind:
einmal – und – nichts – als – und – vom – und – bei – von – in – aber – als – beim – und – dass – Und – als – wenn – zurück – ins – dann – zu

Klassenarbeiten

AUFGABE 1 **S. 21**
Substantive: Fass, Spiel, Kenner, Männer, Schere
Pronomen: ich, mein, wir, ihre
Adjektive: nächtlich, viel, kompliziert, blau
Verben: kennen, unterbieten, raufen, finden

AUFGABE 2 **S. 21**
1. deklinierbar
2. konjugierbar
3. deklinierbar
4. deklinierbar
5. deklinierbar

AUFGABE 3 **S. 21**
siehe unten

AUFGABE 4 **S. 22**
sitzt – lässt – träumt – werde ... planen – bin ... gekommen – waren – geht – erwartet – Fahren – werden sehen – wird ... gefallen – bin ... gereist – schwärme – weiß – hat ... vorgestellt – wird ... werden

AUFGABE 5 **S. 22**
Präsens: er sitzt, er lässt, er träumt, er geht, sie erwartet, ich schwärme, er weiß
Präteritum: sie waren
Perfekt: ich bin gekommen, ich bin gereist, er hat sich vorgestellt
Futur: ich werde planen, Sie werden sehen, es wird gefallen, es wird werden
Imperativ: Fahren Sie

AUFGABE 6 **S. 23**
ich lese – ich las – ich lag – ich habe gelegen – er hat gelegen – sie haben gelegen
der Haut – der Haut – dem Bäcker – den Bäckern – die Bäcker – die Bücher

AUFGABE 7 **S. 23**
1. Der **Komparativ** ist kein Kasus.
2. Der **Genitiv** hat nichts mit Steigerung zu tun.
3. Das **Pronomen** hat nichts mit Konjugation zu tun.
4. Der **Imperativ** hat nichts mit Deklination zu tun.

AUFGABE 8 **S. 24**
dachte er – kann er – hat er – für ihn – überlegte er – war er – er war

AUFGABE 9 **S. 24**
1. Ein Geräusch, ein Schatten oder ein Nichts erschreckten den Hasen.
2. Der Hase hörte ein leises Säuseln.
3. Der Hase sprang auf und lief davon.
4. Der Hase hetzte bis an das Ufer des Teiches.
5. Die Frösche erschraken vor dem Hasen.
6. Die aufgescheuchten Frösche sprangen ins Wasser.

AUFGABE 3

	Personal-pronomen	Präsens (Gegenwart)	Perfekt (1. Vergangenheit)	Präteritum (2. Vergangenheit)	Futur (Zukunft)
1. Pers.	ich	laufe	bin gelaufen	lief	werde laufen
2. Pers.	du	läufst	bist gelaufen	liefst	wirst laufen
3. Pers.	er / sie / es	läuft	ist gelaufen	lief	wird laufen

AUFGABE 10 **S. 24**

Positiv	Komparativ	Superlativ
furchtsam	furchtsamer	am furcht-samsten
unglücklich	unglücklicher	am unglück-lichsten
ungestört	–	–
offen	–	–
ängstlich	ängstlicher	am ängst-lichsten
leise	leiser	am leisesten

AUFGABE 11 **S. 25**

sagte: Verb
mir: Pronomen
Tiere: Substantiv
vor: Präposition
dem: bestimmter Artikel
feige: Adjektiv
findet: Verb
immer: Adverb
ein: unbestimmter Artikel
Feigling: Substantiv

AUFGABE 12 **S. 25**

sagte: 3. Person Singular Präteritum
findet: 3. Person Singular Präsens
Tiere: Akkusativ Plural Neutrum
Feigling: Nominativ Singular Maskulinum

AUFGABE 13 **S. 25**

sägen: schwach, haben
gehen: stark, sein
beißen: stark, haben
lachen: schwach, haben
singen: stark, haben
gähnen: schwach, haben
hüpfen: schwach, sein
spielen: schwach, haben

AUFGABE 14 **S. 25**

1. Der **Genitiv** ist ein Kasus und passt nicht, denn die anderen Begriffe beschreiben ein Verb genauer.
2. Der **Superlativ** ist die zweite Steigerungsform des Adjektivs und passt nicht, denn die anderen Begriffe beschreiben ein Substantiv genauer.
3. Der **Komparativ** ist die erste Steigerungsform des Adjektivs und passt nicht, denn die anderen Begriffe beschreiben ein Verb genauer.
4. Die **Präposition** passt nicht, denn sie ist nicht deklinierbar wie die anderen drei Wortarten.
5. Das **Verb** passt nicht, denn es ist konjugierbar im Gegensatz zu den anderen drei Wortarten.

AUFGABE 15 **S. 26**

ich bin geschwommen: 1. Pers. Singular Perfekt, schwimmen
sie lesen: 3. Pers. Plural Präsens, lesen
sie ließ: 3. Pers. Singular Präteritum, lassen
du wirst feiern: 2. Pers. Singular Futur, feiern

AUFGABE 16 **S. 26**

des Hauses – die Enten – den Händlern – der Bücher – das Kabel – der Kerze

AUFGABE 17 **S. 26**

ihr singt – ihr sangt – du sangst – er / sie sang – er / sie tauchte – er / sie ist getaucht
den Baum – die Hütte – der Hütte – den Hütten – den Autos – der Autos

AUFGABE 18 **S. 27**

reich, reicher, am reichsten
herrlich, herrlicher, am herrlichsten
froh, froher, am frohsten
prächtig, prächtiger, am prächtigsten
groß, größer, am größten
Nicht steigerbar sind: königlich, offen, goldgelockt, unzählig

AUFGABE 19 **S. 27**

Artikel	Substantiv	Verb	Adjektiv
eine	Stadt	verpfändete	anderen
der	Juwelen	verkaufte	ganzen
die	Bedienten	entließ	altes
ein	Eigentum	blieben	tugendsame
eine	Waldschloss		wunderschöne
	Gemahlin		
	Töchter		

Pronomen	Präposition	Konjunktion	Adverb
er	nach	und	nur
seine	von	und	
es			
ihm			
seinem			

AUFGABE 20 **S. 27**

Stadt: Akkusativ Singular Femininum
Juwelen: Akkusativ Plural Neutrum / Maskulinum
Bedienten: Akkusativ Plural Femininum oder Maskulinum
Eigentum: Dativ Singular Neutrum
Waldschloss: Nominativ Singular Neutrum
Gemahlin: Nominativ Singular Femininum
Töchter: Nominativ Plural Femininum
er verpfändete: 3. Pers. Singular Präteritum
er verkaufte: 3. Pers. Singular Präteritum
er entließ: 3. Pers. Singular Präteritum
es blieb: 3. Pers. Singular Präteritum

AUFGABE 21 **S. 27**

siehe unten

AUFGABE 21

Fehler	falscher Kasus	richtiger Kasus	Korrektur
alle Welt	Nominativ / Akkusativ	Dativ	aller Welt
der Gräfin	Genitiv / Dativ	Nominativ	die Gräfin
ihre Töchter	Nominativ / Akkusativ	Dativ	ihren Töchtern
der Küche	Genitiv	Akkusativ	die Küche
die Kochkunst	Nominativ / Akkusativ	Genitiv	der Kochkunst
diesen Mahlzeiten	Dativ	Nominativ	diese Mahlzeiten
den Graf	Akkusativ	Dativ	dem Graf(en)

2 Satzglieder

ÜBUNG 1 S. 29
Lösungsvorschlag: Gestern Abend traf ich Peters Schwester in der Stadt.

ÜBUNG 2 S. 29
Ich traf gestern Abend Peters Schwester in der Stadt. – In der Stadt traf ich gestern Abend Peters Schwester. – Gestern Abend traf ich in der Stadt Peters Schwester. – Peters Schwester traf ich gestern Abend in der Stadt. – Ich traf gestern Abend in der Stadt Peters Schwester.

ÜBUNG 3 S. 29
Gestern Abend / traf / ich / Peters Schwester / in der Stadt.

ÜBUNG 4 S. 29
1. Sie besuchen den Kinderspielplatz in der Kirchstraße. **2.** In den Ferien leihe ich **sie** mir immer aus. **3.** Ohne Zögern beging der Dieb **ihn**. **4. Es** ging ihr auf die Nerven.

ÜBUNG 5 S. 31
1. habe **2.** haben **3.** nagen **4.** macht **5.** bleibt

ÜBUNG 6 S. 31
1. fährt – bleibt **2.** bastelt – nutzt **3.** mag – verbringt **4.** findet – herstellt

ÜBUNG 7 S. 31
Versuchen – benötigen – lohnt – mahlt – verrührt – siebt – gibt hinzu – verknetet – rollt – setzt – werden gebacken

ÜBUNG 8 S. 32
1. Anfang Oktober **schreibt** Pia schon ihren Wunschzettel für Weihnachten. (Aussagesatz)
2. Kann sie es denn gar nicht mehr **abwarten?** (Entscheidungsfrage)
3. Sie **wünscht** sich von ganzem Herzen einen Computer. (Aussagesatz)
4. Den ganzen Tag **liegt** sie ihrer Mutter in den Ohren: „Bitte, Mama, **schenk** mir einen Computer zu Weihnachten!“ (Aufforderungssatz)
5. Mein Mathelehrer **sagt** immer: „**Wünscht** euch zum nächsten Fest einen Computer!“ (Aufforderungssatz)
6. „**Wirst** du meinen Wunsch **erfüllen**?“ (Entscheidungsfrage)
7. „**Gib** endlich Ruhe!“ (Aufforderungssatz), schimpft ihre Mutter. „Schließlich **ist** erst Oktober.“ (Aussagesatz)

ÜBUNG 9 S. 32
Lösungsvorschlag:
Martina ließ den Kopf hängen: Die Haare, die sie so lange hatte wachsen lassen, sahen furchtbar aus! Ihr Traum von einer Frisur, mit der sie Miss Germany werden könnte, war zerplatzt. Sie war verzweifelt. Konnte sie so nach Hause zurückkehren? Als sie dort ankam, stellte sie sich zuerst vor den Spiegel und dachte darüber nach, welche Frisur ihr jetzt noch stehen würde. Sie kam auf die Idee, dass sie die Haare hochstecken könnte.

ÜBUNG 10 S. 34
1. der Hausmeister – Wer baut die große Bühne für das Schulfest schon früh am Morgen auf?
2. die Schüler – Wer trifft wenig später ein, um die Buden und Verkaufsstände aufzustellen?
3. Die Buden – Was wurde gemeinschaftlich von Schülern, Lehrern und Eltern hergestellt?
4. alle – Wer wartet jetzt gespannt auf die ersten Gäste?
5. Der Erlös des Festes – Was soll vollständig der Schule zugutekommen?

ÜBUNG 11 S. 34
1. Die Schüler werden gelobt. **2.** Die Lehrer tanzen. **3.** Das Fest endet.

ÜBUNG 12 S. 34
1. Die Schminkaktion – Was?
2. Die kleinen Kinder – Wer?
3. neue Kunden – Wer?
4. das Clowngesicht – Was?
5. Der Preis von einem Euro pro Kindergesicht – Was?

ÜBUNG 13 **S. 35**

Subjekt	Pronomen
die Sonne	sie
Dunkle Regenwolken	Sie
der erste Ferientag	er
die Kinder	sie
Patrick	er
Das neue Buch von Tante Gabi	Es
Patrick	er
dicke Regentropfen	sie
ein Tag im Bett	es
Patrick	er
der Bücherwurm	er
Das dicke Buch	Es

ÜBUNG 14 **S. 37**

1. dir **2.** dich **3.** seiner **4.** ihr **5.** euch
6. unser **7.** mir

ÜBUNG 15 **S. 37**

1. an **2.** um

ÜBUNG 16 **S. 37**

Lösungsvorschlag:

1. einen Brief **2.** meinem Freund **3.** dir einen Brief **4.** der Opfer **5.** auf dich **6.** dem Lehrer zum Geburtstag **7.** der Frau die Hand

ÜBUNG 17 **S. 38**

1. ihre Großmutter (AO) mit einem Geburtstagskuchen (PO)
2. einen Überraschungsbesuch ihres Sohnes aus Amerika (PO)
3. der Großmutter (DO) eine „kurze“ Rede (AO)
4. dieser Rede (DO)
5. mit ihren Verwandten und Bekannten (PO)
6. bei ihren Gästen (PO)
7. des schönen Tages (GO)

ÜBUNG 18 **S. 38**

Substantiv: Milch, H-Milch, Frischmilch, Frucht-Milchshakes, Shakes, Banane
Pronomen: sie, ihn, sie

ÜBUNG 19 **S. 39**

1. Nominativ **2.** Akkusativ **3.** Genitiv
4. Nominativ **5.** Dativ **6.** Akkusativ **7.** Dativ

ÜBUNG 20 **S. 39**

1. Ich warte auf den Besuch meines Freundes.
2. Er drückte sich gepflegt aus.
3. Wir kümmerten uns um unsere neue Mitschülerin.
4. Auf der Flucht warf der Dieb seine Beute weg.
5. Die Schüler hatten die Hilfe des Lehrers nötig.
6. Dem Kind wurde das Daumenlutschen abgewöhnt.
7. Der Dieb stahl der Passantin die Tasche.

Klassenarbeiten

AUFGABE 1 **S. 40**

Lösungsvorschlag:

1. Heiner nimmt sich des kranken Hundes an.
2. Wir bitten die Lehrerin um die Rückgabe der Arbeit.
3. Ich habe Angst um meinen kranken Hund.
4. Deine beste Freundin Lisa glaubt ihm kein Wort.
5. Der Hund beißt unserem Nachbarn ein Ohr ab.
6. Die armen Menschen bedürfen der Unterstützung.
7. Sie begegnen den Kindern.
8. Die Lehrer achten auf Sauberkeit in der Klasse.

AUFGABE 2 **S. 41**

1. Ich (S) liebe dich (AO).
2. Liebst du (S) mich (AO)?
3. Diese Frage (S) beschäftigte den Jungen (AO) noch lange.
4. Der Junge (S) stellte viele Fragen (AO).
5. Das Mädchen (AO) quälten böse Träume (S).
6. Der Hund (S) verjagt die Katze (AO).

AUFGABE 3 **S. 41**

Lösungsvorschlag:

1. Der Maurer baut.
2. Der Maurer baut ein neues Haus.
3. Der neue MP3-Player gehört meinem Bruder.
4. Ich schenke meiner Oma einen Strauß Blumen.
5. Marco telefoniert mit seiner neuen Freundin.
6. Der Detektiv sucht mit einer Lupe nach einem Beweis.
7. Das Kind fragt den Opa nach einem Eis.

AUFGABE 4 **S. 41**

Ergänzung in dieser Reihenfolge:
herrschte – gibt – vorgeworfen – eingelassen – saß – wartete – erinnerte – verbracht hatte – wurde – eingeführt – Sollte – bedeuten

AUFGABE 5 **S. 42**

Ergänzung in dieser Reihenfolge:
mit ihrer Mutter (PO) – von den Nationalsozialisten (PO) – um ihr Leben (PO) – an die schlimmen Erlebnisse (PO) – den Auftrag (AO) – des völlig verschüchterten Mädchens (GO) – auf diese schwierige Aufgabe (PO) – wenig Zeit (AO) – ihr Vertrauen (AO) – alles (AO)

AUFGABE 6 **S. 42**

Einst / lebte (P) / eine berühmte Heilerin (S). / Viele Menschen / vertrauten (P) / ihrem Wissen (DO). / Drei weise Männer / wollten (P) / den kranken Bürgermeister (AO) / heilen (P). / Jedoch / keiner von ihnen (S) / schaffte (P) / dies. / Da / bat (P) / die Frau des Kranken / die Heilerin / um Hilfe (PO). / Diese / gab (P) / der Frau (DO) / bereitwillig / die rettende Medizin. / Der Bürgermeister (S) / erholte sich (P) / sofort. / Trotzdem / klagten (P) / die Leute des Dorfes / die Heilerin / der Gotteslästerung (GO) / an (gehört zu P). / Nun / bedurfte (P) / sie (S) / der Hilfe des Bürgermeisters. / Doch / er / hatte (P) / sie (AO) / vergessen (P).

AUFGABE 7 **S. 43**

Subjekte (rot): die Menschen – Diese ersten Wehranlagen – man – die Arbeiter – Die mächtigsten Burgherren – sie – Die Bauern dieser Gebiete – Sie – Die einfachen Hütten der Bauern – die Bauernfamilie – Der Burgherr – ein wärmendes Feuer – die Burgherrin
Prädikate (grün): bauten – brannten – errichtete – lebten – besaßen – herrschten – unterstanden – lebten – bestanden – wohnte – besaß – brannte – überwachte

AUFGABE 8 **S. 43**

mit einem Falken – präpositionales Objekt
den Hasen – Akkusativobjekt
zu seinem Herrn – präpositionales Objekt
die Jagdgesellschaft – Subjekt
Rebhühner – Akkusativobjekt
Wildschweine – Akkusativobjekt
auf seinem Jagdpferd – präpositionales Objekt
Eine Hundemeute – Subjekt
dem Hirsch – präpositionales Objekt
Die Jäger – Subjekt
Gelage – Akkusativobjekt
das Wild – Subjekt
ein köstlicher Honigwein – Subjekt

AUFGABE 9 **S. 44**

Subjekt: ein kleiner Junge, Turniere, Der Ehrenpreis
Prädikat: hat angeschaut, bekam
Genitivobjekt: –
Dativobjekt: dem Sieger
Akkusativobjekt: Edelsteine
präpositionales Objekt: für Schwerter

AUFGABE 10 **S. 44**

1. Bei der Ankunft auf dem Parkplatz sah ich in der Ferne schon die Burg.
2. Früher lebten in dieser Burg ein Ritter und seine schöne Frau.
3. In den Burgmauern waren Schießscharten, aus denen haben die Bogenschützen geschossen, wenn Feinde die Leute auf der Burg überfallen wollten.

4. Die Verteidiger der Burg haben Pech auf die Angreifer gegossen.
5. Die Gemächer des Burgherrn hat uns eine Frau gezeigt.
6. Meine Mutter fand die Ritterrüstungen im Museum ganz toll.

AUFGABE 11 **S. 45**
1. klingelt 2. trägt 3. schickt 4. untersuchen 5. fürchten

AUFGABE 12 **S. 45**
1. alle Notrufe der Hilfesuchenden, sie
2. Wer sollte gesund und sportlich sein? – ein Bewerber für den Polizeiberuf, er
3. Wer kontrolliert das Geschehen auf Seen und Flüssen? – Wasserschutzpolizisten, sie
4. Wer begleitet den hohen Besuch? – die Motorradstaffel, sie
5. Wer hat bei Demonstrationen einen guten Überblick? – Polizisten auf Pferden, sie

AUFGABE 13 **S. 46**
1. Wonach suchen Hundeführer mit Hunden? – präpositionales Objekt
2. Wem ist das Einsatzkommando häufig ausgesetzt? – Dativobjekt
3. Wen warnt der herannahende Einsatzwagen mit Martinshorn und Blaulicht? – Akkusativobjekt
4. Wen verfolgt der Streifenwagen? – Akkusativobjekt
5. Wem wird die gestohlene Handtasche nach der Festnahme zurückgegeben? – Dativobjekt
6. Wessen bezichtigt der Kommissar den Verdächtigen? – Genitivobjekt

AUFGABE 14 **S. 46**
1. Der Polizist (S) schreibt für den Staatsanwalt (PO) ein Protokoll über den Diebstahl (AO).
2. Für den Hinweis zur Klärung des Verbrechens (PO) hatte die Kriminalpolizei (S) eine Belohnung von 2000 € (AO) ausgesetzt.
3. Die alte Frau (S) dankt dem Polizisten (DO) für die rasche Klärung des Falls (PO).
4. Sie (S) hat fest mit der Aufklärung des Falls (PO) gerechnet.
5. Zur Aufklärung der Straftat (PO) bedurfte es (S) einer großen Kombinationsgabe der Ermittler (GO).

3 Rechtschreibung

ÜBUNG 1 **S. 49**
die Menge – der Hai – die Reihe – das Meer – der Wal – der Zahn – der Hai – die Reihe – das Schicksal – der Kragen – der Schmerz – das Herz – das Mitleid – der Hai – die Medizin – die Medizin – der Wal – die Qual

ÜBUNG 2 **S. 49**
Ergänzung in dieser Reihenfolge:
Ihnen – Ihr – Sie – Sie – Ihrem – Ihnen – Sie – Ihre – Ihr

ÜBUNG 3 **S. 50**
Trübsal – Freiheit – Wagnis – Untersuchung – Sauberkeit – Achtung – Erlebnis

ÜBUNG 4 **S. 50**
1. Auf der Klassenfahrt haben wir **manches besonders Aufregende** erlebt.
2. Aber **das Beste** war die Nachtwanderung im Schnee.
3. Da gab es in unserer Klasse **einige sehr Ängstliche**.
4. Sie glaubten, dass **etwas besonders Unheimliches** passieren würde.
5. **Das ausdauernde Wandern** durch die kalte Winternacht war für einige von uns ziemlich ermüdend.
6. Umso mehr freuten wir uns auf **das Ausschlafen** am nächsten Morgen.

ÜBUNG 5 **S. 50**
Ergänzung in dieser Reihenfolge:
Zum Herbstanfang – ins Disneyland – Am 20. September – zum ersten Mal – ins Ausland – Beim Kofferpacken – im / vom Amt – Zum Glück – zum Termin – ans Weckerstellen

ÜBUNG 6 **S. 51**

1. Wir gratulieren **Ihrem** Sohn zur bestandenen Prüfung. – Die Eltern gratulieren **ihrem** Sohn zur bestandenen Prüfung.

2. Sarah und Simon fahren nach Fuerteventura, denn **sie** wollen einmal richtig ausspannen. – Fuerteventura ... wenn **Sie** einmal richtig ausspannen wollen!

3. Sparen **Sie** nicht am Futter! Wenn **Sie Ihre** Katze verwöhnen wollen, kaufen **Sie** LOMPIES! – Weil unsere Nachbarn **ihre** Katze verwöhnen wollen, sparen **sie** nicht am Futter, **sie** kaufen LOMPIES.

4. Wahre Genießer wissen, was zählt, **sie** gönnen **sich** ein Bad mit Rosenextrakt. – Machen **Sie sich** das Leben schön, gönnen **Sie sich** ein Bad mit Rosenextrakt von SMOOTHBATH!

5. Sehr geehrter Herr Meier, wir danken **Ihnen** für **Ihre** Teilnahme an unserem Gewinnspiel. – In dem Brief steht, dass sie **mir / dir / ihnen / euch / uns** für **meine / deine / ihre / eure / unsere** Teilnahme an **ihrem** Gewinnspiel danken.

ÜBUNG 7 **S. 51**

1. zum **E**ssen – zu **e**ssen
2. zu **s**tricken – Zum **S**tricken
3. Das **L**esen – zu **l**esen
4. zu **s**pielen – Zum **S**pielen
5. zu **s**chreien – Vom **S**chreien

ÜBUNG 8 **S. 54**

Langer Vokal: Blume – sieben – malen – Rose – rot – nehmen – rufen
Kurzer Vokal: Locke – Witz – List – Sonne – Wand – bunt

ÜBUNG 9 **S. 51**

1. hören: Hörer, Zuhörer, Hörgerät, Hörfunk, Hörspiel, mithören, überhören, hörenswert, schwerhörig
2. nehmen: annehmen, zunehmen, hochnehmen, mitnehmen, unternehmen, Teilnehmer, Benehmen, Mitnahme
3. viel: vielschichtig, vielfältig, Vielzahl, Vielfraß, Vielfalt, vielfach, vielleicht
4. nahe: beinahe, hautnah, sich nähern, näher kommen, Nähe, Nahverkehr, Naherholungsgebiet, unnahbar
5. Segen: segensreich, Segensworte, einsegnen, absegnen, Segnung, gesegnet
6. Haar: Haarband, Haarfarbe, rothaarig, um Haaresbreite, haarsträubend, haarscharf, haarklein, Härchen

ÜBUNG 10 **S. 54**

Beerenobst, Schneeschaufel, Ballsaal, Speerwurf, Moorhuhn, Kaffeetasse, Seemann, Personenwaage

ÜBUNG 11 **S. 55**

R	U	M	M	E	L	F	S	C	H	P	A	S
R	O	O	T	V	Q	D	C	H	X	A	W	T
E	M	T	A	G	I	J	H	S	Q	P	Z	I
S	B	T	R	U	Y	U	W	L	L	P	H	M
T	K	E	T	T	E	I	I	F	N	E	R	M
A	Y	E	A	E	T	I	M	G	R	E	V	E
M	K	L	N	E	B	O	M	U	R	C	E	N
P	E	N	N	R	F	L	B	A	B	U	L	A
E	N	W	E	E	K	K	A	T	Z	E	F	N
L	K	I	N	N	H	L	D	E	R	C	E	L
U	L	E	N	B	E	L	R	I	S	K	I	C
H	T	K	D	E	C	K	E	L	E	E	P	R
V	P	O	Z	K	B	L	E	V	E	X	H	A

ÜBUNG 12 **S. 55**

Lösungsvorschlag:
-atz-: Platz, platzen, Schatz, Katze, Tatzen, Satz, Glatze, Latz
-itz-: sitzen, blitzen, Besitz, flitzen, kritzeln, Ritze, witzig, Hitze, Glitzer, Spitze
-eck-: Decke, Hecke, verstecken, entdecken, Fleck, Zweck, keck, schmecken

ÜBUNG 13 **S. 56**

wiedergeben – wiederholen – wiederherstellen – wiederbringen
widersprechen – widerstehen – widerlegen – widerspiegeln – widerrufen

ÜBUNG 14 **S. 56**
lebten – Diebesbanden – verlegt – Damen – ihre – Juwelen – stehlen – Ohne – Vorwarnung – nahmen – ihre – lieferten – Gefahren – Bestohlenen – wehrten – flehten – Gnade – wohl – Wahl

ÜBUNG 15 **S. 56**
Ur-/ur-: Ursprung, Urchristen, uralt, Urzeit
-tum: Christentum, Altertum
-sal: Mühsal, Schicksal, Rinnsal
-bar: fahrbar, wunderbar, fruchtbar, ehrbar, furchtbar, genießbar
-sam: mühsam, wundersam, seltsam, langsam, furchtsam

ÜBUNG 16 **S. 58**
1. Erbse **2.** Schloss **3.** Straße

ÜBUNG 17 **S. 59**
1 Pass **2** Fluss **3** Los **4** Ast **5** Umriss
6 Maß **7** Essen **8** Nüsse **9** Moos **10** Untertasse
11 Spieß
Lösungswort: Pflaumenmus

ÜBUNG 18 **S. 59**
s: Los, Ast, Moos
ss: Pass, Fluss, Umriss, Essen, Nüsse, Untertasse
ß: Maß, Spieß

ÜBUNG 19 **S. 59**
essen: ich esse, du isst, er / sie / es isst – ich aß, du aßest, er / sie / es aß – ich habe gegessen, du hast gegessen, er / sie / es hat gegessen
lesen: ich lese, du liest, er / sie / es liest – ich las, du lasest, er / sie / es las – ich habe gelesen, du hast gelesen, er / sie / es hat gelesen

ÜBUNG 20 **S. 60**
1. Der Koch misst den Zucker für die Nusssoße ab.
2. Der Mann vergaß zum Schluss, die Rosen zu gießen.
3. Heiße Würstchen isst mein großer Bruder am liebsten.
4. Schließlich raste der Hund auf die Straße und zerriss die Hose.

Klassenarbeiten

AUFGABE 1 **S. 61**
Gestern kam der Zirkus in unsere Stadt. Schon am Vortag hatten die Männer vom Zirkus die Plakate aufgehängt und waren mit Lautsprechern auf dem Dach durch die Stadt gefahren: „Achtung, Achtung! Ab morgen gastiert der Zirkus Luftikus in Ihrer Stadt. Die Vorstellungen sind jeden Abend um 20:00 Uhr, die Kindervorstellungen täglich um 16:00 Uhr.“ Am nächsten Tag begann man sofort damit, das riesige Zelt aufzubauen. Viele Kinder schauten der Mannschaft neugierig bei ihrer Arbeit zu. Das Aufbauen dauerte nur einen Vormittag, dann ging es endlich los. Zur ersten Vorstellung strahlte den Gästen von der Zirkuskuppel ein tiefes Blau entgegen.

AUFGABE 2 **S. 61**
Floß, Flöße – Ergebnis, Ergebnisse – Hindernis, Hindernisse – Kürbis, Kürbisse – Gruß, Grüße – Krokus, Krokusse – Verließ, Verließe – Fluss, Flüsse – Biss, Bisse – Bus, Busse – Kuss, Küsse – Kloß, Klöße – Laus, Läuse – Ausweis, Ausweise

AUFGABE 3 **S. 61**
Mark Twain wurde am 30. November 1835 in einem kleinen Dorf in Florida geboren. Das Dorf zählte nur etwa hundert Einwohner, sodass die Geburt des kleinen Jungen ein großes Ereignis darstellte. Dieses Dorf hatte nur zwei echte Straßen, die übrigen Wege waren lediglich Fußwege entlang der riesigen Maisfelder. Bei nassem Wetter überzog die Wege ein zäher Schlamm, bei trockenem eine feine Staubschicht. Die meisten Häuser waren einfache Blockhäuser.

AUFGABE 4 **S. 62**
Selbst die Kirche war ganz aus Holz. Der Bohlenboden bestand aus groben Stämmen und war nicht mit Teppich bedeckt. Wenn jemand zufällig etwas fallen ließ, so verschwand dies auf Nimmerwiedersehen zwischen den Spalten der großen Bretter. Da die Kirche auf kur-

zen Holzpflöcken stand, hausten dort unten Schweine, Katzen und Hunde, die den Gottesdienst des Pfarrers störten, wenn die Tiere nicht gerade schliefen. An Wochentagen wurde die Kirche als Schule benutzt. Es gab auch ein Geschäft im Dorf, dessen Besitzer mein Onkel war. Es war ein sehr bescheidener Laden, in dem es lediglich ein paar Gerätschaften wie Besen, Äxte, Hacken und dergleichen zu kaufen gab.

AUFGABE 6 **S. 63**

Aufmerksamkeit – Ameise – emsigen – Neues – besonders – was – dass – außerordentlich – besonders – was – etwas – Besseres – bis – gesehen – Fleiß – durchaus – auszusetzen

AUFGABE 7 **S. 63**

rennt – Kieselstein – anstatt – ihn – klettert – zerrt – kugelt – wütend – schüttelt – gierig – ihrem – Eigentum – stößt – schiebt – dreht – zerrt – immer – plötzlich – wieder – fortrennt

AUFGABE 8 **S. 64**

Betrachten wir sie einmal, wenn sie auf Beute ausgeht. Sie hat einen Fang getan; aber was macht sie dann? Geht sie etwa nach Hause? Durchaus nicht, gerade im Gegenteil; sie weiß nicht mehr, wo ihre Wohnung ist, und kann sie nicht finden, wenn sie auch kaum drei Fuß davon entfernt ist. Sie tut einen Fang, habe ich gesagt, aber gewöhnlich ist das Ding zehnmal so groß, als es sein sollte, sie fasst es gerade am unbequemsten Ende an und hebt es mit aller Gewalt in die Höhe; dann trägt sie es fort, nicht nach Hause, sondern in entgegengesetzter Richtung, nicht ruhig und bedächtig, sondern mit rasender Eile, bei der sie ihre Kräfte unnütz vergeudet.

AUFGABE 9 **S. 64**

stößt – Pflanze – fällt – ihr – muss – hinaufklettern – Spitze – Besitz – herziehen – wenn – Mehl – schleppen – hinaufkommt – dass – klettert – kugelt – nimmt – gewöhnlich

AUFGABE 10 **S. 64**

Dass die Ameise sich wieder und wieder mit viel zu großen Beutestücken herumplagt, ist absolut widersinnig. An einem Punkt aber muss ich Mark Twain widersprechen: Dass die Ameise ihre Fehler immer wiederholt, verärgert mich nicht, sondern es amüsiert mich. Meine Freundin findet Ameisen widerlich, ich finde, dass sie niedlich sind.

AUFGABE 11 **S. 65**

Die korrigierten Wörter:
Leichtsinn – Dummheit – unternahm – Haare – dreizehn Jahre – Gedanken – dass (jede) – fliegen – explodieren – (das) Spielen – Straße – Angelegenheit

AUFGABE 12 **S. 65**

Überhaupt war mir damals vieles noch ziemlich unklar. Meine Zensuren waren nicht nur mäßig, sie waren ausgesprochen lausig. Ich tat nur das, was mir Spaß machte, und das waren meistens Dinge, die nicht auf dem Lehrplan unserer Klasse standen. Dass ich auf diese Weise weder zu Hause noch in der Schule große Lorbeeren ernten konnte, merkte ich erst, als es bereits zu spät war und ich mit Pauken und Trompeten, sehr zum Entsetzen meiner Eltern, in der achten Klasse kleben blieb.

AUFGABE 13 **S. 65**

besuchte – damals – Französische – Gymnasium – Berlin – Interessen – schwankten – zwischen – Naturwissenschaft – Musik – Schulorchester – Cello – Mutter – Geduld – vierhändige – Sonaten – mir – versuchte – allerdings – dünnen – schwelgte – Öfteren – erfolgreichen – Komponistenlaufbahn

AUFGABE 14 **S. 66**

Der Held meiner **Jugend** war jedoch Nikolaus Kopernikus. Er war der Erste, der es wagte, im **Widerspruch** zu den damaligen **Anschauungen** der Kirche und der Gelehrtenwelt zu erklären, dass nicht die Erde, sondern die Sonne der Mittelpunkt unseres Weltalls ist. Doch wurde ich von meiner **Bewunderung** für die Geistestaten des Kopernikus bald durch wesentlich näherliegende und aufsehenerregendere **Geschehnisse** abgelenkt. Max Valier, einer der einfallsreichsten und aktivsten Förderer des Einsatzes von Raketen als Antrieb zur **Erzielung** hoher Geschwindigkeiten, hatte sich mit Fritz von Opel, dem bekannten deutschen Automobilfabrikanten, zusammengetan: In gemeinsamer **Anstrengung** entwickelten sie einen durch Raketen angetriebenen Wagen, mit dem sie neue Geschwindigkeitsrekorde aufstellten: Die **Zeitungen** waren voller Schlagzeilen über die kühnen, wenn auch nicht immer von Erfolg gekrönten Versuche, die meine Fantasie und meinen Unternehmungsgeist zu wilden **Überlegungen** anregten.

4 Zeichensetzung

ÜBUNG 1 S. 68

Schenken Sie uns fünf Minuten Aufmerksamkeit**!** Es geht um die Sicherheit Ihrer Kinder**. / !** Immer mehr Kinder sterben im Straßenverkehr**.** Das muss nicht sein**!** Wollen Sie noch länger tatenlos zusehen**?** Durch Geschwindigkeitsbegrenzungen könnten viele Unfälle vermieden werden**.** Wir Eltern setzen uns dafür ein**.** Machen auch Sie bei unserer Aktion mit**!**

ÜBUNG 2 S. 68

Wo warst du gestern**?** Hast du mich gesehen**?**
Wie alt ist sie**?** Das geht dich gar nichts an**! (.)**
Jetzt sag schon**!**
Wer war das**?** Das sage ich dir doch nicht**.**
Bist du sauer**?** Warum fragst du**?** Ich meine nur so**.**

ÜBUNG 3 S. 68

Frage mit Fragewort: Wo warst du gestern? Wie alt ist sie? Wer war das? Warum fragst du?
Satz- bzw. Entscheidungsfrage: Hast du mich gesehen? Bist du sauer?

ÜBUNG 4 S. 69

Das Schiff fuhr nach Ostindien und ich hatte als Arzt nicht viel zu tun**.** Doch eines Nachts gerieten wir in einen fürchterlichen Sturm**.** Welch ein Unwetter zog über uns her**!** Nach mehreren Tagen Kampf gegen die Naturgewalten zerbrach die „Antilope" und versank**.**
Was sollte jetzt aus mir werden**?** Ich trieb ohne Hoffnung auf Rettung allein auf den Wellen durch die dunkle Nacht**.** Da, plötzlich fühlte ich Grund unter meinen Füßen**!** Wo war ich**?** Ich stolperte durch den nassen Sand und betrat festen Boden**.** Nirgends sah ich Zeichen menschlichen Lebens**.** Nun, fürs Erste war ich gerettet**!** Beruhigt schlief ich ein**.**
Am nächsten Morgen wollte ich mich wohlig in der warmen Sonne räkeln**.** Doch was war das**?** Ich konnte mich nicht bewegen**!** Jetzt wollte ich mich aufsetzen**.** Es misslang**!** Noch nicht einmal den Kopf konnte ich bewegen**!** Was war in der Nacht geschehen**?**
Mit langsamen Bewegungen versuchte ich, den Kopf aus seiner seltsamen Starre zu befreien**.** Nach Verlust mehrerer Haare konnte ich den Kopf wenigstens so weit bewegen, dass ich an mir hinunterschauen konnte**.** Ich war von Kopf bis Fuß gefesselt**!** Doch wer hatte dieses Kunststück vollbracht**?**

ÜBUNG 5 S. 69

1. Ist das Auto rot?
2. Frisst die Kuh Gras?
3. Räumt Sebastian auf?
4. Läuft Vanessa schnell?
5. Maria, pass auf!
6. Jan, pack den Koffer!
7. Tobias, hol Kreide!
8. Leni, sprich leise!

ÜBUNG 6 S. 71

Die Hauptsätze sind fett hervorgehoben, alle anderen Sätze sind Nebensätze:
Eine Stadtmaus ging einmal spazieren, als ihr eine Feldmaus begegnete. **Die Feldmaus gab der Stadtmaus gerne von ihren Vorräten ab.** Weil die Feldmaus aber nur Früchte des Waldes und Feldes besaß, **lud die Stadtmaus sie zu**

einem Besuch bei sich ein. Ihre Speisekammer war randvoll mit herrlichen Speisen. Es gab Brot, Schinken, Speck, Kuchen und allerlei mehr. Da kam aber der Besitzer der Speisekammer herein und die Mäuse suchten zitternd vor Angst ein Versteck. Die Stadtmaus floh sofort in ihr Mäuseloch. Da die Feldmaus sich aber nicht auskannte, **fand sie kein sicheres Versteck und wäre beinahe umgekommen. Nach diesem Erlebnis wollte die Feldmaus schnell wieder nach Hause. „Bleibe du ruhig eine feine Stadtmaus. Ich will ein armes Feldmäuschen bleiben und bescheiden meine Eicheln und Nüsse essen. Hier in der Stadt fühle ich mich nicht sicher,** weil es überall Gefahren gibt. **Allein auf dem Land bin ich frei und sicher in meinem Feldlöchlein.“**

ÜBUNG 7 **S. 71**

1. Bis wir uns wiedersehen, werde ich jeden Tag an dich denken.
2. Ich denke an dich(,) und wir sehen uns bestimmt wieder.
3. Wir treffen uns in Hamburg(,) oder wir sehen uns in Berlin wieder.
4. Das erste Mal sahen wir uns in Hamburg**,** das nächste Mal trafen wir uns in Berlin.
5. Für nächstes Jahr planen sie einen gemeinsamen Urlaub(,) oder sie bleiben zu Hause.
6. Obwohl ein gemeinsamer Urlaub geplant war**,** blieben sie lieber zu Hause.

ÜBUNG 8 **S. 72**

Lösungsvorschlag:

Für die Zubereitung einer Biskuitrolle benötigt man 4 Eier, 150 g Zucker, eine Prise Salz, 2 Esslöffel warmes Wasser, 100 g Mehl, 2 Teelöffel Backpulver und 2 Päckchen Vanillepuddingpulver. Zuerst wird steif geschlagenes Eiweiß mit Zucker, Salz und Wasser in eine Schüssel getan und alles mit einem Mixer schaumig gerührt. Dann muss das Eigelb vorsichtig untergezogen werden. Danach werden das Mehl, das Backpulver und das Puddingpulver gesiebt, dadurch gemischt und vorsichtig mit der Eier-Zucker-Masse verrührt. Der Teig wird auf einem Backblech gleichmäßig verteilt und bei 180° C etwa 20 Minuten gebacken. Anschließend wird der noch warme Biskuit auf ein gezuckertes Geschirrtuch gestürzt, gerollt und muss dann auskühlen. Die Rolle kann mit Sahnecreme, Marmelade, Schokocreme oder Fruchtquark gefüllt werden.

ÜBUNG 9 **S. 72**

Kleine**,** quirlige Kinder – zwei unvorsichtige Vierjährige – die unübersichtliche, alte Hauptstraße – süßes kleines Kaninchen – mit schweren, schmerzenden und langwierigen Verletzungen

ÜBUNG 10 **S. 74**

„Das ist nicht recht, Vater“, schimpft er, „dass Ihr ...“ – Er spricht: „Das ist nicht recht, ...“ – „... Steigt sofort ab!“, ruft er. – „Ihr seid drei kuriose Gesellen“, wundert er sich.

ÜBUNG 11 **S. 74**

1. **„**Mir gefällt die Geschichte überhaupt nicht!**“,** schimpft Peter.
2. **„**Warum?**“,** fragt Katrin nach. **„**Ist doch wirklich lustig, dass die beiden sich so durch andere verunsichern lassen.**“**
3. **„**Und außerdem**“,** fügt Maria hinzu**, „**ist die Vorstellung, dass ein Esel getragen wird, echt komisch.**“**
4. **„**Aber die ganze Zeit passiert immer dasselbe: Ein Wanderer kommt, ein Wanderer kommt und so weiter**“,** meckert nun auch Tom.
5. **„**Ach, ihr Jungs habt doch keine Ahnung!**“,** ereifert sich Verena. **„**Ihr wisst doch gar nicht, was eine gute Geschichte [illegible]**“**
6. Die Lehrerin fällt Verena ins Wort: **„**Nun mal langsam. Jeder darf hier sagen, wie ihm die Geschichte gefällt.**“**

ÜBUNG 12 **S. 75**

Lösungsvorschlag:

Paul kommt aufgeregt zu seinen drei Freunden Christopher, Lisa und Laura gelaufen: „Das waren tolle Ferien! Ich war in Amerika und habe dort einen Jungen kennengelernt. Er heißt Steven und wir schreiben uns jede Woche E-Mails.“ „Meine Ferien waren leider ziemlich langweilig“, meint Laura, „denn ich war bloß bei meiner Oma in Kleinniederheuersfeld. Hoffentlich werden die nächsten Ferien besser!“ Betrübt zeigt Lisa auf ihr Bein: „Was soll ich denn sagen? Ich lag vier Wochen mit einem gebrochenen Bein im Krankenhaus. Dort hatte ich nur wenig Besuch, denn es waren ja fast alle im Urlaub. Da habe ich dann eben viel gelesen.“ „Wenigstens hattest du Ruhe. Ich habe einen schreienden Bruder bekommen“, erzählt Christopher. „Ich bin vor dem Geschrei jeden Tag ins Schwimmbad geflohen. Da habe ich auch noch einen schlimmen Sonnenbrand bekommen.“

ÜBUNG 13 S. 75

1. Frieda klagt: „Meine Mutter schimpft immer mit mir: ‚Putz dir die Schuhe ab!‘“
2. In der Zeitung stand: „Die Polizei ließ gestern die frohe Botschaft verkünden ‚Das verschwundene Mädchen ist wieder aufgetaucht‘ und hat dies heute noch einmal bestätigt.“
3. Kalle erzählt seinem Freund: „Der Zugführer rief ‚Einsteigen bitte!‘ und der Zug setzte sich in Bewegung.“

Klassenarbeiten

AUFGABE 1 S. 76

1. **„**Wenn ich das geahnt hätte, hätte ich dich sicher nicht gehen lassen!**“,** schimpfte mein Vater.
2. **„**Nein**“,** betonte der Mann, **„**das Geschäft bleibt geschlossen!**“**
3. **„**Was willst du**?“,** fragte ich sie.
4. **„**Gib mir mal dein Heft**!“,** bat mich Klaus.
5. **„**Ist es wirklich wahr**“,** fragte sie entgeistert, **„**dass ihr heiratet**?“**
6. Die Köchin rief**: „**Das Essen ist fertig**!“**
7. Der Lehrer erklärte**: „**Der Dativ ist der Wem-Fall.**“**
8. **„**Der Akkusativ**“,** fuhr der Lehrer fort, **„**ist der Wen-Fall.**“**
9. **„**Zweifellos werden wir uns morgen sehen**“,** versprach Lisa ihrem Freund.
10. **„**Kannst du mir mal helfen**?“,** bat Helena.

AUFGABE 2 S. 76

Die Hauptsätze sind fett hervorgehoben, alle anderen Sätze sind Nebensätze:
Als ich im November per Kutsche von Frankfurt nach Weimar reisen wollte, **nahm ich eine Kammerfrau mit. Es war ein langer, beschwerlicher Weg,** weil die Straßen schlecht waren und überall Räuberbanden lauerten. **Bis Hanau kamen wir recht gut voran,** obgleich unser Pferd gefallen war. **Gleich dahinter fing das Unglück aber an.** Obwohl die Nacht schon hereingebrochen war, **waren wir an zahlreichen Gasthöfen vorbeigefahren,** da wir die Ausplünderung in diesen Spelunken fürchteten. **Da aber blieben wir in einem Loch stecken und mussten uns nun doch ein Quartier suchen,** wenn wir die Nacht nicht unter freiem Himmel verbringen wollten.

AUFGABE 3 S. 77

Der Wirt eines nahe gelegenen Gasthofes bot uns sofort ein Zimmer an. **„**Der Wagen ist hin. Ihr müsst morgen einen Schmied holen lassen**!“** Doch ich traute dem Wirt nicht. So begab ich mich in der Nacht zu unserem Wagen und sah mit Entsetzen, wie der Wirt mein Gepäck untersuchte. Kaum eine Viertelstunde später saßen wir wieder in unserem Wagen und kämpften uns durch die Nacht.
Doch plötzlich stand der Wagen still. **„**Was ist denn los**?“,** fragte meine Kammerfrau den Kutscher. **„**Das Pferd ist gestürzt und kann nimmermehr aufstehen**“,** antwortete dieser und ging los**,** um Hilfe zu holen. So saßen wir zwei Frauen zwei Stunden nach Mitternacht mutterseelenallein mitten in einem großen, unheimlichen Wald in unserer Kutsche und warteten auf Hilfe. Erst als der Morgen dämmerte, kam unser Kutscher mit einem Pferd zurück und wir konnten die beschwerliche Fahrt fortsetzen.

AUFGABE 5 S. 78

Ein Mann mit einem Frosch auf dem Kopf kommt zum Arzt. <u>Der Arzt **fragt** ihn verwundert:</u> „Wie sind Sie denn daran geraten?“ <u>Da **antwortet** der Frosch:</u> „Ich weiß auch nicht. Gestern bin ich reingetreten und seitdem geht es nicht mehr ab.“

AUFGABE 6 **S. 78**

1. Ein Bauer karrt eine Fuhre Mist durch die Straße. „Wo kommt denn dieser Mist hin?“, fragt die kleine Anna. „Auf die Erdbeeren!“, antwortet der Bauer. „Komisch“, meint Anna, „wir geben immer Sahne auf die Erdbeeren.“
2. „Voriges Jahr haben wir eine Weltreise gemacht“, gibt Stefan an. „Und dieses Jahr?“, fragt Judith. „Dieses Jahr? Da fahren wir woanders hin.“

AUFGABE 7 **S. 78**

Reiten, Schwimmen oder Lesen sind Freizeitbeschäftigungen, die Kinder gerne in ihren Ferien ausüben, während ihre Eltern sich **Burgen, Schlösser und andere bedeutsame Bauwerke** anschauen wollen. Wegen der unterschiedlichen Vorstellungen gibt es oft schon am ersten Morgen des Urlaubs **schlechte Laune, Ärger oder sogar Streit**, weil man sich nicht einigen kann, ob einfach nur **gefaulenzt, gebadet oder aber gewandert** werden soll.

AUFGABE 8 **S. 78**

Familie Müller verbringt die Ferien auf einem familienfreundlichen, kleinen und günstigen Bauernhof, der in der Nähe von Bremen liegt. Sie haben Frühstück, Mittagessen und auch Abendessen gebucht, weil niemand in den Ferien Zeit und Lust zum Kochen hat.

5 Erzählen

ÜBUNG 1 **S. 81**

Erlebniserzählung: 1, 2, 3, 6, 7, 10, 12, 13, 15
Fantasiegeschichte: 4, 5, 8, 9, 11, 14, 16

ÜBUNG 2 **S. 81**

Meine Geburtstagsparty: Einladungskarten schreiben und verteilen – Grill aus dem Keller holen – Hefeteig für Stockbrot ansetzen – Holzkohle kaufen
Begegnung mit einem Drachen: sumpfiges Gelände – Schwefelgeruch – angekohlte Bäume und Sträucher – furchterregende Geräusche
Die Supermaus: dreißig Käselaibe pro Tag verspeisen – allerdings etwas eitel, bürstet sich jeden Tag zwei Stunden lang das Fell – als Blitzkurier einsetzbar – erfolgreiches Studium an der Mäuse-Akademie

ÜBUNG 3 **S. 82**

Lösungsvorschlag:
Meine Geburtstagsparty: Musik auflegen – Krach mit den Nachbarn – Lagerfeuer
Begegnung mit einem Drachen: aufsteigender Rauch – grüne Schwanzflosse – Kampf
Die Supermaus: Kamm verloren – noch schnell einen kaufen – Ladenschluss

ÜBUNG 4 **S. 82**

Die richtige Reihenfolge: 3, 7, 1, 5, 4, 6, 2

ÜBUNG 5 **S. 82**

Lösungsvorschlag:
Gestern habe ich davon geträumt, über Nacht im Zoo eingeschlossen worden zu sein. Um mich vor dem großen Gewitter zu schützen, ging ich ins Tropenhaus. Als ich kurz nach dem Gewitter das Tropenhaus verlassen wollte, war die Ausgangstür verschlossen. Sofort eilte ich zur Eingangstür, aber dort konnte ich auch nicht hinaus. Eindringliches Pochen gegen die Tür und lautes Rufen halfen nichts. Niemand reagierte. Ein Blick auf die Uhr zeigte, dass es bereits nach 18:00 Uhr und ich offensichtlich bis zum nächsten Morgen hier gefangen war. Ich

machte mich auf die Suche nach einem Schlafplatz. Hinter dem Kaimanbecken gab es eine kleine, trockene Höhle, zu der ich über einen schmalen Steg gelangen konnte. Als Abendbrot nahm ich drei Orangen mit, die ich von der Aussichtsplattform aus pflücken konnte. Ich tastete mich vorsichtig über den Steg und versuchte, die Aufmerksamkeit der Kaimane nicht auf mich zu lenken. Solange die Beleuchtung noch an war, hielten die grünen Tiere immer wieder ihre Schnauze aus dem Wasser und beobachteten mich neugierig. Nachdem dann das Licht ausgegangen war, wurden die Fledermäuse in der Höhle aktiv. Vor Aufregung konnte ich kaum schlafen. Offensichtlich sind mir dann doch irgendwann die Augen zugefallen, denn am nächsten Morgen wurde ich durch sanftes Rütteln geweckt. Die Tierpflegerin staunte nicht schlecht, als ich ihr von meinen Erlebnissen berichtete.

ÜBUNG 6 **S. 84**

sagen: rufen, meinen, antworten, erwidern, entgegnen, flüstern, schreien ...
denken: sich etwas durch den Kopf gehen lassen, überlegen, meinen, erwarten, für richtig halten ...
lustig: fröhlich, fidel, munter, komisch, merkwürdig, lächerlich, witzig, heiter, gut gelaunt ...
sich freuen: glücklich sein, zufrieden sein, lachen, strahlen über das ganze Gesicht ...
helfen: behilflich sein, Hilfe leisten, unter die Arme greifen, unterstützen, zur Hand gehen ...
Hindernis: Sperre, Blockade, Hürde, Barriere, Schwierigkeit, Problem ...

ÜBUNG 7 **S. 84**

Christian: „Mist, so 'ne Chance ...!"
Pia: „Super, jetzt kann ich ...!"
Laura: „Da ist bestimmt ...!"
Chiara: „Hoffentlich ist ihr ...!"
Jonas: „Eigentlich hatte ich ...!"

ÜBUNG 8 **S. 85**

Lösungsvorschlag:

1. „Oh Mist, das wird bestimmt teuer!"
2. „Oje, die Rutsche ist aber hoch!"
3. „Da habe ich mich wohl verschnitten."
4. „Das muss es jetzt sein, sonst ist das Spiel gelaufen!"

ÜBUNG 9 **S. 85**

vor lauter Enttäuschung heulen – mit schlotternden Knien ankommen – in klirrender Kälte erfrieren – mit rasanter Geschwindigkeit brausen – vor Aufregung starkes Herzklopfen haben – keinen Finger krumm machen – mit ohrenbetäubender Lautstärke kreischen – einen bitterbösen Blick zuwerfen

ÜBUNG 10 **S. 86**

1. sich täuschen lassen, auf jemanden hereinfallen
2. genau hinhören, neugierig lauschen
3. schlagfertig sein, gut reden können
4. etwas plötzlich, unerwartet und heimlich tun
5. jemanden haben, dem man vertrauen kann

ÜBUNG 11 **S. 86**

Wetter: diesig, neblig, frostig, heiter, schön, heiß
Mittagessen: lecker, würzig, teuer, reichhaltig, heiß, versalzen
Briefträger: fröhlich, heiter, munter, spindeldürr, schön, blond
Traumschloss: prachtvoll, schön, teuer, gläsern
Radrennen: heiter, munter, schön, spannend, umkämpft, anstrengend

ÜBUNG 12 **S. 87**

Lösungsvorschlag:

Gestern ist mir etwas ganz Seltsames passiert. Als ich kurz vor dem Mittagessen zu unserem Briefkasten ging, traf ich unseren Briefträger. Er, der sonst immer so griesgrämig und böse blickte, war fröhlich und heiter und lud mich ein, mit ihm an einem Radrennen teilzunehmen. Trotz des frostigen und nebligen Wetters schwang ich mich auf mein Rad und wir fuhren zum Start an der Colman-Brücke. Dort waren bestimmt schon hundert Teilnehmerinnen und Teilnehmer versammelt. Das Ziel war die Olfrather Mühle am anderen Ende der Stadt. Für die ersten zehn Gewinner gab es einen teuren MP3-Player. Nach dem Startschuss radelten wir in hohem Tempo in Richtung Stadtwald. Wegen des diesigen Wetters mussten wir darauf achten, nicht auf die vor uns Radelnden aufzufahren. Als wir den Stadtwald erreichten, hatten wir den Anschluss an die haushoch überlegenen Spitzenreiter bereits verloren. Bis dahin schien mir noch alles ganz normal, aber als wir dann an einem gläsernen Traumschloss vorbeikamen, war ich schon nicht mehr ganz so munter. Ein spindeldürrer Mann winkte uns aus einem Fenster zu und machte uns klar, dass wir ins Schloss kommen sollten. Wir betraten eine prachtvolle Eingangshalle und wurden in einen schönen Speisesaal geleitet. Dort erwartete uns eine ausgelassene Gesellschaft, die an einem reichhaltig gedeckten Tisch saß. Wir nahmen an der Tafel Platz. Eine Schar von Dienern reichte uns würzige und wohlriechende Speisen. Mein Tischnachbar erzählte mir Witze, über die wir schallend lachen mussten. Wir amüsierten uns prächtig. Satt und zufrieden verließen wir das Schloss nach mehreren Stunden. Gestärkt machten wir uns auf zum Ziel. Dabei gelang es uns, die übrigen Teilnehmer und Teilnehmerinnen zu überholen und unter den ersten zehn ins Ziel zu gelangen. So erhielten wir auch noch einen Preis und fuhren danach beschwingt und glücklich heim. Das war sicher eines meiner spannendsten Abenteuer! Nach diesem anstrengenden Wettkampf hatte ich so einen Hunger, dass mir sogar der Erbseneintopf meiner Mutter schmeckte, obwohl er – wie immer – unglaublich heiß und dazu noch versalzen war.

ÜBUNG 13 **S. 89**

1. Sommer
2. Willi
3. Mann

ÜBUNG 14 **S. 89**

Bild a: der Hund ist nicht da; alle fragen sich, wo er geblieben ist
Bild b: er bekommt eine Wurst
Bilder c und d: die Familie ist an ihrem Urlaubsort angekommen, der Vater holt die Koffer aus dem Wagen und bringt sie ins Zimmer, der Hund läuft weg
Bild e: an einen Würstchenstand, weil er Hunger hatte
Bild f: sie läuft durch den Ort, am Straßenrand steht ein Mann

ÜBUNG 15 **S. 89**

d → c → a → f → e → b

ÜBUNG 16 **S. 90**

Sorge: Kummer, Last, Not, Traurigkeit, Ratlosigkeit, Verzweiflung, Unruhe
Freude: Glücksgefühl, Begeisterung, Jubel, Zufriedenheit, Erleichterung
Hunger: Magenknurren, Appetit, Heißhunger, Kohldampf

ÜBUNG 17 **S. 90**

Bild c: „Super, endlich angekommen!“ – „Und wo ist mein Bett?“ – „Das wird bestimmt ein toller Urlaub!“
Bild a: „Wo ist Bello?“ – „Ist er nach der letzten Rast wieder eingestiegen?“ – „Wir müssen ihn unbedingt suchen!“
Bild f: „Jetzt sind wir schon durch den ganzen Ort gelaufen, wo ist er nur?“ – „Sollen wir nicht mal bei der Polizei nachfragen?“ – „Hoffentlich finden wir ihn wieder!“

ÜBUNG 18 **S. 90**

Lösungsvorschlag:
Unser Bello ist wirklich ein schlaues Hündchen. Als wir letzten Sommer nach siebenstündiger Fahrt endlich in den Bergen ankamen, waren wir alle froh, am Ziel zu sein. Während wir die Koffer ins Hotelzimmer trugen, hüpfte Bello unbemerkt aus dem Wagen. Wir waren glücklich über unser schönes Zimmer und räumten sofort unsere Koffer aus. Plötzlich fiel mir auf, dass Bello gar nicht da war. Ich erschrak und

auch meine Eltern waren besorgt. Sofort ließen wir alles stehen und liegen und gingen ins Dorf. Wir riefen nach Bello und guckten überall nach. Da entdeckte ich unseren Hund, der mit hängender Zunge an der Theke von Willis Wurstbude stand. Offensichtlich war er durch den Geruch angezogen worden und bettelte nun um ein Würstchen. Unser Bello hatte so großen Hunger, dass er sich selbstständig auf die Suche nach etwas Essbarem gemacht hatte. Ich kaufte ihm also ein Würstchen und Bello war erst einmal zufrieden.

Klassenarbeiten

AUFGABE 1 **S. 91**

1. c **2.** f **3.** d **4.** b **5.** g **6.** a **7.** e **8.** h
Einleitung: c, f
Hauptteil: d, b, g, a
Schluss: e, h

AUFGABE 2 **S. 91**

Ergänzung in dieser Reihenfolge:
orangefarbenen – freudig – wunderschön blühenden – neu verputzte – gelb gestrichenen – durchdringende – großer – besorgt blickende – zügig

AUFGABE 3 **S. 92**

Die richtige Reihenfolge:
2, 7, 13, 9, 11, 15, 14, 4, 12, 6, 1, 5, 10, 3, 8

AUFGABE 4 **S. 92**

Lösungsvorschlag:
Gestern Abend spielten wir „Das Nilpferd auf der Achterbahn". Als Erstes bekam meine Schwester die Aufgabe, rückwärts auf einem Bein um den Tisch zu hüpfen. Danach musste mein Vater ein Lied gurgeln, das wir erraten sollten. Er sang „Hänschen klein", sodass wir aus vollem Halse lachten. Schließlich war ich an der Reihe und musste fünf Begriffe nennen, die mit C anfangen. Die anderen rieten mit und halfen mir. Nachdem wir gegen 22 Uhr aufgehört hatten zu spielen, fielen wir todmüde ins Bett.

AUFGABE 5 **S. 93**

Lösungsvorschlag:
2. ziemlich: Ich bekam riesige Angst. **sehr:** Ich dachte, gleich ist es vorbei!
3. ziemlich: Er begann mir wirklich auf die Nerven zu gehen. **sehr:** Ich wurde rasend vor Zorn!
4. ziemlich: Mir rutschte das Herz in die Hose! **sehr:** Ich wusste nicht mehr ein noch aus vor lauter Angst.

AUFGABE 6 **S. 93**

Lösungsvorschlag:
So etwas ist mir aber auch noch nie passiert. Obwohl ich gut auf die Mathearbeit vorbereitet war und mich auch putzmunter gefühlt habe, muss ich die Arbeit in der nächsten Woche nachschreiben. Warum?
Als gestern der Wecker klingelte, bin ich wie immer nur mühsam aus dem Bett gekommen, aber nach dem Duschen war ich endlich wach. Ich ging also zum Frühstück und machte mich dann fertig für den Schulweg. Kaum war ich auf der Straße, fiel mir ein, dass ich noch einmal nach meinem Matheheft gucken sollte. Die Beyer, meine Lehrerin, ist nämlich eine ganz Strenge, mit der verscherzt man es sich am besten nicht. Und klar: Ich hatte es vergessen! Ich bin also wie der Blitz zurück ins Haus gerannt, in mein Zimmer. Zum Glück fand ich das Heft gleich! Aber dann entdeckte ich noch etwas: Pucki, mein Meerschweinchen, war irgendwie entwischt! Ich suchte es im ganzen Zimmer, bis ich es gefunden hatte. Es hatte sich unter meinem Bett verkrochen, dort konnte es unmöglich den ganzen Tag bleiben. Aber wie sollte ich es wieder aus seinem Versteck locken? Ich probierte es mit Pfeifen und mit gutem Zureden: „Pucki-Schnucki, komm doch raus!" Aber nichts half. Dann kam ich darauf, es mit seiner Lieblingsspeise zu füttern. Ich rannte also in den Keller und schaute nach, ob im Gemüseregal noch eine Schlangengurke zu bekommen war, Glück gehabt! Dann in die Küche: Es musste etwas von dem Nussmus, das meine Mutter immer zum Backen gebrauchte, her. Auch das war zum Glück kein Problem. Kaum hatte ich Pucki die Gurkenscheiben mit Nussmus angebo-

ten, kam er auch schon unter dem Bett hervor, ließ sich ohne Weiteres einfangen und zurück in den Käfig sperren.
Ich bin wieder losgerannt, um noch einigermaßen pünktlich zur Mathearbeit zu kommen, aber natürlich war der Bus schon weg. Das hieß: eine halbe Stunde Verspätung. Sicher hat Frau Beyer meine Geschichte für eine ganz faule Ausrede gehalten und wahrscheinlich wird die Arbeit nächste Woche viel schwieriger. Aber hätte ich Pucki im Stich lassen sollen?

AUFGABE 7 **S. 94**
Lösungsvorschlag (als Beispiel ist eine Idee ganz ausgearbeitet):
Ich war in einer Backstube. Plötzlich sah ich ein menschliches Wesen, das sich auf ganz seltsame Art bewegte. Dieses begann zu sprechen, denn es wollte mit mir spielen.

AUFGABE 8 **S. 94**
1. Ich rieche den Duft von frisch gebackenen Mandelhörnchen.
2. Ich höre, wie das menschliche Wesen einen Fuß nachzieht und wie das Feuer im Backofen knistert.
3. Ich sehe eine gebückte Gestalt, die sich im Halbdunkel der nur von wenigen Kerzen erleuchteten Backstube mühsam bewegt.

AUFGABE 9 **S. 94**
grün lackierte Fingernägel, pinkfarbene Augen, bis zur Erde reichende Arme

AUFGABE 10 **S. 95**
„Gut, dass du kommst! Ich hatte schon lange keinen Besuch mehr und freue mich, dass ich endlich mal wieder jemanden habe, mit dem ich Mühle spielen kann."

AUFGABE 11 **S. 95**
rufen, stammeln, drohen, jammern, entgegnen, zurechtweisen

AUFGABE 12 **S. 95**
Mühlespiel wird zum Duell über Leben und Tod.

AUFGABE 13 **S. 95**
Als er gerade den drittletzten Mühlestein nahm, klingelte ...

AUFGABE 14 **S. 95**
Lösungsvorschlag:
Langsam öffnete ich die knarrende Tür zur Backstube. Ich war sehr gespannt, was sich dahinter verbarg. Der schiefe Schornstein und die Spinnweben im Türrahmen ließen darauf schließen, dass schon lange niemand mehr durch diese Tür gegangen war. Meine Augen mussten sich erst an das Halbdunkel des Raumes gewöhnen. Als Erstes sah ich das hell lodernde Feuer im Backofen, dessen Tür ein wenig offen stand. Dann nahm ich plötzlich diesen wohlriechenden Duft von frisch gebackenen Mandelhörnchen wahr. Das Wasser lief mir im Mund zusammen. Doch plötzlich kam hinter einem Holzregal ein gebückter Mensch hervor, der die Kapuze seines braunen Umhangs tief ins Gesicht gezogen hatte. Das Wesen begann zu sprechen und zog die Kapuze ein wenig zurück. Ein hintergründiges Lächeln huschte über sein Gesicht: „Gut, dass du kommst!", rief es. „Ich hatte schon lange keinen Besuch mehr und freue mich, dass ich endlich mal wieder jemanden habe, mit dem ich Mühle spielen kann. Du kennst das Spiel doch, oder? Ich bringe es dir sonst schnell bei!", drohte es. Erfreut, seine Arbeit unterbrechen zu können, setzte sich das Wesen, dessen Fingernägel grün lackiert waren, an einen einfachen Holztisch, an dem zwei dreibeinige Hocker standen. Erst jetzt konnte ich in seine pinkfarbenen Augen blicken.
Mit seinen bis zur Erde reichenden Armen holte es ein Mühlespiel aus einer Schublade und wies mir die schwarzen Steine zu. Ich muss ratlos ausgesehen haben, denn sofort fragte es mich drohend: „Muss ich dir die Spielregeln etwa erst erklären?" „Nein, nein!", entgegnete ich

sofort. Irgendwie würde ich mich an die Regeln, die mir meine Großmutter beigebracht hatte, schon schnell erinnern. Nachdem mein Gegner die ersten Steine gesetzt hatte, murmelte er beiläufig, dass ich im Falle eines Sieges das Backhaus wieder lebendig verlassen könnte, andernfalls sei mein Tod vorherbestimmt.
Mir lief es kalt über den Rücken. War ich gut genug, um in diesem Spiel bestehen zu können? Nachdem wir alle unsere Steine gesetzt hatten, schlossen wir schnell immer wieder Mühlen und nahmen dem Gegner Spielsteine ab. Die pinkfarbenen Augen meines Gegenspielers blickten mich dabei funkelnd und siegessicher an. Schließlich hatte er die Gelegenheit, mit seinen drei Steinen zu springen, meine Spielzüge zu blockieren und mir Steine abzunehmen. Wie würde ich es schaffen, doch noch zu siegen? Ich schwitzte und meine Haare klebten im Nacken. Meine Lage wurde immer aussichtsloser. Ich hatte noch drei Spielsteine. Noch einer weniger und mein Leben wäre zu Ende. Ich wagte es kaum mehr, auf das Spielbrett zu sehen. Als er gerade meinen drittletzten Mühlestein nahm, klingelte der Wecker und ich musste schleunigst aufstehen, damit ich nicht zu spät zur Schule kam.

6 Beschreiben

ÜBUNG 1 S. 97

1. Satz b) ist der beste, weil der Gegenstand genau benannt wird.
2. Satz b) ist der beste, weil erwähnt wird, dass eine Bratpfanne beschrieben wird.
3. Satz c) ist der beste, weil die Inlineskates genauer beschrieben werden.

ÜBUNG 2 S. 98

Reihenfolge (von oben nach unten):
1=f, 2=i, 3=d, 4=h, 5=a, 6=c, 7=g, 8=b, 9=e

ÜBUNG 3 S. 98

Lösungsvorschlag:
ein Paar Inlineskates. – Die Inlineskates sind rot, orange und blau. – Jeder Inlineskate hat vier hintereinander angeordnete Rollen. – Durch einen hinten angebrachten Stopper kann man bremsen. – Mit der weißen Schnalle lässt sich die Weite des Schuhs verstellen.

ÜBUNG 4 S. 99

eine Schreib- und Maltafel für Kinder – Die Tafel ist insgesamt 1,19 Meter hoch ... – Sie ist aus schwarzem Material hergestellt ... – Um die Schreibfläche herum ist ein Rahmen ... – Auf der Höhe von etwa 45 Zentimetern befindet sich ... – Diese Schreib- und Maltafel scheint besonders für Kinder ...

ÜBUNG 5 S. 100

1. Ein reguläres Fußballfeld ist 105 m lang und 75 m breit.
2. Opa Hermann hat drei Töchter, zwei Söhne und 14 (vierzehn) Enkelkinder.
3. Zu meinem Geburtstag kommen sieben Gäste und es gibt vier Torten.

ÜBUNG 6 S. 100

a) **1.** Schreibtischleuchte **2.** lackiertes Metall
3. schwarz **4.** verstellbarer Leuchtenkopf
5. Beleuchtung von Arbeitsplätzen, insbesondere Schreibtischen

b) **1.** rund (flacher Zylinder) – Durchmesser: ca. 20 cm, Höhe: etwa 3 cm **2.** flexibler Arm aus Metall, Länge: ca. 40 cm **3.** glockenförmig – Durchmesser an der breitesten Stelle: ca. 15 cm

c) Ich beschreibe eine Schreibtischleuchte.
Sie besteht aus dunkel lackiertem Metall. Die Lampe steht auf einem zylindrischen Fuß, der einen Durchmesser von etwa 20 cm und eine Höhe von etwa 3 cm hat. Auf diesem Fuß ist auf einer Erhöhung ein flexibler, ca. 40 cm langer Arm befestigt. Der Leuchtenkopf kann damit in viele verschiedene Positionen gebracht werden. Der Arm trägt den eigentlichen Leuchtenkopf, der in derselben Farbe lackiert ist und die Form einer Glocke hat. An seiner breitesten Stelle misst der Leuchtenkopf etwa 15 cm.

ÜBUNG 7 **S. 102**

1. Vor mir liegt ein Bild, auf dem eine etwa 45 Jahre alte Frau abgebildet ist, die mit ihrer rechten Hand einen Schrankgriff anfasst.

2. Auf dem Bild ist eine Frau im Alter von etwa 35 Jahren zu sehen, die auf einem Rasen vor einem Holzgebäude steht.

3. Auf dem vor mir liegenden Bild sehe ich einen etwa 30-jährigen Mann, der einen blauen Wanderanorak und einen Rucksack trägt.

4. Im Folgenden beschreibe ich einen etwa zwölfjährigen Jungen, der ein buntes Ringel-T-Shirt und eine blaue Jeans trägt.

ÜBUNG 8 **S. 102**

Bild 1 (Frau, 45 Jahre): 1, 5, 6, 16
Bild 2 (Frau, 35 Jahre): 3, 10, 11, 13, 15
Bild 3 (Wanderer): 4, 8, 9, 14
Bild 4 (Junge): 2, 7, 12

ÜBUNG 9 **S. 103**

Bild 1: elegante Kleidung – ein kleines Lächeln
Bild 2: sommerliche Freizeitkleidung – dunkle, nackenlange Haare, die ins Gesicht fallen
Bild 3: trägt großen, rotbraunen Rucksack mit Metallgestell – blauer Wanderanorak

ÜBUNG 10 **S. 103**

Bild 1: elegante Dame
Bild 2: Frau im Sommerurlaub
Bild 3: junger Mann auf Wanderung

ÜBUNG 11 **S. 104**

a) männlich; 13 Jahre; 1,60 m; längliches Gesicht, breite Nase, schmale Lippen; etwas missmutiger Gesichtsausdruck, Mütze tief ins Gesicht gezogen; schlanker Körperbau, rechter Arm angewinkelt, Handfläche nach oben ausgestreckt, rechtes Bein angewinkelt; hellgraue Mütze, roter Anorak mit weißen Streifen und grauen Schultern, hellgraue Hose, dunkle Lederschuhe, an der rechten Hand einen hellgrauen Handschuh, in der linken Hand den anderen Handschuh

b) Im Folgenden beschreibe ich einen 13-jährigen Jungen.

c) Der Junge hat einen länglichen Kopf. Da er ernst schaut, wirken die Lippen eher schmal. Der Mund ist geschlossen. Er trägt eine graue Wollmütze, die Haare und Ohren vollständig bedeckt.
Der Gesichtsausdruck lässt vermuten, dass der Junge im Moment nicht gut gelaunt ist. Er geht auf einer regennassen Straße. Der Junge hat die rechte Hand nach vorne gestreckt. An dieser Hand trägt er einen grauen Fingerhandschuh. Die linke Hand lässt er baumeln, sie hält den anderen Handschuh. Der Junge ist mit einem roten Anorak bekleidet, der im Brustbereich und an den Amen einen weißen Streifen hat. Im Schulterbereich ist er grau abgesetzt. Der Reißverschluss ist geschlossen und durch eine Leiste verdeckt. Er hat eine hellgraue Hose an mit einer aufgesetzten Tasche in Höhe des linken Knies. An den Füßen trägt er feste dunkle Schuhe.

d) Der Junge wirkt auf mich ein wenig missmutig, vielleicht wegen des Regens.

ÜBUNG 12 **S. 106**

Ergänzung z. B. in dieser Reihenfolge:
Zuerst – anschließend – daraufhin – nun – danach – im Anschluss daran – zum Schluss

ÜBUNG 13 **S. 106**

1. Einkaufen: Ware bezahlen
2. CD abspielen: Musik hören
3. Schuhe putzen: Schuhputzzeug holen
4. Hausaufgaben machen: Aufgabenstellung lesen / Aufgabe gründlich vorbereiten

ÜBUNG 14 **S. 106**

Lösungsvorschlag:
Zunächst werden die Kartoffeln gewaschen, geschält und in Scheiben geschnitten. Im Anschluss daran gibt man etwa 20 ml Öl in eine Pfanne und erhitzt dieses. Die Kartoffelscheiben werden in das heiße Öl gegeben. Danach kann man die Kartoffeln nach Geschmack mit Salz, Pfeffer oder Paprika würzen. Die Kartoffelscheiben werden bei großer Hitze goldgelb gebraten, dabei sollten sie von Zeit zu Zeit gewendet werden.

ÜBUNG 15 **S. 107**

1. Kuchen backen: **a)** Zutaten einkaufen **b)** Teig rühren **c)** Ofen vorheizen **d)** Form in den Ofen schieben
2. Halma spielen: **a)** Spielesammlung hervorholen **b)** Spielbrett auf den Tisch legen **c)** Spielfiguren aufstellen **d)** den ersten Spielzug machen
3. Elfmeter schießen: **a)** Ball nehmen **b)** auf den Punkt legen **c)** Anlauf nehmen **d)** Tor erzielen
4. Fahrkarte kaufen: **a)** zum Automaten gehen **b)** Fahrziel aussuchen, Fahrpreis herausfinden **c)** Geld einwerfen **d)** Fahrkarte entnehmen

ÜBUNG 16 **S. 108**

Ergänzung in dieser Reihenfolge:
Bodenstaubsauger – 80 cm langen, 40 cm breiten und 30 cm hohen – durch drei an der Unterseite angebrachte Rollen – durch einen etwa 1 m langen, biegsamen Schlauch – auf der Oberseite des Saugfußes – zum Tragen des Staubsaugers – zum Einstellen der Saugkraft – ein Motor im Inneren des Bodengeräts – auf den Bodensauger – durch das Saugrohr und den Saugschlauch

ÜBUNG 17 **S. 109**

Lösungsvorschlag:
Um das Gerät ein- und auszuschalten, drücken Sie lange (mindestens drei Sekunden) auf Knopf A. Mit einem kurzen Druck auf Knopf A starten Sie den ersten Titel. Um das Gerät auf „Pause" zu stellen, drücken Sie erneut kurz auf Knopf A. Um zum vorhergehenden Titel zu gelangen, drücken sie kurz auf Knopf B. Drücken Sie lange (mindestens drei Sekunden) auf Knopf B, um die Lautstärke zu senken.
Mit einem kurzen Druck auf Knopf C gelangen Sie zum nächsten Titel. Drücken Sie lange (mindestens drei Sekunden) auf Knopf C, um die Lautstärke zu steigern.
Mit einem kurzen Druck auf Knopf D starten Sie das Gerät neu.
Es könnte sich z. B. um einen CD-Player oder einen MP3-Player handeln.

Klassenarbeiten

AUFGABE 1 **S. 110**

1. Gefühle vermitteln gehört zum Erzählen, nicht zum Beschreiben.
2. Handlungsanweisung gehört zur Vorgangsbeschreibung, die anderen Begriffe zur Personenbeschreibung.
3. Allgemeine Erscheinung gehört zur Personenbeschreibung, die anderen Begriffe zur Vorgangsbeschreibung.
4. Spannung erzeugen gehört zum Erzählen, nicht zum Beschreiben.

AUFGABE 2 **S. 110**

Vorbereitung: Reise buchen, Grammatik pauken, Gitarre stimmen
Durchführung: mit der Bahn fahren, bei der Klassenarbeit konzentriert arbeiten, vor großem Publikum auftreten
Ergebnis: Erholung pur genießen, gute Note bekommen, riesigen Applaus ernten

AUFGABE 3 **S. 111**

1. Der abgebildete Mann trägt außergewöhnlich bunte Kleidung.
2. Zunächst gibt man 20 Gramm Butter in die Pfanne und lässt sie bei geringer Hitze zergehen.
3. Das Handy ist sehr vielseitig, außer als Telefon kann man es auch als Digitalkamera, Notizbuch und Terminkalender sowie für SMS nutzen.

AUFGABE 4 **S. 111**

Diese Sätze gehören zusammen:
Ich beschreibe jetzt mein **Lineal**. Es ist ein flacher Gegenstand aus Holz. Es ist 4 mm hoch, etwa 3 cm breit und etwas mehr als 20 cm lang. An der oberen Seite hat es eine Skala, auf die im Abstand von je 1 cm etwa 9 mm lange senkrechte Striche, im Abstand von je 0,5 cm etwa 0,6 mm lange senkrechte Striche und im Abstand von je 1 mm etwa 3 mm lange senkrechte Striche markiert sind. Außerdem sind die Zahlen von 0 bis 20 aufgedruckt. Man benutzt diesen Gegenstand zum Abmessen kleinerer Strecken bzw. zum Ziehen gerader Linien.

Vor mir liegt ein **Tintenlöscher**. Er ist ein zylinderförmiger Gegenstand aus Kunststoff. Er hat einen Durchmesser von etwa 1 cm und eine Länge von etwa 18 cm. An der einen Seite ist eine etwa 2,5 cm lange blaue Kappe auf diesen Gegenstand gesteckt, auf der anderen Seite steckt eine weiße Kappe, die in Form und Größe mit der blauen identisch ist. Man benutzt diesen Gegenstand zum Korrigieren von Schreibfehlern.

AUFGABE 5 **S. 112**

Lösungsvorschlag:
Im Folgenden beschreibe ich eine Schreib- und Maltafel aus Kunststoff für Kinder. Die Tafel ist etwa 1,30 m hoch und 65 cm breit. Die vier weißen Stützen sind in der Höhe verstellbar. Das blau umrandete weiße Mal- und Schreibfeld misst 50 x 55 cm. Auf ihm kann nur mit einem Spezialstift geschrieben werden. Unter dem Schreibfeld befindet sich auf der ganzen Breite eine Plastiktasche zur Aufbewahrung des Stiftes und des Putztuches.
Diese Schreibtafel eignet sich für Kinder im Vor- und Grundschulalter.

AUFGABE 6 **S. 112**

Zutaten: 500 ml Milch, 3 Esslöffel Zucker, 1 Päckchen Puddingpulver, 1 Ei
Arbeitsschritte: **1.** 50 ml Milch abmessen – **2.** Ei trennen – **3.** Puddingpulver mit der Milch und dem Eigelb glattrühren – **4.** Eiweiß zu Eischnee schlagen – **5.** die restliche Milch mit dem Zucker aufkochen – **6.** wenn sie kocht, von der Herdplatte nehmen – **7.** das in der Milch gelöste Puddingpulver hinzugeben – **8.** gründlich umrühren – **9.** wieder auf die Herdplatte stellen und kurz aufkochen (bis der Pudding blubbert) – **10.** in eine Schüssel geben – **11.** Eischnee vorsichtig unterheben – **12.** Pudding abkühlen lassen
Lösungsvorschlag Vorgangsbeschreibung:
Um einen leckeren Pudding zu kochen, braucht man 500 ml Milch, 3 Esslöffel Zucker, 1 Päckchen Puddingpulver und 1 Ei.
Zunächst misst man von der Milch 50 ml ab und trennt das Ei. Das Puddingpulver rührt man mit dem kleinen Teil der Milch und dem Eigelb glatt. Das Eiweiß schlägt man zu Eischnee. In die restliche Milch rührt man den Zucker und kocht sie auf. Wenn die Milch kocht, nimmt man sie von der Herdplatte, gibt das gelöste Puddingpulver dazu und rührt es gründlich um. Nun stellt man den Topf wieder auf den Herd und kocht den Pudding noch einmal kurz auf. Wenn der Pudding blubbert, kann man ihn von der Herdplatte nehmen und in eine Schüssel geben. Zum Schluss hebt man den Eischnee unter und lässt den Pudding abkühlen.

7 Umgang mit Texten

ÜBUNG 1 **S. 114**

1. die Schafe **2.** am Bart und am Geruch **3.** in den Gebirgen des Mittleren Ostens und Zentralasiens **4.** von den Wildziegen **5.** für ihr weiches, warmes Fell **6.** der Steinbock

ÜBUNG 2 **S. 114**

Madrid: Hauptstadt, Millionen, Einwohner, Jahrhundert
Kästner: geboren, gestorben, deutsch, humorvoll
Märchen: fantastische, Erzählung, zwischen, Wirklichkeit

ÜBUNG 3 **S. 115**

Moslem: Gläubiger des Islam
Palästina: Bezeichnung für das Gebiet zwischen Mittelmeer und Jordan, in dem der Geburts- und der Sterbeort Jesu liegen
organisieren: einrichten, aufbauen, planen, gestalten
Orient: zu lateinisch oriens = aufgehende Sonne; der Raum der vorderasiatischen Hochkulturen und die islamischen Länder im Nahen Osten und in Nordafrika

ÜBUNG 4 **S. 115**

Die beiden wichtigsten Sätze:
Die Katzen oder Raubkatzen umfassen ungefähr 40 Arten aus der Gruppe der Raubtiere. – Die europäische Wildkatze unterscheidet sich besonders durch ihre Größe von der Hauskatze, die von der afrikanischen Falbkatze abstammt.

ÜBUNG 5 **S. 116**

1. falsch **2.** falsch **3.** falsch **4.** richtig **5.** richtig **6.** falsch **7.** richtig

ÜBUNG 6 **S. 117**

Breiten- und Längengrad – Lage eines Ortes – Breitengrad, wie weit vom Äquator entfernt – Äquator, null Grad – südliche, nördliche Halbkugel

ÜBUNG 8 **S. 117**

... um sich einmal um die **Sonne** zu drehen. Ein Jahr umfasst genau **365 Tage** und sechs Stunden. ... Alle vier Jahre fassen wir die **Vierteltage** zu einem **Schalttag** zusammen. Ein solches **Schaltjahr** hat dann 366 Tage. Darum gibt es nur alle vier Jahre einen **29. Februar.**
... die Ägypter die **Länge** des Jahres. Sie bemerkten, dass der **Nil** stets die Ufer überflutete ...

ÜBUNG 10 **S. 120**

1. Ankerplatz vor dem Hafen
2. festmachen
3. Gesichtsausdruck
4. Wohn- und Aufenthaltsraum auf Schiffen
5. Mastbäume absägen
6. ein Loch haben
7. rückwärts
8. tapfer
9. Seil mit einem Schwimmkörper
10. das Seil fangen

ÜBUNG 11 **S. 120**

Ergänzung in dieser Reihenfolge:
ihren Segelmast abschneiden mussten – unterging – ziellos auf dem Meer herumtrieben – schlug leck – alle Matrosen – das eindringende Wasser – Kapitän – ein Hilfsboot – durch energisches Rudern – gesamten Mannschaft des Schiffs – nur kurze Zeit – sank

ÜBUNG 12 **S. 121**

Lösungsvorschlag:
Zu Beginn liegt das Schiff an einem Ankerplatz. Dann kommt ein Sturm auf. Dabei werden mehrere andere Schiffe beschädigt. Auch das Schiff von Robinson Crusoe schlägt leck. Deshalb werden alle Matrosen an die Pumpen gerufen. Außerdem lässt der Kapitän Notschüsse abfeuern. Glücklicherweise kommt ein Hilfsboot, das die Mannschaft des Schiffes aufnimmt. Zum Schluss sinkt das Schiff, auf dem Robinson sich vorher befunden hat.

ÜBUNG 13 **S. 121**

1. wie ein wild gewordenes Tier.
2. waren auf das Schlimmste gefasst.
3. forderte seine ganze Kraft.
4. als Rettung in letzter Minute.
5. mit schwankenden Knien

ÜBUNG 14 **S. 121**

Lösungsvorschlag:

Robinson Crusoe lag mit dem Schiff, auf dem er arbeitete, vor dem Hafen von Yarmouth fest, als ein Sturm aufkam, der immer heftiger und bedrohlicher wurde. Der Kapitän hatte solche Angst um das Schiff, dass er leise vor sich hin sagte: „Gott sei uns gnädig! Wir sind alle verloren." Die Schiffe in der näheren Umgebung waren bereits schwer beschädigt oder untergegangen, als auch auf seinem Schiff ein Leck und vier Fuß Wasser im Schiffsraum entdeckt wurden. Schnell wurde ein Pumpdienst organisiert und auch Crusoe, der sonst nicht viel helfen konnte, wurde geweckt, um mit anzupacken. Als klar wurde, dass das Wasser immer weiter stieg, wurden Notschüsse als Hilferuf abgegeben. Trotz äußerster Gefahr kam ein Boot so nahe an das Schiff heran, dass die gesamte Mannschaft gerettet werden konnte. Schon eine Viertelstunde nachdem sie in das zur Hilfe geeilte Boot umgestiegen waren, ging das eigene Schiff unter.

ÜBUNG 15 **S. 123**

Lösungsvorschlag:

Leben: geben, Streben, vergeben, weben, erleben, Reben
Wellen: Stellen, fällen, Fellen, Quellen, hellen, Rebellen
sagen: fragen, tragen, plagen, klagen, Behagen, Magen
reisen: weisen, speisen, leisen, umkreisen, Meisen, Ameisen
auch: Bauch, Schlauch, Rauch, Lauch, Strauch, Hauch

ÜBUNG 16 **S. 123**

Frühling lässt sein blaues Band
Wieder flattern durch die Lüfte;
Süße, wohlbekannte Düfte
Streifen ahnungsvoll das Land.
Veilchen träumen schon,
Wollen balde kommen.
– Horch, von fern ein leiser Harfenton!
Frühling, ja du bist's!
Dich hab ich vernommen!

Im Nebel ruhet noch die Welt,
Noch träumen Wald und Wiesen:
Bald siehst du, wenn der Schleier fällt,
Den blauen Himmel unverstellt,
Herbstkräftig die gedämpfte Welt
In warmem Golde fließen.

ÜBUNG 17 **S. 123**

Lösungsvorschlag:

Yanik: „Ich möchte gerne Hähnchen, das kann ich dann mit den Fingern essen."
Angie: „Oh nee, ich möchte es lieber etwas feiner haben. Wie wärs mit dem Italiener um die Ecke?"
Pia: „Aber ich möchte am liebsten Pommes mit Ketchup!"
Vater: „Dann gehen wir also in die Pommesbude?"
Mutter: „Nein, das kommt nicht infrage! Ich möchte einmal gemütlich mit euch zusammensitzen und nicht im Stehen essen."

ÜBUNG 18 **S. 124**

Reihenfolge der Regieanweisungen:
2, 6, 7, 1, 4, 8, 5, 3

ÜBUNG 19 **S. 124**

Lösungsvorschlag:

Es war einmal ein Bauerntölpel, der in einem kleinen Dorf lebte. Eines Tages sagte er zu sich selbst, dass ihm sein Leben zu langweilig sei und er sich auf die Suche nach einem großen Schatz machen wolle. Darum ging er in die weite Welt hinaus. Jedem, den er auf seinem Weg traf, erzählte er von seinem Vorhaben und alle blickten ihn ungläubig an. Dann aber traf er einen Schornsteinfegergesellen, der ihm von einem verschlafenen Dorf erzählte, durch das er soeben gezogen war. In diesem Dorf, so sagte der Geselle, gebe es eine alte Mühle und unter dem Mühlstein sei ein Schatz versteckt. Frohen Mutes machte sich der Bauernjunge auf den Weg ins Dorf und schon von Weitem sah er die Mühle. Tölpelhaft wie er war, hatte er jedoch nicht daran gedacht, dass man einen Mühlstein nicht einfach bewegen kann. Also saß er auf dem Mühlstein und ließ seinen Blick schweifen. Dabei entdeckte er einen goldenen Schlüssel, der jedoch ganz von Staub bedeckt war. Als er diesen Schlüssel in das Schloss am Mühlstein steckte, hob sich dieser wie von Zauberhand ein wenig in die Höhe und er konnte den Schatzbeutel herausziehen. Und tatsächlich befanden sich darin hundert Dukaten! Mit seinem Schatz machte er sich wieder auf zu seinen Eltern. Unterwegs kaufte er von dem Geld Milch, Brot und Honig, denn er wusste, dass seine alte Mutter sich sehr darüber freuen würde. Er bemerkte bald, dass sich auch nach seinem Einkauf noch hundert Dukaten in seinem Schatzbeutel befanden, also kaufte er auch noch Fleisch und Wein für seinen Vater. Als er nach Hause kam, gab es ein großes Festmahl. Der Bauerntölpel aber behielt sein Geheimnis für sich, niemandem erzählte er von seinem wunderbaren Schatzbeutel. Und so waren er und seine Eltern glücklich und zufrieden bis ans Ende ihrer Tage. Und wenn sie nicht gestorben sind, dann leben sie noch heute.

Klassenarbeiten

AUFGABE 1 **S. 125**

Mai – Bienen – Bienenhaus – Tageszeit
Tal – wiege – nieder – Ruh

AUFGABE 2 **S. 125**

1. für die Biegsamkeit der Knochen
2. wenn er richtig zusammengefügt ist
3. in den Röhrenknochen

AUFGABE 3 **S. 126**

Lösungsvorschlag:

Es war einmal ein armes Mädchen, das seine Eltern verloren hatte. Deshalb machte es sich auf den Weg zu seiner Großmutter. Auf dem Weg traf es einen armen Mann, der es bat, ihm sein Brot zu geben, weil er so großen Hunger hatte. Dann traf es eine Frau, der es sein Kopftuch gab, weil sie so kalte Ohren hatte. Nach einiger Zeit lief ihm ein Kind über den Weg, dem gab es sein Kleidchen, weil das Kind gar nichts anzuziehen hatte. Danach trug das arme Mädchen also nichts mehr als sein Hemdchen. So war es auch am späten Abend noch unterwegs, als plötzlich vom Himmel goldene Sterntaler fielen, die es in seinem Hemdchen aufsammelte. Und so lebte es zusammen mit seiner Großmutter noch viele Jahre glücklich.

AUFGABE 4 **S. 126**

Sachtext: klare, sachliche Sprache – knapper Text – Informationen – Fremdwörter nachschlagen – Abkürzungen auflösen – Schlüsselwörter herausschreiben
Gedicht: Reimwörter – gleich klingende Schlusssilbe – bildliche Sprache – Strophen – Verse
Dialogtext: Rollenspiel – Szene – Bühnenbild – Regieanweisungen – Kostüme
Märchen: große Aufgabe – Prinzessin – große Belohnung – wunderbare Fähigkeiten – drei Wünsche freihaben –
Fabel: Fuchs – recht kurzer Erzähltext – sprechende Tiere – Lehre vermitteln

DUDEN

ISBN 978-3-411-72153-5